AF334459

PERSPECTIVES ON
SUPERSYMMETRY

ADVANCED SERIES ON DIRECTIONS IN HIGH ENERGY PHYSICS

Advanced Series on
Directions in High Energy Physics — Vol. 18

PERSPECTIVES ON SUPERSYMMETRY

Editor

Gordon L. Kane

University of Michigan

World Scientific
Singapore • New Jersey • London • Hong Kong

Published by

World Scientific Publishing Co. Pte. Ltd.
P O Box 128, Farrer Road, Singapore 912805
USA office: Suite 1B, 1060 Main Street, River Edge, NJ 07661
UK office: 57 Shelton Street, Covent Garden, London WC2H 9HE

British Library Cataloguing-in-Publication Data
A catalogue record for this book is available from the British Library.

PERSPECTIVES ON SUPERSYMMETRY

ISBN 981-02-3553-4

Printed in Singapore by Eurasia Press Pte Ltd

CONTENTS

INTRODUCTION

Supersymmetry is a hypothetical symmetry of the basic laws of nature. It proposes that the basic laws are symmetric under interchanging bosons and fermions in the appropriate manner. We do not yet know for sure that nature is actually supersymmetric at the scale of weak interactions, but there is considerable indirect evidence that it is.

One of the remarkable things about supersymmetry is that it was not invented for typical reasons. Most physical theories are invented in response to a puzzle, or to explain data, or to improve the consistency of previous theories. Yang-Mills theories are one exception – they were an extension of Abelian gauge theories to non-Abelian ones, and have turned out to be of great importance for understanding nature. Supersymmetry was also an extension, found [a] in the early 1970's both by examining certain field theories and also by trying to understand if a boson-fermion symmetry could be consistent with relativistic quantum field theory. The key point for our purposes is that supersymmetry was not invented to solve any puzzle or explain any observation.

Over about a decade after the discovery of supersymmetry, it was slowly realized by a number of people that it had the potential to solve an astonishing number of major physical problems. Even though it was not invented to do so, supersymmetry could provide a solution to the hierarchy problem (this and other terminology are clearly explained in Steve Martin's fine pedagogical Chapter), could provide a derivation of the Higgs mechanism (which was technically satisfactory for the Standard Model (SM) but lacked a physical basis there) and in the process successfully predict a heavy top quark in the early 1980's, allowed the unification of the SM forces (that supersymmetry was needed to achieve this was confirmed in the early 1990's, a decade after it was predicted), provided a connection between the SM forces and gravity, and provided a candidate for the cold dark matter of the universe (the lightest superpartner (LSP)) before astronomy demonstrated the need for non-baryonic cold dark matter. All of this was in place by the early 1980's.

More recently, it has been recognized that the scalar potential of supersymmetry, the potential energy from the scalars and their interactions, provides the potential energy that determines the course of inflation(s), and the scalars themselves can be the inflaton(s) (see Randall's chapter). The parameters in that scalar potential also determine how scalars will behave at colliders, and can affect neutrino masses, baryogenesis and other phenomena. The study of fermion masses becomes a serious research problem in a supersymmetric frame-

---

[a] The history of supersymmetry is complicated and interesting. Rather than treat it superficially here, I will not address it.

work, and supersymmetry affects our approach to proton decay, baryogenesis, and CP violation. It also suggests approaches to understanding the cosmological constant. Given all of these successes and opportunities, it is clear why so many physicists expect the world to be supersymmetric.

Is nature actually supersymmetric on the electroweak scale? Although that is a statement about the theory that describes nature, the way people will finally be convinced is by the explicit detection of superpartners of the SM particles. Contrary to what is sometimes stated, it is not surprising that superpartners have not yet been detected. First, theoretically all particles can get masses from two sources, the breaking of the electroweak symmetry and the breaking of supersymmetry. One set of particles only gets mass from the electroweak breaking, and all of those except one (the Higgs boson) have been observed. The masses of all those particles would vanish if the electroweak symmetry were restored. It is entirely reasonable that the particles that get mass from both sources should be somewhat heavier and not yet observed. The one exception, the Higgs boson, is very difficult to produce and to detect at hadron colliders – it should be observed at LEP soon or at FNAL in a few years. Second, phenomenologically there are no general bounds on superpartners masses that make the non-colored ones heavier than M_W, or colored ones heavier than M_{top}, so searches have not yet reached levels where one might have expected to find the superpartners. (All published limits depend on guesses for parameters and models.)

It would be very nice if clean, unambiguous experimental signals could appear one day. But a little thought tells us that is unlikely — probably impossible. Consider colliders. At least until the LHC, which is unlikely to produce its first paper relevant to supersymmetry until about a decade from now, what will happen is that as energy and/or luminosity increases at LEP and FNAL a few events of superpartner production will occur. Perhaps such events have already occurred. Each event has two escaping LSP's , so it is never possible to find a dramatic Z-like two body peak, or even a W-like peak with one escaping particle. Even worse, often several channels look similar to detectors so simple features can be obscured. And usually there are SM processes that can fake any particular signature, as well as ways to fake signatures because detectors are imperfect.

Thus to make progress it is essential to proceed with limited amounts of incomplete information. Without theory input it is entirely possible that signals would not be noticed, hidden under backgrounds since there was no guide to what cuts to use (an example is discussed below). Further, a particular signal might be encouraging but not convincing – only when combined with other signals that were related by the theory but not directly experimentally could a strong case be made.

What about information from decays? In the best cases, such as $b \to s\gamma$, where there is no tree level decay, the SUSY contribution could be comparable to the SM one, say a large effect of order 30%. Then to get a significant effect the combined errors of the theoretical calculation of the SM value and the experiment have to be below 10%. Only $b \to s\gamma$ of known decays can approach that (see Chankowski and Pokorski's chapter); presently the theoretical error is at about that level, and the experimental error about 20%. At best, in a couple of years this could provide evidence of new physics. If so, taken alone several interpretations would be possible, but with theoretical input it could be combined with collider data and other decays to determine which interpretation was consistent.

When there is a tree level SM contribution, such as for $R_b = \Gamma\left(Z \to b\bar{b}\right) / \Gamma\left(Z \to \text{hadrons}\right)$ the SUSY effect has to be a loop and can be $\sim \frac{1}{2}\%$, so experimental errors have to be several times smaller. Just from statistics that requires $\sim 10^6$ events, which is unlikely. Sometimes several decays are related in a particular model, in which case the combined predictions can be tested and the results are somewhat more significant (that is the situation for $b \to s\gamma$ plus R_b plus α_s). Further input could come from proton decay to channels favored by SUSY, from occurrence of decays forbidden in the SM such as $\mu \to e\gamma$ or $K \to \mu e$, from neutron or electron electric dipole moments, from non-SM CP violation, or other rare phenomena.

If we can discover superpartners before LHC, it will be necessary to proceed with fragments of information, check their consistency, make predictions to test them, and slowly build a case. We will see below that there are things to work with (which need not have happened – most fluctuations from the SM could never be interpreted as SUSY). The first goal, of course, is to establish that superpartners indeed exist at the EW scale. Given that, then the goal becomes the mesurement of the parameters of the Lagrangian, so that theories implying a particular Lagrangian can be tested, and the form the Lagrangian takes will suggest how SUSY is broken. Measuring the basic parameters will be difficult or impossible at LEP and hard at FNAL, but should be feasible at LHC combined with a lepton collider with polarized beams. For example, $\tan\beta$ cannot be measured at LEP.

The way superpartners will behave — particularly their experimental signatures — is largely determined by the quantum numbers of the LSP. Theoretical and phenomenological arguments probably combine to imply the LSP is either the lightest neutralino (the superpartners of γ, Z, and the neutral Higgs fields of supersymmetry can quantum mechanically mix — the mass eigenstates are "neutralinos") or the gravitino. If it is the lightest neutralino over much of the parameter space it turns out to be mainly bino (the partner of the $U(1)$ B_μ gauge field), or in a smaller but robust part of the parameter space, higgsino, a particular combination of partners of Higgs fields, approximately the partner of the Higgs boson.

Almost all of the studies of supersymmetric models can be classified according to what is the LSP, and fall in three categories, as show in Table 1. A fourth category is "unstable LSP"; see Dreiner's chapter. On the left several criteria are listed that are often used to compare and test models. The first world listed has a light gravitino LSP ($\widetilde{G}$LSP), the second an LSP that is mainly bino ($\widetilde{B}$LSP) and the third mainly higgsino ($\tilde{h}$LSP). The implications for SUSY-breaking and experimental signature are very different for the three cases – it will be easy to recognize which is being observed once one is detected. ($\widetilde{G}$LSP) corresponds to gauge-mediated SUSY-breaking, and the other two gravity-mediated SUSY-breaking.)

In a $\widetilde{G}LSP$ world there is a small window for a signal at LEP but no phenomenological reason to expect one there. If the CDF event were interpreted as evidence for a $\widetilde{G}LSP$ world, which is difficult given constraints but not excluded, many such events world occur at FNAL after the collider starts running again in 1999; otherwise there is only a small window at FNAL. Such a world could probably be detected at LHC or a lepton collider about 2010.

In a $\widetilde{B}LSP$ world there is no evidence today for sparticles, and all hints must disappear (no more $ee\gamma\gamma\not{E}_T$ events at FNAL; $BR(b \to s\gamma) \to$ SM, $R_b \to$ SM, $\alpha_s^{\Gamma z} - \alpha_s^{\text{other}} \to 0$; no future excess of $\gamma\gamma\not{E}$ events at LEP; baryogenesis not at EW scale; etc.). There are small windows at LEP and FNAL but no phenomenological reason for sparticles to be there. Such a world probably be detected at LHC or a lepton collider about 2010.

In a $\tilde{h}LSP$ world sparticles may have already been observed. Confirmation will occur at LEP once 50 pb^{-1}/detector at $\sqrt{s} \gtrsim$ 190 GeV has been accumulated or before. In my short chapter the hints for an $\tilde{h}LSP$ world and some of the tests are described in a little more detail.

Once superpartners are detected it will be a delightful challenge to extract physics information from the raw data. As discussed above, often signals will come from several channels. The relationship of what is observed by experimenters to the masses and couplings of the superpartners will be complicated and nonlinear, and the relationship to the parameters of the effective Lagrangian in the electroweak symmetry basis is even more difficult to extract. To connect with theoretical work it is necessary to measure $\tan\beta, \mu$, and the soft-breaking masses, which are only indirectly related to the particle masses.

Actual measurements of effects of superpartners will produce cross sections and distributions, excesses of events with some set of particles such as gammas

Table 1.

Evidence/Criteria	$\widetilde{G}LSP$	$\widetilde{B}LSP$	$\widetilde{h}LSP$
Absence of FCNC	Yes, if messenger scale low	Mechanisms exist, but don't know if they are applicable	
Number of parameters	Presently all about same – $\widetilde{G}LSP$ somewhat more than others now, but will be more predictive after squarks and sleptons observed.		
prompt γs	all events or none	no	events with 0,1, and 2
trilepton + $\not{E}_T$ events	no	yes	no
CDF e "e" $\gamma\gamma\not{E}_T$	maybe	no	yes
$R_b, BR(b \to s\gamma), \alpha_s$	maybe bsγ no for R_b, α_s	no	yes
LEP $\gamma\gamma\not{E}$ events	no	no	yes
Cold Dark Matter	not LSP	ok	yes
$\tilde{g}$, light $\tilde{t}$ at FNAL?	no	no	ok
EW baryogenesis	no	no	ok
$m_{h^\circ} \lesssim M_Z$	no reason	no reason	yes

and perhaps with missing energy. They do not produce measurements of the masses or couplings of superpartners, and determination of the soft-breaking parameters or μ or $\tan\beta$ is even less likely. Most distributions get contributions from several processes as well. How can we proceed to extract the physics parameters of interest from data in such a nonlinear situation?

Most analyses have proceeded by assuming a model, and perhaps also assuming values for some parameters, until the problem is reduced to a simpler one. That is of use since there is no guarantee a solution can be obtained, so if a consistent solution can be found it may be a physical one.

In fact, the general problem has been addressed and a procedure ("carving") given [1] to extract the sparticle masses and/or the parameters of $\mathcal{L}_{EFF}$. Every measurement provides information. There is an excess of events at a certain cross section level in one process, none in another. The optimum procedure is to randomly select values for all parameters, calculate all observables, and discard values of parameters that give observables in disagreement with data. While that sounds like a big task it is not so hard in practice since any given observable only depends on a few parameters. Obviously to determine N parameters one will need at least N observables sensitive to the parameters. Also, often it is easy after a little thought to put limits on parameters that reduce the size of the problem . This method could already be used to get general limits on sparticle masses, though it has not been used much.

As data about superpartners is increasingly available, more and more of the parameters of the effective Lagrangians $\mathcal{L}_{EFF}$ of the theory at the electroweak scale will be measured. Constraints from rare decays, CP violation, baryogenesis etc. will be included. Then, assuming the theory to be perturbative to the scale where the gauge coupling unify, the effective Lagrangian at that scale will be calculated by using renormalization group equations. It is not necessary (nor expected) that there be a desert in between, but only that the theory be perturbative. Intermediate matter and scales are expected. There will be consistency checks that allow the perturbativity to be confirmed. Since constraints occur at both ends it is not just an extrapolation. Unification may or may not involve a unified gauge group.

String theory, on the other hand, starts somewhat above the unification scale, at the Planck scale. If the way to select the vacuum was known, and also how SUSY was broken (perhaps once the vacuum is known the latter will be determined), then the $\mathcal{L}_{EFF}$ could be predicted, and compared with the $\mathcal{L}_{EFF}$ deduced from data. In practice I expect it to be the other way, as it has been throughout the history of physics – once we know the experimental $\mathcal{L}_{EFF}$ deduced from data the patterns of parameters will be recognizable and will tell someone how SUSY is broken and how the vacuum is selected. After that it will be possible to derive it from string theory. String theorists and sphenomenologists will meet at the unification scale.

The present book is designed to fill several gaps in a coherent way. Anyone with a basic grasp of the Standard Model of particle physics, such as can be obtained from ref. 2, can learn to understand the achievements and goals and issues of supersymmetry from the initial fine chapters of S. Martin, and of K. Dienes and C. Kolda. Martin's chapter is the best pedagogical introduction to supersymmetry anywhere. (A number of chapters had to be shortened for space reasons. Original longer versions are sometimes available elsewhere, and are high quality and useful.)

Supersymmetry is an active, ongoing area of research. The chapter from Dienes and Kolda is a masterful introduction to most of the open theoretical questions in supersymmetry — it can be viewed as a list of opportunities for theoretical researchers.

A number of chapters focus on perhaps the most important question at present — how do we detect the superpartners if they are there. The chapter on Fermilab will be particularly useful since new results will not come from Fermilab for about three years. Others look in more detail at theoretical questions. Some focus on the connections of experiment and theory, or aspects of the connection of unification or Planck scale theory with low energy data. All of the chapters have introductions that are useful to non-experts, and references that point to places to learn more, as well as summaries of the current status of their subject.

Some topics are not included in this book because they are well covered recently in what can be thought of as a companion volume, "Perspectives on Higgs Physics II", that I edited for World Scientific. These include the supersymmetric Higgs sector properties and ways to detect the supersymmetric Higgs bosons, the general upper limit of about 150 GeV on the mass of the lightest Higgs boson that is the only quantitative upper limit that is a test of supersymmetry, the role of supersymmetry in electroweak baryogenesis, and more.

Supersymmetry is an adolescent field. The situation with supersymmetry today is actually quite analogous to the situation of the Standard Model in 1973. The theory was attractive and convincing, and all experimental evidence was indirect. The field is poised to move rapidly once data on superpartners begins to come. The main challenge will be to go from the raw data to measure the basic parameters of the theory – as I said above much is understood about how to do that, though not all.

Then the central problem of the field for theorists is to understand how supersymmetry is broken. The result of the breaking is a set of soft-breaking masses and interaction parameters. Measuring these soft parameters (which are not the masses of the physical particles) will be the central problem for experimenters. Soon after these are measured it is likely that their pattern will be recognized as one generated by a particular physical mechanism. Then the cosmological implications of supersymmetry and its connections to formulating a fundamental Planck scale theory will become the central topics of a mature field.

REFERENCES

1. G.L. Kane, C. Kolda, L. Roszkowski, and J. Wells, Phys. Rev. **D49**, (1994) 6173, Section G.

2. "Modern Elementary Particle Physics", G.L. Kane, 1993, updated edition, Addison-Wesley.

Gordon Kane

A SUPERSYMMETRY PRIMER

STEPHEN P. MARTIN

Randall Physics Laboratory, University of Michigan
Ann Arbor MI 48109-1120 USA

I provide a pedagogical introduction to supersymmetry. The level of discussion is aimed at readers who are familiar with the Standard Model and quantum field theory, but who have little or no prior exposure to supersymmetry. Topics covered include: motivations for supersymmetry; the construction of supersymmetric Lagrangians; supersymmetry-breaking interactions; the Minimal Supersymmetric Standard Model (MSSM); R-parity and its consequences; the origins of supersymmetry breaking; and the mass spectrum of the MSSM.

1 Introduction

The Standard Model of high energy physics provides a remarkably successful description of presently known phenomena. Still, it will have to be extended to describe physics at arbitrarily high energies. Certainly a new framework will be required at the reduced Planck scale $M_{\rm P} = (8\pi G_{\rm Newton})^{-1/2} = 2.4 \times 10^{18}$ GeV, where quantum gravitational effects become important.

The mere fact that the ratio $M_{\rm P}/M_W$ is so huge is already a powerful clue to the character of physics beyond the Standard Model, because of the infamous "hierarchy problem".[1] This is not really a difficulty with the Standard Model itself, but rather a disturbing sensitivity of the Higgs potential to new physics in almost any imaginable extension of the Standard Model. The electrically neutral part of the Standard Model Higgs field is a complex scalar H with a classical potential given by

$$V = m_H^2 |H|^2 + \lambda |H|^4 \,. \tag{1.1}$$

The Standard Model requires a non-vanishing vacuum expectation value (VEV) for H at the minimum of the potential. This will occur if $m_H^2 < 0$, resulting in $\langle H \rangle = \sqrt{-m_H^2/2\lambda}$. Since we know experimentally that $\langle H \rangle = 174$ GeV from measurements of the properties of the weak interactions, it must be that m_H^2 is very roughly of order $-(100 \text{ GeV})^2$. However, m_H^2 receives enormous quantum corrections from the virtual effects of every particle which couples, directly or indirectly, to the Higgs field.

For example, in Fig. 1a we have a correction to m_H^2 from a loop containing a Dirac fermion f with mass m_f. If the Higgs field couples to f with a term in the lagrangian $-\lambda_f H \overline{f} f$, then the Feynman diagram in Fig. 1a yields a

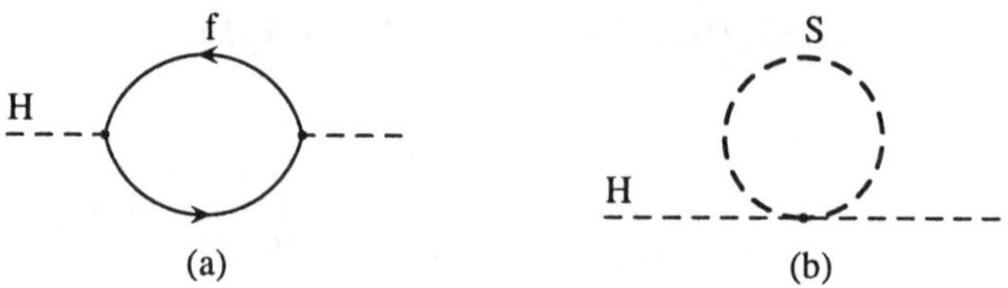

Figure 1: Quantum corrections to the Higgs (mass)^2.

correction

$$\Delta m_H^2 = \frac{|\lambda_f|^2}{16\pi^2} \left[-2\Lambda_{\mathrm{UV}}^2 + 6m_f^2 \ln(\Lambda_{\mathrm{UV}}/m_f) + \ldots \right]. \qquad (1.2)$$

Here Λ_{UV} is an ultraviolet momentum cutoff used to regulate the loop integral; it should be interpreted as the energy scale at which new physics enters to alter the high-energy behavior of the theory. The ellipses represent terms which depend on the precise manner in which the momentum cutoff is applied, and which do not get large as Λ_{UV} does. Each of the leptons and quarks of the Standard Model can play the role of f; for quarks, eq. (1.2) should be multiplied by 3 to account for color. The largest correction comes when f is the top quark with $\lambda_f \approx 1$. The problem is that if Λ_{UV} is of order M_{P}, say, then this quantum correction to m_H^2 is some 30 orders of magnitude larger than the aimed-for value of $m_H^2 \sim -(100 \text{ GeV})^2$. This is only directly a problem for corrections to the Higgs scalar boson (mass)^2, because quantum corrections to fermion and gauge boson masses do not have the quadratic sensitivity to Λ_{UV} found in eq. (1.2). However, the quarks and leptons and the electroweak gauge bosons Z^0, $W^\pm$ of the Standard Model all owe their masses to $\langle H \rangle$, so that the entire mass spectrum of the Standard Model is directly or indirectly sensitive to the cutoff Λ_{UV}.

One could imagine that the solution is to simply pick an ultraviolet cutoff Λ_{UV} which is not too large. However, one still has to concoct some new physics at the scale Λ_{UV} which not only alters the propagators in the loop, but actually cuts off the loop integral. This is not easy to do in a theory whose lagrangian does not contain more than two derivatives, and higher derivative theories generally suffer from a loss of unitarity. In string theories, loop integrals are cut off at high Euclidean momentum p by factors $e^{-p^2/\Lambda_{\mathrm{UV}}^2}$, but then Λ_{UV} is a string scale which for other reasons is thought to be not very far below M_{P}. Furthermore, there is a contribution similar to eq. (1.2) from the virtual effects of any arbitrarily heavy particles which might exist. For example, suppose there exists a heavy complex scalar particle S with mass m_S which couples to

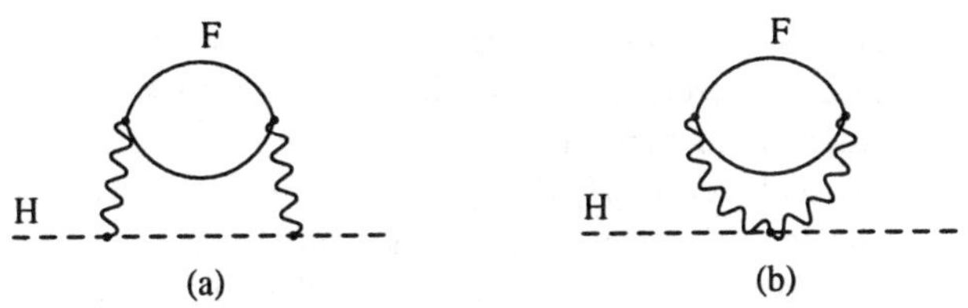

Figure 2: Two-loop corrections to the Higgs (mass)^2 due to a heavy fermion.

the Higgs with a lagrangian term $-\lambda_S |H|^2 |S|^2$. Then the Feynman diagram in Fig. 1b gives a correction

$$\Delta m_H^2 = \frac{\lambda_S}{16\pi^2} \left[\Lambda_{\mathrm{UV}}^2 - 2m_S^2 \ln(\Lambda_{\mathrm{UV}}/m_S) + \ldots \right]. \qquad (1.3)$$

If one rejects a physical interpretation of Λ_{UV} and uses dimensional regularization on the loop integral instead of a momentum cutoff, then there will be no Λ_{UV}^2 piece. However, even then the term proportional to m_S^2 cannot be eliminated without the physically unjustifiable tuning of a counter-term specifically for that purpose. So m_H^2 is sensitive to the masses of the *heaviest* particles that H couples to; if m_S is very large, its effects on the Standard Model do not decouple, but instead make it very difficult to understand why m_H^2 is so small.

This problem arises even if there is no direct coupling between the Standard Model Higgs boson and the heavy particles. For example, suppose that there exists a heavy fermion F which, unlike the quarks and leptons of the Standard Model, has vector-like quantum numbers and therefore gets a large mass m_F without coupling to the Higgs field [i.e., a mass term of the form $m_F \overline{F} F$ is not forbidden by any symmetry, including $SU(2)_L$]. In that case, no diagram like Fig. 1a exists for F. Nevertheless there will be a correction to m_H^2 as long as F shares some gauge interactions with the Standard Model Higgs field; these may be the familiar electroweak interactions, or some unknown gauge forces which are broken at a very high energy scale inaccessible to experiment. In any case, the two-loop Feynman diagrams in Fig. 2 yield a correction

$$\Delta m_H^2 = x \left(\frac{g^2}{16\pi^2} \right)^2 \left[a\Lambda_{\mathrm{UV}}^2 + 48m_F^2 \ln(\Lambda_{\mathrm{UV}}/m_F) + \ldots \right], \qquad (1.4)$$

where g is the gauge coupling in question, and x is a group theory factor of order 1 (specifically, the product of the quadratic Casimir invariant of H and the Dynkin index of F for the gauge group in question). The coefficient a

depends on the precise method of cutting off the momentum integrals. It does not arise at all if one rejects the possibility of a physical interpretation for Λ_{UV} and uses dimensional regularization, but the m_F^2 contribution is always present. The numerical factor $(g^2/16\pi^2)^2$ may be quite small (of order 10^{-5} for electroweak interactions) but the important point is that these contributions to Δm_H^2 are sensitive to the largest masses and/or ultraviolet cutoff in the theory, presumably of order M_{P}. The "natural" (mass)^2 of a fundamental Higgs scalar, including quantum corrections, seems to be more like M_{P}^2 than the experimentally favored value! Even very indirect contributions from Feynman diagrams with three or more loops can give unacceptably large contributions to Δm_H^2. If the Higgs boson is a fundamental particle, we have two options: either we must make the rather bizarre assumption that there do not exist *any* heavy particles which couple (even indirectly or extremely weakly) to the Higgs scalar field, or some rather striking cancellation is needed between the various contributions to Δm_H^2.

The systematic cancellation of the dangerous contributions to Δm_H^2 can only be brought about by the type of conspiracy which is better known to physicists as a symmetry. It is apparent from comparing eqs. (1.2), (1.3) that the new symmetry ought to relate fermions and bosons, because of the relative minus sign between fermion loop and boson loop contributions to Δm_H^2. (Note that λ_S must be positive if the scalar potential is to be bounded from below.) If each of the quarks and leptons of the Standard Model is accompanied by two complex scalars with $\lambda_S = |\lambda_f|^2$, then the Λ_{UV}^2 contributions of Figs. 1a and 1b will neatly cancel.[2] Clearly, more restrictions on the theory will be necessary to ensure that this success persists to higher orders, so that, for example, the contributions in Fig. 2 and eq. (1.4) from a very heavy fermion are cancelled by the two-loop effects of some very heavy bosons. Fortunately, conditions for cancelling all such contributions to scalar masses are not only possible, but are actually unavoidable once we merely assume that a symmetry relating fermions and bosons, called a *supersymmetry*, should exist.

A supersymmetry transformation turns a bosonic state into a fermionic state, and vice versa. The operator Q which generates such transformations must be an anticommuting spinor, with

$$Q|\text{Boson}\rangle = |\text{Fermion}\rangle; \qquad Q|\text{Fermion}\rangle = |\text{Boson}\rangle. \qquad (1.5)$$

Spinors are intrinsically complex objects, so $Q^\dagger$ (the hermitian conjugate of Q) is also a symmetry generator. Because Q and $Q^\dagger$ are fermionic operators, they carry spin angular momentum $1/2$, so it is clear that supersymmetry must be a spacetime symmetry. The possible forms for such symmetries in an interacting quantum field theory are highly restricted by the Haag-Lopuszanski-Sohnius extension of the Coleman-Mandula theorem.[3] For realistic theories which, like the Standard Model, have chiral fermions (i.e., fermions whose left- and right-handed pieces transform differently under the gauge group) and thus the possibility of parity-violating interactions, this theorem implies that the generators Q and $Q^\dagger$ must satisfy an algebra of anticommutation and commutation relations with the schematic form

$$\{Q, Q^\dagger\} = P^\mu \tag{1.6}$$

$$\{Q, Q\} = \{Q^\dagger, Q^\dagger\} = 0 \tag{1.7}$$

$$[P^\mu, Q] = [P^\mu, Q^\dagger] = 0 \tag{1.8}$$

where P^μ is the momentum generator of spacetime translations. Here we have ruthlessly suppressed the spinor indices on Q and $Q^\dagger$; after developing some notation we will, in section 3.1, derive the precise version of eqs. (1.6)-(1.8) with indices restored. In the meantime, we simply note that the appearance of P^μ on the right-hand side of eq. (1.6) is unsurprising, since it transforms under Lorentz boosts and rotations as a spin-1 object while Q and $Q^\dagger$ on the left-hand side each transform as spin-1/2 objects.

The single-particle states of a supersymmetric theory fall naturally into irreducible representations of the supersymmetry algebra which are called *supermultiplets*. Each supermultiplet contains both fermion and boson states, which are commonly known as *superpartners* of each other. By definition, if $|\Omega\rangle$ and $|\Omega'\rangle$ are members of the same supermultiplet, then $|\Omega'\rangle$ is proportional to some combination of Q and $Q^\dagger$ operators acting on $|\Omega\rangle$, up to a spacetime translation or rotation. The (mass)^2 operator $-P^2$ commutes with the operators Q, $Q^\dagger$, and with all spacetime rotation and translation operators, so it follows immediately that particles which inhabit the same irreducible supermultiplet must have equal eigenvalues of $-P^2$, and therefore equal masses.

The supersymmetry generators $Q, Q^\dagger$ also commute with the generators of gauge transformations. Therefore particles in the same supermultiplet must also be in the same representation of the gauge group, and so must have the same electric charges, weak isospin, and color degrees of freedom.

Each supermultiplet contains an equal number of fermion and boson degrees of freedom. To prove this, consider the operator $(-1)^{2s}$ where s is the spin angular momentum. By the spin-statistics theorem, this operator has eigenvalue $+1$ acting on a bosonic state and eigenvalue -1 acting on a fermionic state. Any fermionic operator will turn a bosonic state into a fermionic state and vice versa. Therefore $(-1)^{2s}$ must anticommute with every fermionic operator in the theory, and in particular with Q and $Q^\dagger$. Now consider the subspace of states $|i\rangle$ in a supermultiplet which have the same eigenvalue p^μ of the four-momentum operator P^μ. In view of eq. (1.8), any combination of Q or $Q^\dagger$

6

acting on $|i\rangle$ will give another state $|i'\rangle$ which has the same four-momentum eigenvalue. Therefore one has a completeness relation $\sum_i |i\rangle\langle i| = 1$ within this subspace of states. Now one can take a trace over all such states of the operator $(-1)^{2s} P^\mu$ (including each spin helicity state separately):

$$
\begin{aligned}
\sum_i \langle i|(-1)^{2s} P^\mu |i\rangle &= \sum_i \langle i|(-1)^{2s} Q Q^\dagger |i\rangle + \sum_i \langle i|(-1)^{2s} Q^\dagger Q |i\rangle \\
&= \sum_i \langle i|(-1)^{2s} Q Q^\dagger |i\rangle + \sum_i \sum_j \langle i|(-1)^{2s} Q^\dagger |j\rangle\langle j|Q|i\rangle \\
&= \sum_i \langle i|(-1)^{2s} Q Q^\dagger |i\rangle + \sum_j \langle j|Q(-1)^{2s} Q^\dagger |j\rangle \\
&= \sum_i \langle i|(-1)^{2s} Q Q^\dagger |i\rangle - \sum_j \langle j|(-1)^{2s} Q Q^\dagger |j\rangle \\
&= 0.
\end{aligned}
\tag{1.9}
$$

The first equality follows from the supersymmetry algebra relation eq. (1.6); the second and third from use of the completeness relation; and the fourth from the fact that $(-1)^{2s}$ must anticommute with Q. Now $\sum_i \langle i|(-1)^{2s} P^\mu |i\rangle = p^\mu \mathrm{Tr}[(-1)^{2s}]$ is just proportional to the number of bosonic degrees of freedom n_B minus the number of fermionic degrees of freedom n_F in the trace, so that

$$
n_B = n_F \tag{1.10}
$$

must hold for a given $p^\mu \neq 0$ in each supermultiplet.

The simplest possibility for a supermultiplet has a single Weyl fermion (with two helicity states, so $n_F = 2$) and two real scalars (each with $n_B = 1$). It is natural to assemble the two real scalar degrees of freedom into a complex scalar field; as we will see below this provides for convenient formulation of the supersymmetry algebra, Feynman rules, supersymmetry violating effects, etc. This combination of a two-component Weyl fermion and a complex scalar field is called a *chiral* or *matter* or *scalar* supermultiplet.

The next simplest possibility for a supermultiplet contains a spin-1 vector boson. If the theory is to be renormalizable this must be a gauge boson which is massless, at least before the gauge symmetry is spontaneously broken. A massless spin-1 boson has two helicity states, so the number of bosonic degrees of freedom is $n_B = 2$. Its superpartner is therefore a massless spin-1/2 Weyl fermion, again with two helicity states, so $n_F = 2$. (If one tried instead to use a massless spin-3/2 fermion, the theory would not be renormalizable.) Gauge bosons must transform as the adjoint representation of the gauge group, so their fermionic partners must also. Since the adjoint representation of a gauge group is always its own conjugate, this means that these fermions must have the same gauge transformation properties for left-handed and for right-handed components. Such a combination of spin-1/2 and spin-1 particles is called a *gauge* or *vector* supermultiplet.

There are other possible combinations of particles with spins which can satisfy $n_B = n_F$. However, these are always reducible to combinations of chiral and gauge supermultiplets if they have renormalizable interactions, except in certain theories with "extended" supersymmetry which means that they have more than one distinct copy of the supersymmetry generators $Q, Q^\dagger$. Such theories are mathematically amusing, but evidently do not have any phenomenological prospects. The reason is that extended supersymmetry in four-dimensional field theories cannot allow for chiral fermions or parity violation as observed in the Standard Model. So we will not discuss such possibilities further, although extended supersymmetry in higher dimensional field theories might describe the real world if the extra dimensions are compactified, and extended supersymmetry in four dimensions provides interesting toy models. The ordinary, non-extended, phenomenologically-viable type of supersymmetric model is sometimes called $N = 1$ supersymmetry, with N referring to the number of supersymmetries (the number of distinct copies of $Q, Q^\dagger$).

In a supersymmetric extension of the Standard Model,[4,5,6] each of the known fundamental particles must therefore be in either a chiral or gauge supermultiplet and have a superpartner with spin differing by 1/2 unit. The first step in understanding the exciting phenomenological consequences of this prediction is to decide how the known particles fit into supermultiplets, and to give them appropriate names. A crucial observation here is that only chiral supermultiplets can contain fermions whose left-handed parts transform differently under the gauge group than their right-handed parts. All of the Standard Model fermions (the known quarks and leptons) have this property, so they must be members of chiral supermultiplets.[a] The names for the spin-0 partners of the quarks and leptons are constructed by prepending an "s", which is short for scalar. Thus generically they are called *squarks* and *sleptons* (short for "scalar quark" and "scalar lepton"). The left-handed and right-handed pieces of the quarks and leptons are separate two-component Weyl fermions with different gauge transformation properties in the Standard Model, so each must have its own complex scalar partner. The symbols for the squarks and sleptons are the same as for the corresponding fermion, but with a tilde used to denote the superpartner of a Standard Model particle. For example, the superpartners of the left-handed and right-handed parts of the electron Dirac

---

[a]In particular, one cannot attempt to make a spin-1/2 neutrino be the superpartner of the spin-1 photon; the neutrino is in a doublet, and the photon neutral, under weak isospin.

field are called left- and right-handed selectrons, and are denoted $\tilde{e}_L$ and $\tilde{e}_R$. It is important to keep in mind that the "handedness" here does not refer to the helicity of the selectrons (they are spin-0 particles) but to that of their superpartners. A similar nomenclature applies for smuons and staus: $\tilde{\mu}_L$, $\tilde{\mu}_R$, $\tilde{\tau}_L$, $\tilde{\tau}_R$. In the Standard Model the neutrinos are always left-handed, so the sneutrinos are denoted generically by $\tilde{\nu}$, with a possible subscript indicating which lepton flavor they carry: $\tilde{\nu}_e$, $\tilde{\nu}_\mu$, $\tilde{\nu}_\tau$. Finally, a complete list of the squarks is $\tilde{q}_L$, $\tilde{q}_R$ with $q = u, d, s, c, b, t$. The gauge interactions of each of these squark and slepton field are the same as for the corresponding Standard Model fermion; for instance, a left-handed squark like $\tilde{u}_L$ will couple to the W boson while $\tilde{u}_R$ will not.

It seems clear that the Higgs scalar boson must reside in a chiral supermultiplet, since it has spin 0. Actually, it turns out that one chiral supermultiplet is not enough. One way to see this is to note that if there were only one Higgs chiral supermultiplet, the electroweak gauge symmetry would suffer a triangle gauge anomaly, and would be inconsistent as a quantum theory. This is because the conditions for cancellation of gauge anomalies include $\mathrm{Tr}[Y^3] = \mathrm{Tr}[T_3^2 Y] = 0$, where T_3 and Y are the third component of weak isospin and the weak hypercharge, respectively, in a normalization where the ordinary electric charge is $Q_{\mathrm{EM}} = T_3 + Y$. The traces run over all of the left-handed Weyl fermionic degrees of freedom in the theory. In the Standard Model, these conditions are already satisfied, somewhat miraculously, by the known quarks and leptons. Now, a fermionic partner of a Higgs chiral supermultiplet must be a weak isodoublet with weak hypercharge $Y = 1/2$ or $Y = -1/2$. In either case alone, such a fermion will make a non-zero contribution to the traces and spoil the anomaly cancellation. This can be avoided if there are two Higgs supermultiplets, one with each of $Y = \pm 1/2$. In that case the total contribution to the anomaly traces from the two fermionic members of the Higgs chiral supermultiplets will vanish. It turns out (as we will see in section 5.1) that both of these are also necessary for another completely different reason: because of the structure of supersymmetric theories, only a $Y = +1/2$ Higgs chiral supermultiplet can have the Yukawa couplings necessary to give masses to charge $+2/3$ up-type quarks (up, charm, top), and only a $Y = -1/2$ Higgs can have the Yukawa couplings necessary to give masses to charge $-1/3$ down-type quarks (down, strange, bottom) and to charged leptons. We will call the $SU(2)_L$-doublet complex scalar fields corresponding to these two cases H_u and H_d respectively.[b] The weak isospin components of H_u

---

[b]Other notations which are popular in the literature have $H_d, H_u \to H_1, H_2$ or $H, \overline{H}$. The one used here has the virtue of making it easy to remember which Higgs is responsible for giving masses to which quarks.

Table 1: Chiral supermultiplets in the Minimal Supersymmetric Standard Model.

Names		spin 0	spin 1/2	$SU(3)_C,\, SU(2)_L,\, U(1)_Y$
squarks, quarks	Q	$(\widetilde{u}_L \ \ \widetilde{d}_L)$	$(u_L \ \ d_L)$	$(\mathbf{3},\, \mathbf{2},\, \frac{1}{6})$
($\times 3$ families)	$\overline{u}$	$\widetilde{u}_R^*$	$u_R^\dagger$	$(\overline{\mathbf{3}},\, \mathbf{1},\, -\frac{2}{3})$
	$\overline{d}$	$\widetilde{d}_R^*$	$d_R^\dagger$	$(\overline{\mathbf{3}},\, \mathbf{1},\, \frac{1}{3})$
sleptons, leptons	L	$(\widetilde{\nu} \ \ \widetilde{e}_L)$	$(\nu \ \ e_L)$	$(\mathbf{1},\, \mathbf{2},\, -\frac{1}{2})$
($\times 3$ families)	$\overline{e}$	$\widetilde{e}_R^*$	$e_R^\dagger$	$(\mathbf{1},\, \mathbf{1},\, 1)$
Higgs, higgsinos	H_u	$(H_u^+ \ \ H_u^0)$	$(\widetilde{H}_u^+ \ \ \widetilde{H}_u^0)$	$(\mathbf{1},\, \mathbf{2},\, +\frac{1}{2})$
	H_d	$(H_d^0 \ \ H_d^-)$	$(\widetilde{H}_d^0 \ \ \widetilde{H}_d^-)$	$(\mathbf{1},\, \mathbf{2},\, -\frac{1}{2})$

with $T_3 = (+1/2, -1/2)$ have electric charges 1, 0 respectively, and are denoted (H_u^+, H_u^0). Similarly, the $SU(2)_L$-doublet complex scalar H_d has $T_3 = (+1/2, -1/2)$ components (H_d^0, H_d^-). The neutral scalar that corresponds to the physical Standard Model Higgs boson is in a linear combination of H_u^0 and H_d^0; we will discuss this further in section 7.2. The generic nomenclature for a spin-1/2 superpartner is to append "-ino" to the name of the Standard Model particle, so the fermionic partners of the Higgs scalars are called higgsinos. They are denoted by $\widetilde{H}_u$, $\widetilde{H}_d$ for the $SU(2)_L$-doublet left-handed Weyl spinor fields, with weak isospin components $\widetilde{H}_u^+$, $\widetilde{H}_u^0$ and $\widetilde{H}_d^0$, $\widetilde{H}_d^-$.

We have now found all of the chiral supermultiplets of a minimal phenomenologically viable extension of the Standard Model. They are summarized in Table 1, classified according to their transformation properties under the Standard Model gauge group $SU(3)_C \times SU(2)_L \times U(1)_Y$, which combines u_L, d_L and ν, e_L degrees of freedom into $SU(2)_L$ doublets. Here we have followed the standard convention that all chiral supermultiplets are defined in terms of left-handed Weyl spinors, so that the *conjugates* of the right-handed quarks and leptons (and their superpartners) appear in Table 1. This protocol for defining chiral supermultiplets turns out to be very useful for constructing supersymmetric lagrangians, as we will see in section 3. It is useful also to have a symbol for each of the chiral supermultiplets as a whole; these are indicated in the second column of Table 1. Thus for example Q stands for the $SU(2)_L$-doublet chiral supermultiplet containing $\widetilde{u}_L, u_L$ (with weak isospin component $T_3 = +1/2$), and $\widetilde{d}_L, d_L$ (with $T_3 = -1/2$), while $\overline{u}$ stands for the $SU(2)_L$-

Table 2: Gauge supermultiplets in the Minimal Supersymmetric Standard Model.

Names	spin 1/2	spin 1	$SU(3)_C,\ SU(2)_L,\ U(1)_Y$
gluino, gluon	$\widetilde{g}$	g	(**8**, **1** , 0)
winos, W bosons	$\widetilde{W}^{\pm}\ \widetilde{W}^0$	$W^{\pm}\ W^0$	(**1**, **3** , 0)
bino, B boson	$\widetilde{B}^0$	B^0	(**1**, **1** , 0)

singlet supermultiplet containing $\widetilde{u}_R^*,\, u_R^\dagger$. There are three families for each of the quark and lepton supermultiplets, but we have used first-family representatives in Table 1. Below, a family index $i = 1, 2, 3$ will be affixed to the chiral supermultiplet names ($Q_i,\ \overline{u}_i, \ldots$) when needed, e.g. $(\overline{e}_1, \overline{e}_2, \overline{e}_3) = (\overline{e}, \overline{\mu}, \overline{\tau})$. The bar on $\overline{u}$, $\overline{d}$, $\overline{e}$ fields is part of the name, and does not denote any kind of conjugation. It is interesting to note that the Higgs chiral supermultiplet H_d (containing H_d^0, H_d^-, $\widetilde{H}_d^0$, $\widetilde{H}_d^-$) has exactly the same Standard Model gauge quantum numbers as the left-handed sleptons and leptons L_i, e.g. $(\widetilde{\nu}, \widetilde{e}_L, \nu, e_L)$. Naively one might therefore suppose that we could have been more economical in our assignment by taking a neutrino and a Higgs scalar to be superpartners, instead of putting them in separate supermultiplets. This would amount to the proposal that the Higgs boson and a sneutrino should be the same particle. This is a nice try which played a key role in some of the first attempts to connect supersymmetry to phenomenology,[4] but it is now known not to work. Even ignoring the anomaly cancellation problem mentioned above, many insoluble phenomenological problems would result, including lepton number violation and a mass for at least one of the neutrinos in gross violation of experimental bounds. Therefore, all of the superpartners of Standard Model particles are really new particles, and cannot be identified with some other Standard Model state.

The vector bosons of the Standard Model clearly must reside in gauge supermultiplets. Their fermionic superpartners are generically referred to as gauginos. The $SU(3)_C$ color gauge interactions of QCD are mediated by the gluon, whose spin-1/2 color-octet supersymmetric partner is the gluino. As usual, a tilde is used to denote the supersymmetric partner of a Standard Model state, so the symbols for the gluon and gluino are g and $\widetilde{g}$ respectively. The electroweak gauge symmetry $SU(2)_L \times U(1)_Y$ has associated with it spin-1 gauge bosons W^+, W^0, W^- and B^0, with spin-1/2 superpartners $\widetilde{W}^+, \widetilde{W}^0, \widetilde{W}^-$ and $\widetilde{B}^0$, called *winos* and *bino*. After electroweak symmetry breaking, the W^0,

B^0 gauge eigenstates mix to give mass eigenstates Z^0 and γ. The corresponding gaugino mixtures of $\widetilde{W}^0$ and $\widetilde{B}^0$ are called zino ($\widetilde{Z}^0$) and photino ($\widetilde{\gamma}$); if supersymmetry were unbroken, they would be mass eigenstates with masses m_Z and 0. Table 2 summarizes the gauge supermultiplets of a minimal supersymmetric extension of the Standard Model.

The chiral and gauge supermultiplets in Tables 1 and 2 make up the particle content of the Minimal Supersymmetric Standard Model (MSSM). The most obvious and interesting feature of this theory is that none of the superpartners of the Standard Model particles has been discovered as of this writing. If supersymmetry were unbroken, then there would have to be selectrons $\widetilde{e}_L$ and $\widetilde{e}_R$ with masses exactly equal to $m_e = 0.511...$ MeV. A similar statement applies to each of the other sleptons and squarks, and there would also have to be a massless gluino and photino. These particles would be extraordinarily easy to detect. Clearly, therefore, *supersymmetry is a broken symmetry* in the vacuum state chosen by nature.

A very important clue as to the nature of supersymmetry breaking can be obtained by returning to the motivation provided by the hierarchy problem. Supersymmetry forced us to introduce two complex scalar fields for each Standard Model Dirac fermion, which is just what is needed to enable a cancellation of the quadratically divergent ($\Lambda_{\rm UV}^2$) pieces of eqs. (1.2) and (1.3). This sort of cancellation also requires that the associated dimensionless couplings should be related (e.g. $\lambda_S = |\lambda_f|^2$). The necessary relationships between couplings indeed occur in unbroken supersymmetry, as we will see in section 3. In fact, unbroken supersymmetry guarantees that the quadratic divergences in scalar squared masses must vanish to all orders in perturbation theory.[c] Now, if broken supersymmetry is to provide a solution to the hierarchy problem, then the relationships between dimensionless couplings which hold in an unbroken supersymmetric theory must be maintained. Otherwise, there would be quadratically divergent radiative corrections to the Higgs scalar masses of the form

$$\Delta m_H^2 = \frac{1}{8\pi^2}(\lambda_S - |\lambda_f|^2)\Lambda_{\rm UV}^2 + \cdots. \tag{1.11}$$

We are therefore led to consider "soft" supersymmetry breaking. This means that the effective lagrangian of the MSSM can be written in the form

$$\mathcal{L} = \mathcal{L}_{\rm SUSY} + \mathcal{L}_{\rm soft}, \tag{1.12}$$

---

[c]A simple way to understand this is to note that unbroken supersymmetry requires the degeneracy of scalar and fermion masses. Radiative corrections to fermion masses are known to diverge at most logarithmically, so the same must be true for scalar masses in unbroken supersymmetry.

where $\mathcal{L}_{\text{SUSY}}$ preserves supersymmetry invariance, and $\mathcal{L}_{\text{soft}}$ violates supersymmetry but contains only mass terms and couplings with *positive* mass dimension. Without further justification, soft supersymmetry breaking might seem like a rather arbitrary requirement. Fortunately, theoretical models for supersymmetry breaking can indeed yield effective lagrangians with just such terms for $\mathcal{L}_{\text{soft}}$. If the largest mass scale associated with the soft terms is denoted m_{soft}, then the additional non-supersymmetric corrections to the Higgs scalar (mass)^2 must vanish in the $m_{\text{soft}} \to 0$ limit, so by dimensional analysis they cannot be proportional to Λ_{UV}^2. More generally, these models maintain the cancellation of quadratically divergent terms in the radiative corrections of all scalar masses, to all orders in perturbation theory. The corrections also cannot go like $\Delta m_H^2 \sim m_{\text{soft}} \Lambda_{\text{UV}}$, because in general the loop momentum integrals always diverge either quadratically or logarithmically, not linearly, as $\Lambda_{\text{UV}} \to \infty$. So they must be of the form

$$\Delta m_H^2 = m_{\text{soft}}^2 \left[\frac{\lambda}{16\pi^2} \ln(\Lambda_{\text{UV}}/m_{\text{soft}}) + \ldots \right]. \qquad (1.13)$$

Here λ is schematic for various dimensionless couplings, and the ellipses stand both for terms which are independent of Λ_{UV} and for higher loop corrections (which depend on Λ_{UV} through powers of logarithms).

Since the mass splittings between the known Standard Model particles and their superpartners are just determined by the parameters m_{soft} appearing in $\mathcal{L}_{\text{soft}}$, eq. (1.13) tells us that the superpartner masses cannot be too huge. Otherwise, we would lose our successful cure for the hierarchy problem since the m_{soft}^2 corrections to the Higgs scalar (mass)^2 would be unnaturally large compared to the electroweak breaking scale of 174 GeV. The top and bottom squarks and the winos and bino give especially large contributions to $\Delta m_{H_u}^2$ and $\Delta m_{H_d}^2$, but the gluino mass and all the other squark and slepton masses also feed in indirectly, through radiative corrections to the top and bottom squark masses. Furthermore, in most viable models of supersymmetry breaking that are not unduly contrived, the superpartner masses do not differ from each other by more than about an order of magnitude. Using $\Lambda_{\text{UV}} \sim M_{\text{P}}$ and $\lambda \sim 1$ in eq. (1.13), one finds that roughly speaking m_{soft}, and therefore the masses of at least the lightest few superpartners, should be at the most about 1 TeV or so, in order for the MSSM scalar potential to provide a Higgs VEV resulting in $m_W, m_Z = 80.4, 91.2$ GeV without miraculous cancellations. This is the best reason for the optimism among many theorists that supersymmetry will be discovered at LEP2, the Tevatron, the LHC, or a next generation lepton linear collider.

However, it is useful to keep in mind that the hierarchy problem was *not* the historical motivation for the development of supersymmetry in the early 1970's. The supersymmetry algebra and supersymmetric field theories were originally concocted independently in various disguises [7,8,9,10] which bear little resemblance to the MSSM. It is quite impressive that a theory which was developed for quite different reasons, including purely aesthetic ones, can later be found to provide a solution for the hierarchy problem.

One might also wonder if there is any good reason why all of the superpartners of the Standard Model particles should be heavy enough to have avoided discovery so far. There is. All of the particles in the MSSM which have been discovered so far have something in common; they would necessarily be massless in the absence of electroweak symmetry breaking. In particular, the masses of the $W^\pm, Z^0$ bosons and all quarks and leptons are equal to dimensionless coupling constants times the Higgs VEV $\sim$ 174 GeV, while the photon and gluon are required to be massless by electromagnetic and QCD gauge invariance. Conversely, all of the undiscovered particles in the MSSM have exactly the opposite property, since each of them can have a lagrangian mass term in the absence of electroweak symmetry breaking. For the squarks, sleptons, and Higgs scalars this follows from a general property of complex scalar fields that a mass term $m^2|\phi|^2$ is always allowed by all gauge symmetries. For the higgsinos and gauginos, it follows from the fact that they are in a real representation of the gauge group. From the point of view of the MSSM, then, the discovery of the top quark in 1995 marked a quite natural milestone; the already-discovered particles are precisely those which had to be light, based on the principle of electroweak gauge symmetry. There is a single exception: one neutral Higgs scalar boson should be lighter than about 150 GeV if supersymmetry is correct, for reasons to be discussed in section 7.2.

A very important feature of the MSSM is that the superpartners listed in Tables 1 and 2 are not necessarily the mass eigenstates of the theory. This is because after electroweak symmetry breaking and supersymmetry breaking effects are included, there can be mixing between the electroweak gauginos and the higgsinos, and between the various squarks and sleptons and Higgs scalars which have the same electric charge. The lone exception is the gluino, which is a color octet fermion and therefore does not have the appropriate quantum numbers to mix with any other particle. The masses and mixings of the superpartners are obviously of paramount importance to experimentalists. It is perhaps slightly less obvious that these phenomenological issues are all quite directly related to one central question which is also the focus of much of the theoretical work in supersymmetry: "How is supersymmetry broken?" The reason for this is that most of what we do not already know about the MSSM has to do with $\mathcal{L}_{\text{soft}}$. The structure of supersymmetric lagrangians allows very little arbitrariness, as we will see in section 3. In fact, all of the dimensionless couplings and all but one mass term in the supersymmetric part of the MSSM lagrangian correspond directly to some parameter in the ordinary Standard Model which has already been measured by experiment. For example, we will find out that the supersymmetric coupling of a gluino to a squark and a quark is determined by the QCD coupling constant α_S. In contrast, the supersymmetry-breaking part of the lagrangian apparently contains many unknown parameters and a considerable amount of arbitrariness. Each of the mass splittings between Standard Model particles and their superpartners correspond to terms in the MSSM lagrangian which are purely supersymmetry-breaking in their origin and effect. These soft supersymmetry-breaking terms can also introduce a large number of mixing angles and CP-violating phases not found in the Standard Model. Fortunately, as we will see in section 5.4, there is already rather strong evidence that the supersymmetry-breaking terms in the MSSM are actually not arbitrary at all. Furthermore, the additional parameters will be measured and constrained as the superpartners are detected. From a theoretical perspective, the challenge is to explain all of these parameters with a model for supersymmetry breaking.

The rest of our discussion is organized as follows. Section 2 provides a list of important notations. In section 3, we will learn how to construct lagrangians for supersymmetric field theories. Soft supersymmetry-breaking couplings are described in section 4. In section 5, we will apply the preceding general results to the special case of the MSSM, introduce the concept of R-parity, and emphasize the importance of the structure of the soft terms. In section 7, we will study the mass and mixing angle patterns of the new particles predicted by the MSSM. The discussion will be lacking in historical accuracy or perspective, for which the author apologizes in advance. The reader is encouraged to consult the many outstanding textbooks,[11–18] review articles,[19–34] and the reprint volume,[35] which contain a much more consistent guide to the original literature.

2 Interlude: Notations and Conventions

Before proceeding to discuss the construction of supersymmetric lagrangians, we need to specify our notations. It is overwhelmingly convenient to employ two-component Weyl notation for fermions, rather than four component Dirac or Majorana spinors. The lagrangian of the Standard Model (and supersymmetric extensions of it) violates parity; each Dirac fermion has left-handed and right-handed parts with completely different electroweak gauge interactions. If one used four-component notation, one would therefore have to include clumsy left- and right-handed projection operators

$$P_{L,R} = (1 \pm \gamma_5)/2 \qquad (2.1)$$

all over the place. The two-component Weyl fermion notation has the advantage of treating fermionic degrees of freedom with different gauge quantum numbers separately from the start (as Nature intended for us to do). But an even better reason for using two-component notation here is that in supersymmetric models the minimal building blocks of matter are chiral supermultiplets, each of which contains a single two-component Weyl fermion.

Since two-component fermion notation may be unfamiliar to some readers, we will specify our conventions[d] by showing how they correspond to the four-component fermion language. A four-component Dirac fermion Ψ_D with mass M is described by the lagrangian

$$\mathcal{L}_{\text{Dirac}} = -i\overline{\Psi}_D \gamma^\mu \partial_\mu \Psi_D - M\overline{\Psi}_D \Psi_D . \qquad (2.2)$$

We use a spacetime metric $\eta_{\mu\nu} = \text{diag}(-1,1,1,1)$. For our purposes it is convenient to use the specific representation of the 4×4 gamma matrices given in 2×2 blocks by

$$\gamma_\mu = \begin{pmatrix} 0 & \sigma_\mu \\ \overline{\sigma}_\mu & 0 \end{pmatrix}; \qquad \gamma_5 = \begin{pmatrix} 1 & 0 \\ 0 & -1 \end{pmatrix}, \qquad (2.3)$$

where

$$\sigma_0 = \overline{\sigma}_0 = \begin{pmatrix} 1 & 0 \\ 0 & 1 \end{pmatrix}; \qquad \sigma_1 = -\overline{\sigma}_1 = \begin{pmatrix} 0 & 1 \\ 1 & 0 \end{pmatrix};$$

$$\sigma_2 = -\overline{\sigma}_2 = \begin{pmatrix} 0 & -i \\ i & 0 \end{pmatrix}; \qquad \sigma_3 = -\overline{\sigma}_3 = \begin{pmatrix} 1 & 0 \\ 0 & -1 \end{pmatrix}. \qquad (2.4)$$

In this basis, a four component Dirac spinor is written in terms of 2 two-component, complex, anticommuting objects $(\xi)_\alpha$ with $\alpha = 1,2$ and $(\chi^\dagger)^{\dot\alpha}$ with $\dot\alpha = 1,2$:

$$\Psi_D = \begin{pmatrix} \xi_\alpha \\ \chi^{\dagger\dot\alpha} \end{pmatrix}; \qquad \overline{\Psi}_D = \begin{pmatrix} \chi^\alpha & \xi^\dagger_{\dot\alpha} \end{pmatrix}. \qquad (2.5)$$

The undotted (dotted) indices are used for the first (last) two components of a Dirac spinor. The heights of these indices are important; for example,

---

[d]The conventions used here are the same as in Ref.[11], except that we use a dagger rather than a bar to indicate hermitian conjugation for Weyl spinors.

comparing eqs. (2.2)-(2.5), we observe that the matrices $(\sigma^\mu)_{\alpha\dot\alpha}$ and $(\overline\sigma^\mu)^{\dot\alpha\alpha}$ defined by eq. (2.4) carry indices with the heights as indicated. The spinor indices are raised and lowered using the antisymmetric symbol $\epsilon^{12} = -\epsilon^{21} = \epsilon_{21} = -\epsilon_{12} = 1$; $\epsilon_{11} = \epsilon_{22} = \epsilon^{11} = \epsilon^{22} = 0$, according to

$$\xi_\alpha = \epsilon_{\alpha\beta}\xi^\beta; \qquad \xi^\alpha = \epsilon^{\alpha\beta}\xi_\beta; \qquad \chi^\dagger_{\dot\alpha} = \epsilon_{\dot\alpha\dot\beta}\chi^{\dagger\dot\beta}; \qquad \chi^{\dagger\dot\alpha} = \epsilon^{\dot\alpha\dot\beta}\chi^\dagger_{\dot\beta}. \qquad (2.6)$$

This is consistent since $\epsilon_{\alpha\beta}\epsilon^{\beta\gamma} = \epsilon^{\gamma\beta}\epsilon_{\beta\alpha} = \delta^\gamma_\alpha$ and $\epsilon_{\dot\alpha\dot\beta}\epsilon^{\dot\beta\dot\gamma} = \epsilon^{\dot\gamma\dot\beta}\epsilon_{\dot\beta\dot\alpha} = \delta^{\dot\gamma}_{\dot\alpha}$. The field ξ is called a "left-handed Weyl spinor" and $\chi^\dagger$ is a "right-handed Weyl spinor". The names fit, because

$$P_L\Psi_D = \begin{pmatrix} \xi_\alpha \\ 0 \end{pmatrix}; \qquad P_R\Psi_D = \begin{pmatrix} 0 \\ \chi^{\dagger\dot\alpha} \end{pmatrix}. \qquad (2.7)$$

The hermitian conjugate of a left-handed Weyl spinor is a right-handed Weyl spinor $(\psi_\alpha)^\dagger = (\psi^\dagger)_{\dot\alpha}$ and vice versa $(\psi^{\dagger\dot\alpha})^\dagger = \psi^\alpha$. Therefore any particular fermionic degrees of freedom can be described equally well using a Weyl spinor which is left-handed (with an undotted index) or one which is right-handed (with a dotted index). By convention all names of fermion fields are chosen so that left-handed Weyl spinors do not carry daggers and right-handed Weyl spinors do carry daggers, as in eq. (2.5).

It is useful to abbreviate expressions with two spinor fields by suppressing undotted indices contracted like $^\alpha_{\ \alpha}$ and dotted indices contracted like $_{\dot\alpha}^{\ \dot\alpha}$. In particular,

$$\xi\chi \equiv \xi^\alpha\chi_\alpha = \xi^\alpha\epsilon_{\alpha\beta}\chi^\beta = -\chi^\beta\epsilon_{\alpha\beta}\xi^\alpha = \chi^\beta\epsilon_{\beta\alpha}\xi^\alpha = \chi^\beta\xi_\beta \equiv \chi\xi \qquad (2.8)$$

with, conveniently, no minus sign in the end. [A minus sign appeared in eq. (2.8) from exchanging the order of anticommuting spinors, but it disappeared due to the antisymmetry of the ϵ symbol.] Likewise, $\xi^\dagger\chi^\dagger$ and $\chi^\dagger\xi^\dagger$ are equivalent abbreviations for $\chi^\dagger_{\dot\alpha}\xi^{\dagger\dot\alpha} = (\xi\chi)^*$, the complex conjugate of $\xi\chi$. In a similar way,

$$\xi^\dagger\overline\sigma^\mu\chi = -\chi\sigma^\mu\xi^\dagger = (\chi^\dagger\overline\sigma^\mu\xi)^* = -(\xi\sigma^\mu\chi^\dagger)^* \qquad (2.9)$$

stands for $\xi^\dagger_{\dot\alpha}(\overline\sigma^\mu)^{\dot\alpha\alpha}\chi_\alpha$, etc. With these conventions, the Dirac lagrangian eq. (2.2) can now be rewritten:

$$\begin{aligned} \mathcal{L}_{\text{Dirac}} &= -i\overline\Psi_D\gamma^\mu\partial_\mu\Psi_D - M\overline\Psi_D\Psi_D & (2.10) \\ &= -i\xi^\dagger\overline\sigma^\mu\partial_\mu\xi - i\chi^\dagger\overline\sigma^\mu\partial_\mu\chi - M(\xi\chi + \xi^\dagger\chi^\dagger) & (2.11) \end{aligned}$$

where we have dropped a total derivative piece $i\partial_\mu(\chi^\dagger\overline\sigma^\mu\chi)$ which does not affect the action.

A four-component Majorana spinor can be obtained from the Dirac spinor of eq. (2.5) by imposing the constraint $\chi = \xi$, so that

$$\Psi_{\mathrm{M}} = \begin{pmatrix} \xi_\alpha \\ \xi^{\dagger\dot\alpha} \end{pmatrix}; \qquad \overline{\Psi}_{\mathrm{M}} = \begin{pmatrix} \xi^\alpha & \xi^\dagger_{\dot\alpha} \end{pmatrix}. \tag{2.12}$$

The lagrangian for a Majorana fermion with mass M

$$\mathcal{L}_{\mathrm{Majorana}} = -\frac{i}{2}\overline{\Psi}_{\mathrm{M}}\gamma^\mu\partial_\mu\Psi_{\mathrm{M}} - \frac{1}{2}M\overline{\Psi}_{\mathrm{M}}\Psi_{\mathrm{M}} \tag{2.13}$$

in the four-component Majorana spinor form can therefore be rewritten

$$\mathcal{L}_{\mathrm{Majorana}} = -i\xi^\dagger\overline{\sigma}^\mu\partial_\mu\xi - \frac{1}{2}M(\xi\xi + \xi^\dagger\xi^\dagger) \tag{2.14}$$

in the more economical two-component Weyl spinor representation. [Note that even though ξ_α is anticommuting, $\xi\xi$ and its complex conjugate $\xi^\dagger\xi^\dagger$ do not vanish, because of the suppressed ϵ symbol, see eq. (2.8).]

More generally, any theory involving spin-1/2 fermions can be written down in terms of a collection of left-handed Weyl spinors ψ_i with

$$\mathcal{L} = -i\psi^{\dagger i}\overline{\sigma}^\mu\partial_\mu\psi_i + \ldots \tag{2.15}$$

where the ellipses represent possible mass terms, gauge interactions, and Yukawa interactions with scalar fields. Here the index i runs over the appropriate gauge and flavor indices of the fermions; it is raised or lowered by hermitian conjugation. There is a different ψ_i for the left-handed piece and for the hermitian conjugate of the right-handed piece of a Dirac fermion. If one has any expression involving bilinears in four-component spinors

$$\Psi_1 = \begin{pmatrix} \xi_1 \\ \chi_1^\dagger \end{pmatrix} \quad \text{and} \quad \Psi_2 = \begin{pmatrix} \xi_2 \\ \chi_2^\dagger \end{pmatrix}, \tag{2.16}$$

then one can translate into two-component Weyl spinor language (or vice versa) using the dictionary:

$$\overline{\Psi}_1 P_L \Psi_2 = \chi_1\xi_2; \qquad\qquad \overline{\Psi}_1 P_R \Psi_2 = \xi_1^\dagger\chi_2^\dagger; \tag{2.17}$$

$$\overline{\Psi}_1 \gamma^\mu P_L \Psi_2 = \xi_1^\dagger\overline{\sigma}^\mu\xi_2; \qquad\qquad \overline{\Psi}_1 \gamma^\mu P_R \Psi_2 = \chi_1\sigma^\mu\chi_2^\dagger \tag{2.18}$$

etc. We will introduce a few other Weyl spinor identities in the following as they are needed.

Let us now see how the Standard Model quarks and leptons are described in this notation. The complete list of left-handed Weyl spinors can be given names corresponding to the chiral supermultiplets in Table 1:

$$Q_i \;=\; (u\,d),\; (c\,s),\; (t\,b) \qquad\qquad\qquad (2.19)$$

$$\overline{u}_i \;=\; \overline{u},\, \overline{c},\, \overline{t} \qquad\qquad \overline{d}_i \;=\; \overline{d},\, \overline{s},\, \overline{b} \qquad (2.20)$$

$$L_i \;=\; (\nu_e\, e),\; (\nu_\mu\, \mu),\; (\nu_\tau\, \tau) \qquad\qquad (2.21)$$

$$\overline{e}_i \;=\; \overline{e},\, \overline{\mu},\, \overline{\tau}. \qquad\qquad\qquad (2.22)$$

Here $i = 1, 2, 3$ is a family index. The bars on these fields are part of the names of the fields, and do *not* denote any kind of conjugation. Rather, the unbarred fields are the left-handed pieces of a Dirac spinor, while the barred fields are the names given to the conjugates of the right-handed piece of a Dirac spinor. For example, e is the same thing as e_L in Table 1, and $\overline{e}$ is the same as $e_R^\dagger$. Together they form a Dirac spinor:

$$\begin{pmatrix} e \\ \overline{e}^\dagger \end{pmatrix} \equiv \begin{pmatrix} e_L \\ e_R \end{pmatrix} \qquad\qquad (2.23)$$

with similar equations for all of the other quark and charged lepton Dirac spinors. (The neutrinos of the Standard Model are not part of a Dirac spinor.) The fields Q_i and L_i are weak isodoublets which always go together when one is constructing interactions invariant under the full Standard Model gauge group $SU(3)_C \times SU(2)_L \times U(1)_Y$. Suppressing all color and weak isospin indices, the purely kinetic part of the Standard Model fermion lagrangian density is then

$$\begin{aligned} \mathcal{L} \;=\; & -iQ^{\dagger i}\overline{\sigma}^\mu \partial_\mu Q_i - i\overline{u}^{\dagger i}\overline{\sigma}^\mu \partial_\mu \overline{u}_i - i\overline{d}^{\dagger i}\overline{\sigma}^\mu \partial_\mu \overline{d}_i \\ & -iL^{\dagger i}\overline{\sigma}^\mu \partial_\mu L_i - i\overline{e}^{\dagger i}\overline{\sigma}^\mu \partial_\mu \overline{e}_i \end{aligned} \qquad (2.24)$$

with the family index $i = 1, 2, 3$ summed over.

3 Supersymmetric lagrangians

In this section we will describe the construction of supersymmetric lagrangians. Our aim is to arrive at a sort of recipe which will allow us to write down the allowed interactions and mass terms of a general supersymmetric theory, so that later we can apply the results to the special case of the MSSM. We will not use the superfield language,[36] which is often more elegant and efficient for those who know it, but which might seem rather cabalistic to some readers. Our approach is therefore intended to be rather complementary to the superfield derivations given in Refs.[11-18] We begin by considering the simplest example of a supersymmetric theory in four dimensions.

3.1 The simplest supersymmetric model: a free chiral supermultiplet

The minimum fermion content of any theory in four dimensions consists of a single left-handed two-component Weyl fermion ψ. Since this is an intrinsically complex object, it seems sensible to choose as its superpartner a complex scalar field ϕ. The simplest action we can write down for these fields just consists of kinetic energy terms for each:

$$S = \int d^4x \; (\mathcal{L}_{\text{scalar}} + \mathcal{L}_{\text{fermion}}) \tag{3.1}$$

$$\mathcal{L}_{\text{scalar}} = -\partial^\mu \phi^* \partial_\mu \phi; \qquad \mathcal{L}_{\text{fermion}} = -i\psi^\dagger \overline{\sigma}^\mu \partial_\mu \psi. \tag{3.2}$$

This is called the massless, non-interacting *Wess-Zumino model*,[9] and it corresponds to a single chiral supermultiplet as discussed in the Introduction.

A supersymmetry transformation should turn the scalar boson ϕ into something involving the fermion ψ_α. The simplest possibility for the transformation of the scalar field is

$$\delta\phi = \epsilon\psi; \qquad \delta\phi^* = \epsilon^\dagger \psi^\dagger \tag{3.3}$$

where ϵ^α is an infinitesimal, anticommuting, two-component Weyl fermion object which parameterizes the supersymmetry transformation. Here, we are discussing global supersymmetry, which means that ϵ^α is a constant, satisfying $\partial_\mu \epsilon^\alpha = 0$. Since ψ has dimensions of $(\text{mass})^{3/2}$ and ϕ has dimensions of (mass), it must be that ϵ has dimensions of $(\text{mass})^{-1/2}$. Using eq. (3.3), we find that the scalar part of the lagrangian transforms as

$$\delta\mathcal{L}_{\text{scalar}} = -\epsilon\partial^\mu\psi \, \partial_\mu\phi^* - \epsilon^\dagger \partial^\mu \psi^\dagger \, \partial_\mu\phi. \tag{3.4}$$

We would like for this to be cancelled by $\delta\mathcal{L}_{\text{fermion}}$, at least up to a total derivative, so that the action will be invariant under the supersymmetry transformation. Comparing eq. (3.4) with $\mathcal{L}_{\text{fermion}}$, we see that for this to have any chance of happening, $\delta\psi$ should be linear in $\epsilon^\dagger$ and in ϕ and contain one spacetime derivative. Up to a multiplicative constant, there is only one possibility to try:

$$\delta\psi_\alpha = i(\sigma^\mu \epsilon^\dagger)_\alpha \, \partial_\mu\phi; \qquad \delta\psi_{\dot\alpha}^\dagger = -i(\epsilon\sigma^\mu)_{\dot\alpha} \, \partial_\mu\phi^*. \tag{3.5}$$

With this guess, one immediately obtains

$$\delta\mathcal{L}_{\text{fermion}} = -\epsilon\sigma^\mu\overline{\sigma}^\nu \partial_\nu\psi \, \partial_\mu\phi^* + \psi^\dagger\overline{\sigma}^\nu\sigma^\mu\epsilon^\dagger \, \partial_\mu\partial_\nu\phi \,. \tag{3.6}$$

This can be put in a slightly more useful form by employing the Pauli matrix identities

$$\left[\sigma^\mu\overline{\sigma}^\nu + \sigma^\nu\overline{\sigma}^\mu\right]_\alpha^{\;\beta} = -2\eta^{\mu\nu}\delta_\alpha^\beta; \qquad \left[\overline{\sigma}^\mu\sigma^\nu + \overline{\sigma}^\nu\sigma^\mu\right]_{\dot\alpha}^{\;\dot\beta} = -2\eta^{\mu\nu}\delta_{\dot\alpha}^{\dot\beta} \tag{3.7}$$

and using the fact that partial derivatives commute ($\partial_\mu \partial_\nu = \partial_\nu \partial_\mu$). Equation (3.6) then becomes

$$\delta\mathcal{L}_{\text{fermion}} = \epsilon\partial^\mu\psi\,\partial_\mu\phi^* + \epsilon^\dagger\partial^\mu\psi^\dagger\,\partial_\mu\phi$$
$$-\partial_\mu\left(\epsilon\sigma^\nu\overline{\sigma}^\mu\psi\,\partial_\nu\phi^* + \epsilon\psi\,\partial^\mu\phi^* + \epsilon^\dagger\psi^\dagger\,\partial^\mu\phi\right). \qquad (3.8)$$

The first two terms here just cancel against $\delta\mathcal{L}_{\text{scalar}}$, while the remaining contribution is a total derivative. So we arrive at

$$\delta S = \int d^4x \;\left(\delta\mathcal{L}_{\text{scalar}} + \delta\mathcal{L}_{\text{fermion}}\right) = 0, \qquad (3.9)$$

justifying our guess of the numerical multiplicative factor made in eq. (3.5).

We are not quite finished in demonstrating that the theory described by eq. (3.1) is supersymmetric. We must also show that the supersymmetry algebra closes; in other words, that the commutator of two supersymmetry transformations is another symmetry of the theory. Using eq. (3.5) in eq. (3.3), one finds

$$(\delta_{\epsilon_2}\delta_{\epsilon_1} - \delta_{\epsilon_1}\delta_{\epsilon_2})\phi = i(\epsilon_1\sigma^\mu\epsilon_2^\dagger - \epsilon_2\sigma^\mu\epsilon_1^\dagger)\,\partial_\mu\phi. \qquad (3.10)$$

This is a remarkable result; in words, we have found that the commutator of two supersymmetry transformations gives us back the derivative of the original field. Since ∂_μ just corresponds to the generator of spacetime translations P_μ, we have reproduced the form of the supersymmetry algebra which was foreshadowed in eq. (1.6) of the Introduction. All of this will be for naught if we do not find the same result for the fermion ψ, however. Using eq. (3.3) in eq. (3.5), we find

$$(\delta_{\epsilon_2}\delta_{\epsilon_1} - \delta_{\epsilon_1}\delta_{\epsilon_2})\psi_\alpha = i(\sigma^\mu\epsilon_1^\dagger)_\alpha\,\epsilon_2\partial_\mu\psi - i(\sigma^\mu\epsilon_2^\dagger)_\alpha\,\epsilon_1\partial_\mu\psi. \qquad (3.11)$$

We can put this into a more useful form by applying the Fierz identity

$$\chi_\alpha\,(\xi\eta) = -\xi_\alpha\,(\eta\chi) - \eta_\alpha\,(\chi\xi) \qquad (3.12)$$

with $\chi = \sigma^\mu\epsilon_1^\dagger$, $\xi = \epsilon_2$, $\eta = \partial_\mu\psi$, and again with $\chi = \sigma^\mu\epsilon_2^\dagger$, $\xi = \epsilon_1$, $\eta = \partial_\mu\psi$, followed in each case by an application of the identity eq. (2.9). The result is

$$(\delta_{\epsilon_2}\delta_{\epsilon_1} - \delta_{\epsilon_1}\delta_{\epsilon_2})\psi_\alpha = i(\epsilon_1\sigma^\mu\epsilon_2^\dagger - \epsilon_2\sigma^\mu\epsilon_1^\dagger)\,\partial_\mu\psi_\alpha$$
$$-i\epsilon_{1\alpha}\,\epsilon_2^\dagger\overline{\sigma}^\mu\partial_\mu\psi + i\epsilon_{2\alpha}\,\epsilon_1^\dagger\overline{\sigma}^\mu\partial_\mu\psi. \qquad (3.13)$$

The last two terms in (3.13) vanish on-shell; that is, if the equation of motion $\overline{\sigma}^\mu\partial_\mu\psi = 0$ following from the action is enforced. The remaining piece is exactly the same spacetime translation that we found for the scalar field.

The fact that the supersymmetry algebra only closes on-shell (when the classical equations of motion are satisfied) might be somewhat worrisome, since we would like the symmetry to hold even quantum mechanically. This can be fixed by a trick. We invent a new complex scalar field F which does not have a kinetic term. Such fields are called *auxiliary*, and they are really just book-keeping devices which allow the symmetry algebra to close off-shell. The lagrangian density for F and its complex conjugate is just

$$\mathcal{L}_{\text{auxiliary}} = F^* F. \tag{3.14}$$

The dimensions of F are $(\text{mass})^2$, unlike an ordinary scalar field which has dimensions of (mass). Equation (3.14) leads to the not-very-exciting equations of motion $F = F^* = 0$. However, we can use the auxiliary fields to our advantage by including them in the supersymmetry transformation rules. In view of eq. (3.13), a plausible thing to do is to make F transform into a multiple of the equation of motion for ψ:

$$\delta F = i\epsilon^\dagger \overline{\sigma}^\mu \partial_\mu \psi; \qquad \delta F^* = -i\partial_\mu \psi^\dagger \overline{\sigma}^\mu \epsilon. \tag{3.15}$$

Once again we have chosen the overall factor on the right hand side by virtue of foresight. Now the auxiliary part of the lagrangian density transforms as

$$\delta\mathcal{L}_{\text{auxiliary}} = i\epsilon^\dagger \overline{\sigma}^\mu \partial_\mu \psi \, F^* - i\partial_\mu \psi^\dagger \overline{\sigma}^\mu \epsilon \, F \tag{3.16}$$

which vanishes on-shell, but not for arbitrary off-shell field configurations. It is easy to see that by adding an extra term to the transformation law for ψ and $\psi^\dagger$:

$$\delta\psi_\alpha = i(\sigma^\mu \epsilon^\dagger)_\alpha \, \partial_\mu \phi + \epsilon_\alpha F; \qquad \delta\psi^\dagger_{\dot\alpha} = -i(\epsilon\sigma^\mu)_{\dot\alpha} \, \partial_\mu \phi^* + \epsilon^\dagger_{\dot\alpha} F^* \tag{3.17}$$

one obtains an additional contribution to $\delta\mathcal{L}_{\text{fermion}}$ which just cancels with $\delta\mathcal{L}_{\text{auxiliary}}$, up to a total derivative term. So our "modified" theory with $\mathcal{L} = \mathcal{L}_{\text{scalar}} + \mathcal{L}_{\text{fermion}} + \mathcal{L}_{\text{auxiliary}}$ is still invariant under supersymmetry transformations. Proceeding as before, one now obtains for each of the fields $X = \phi, \phi^*, \psi, \psi^\dagger, F, F^*$,

$$(\delta_{\epsilon_2}\delta_{\epsilon_1} - \delta_{\epsilon_1}\delta_{\epsilon_2})X = i(\epsilon_1 \sigma^\mu \epsilon^\dagger_2 - \epsilon_2 \sigma^\mu \epsilon^\dagger_1) \, \partial_\mu X \tag{3.18}$$

using eqs. (3.3), (3.15), and (3.17), but without resorting to any of the equations of motion. So we have succeeded in showing that supersymmetry is a valid symmetry of the lagrangian off-shell.

In retrospect, one can see why we needed to introduce the auxiliary field F in order to get the supersymmetry algebra to work off-shell. On shell, the complex scalar field ϕ has two real propagating degrees of freedom, which match with the two spin polarization states of ψ. Off-shell, however, the Weyl fermion ψ is a complex two-component object, so it has four real degrees of freedom. (Going on-shell eliminates half of the propagating degrees of freedom for ψ, because the lagrangian is linear in time derivatives, so that the canonical momenta can be reexpressed in terms of the configuration variables without time derivatives and are not independent phase space coordinates.) To make the numbers of degrees of freedom match off-shell as well as on-shell, we had to introduce two more real scalar degrees of freedom in the complex field F, which are eliminated when one goes on-shell.

Invariance of the action under a symmetry transformation always implies the existence of a conserved current, and supersymmetry is no exception. The *supercurrent* J_α^μ is an anticommuting four-vector which also carries a spinor index, as befits the current associated with a symmetry with fermionic generators.[37] By the usual Noether procedure, one finds for the supercurrent (and its hermitian conjugate) in terms of the variations of the fields $X = \phi, \phi^*, \psi, \psi^\dagger, F, F^*$:

$$\epsilon J_\mu + \epsilon^\dagger J_\mu^\dagger \equiv \sum_X \delta X \, \frac{\delta \mathcal{L}}{\delta(\partial^\mu X)} - K_\mu, \tag{3.19}$$

where K_μ is the object whose divergence is the variation of the lagrangian density under the supersymmetry transformation, $\partial^\mu K_\mu = \delta \mathcal{L}$. A little work reveals that

$$J_\alpha^\mu = (\sigma^\nu \overline{\sigma}^\mu \psi)_\alpha \, \partial_\nu \phi^*; \qquad J_{\dot{\alpha}}^{\dagger\mu} = (\psi^\dagger \overline{\sigma}^\mu \sigma^\nu)_{\dot{\alpha}} \, \partial_\nu \phi. \tag{3.20}$$

The supercurrent and its hermitian conjugate are separately conserved:

$$\partial_\mu J_\alpha^\mu = 0; \qquad \partial_\mu J_{\dot{\alpha}}^{\dagger\mu} = 0 \tag{3.21}$$

as can be verified by use of the equations of motion. From these currents one constructs the conserved charges

$$Q_\alpha = \sqrt{2} \int d^3x \, J_\alpha^0; \qquad Q_{\dot{\alpha}}^\dagger = \sqrt{2} \int d^3x \, J_{\dot{\alpha}}^{\dagger 0} \tag{3.22}$$

which are the generators of supersymmetry transformations. (The factor of $\sqrt{2}$ normalization is included to agree with an arbitrary historical convention.) As quantum mechanical operators, they satisfy

$$\left[\epsilon Q + \epsilon^\dagger Q^\dagger, X\right] = -i\sqrt{2}\, \delta X \tag{3.23}$$

for any field X, up to terms which vanish on-shell. This can be verified explicitly by using the canonical equal-time commutation and anticommutation relations

$$[\phi(x), \pi(y)] = [\phi^*(x), \pi^*(y)] = i\delta^{(3)}(x - y); \qquad (3.24)$$

$$\{\psi_{\dot\alpha}^\dagger(x), \psi_\alpha(y)\} = -\sigma^0_{\alpha\dot\alpha}\,\delta^{(3)}(x - y) \qquad (3.25)$$

derived from the free field theory lagrangian eq. (3.1). Here $\pi = \partial_0\phi^*$ and $\pi^* = \partial_0\phi$ are the momenta conjugate to ϕ and ϕ^* respectively. Now the content of eq. (3.18) can be expressed in terms of canonical commutators as

$$\left[\epsilon_2 Q + \epsilon_2^\dagger Q^\dagger,\ [\epsilon_1 Q + \epsilon_1^\dagger Q^\dagger, X]\right] - \left[\epsilon_1 Q + \epsilon_1^\dagger Q^\dagger,\ [\epsilon_2 Q + \epsilon_2^\dagger Q^\dagger, X]\right] =$$
$$2(\epsilon_2\sigma^\mu\epsilon_1^\dagger - \epsilon_1\sigma^\mu\epsilon_2^\dagger)\,i\partial_\mu X \qquad (3.26)$$

up to terms which vanish on-shell. The spacetime momentum operator P^μ is given in terms of the canonical variables by $P^0 = \pi\pi^* + \partial_j\phi\partial^j\phi^* + i\psi^\dagger\overline\sigma^j\partial_j\psi$ and $P^j = -\pi\partial^j\phi - \pi^*\partial^j\phi^* + i\psi^\dagger\overline\sigma^0\partial^j\psi$, where j is the spacetime vector index restricted to the three spatial dimensions. It generates spacetime translations on the fields X according to

$$[P_\mu, X] = i\partial_\mu X. \qquad (3.27)$$

By rearranging the terms in eq. (3.26) using the Jacobi identity, we therefore have

$$\left[[\epsilon_2 Q + \epsilon_2^\dagger Q^\dagger,\ \epsilon_1 Q + \epsilon_1^\dagger Q^\dagger],\ X\right] = 2(\epsilon_2\sigma^\mu\epsilon_1^\dagger - \epsilon_1\sigma^\mu\epsilon_2^\dagger)\,[P_\mu, X], \qquad (3.28)$$

for any X, so it must be that

$$[\epsilon_2 Q + \epsilon_2^\dagger Q^\dagger,\ \epsilon_1 Q + \epsilon_1^\dagger Q^\dagger] = 2(\epsilon_2\sigma^\mu\epsilon_1^\dagger - \epsilon_1\sigma^\mu\epsilon_2^\dagger)\,P_\mu \qquad (3.29)$$

up to terms which vanish on-shell. Now by expanding out eq. (3.29), one obtains the non-schematic form of the supersymmetry algebra relations

$$\{Q_\alpha, Q_{\dot\alpha}^\dagger\} = 2\sigma^\mu_{\alpha\dot\alpha}P_\mu, \qquad (3.30)$$

$$\{Q_\alpha, Q_\beta\} = \{Q_{\dot\alpha}^\dagger, Q_{\dot\beta}^\dagger\} = 0 \qquad (3.31)$$

as promised in the Introduction. [The commutator in eq. (3.29) turns into anticommutators in eqs. (3.30) and (3.31) in the process of extracting the anticommuting spinors ϵ_1 and ϵ_2.] The results $[Q_\alpha, P_\mu] = 0$ and $[Q_{\dot\alpha}^\dagger, P_\mu] = 0$ follow immediately from eq. (3.27) and the fact that the supersymmetry transformations are global (independent of position in spacetime). This demonstration of the supersymmetry algebra in terms of the canonical generators Q and $Q^\dagger$ requires the use of the Hamiltonian equations of motion, but the symmetry itself is valid off-shell at the level of the lagrangian, as we have already shown.

3.2 Interactions of chiral supermultiplets

In a realistic theory like the MSSM, there are many chiral supermultiplets which have both gauge and non-gauge interactions. In this subsection, our task is to construct the most general possible theory of masses and non-gauge interactions for particles that live in chiral supermultiplets. In the MSSM these are the quarks, squarks, leptons, sleptons, Higgs scalars and higgsino fermions. We will find that the form of the non-gauge couplings, including mass terms, is highly restricted by the requirement that the action is invariant under supersymmetry transformations.

Our starting point is the lagrangian density for a collection of free chiral supermultiplets labelled by an index i which runs over all gauge and flavor degrees of freedom. Since we will want to construct an interacting theory which closes off-shell, each supermultiplet contains a complex scalar ϕ_i and a left-handed Weyl fermion ψ_i as physical degrees of freedom, plus a complex auxiliary field F_i which does not propagate. The results of the previous subsection tell us that the free part of the Lagrangian is

$$\mathcal{L}_{\text{free}} = -\partial^\mu \phi^{*i} \partial_\mu \phi_i - i\psi^{\dagger i}\overline{\sigma}^\mu \partial_\mu \psi_i + F^{*i}F_i \qquad (3.32)$$

where we sum over repeated indices i (not to be confused with the suppressed spinor indices), with the convention that fields ϕ_i and ψ_i always carry lowered indices, while their conjugates always carry raised indices. It is invariant under the supersymmetry transformation

$$\delta\phi_i = \epsilon\psi_i \qquad\qquad \delta\phi^{*i} = \epsilon^\dagger \psi^{\dagger i} \qquad\qquad (3.33)$$

$$\delta(\psi_i)_\alpha = i(\sigma^\mu \epsilon^\dagger)_\alpha\, \partial_\mu \phi_i + \epsilon_\alpha F_i \qquad \delta(\psi^{\dagger i})_{\dot{\alpha}} = -i(\epsilon\sigma^\mu)_{\dot{\alpha}}\, \partial_\mu \phi^{*i} + \epsilon^\dagger_{\dot{\alpha}} F^{*i} \qquad (3.34)$$

$$\delta F_i = i\epsilon^\dagger\overline{\sigma}^\mu \partial_\mu \psi_i \qquad\qquad \delta F^{*i} = -i\partial_\mu \psi^{\dagger i}\overline{\sigma}^\mu\epsilon\,. \qquad\qquad (3.35)$$

As we will now show, the most general set of renormalizable interactions for these fields can be written in the simple form

$$\mathcal{L}_{\text{int}} = -\frac{1}{2}W^{ij}\psi_i\psi_j + W^i F_i + \text{c.c.}, \qquad (3.36)$$

where W^{ij} and W^i are some functions of the scalar fields with dimensions of (mass) and (mass)^2 respectively, and "c.c." henceforth stands for complex conjugate. At this point, we are not assuming that W^{ij} and W^i are related to each other in any way whatsoever. However, soon we will find out that they *are* related, which is why we have chosen the same letter for them. Notice that eq. (2.8) tells us that W^{ij} is symmetric under $i \leftrightarrow j$. We do not need to consider the possibility of W^{ij} or W^i being functions of the fermionic or auxiliary fields, because the lagrangian density would have mass dimension greater than 4, and so could not be renormalizable by power counting. For the same reason, W^i is at most a quadratic polynomial, and W^{ij} linear, in the fields ϕ_i and ϕ^{*i}. Also, we did not include in $\mathcal{L}_{\rm int}$ any term which is a function of the scalar fields ϕ_i, ϕ^{*i} only. If there were such a term, then under a supersymmetry transformation (3.33) it would go into another function of the scalar fields only, multiplied by $\epsilon\psi_i$ or $\epsilon^\dagger\psi^{\dagger i}$, and with no spacetime derivatives or F_i, F^{*i} fields. It is easy to see from eqs. (3.33)-(3.36) that nothing of this form can possibly be cancelled by the supersymmetry transformation of any other term in the lagrangian, so eq. (3.36) is the most general possibility!

We must now require that $\mathcal{L}_{\rm int}$ is invariant under the supersymmetry transformations, since $\mathcal{L}_{\rm free}$ was already invariant by itself. It is easiest to divide the variation of $\mathcal{L}_{\rm int}$ into several parts which must cancel separately. First, we consider the part which contains four spinors:

$$\delta\mathcal{L}_{\rm int}|_{4-\rm spinor} \;=\; -\frac{1}{2}\frac{\delta W^{ij}}{\delta\phi_k}(\epsilon\psi_k)(\psi_i\psi_j) - \frac{1}{2}\frac{\delta W^{ij}}{\delta\phi^{*k}}(\epsilon^\dagger\psi^{\dagger k})(\psi_i\psi_j) + {\rm c.c.} \quad (3.37)$$

The term proportional to $(\epsilon\psi_k)(\psi_i\psi_j)$ cannot cancel against any other term, but the Fierz identity eq. (3.12) implies

$$(\epsilon\psi_i)(\psi_j\psi_k) + (\epsilon\psi_j)(\psi_k\psi_i) + (\epsilon\psi_k)(\psi_i\psi_j) = 0, \qquad (3.38)$$

which allows this contribution to $\delta\mathcal{L}_{\rm int}$ to vanish identically if and only if $\delta W^{ij}/\delta\phi_k$ is totally symmetric under interchange of i,j,k. There is no such identity available for the term proportional to $(\epsilon^\dagger\psi^{\dagger k})(\psi_i\psi_j)$. Since it cannot cancel with any other term, requiring it to vanish tells us that W^{ij} cannot contain ϕ^{*k}, in other words it is *analytic* (or *holomorphic*) in the complex fields ϕ_k.

So far, what we have learned is that we can write

$$W^{ij} = M^{ij} + y^{ijk}\phi_k \qquad (3.39)$$

where M^{ij} is a (symmetric) mass matrix for the fermion fields, and y^{ijk} is a Yukawa coupling of a scalar ϕ_k and two fermions $\psi_i\psi_j$ which must be totally symmetric under interchange of i,j,k. It is convenient to write

$$W^{ij} = \frac{\delta^2}{\delta\phi_i\delta\phi_j}W \qquad (3.40)$$

where we have introduced a very useful object

$$W = \frac{1}{2}M^{ij}\phi_i\phi_j + \frac{1}{6}y^{ijk}\phi_i\phi_j\phi_k \qquad (3.41)$$

which is called the *superpotential*. This is not a scalar potential in the ordinary sense; in fact, it is not even real. It is instead an analytic function of the scalar fields ϕ_i treated as complex variables.

Continuing on our vaunted quest, we next consider the parts of $\delta\mathcal{L}_{\text{int}}$ which contain a spacetime derivative:

$$\delta\mathcal{L}_{\text{int}}|_{\partial} = -iW^{ij}\partial_\mu\phi_j\,\psi_i\sigma^\mu\epsilon^\dagger - iW^i\,\partial_\mu\psi_i\sigma^\mu\epsilon^\dagger + \text{c.c.} \tag{3.42}$$

where we have used the identity eq. (2.9) on the second term, which comes from $(\delta F_i)W^i$. Now we can use eq. (3.40) to observe that

$$W^{ij}\partial_\mu\phi_j = \partial_\mu\left(\frac{\delta W}{\delta\phi_i}\right). \tag{3.43}$$

Then it is clear that eq. (3.42) will be a total derivative if and only if

$$W^i = \frac{\delta W}{\delta\phi_i} = M^{ij}\phi_j + \frac{1}{2}y^{ijk}\phi_j\phi_k\,, \tag{3.44}$$

which explains why we chose its name the way we did. The remaining terms in $\delta\mathcal{L}_{\text{int}}$ are all linear in F_i or F^{*i}, and it is easy to show that they cancel, given the results for W^i and W^{ij} that we have already found.

To recap, we have found that the most general non-gauge interactions for chiral supermultiplets are determined by a single analytic function of the complex scalar fields, the superpotential W. The auxiliary fields F_i and F^{*i} can be eliminated using their classical equations of motion. The part of $\mathcal{L}_{\text{free}}+\mathcal{L}_{\text{int}}$ that contains the auxiliary fields is $F_iF^{*i} + W^iF_i + W_i^*F^{*i}$, leading to the equations of motion

$$F_i = -W_i^*; \qquad\qquad F^{*i} = -W^i\,. \tag{3.45}$$

Thus the auxiliary fields are expressible algebraically (without any derivatives) in terms of the scalar fields. After making the replacement eq. (3.45) in $\mathcal{L}_{\text{free}} + \mathcal{L}_{\text{int}}$, we obtain the lagrangian density

$$\mathcal{L} = -\partial^\mu\phi^{*i}\partial_\mu\phi_i - i\psi^{\dagger i}\overline{\sigma}^\mu\partial_\mu\psi_i - \frac{1}{2}\left(W^{ij}\psi_i\psi_j + W^{*ij}\psi^{\dagger i}\psi^{\dagger j}\right) - W^iW_i^*.\tag{3.46}$$

(Since F_i and F^{*i} appear only quadratically in the action, the result of instead doing a functional integral over them at the quantum level has precisely the same effect.) Now that the non-propagating fields F_i, F^{*i} have been eliminated, it is clear from eq. (3.46) that the scalar potential for the theory is just given in terms of the superpotential by (recall $\mathcal{L}$ contains $-V$):

$$V(\phi, \phi^*) = W^i W_i^* = F_i F^{*i} = M_{ik}^* M^{kj} \phi^{*i} \phi_j \tag{3.47}$$

$$+ \frac{1}{2} M^{in} y_{jkn}^* \phi_i \phi^{*j} \phi^{*k} + \frac{1}{2} M_{in}^* y^{jkn} \phi^{*i} \phi_j \phi_k + \frac{1}{4} y^{ijn} y_{kln}^* \phi_i \phi_j \phi^{*k} \phi^{*l} \, .$$

This scalar potential is automatically bounded from below; in fact, since it is a sum of squares of absolute values (of the W^i), it is always non-negative. If we substitute the general form for the superpotential eq. (3.41) into eq. (3.46), we obtain for the full lagrangian density

$$\begin{aligned}
\mathcal{L} &= -\partial^\mu \phi^{*i} \partial_\mu \phi_i - i\psi^{\dagger i} \overline{\sigma}^\mu \partial_\mu \psi_i \\
&\quad - \frac{1}{2} M^{ij} \psi_i \psi_j - \frac{1}{2} M_{ij}^* \psi^{\dagger i} \psi^{\dagger j} - V(\phi, \phi^*) \\
&\quad - \frac{1}{2} y^{ijk} \phi_i \psi_j \psi_k - \frac{1}{2} y_{ijk}^* \phi^{*i} \psi^{\dagger j} \psi^{\dagger k} .
\end{aligned} \tag{3.48}$$

Now we can compare the masses of the fermions and scalars by looking at the linearized equations of motion:

$$\partial^\mu \partial_\mu \phi_i = M_{ik}^* M^{kj} \phi_j + \ldots ; \tag{3.49}$$

$$-i\overline{\sigma}^\mu \partial_\mu \psi_i = M_{ij}^* \psi^{\dagger j} + \ldots ; \qquad -i\sigma^\mu \partial_\mu \psi^{\dagger i} = M^{ij} \psi_j + \ldots . \tag{3.50}$$

One can eliminate ψ in terms of $\psi^\dagger$ and vice versa in eq. (3.50), obtaining [after use of the identity eq. (3.7)]

$$\partial^\mu \partial_\mu \psi_i = M_{ik}^* M^{kj} \psi_j + \ldots ; \qquad \partial^\mu \partial_\mu \psi^{\dagger j} = \psi^{\dagger i} M_{ik}^* M^{kj} + \ldots . \tag{3.51}$$

Therefore, the fermions and the bosons satisfy the same wave equation with exactly the same $(\text{mass})^2$ matrix with real non-negative eigenvalues, namely $(M^2)_i{}^j = M_{ik}^* M^{kj}$. It follows that diagonalizing this matrix gives a collection of chiral supermultiplets each of which contains a mass-degenerate complex scalar and Weyl fermion, in agreement with the general argument in the Introduction.

3.3 Lagrangians for gauge supermultiplets

The propagating degrees of freedom in a gauge supermultiplet are a massless gauge boson field A_μ^a and a two-component Weyl fermion gaugino λ^a. The index a here runs over the adjoint representation of the gauge group ($a = 1 \ldots 8$ for $SU(3)_C$ color gluons and gluinos; $a = 1, 2, 3$ for $SU(2)_L$ weak isospin;

$a = 1$ for $U(1)_Y$ weak hypercharge). The gauge transformations of the vector supermultiplet fields are then

$$\delta_{\text{gauge}} A_\mu^a = -\partial_\mu \Lambda^a + g f^{abc} A_\mu^b \Lambda^c \tag{3.52}$$

$$\delta_{\text{gauge}} \lambda^a = g f^{abc} \lambda^b \Lambda^c \tag{3.53}$$

where Λ^a is an infinitesimal gauge transformation parameter, g is the gauge coupling, and f^{abc} are the totally antisymmetric structure constants which define the gauge group. (The special case of an abelian group like $U(1)_Y$ is obtained by just setting $f^{abc} = 0$; in particular the corresponding gaugino is a gauge singlet in that case.)

The on-shell degrees of freedom for A_μ^a and λ_α^a amount to two bosonic and two fermionic helicity states (for each a), as required by supersymmetry. However, off-shell λ_α^a consists of two complex, or four real, fermionic degrees of freedom, while A_μ^a only has three real degrees of freedom; one degree of freedom is removed by the inhomogeneous gauge transformation eq. (3.52). So, we will need one real auxiliary field, traditionally called D^a, in order for supersymmetry to be consistent off-shell. This field also transforms as an adjoint of the gauge group [i.e., like eq. (3.53) with $\lambda \rightarrow D$] and satisfies $(D^a)^* = D^a$. Like the chiral auxiliary fields F_i, it has dimensions of (mass)^2 and so no kinetic term, so that it can be eliminated on-shell using its algebraic equation of motion.

Therefore, the lagrangian density for a gauge supermultiplet ought to be

$$\mathcal{L}_{\text{gauge}} = -\frac{1}{4} F_{\mu\nu}^a F^{\mu\nu a} - i\lambda^{\dagger a} \overline{\sigma}^\mu D_\mu \lambda^a + \frac{1}{2} D^a D^a \tag{3.54}$$

where

$$F_{\mu\nu}^a = \partial_\mu A_\nu^a - \partial_\nu A_\mu^a - g f^{abc} A_\mu^b A_\nu^c \tag{3.55}$$

is the usual Yang-Mills field strength, and

$$D_\mu \lambda^a = \partial_\mu \lambda^a - g f^{abc} A_\mu^b \lambda^c \tag{3.56}$$

is the covariant derivative of the gaugino field. One can guess the appropriate form for the supersymmetry transformation of the fields, up to multiplicative constants, from the requirements that they should be linear in the infinitesimal parameters $\epsilon, \epsilon^\dagger$ with dimensions of (mass)$^{-1/2}$; that δA_μ^a is real; and that δD^a should be real and proportional to the field equations for the gaugino, in analogy with the role of the auxiliary field F in the chiral supermultiplet case.

Thus one can guess, up to multiplicative factors,

$$\delta A_\mu^a = -\frac{1}{\sqrt{2}}\left[\epsilon^\dagger \overline{\sigma}_\mu \lambda^a + \lambda^{\dagger a}\overline{\sigma}_\mu \epsilon\right] \tag{3.57}$$

$$\delta \lambda_\alpha^a = -\frac{i}{2\sqrt{2}}(\sigma^\mu \overline{\sigma}^\nu \epsilon)_\alpha\, F_{\mu\nu}^a + \frac{1}{\sqrt{2}}\epsilon_\alpha\, D^a \tag{3.58}$$

$$\delta D^a = \frac{i}{\sqrt{2}}\left[\epsilon^\dagger \overline{\sigma}^\mu D_\mu \lambda - D_\mu \lambda^\dagger \overline{\sigma}^\mu \epsilon\right]. \tag{3.59}$$

The factors of $\sqrt{2}$ are chosen[e] so that the action obtained by integrating $\mathcal{L}_{\text{gauge}}$ is invariant. It is now a little bit tedious, but straightforward, to check that eq. (3.18) is modified to

$$(\delta_{\epsilon_2}\delta_{\epsilon_1} - \delta_{\epsilon_1}\delta_{\epsilon_2})X = i(\epsilon_1 \sigma^\mu \epsilon_2^\dagger - \epsilon_2 \sigma^\mu \epsilon_1^\dagger)D_\mu X \tag{3.60}$$

for X equal to any of the gauge-covariant fields $F_{\mu\nu}^a$, λ^a, $\lambda^{\dagger a}$, D^a, as well as arbitrary covariant derivatives acting on them. This ensures that the supersymmetry algebra eqs. (3.30)-(3.31) is realized on gauge-invariant combinations of fields in gauge supermultiplets, as they were on the chiral supermultiplets.[f] These calculations require the use of identities

$$\xi \sigma^\mu \overline{\sigma}^\nu \chi = \chi \sigma^\nu \overline{\sigma}^\mu \xi = (\chi^\dagger \overline{\sigma}^\nu \sigma^\mu \xi^\dagger)^* = (\xi^\dagger \overline{\sigma}^\mu \sigma^\nu \chi^\dagger)^*; \tag{3.61}$$

$$\overline{\sigma}^\mu \sigma^\nu \overline{\sigma}^\rho = \eta^{\mu\rho}\overline{\sigma}^\nu - \eta^{\nu\rho}\overline{\sigma}^\mu - \eta^{\mu\nu}\overline{\sigma}^\rho - i\epsilon^{\mu\nu\rho\kappa}\overline{\sigma}_\kappa; \tag{3.62}$$

$$\sigma_{\alpha\dot\alpha}^\mu \overline{\sigma}_\mu^{\dot\beta\beta} = -2\delta_\alpha^\beta \delta_{\dot\alpha}^{\dot\beta}. \tag{3.63}$$

If we had not included the auxiliary field D^a, then the supersymmetry algebra eq. (3.60) would hold only after using the equations of motion for λ^a and $\lambda^{\dagger a}$. The auxiliary fields just satisfy the equations of motion $D^a = 0$, but this is no longer true if one couples the gauge supermultiplets to chiral supermultiplets, as we now do.

3.4 Supersymmetric gauge interactions

Finally we are ready to consider a general lagrangian density for a supersymmetric theory with both chiral and gauge supermultiplets. Suppose that the

---

[e] For future convenience in treating the MSSM, we have chosen complex phases so that our λ^a, $\lambda^{\dagger a}$ are equal to $-i$, i times the gaugino spinors in Ref.[11]

[f] The supersymmetry transformations eqs. (3.57)-(3.59) are non-linear for non-abelian gauge symmetries, because of the gauge fields contained in the covariant derivatives acting on the gaugino fields and in the field strength $F_{\mu\nu}^a$. By adding even more auxiliary fields besides D^a, one can make the supersymmetry transformations linear in the fields. The version given here in which those extra auxiliary fields have been removed by gauge transformations is called "Wess-Zumino gauge".[38]

chiral supermultiplets transform under the gauge group in a representation with hermitian matrices $(T^a)_i{}^j$ satisfying $[T^a, T^b] = if^{abc}T^c$. [For example, if the gauge group is $SU(2)$, then $f^{abc} = \epsilon^{abc}$, and the T^a are $1/2$ times the Pauli matrices for a chiral supermultiplet transforming in the fundamental representation.] Thus

$$\delta_{\text{gauge}} X_i = ig\Lambda^a (T^a X)_i \tag{3.64}$$

for $X_i = \phi_i, \psi_i, F_i$; since supersymmetry and gauge transformations commute, the scalar, fermion, and auxiliary fields must be in the same representation of the gauge group. To have a gauge-invariant lagrangian, we need to turn the ordinary derivatives in eq. (3.32) into covariant derivatives:

$$\partial_\mu \phi_i \rightarrow D_\mu \phi_i = \partial_\mu \phi_i + igA_\mu^a (T^a \phi)_i \tag{3.65}$$

$$\partial_\mu \phi^{*i} \rightarrow D_\mu \phi^{*i} = \partial_\mu \phi^{*i} - igA_\mu^a (\phi^* T^a)^i \tag{3.66}$$

$$\partial_\mu \psi_i \rightarrow D_\mu \psi_i = \partial_\mu \psi_i + igA_\mu^a (T^a \psi)_i. \tag{3.67}$$

Naively, this simple procedure achieves the goal of coupling the vector bosons in the gauge supermultiplet to the scalars and fermions in the chiral super-multiplets. However, we also have to consider whether there are any other interactions allowed by gauge invariance involving the gaugino and D^a fields which might have to be included in a supersymmetric lagrangian.

In fact, there are three such possibilities, namely

$$(\phi^* T^a \psi)\lambda^a, \qquad \lambda^{\dagger a}(\psi^\dagger T^a \phi) \qquad \text{and} \qquad (\phi^* T^a \phi)D^a, \tag{3.68}$$

which are renormalizable. Now one can add them, with arbitrary coefficients, to the lagrangians for the chiral and gauge supermultiplets and demand that the whole mess be real and invariant under supersymmetry transformations, up to a total derivative. Not surprisingly, this is possible only if one modifies the supersymmetry transformation laws for the matter fields to include gauge-covariant rather than ordinary derivatives (and to include one strategically-chosen extra term in δF_i):

$$\delta \phi_i = \epsilon \psi_i \tag{3.69}$$

$$\delta(\psi_i)_\alpha = i(\sigma^\mu \epsilon^\dagger)_\alpha D_\mu \phi_i + \epsilon_\alpha F_i \tag{3.70}$$

$$\delta F_i = i\epsilon^\dagger \bar{\sigma}^\mu D_\mu \psi_i + \sqrt{2}g(T^a \phi)_i \epsilon^\dagger \lambda^{\dagger a}. \tag{3.71}$$

After some algebra one fixes the coefficients for the terms in eq. (3.68), so that the full lagrangian density for a renormalizable supersymmetric theory is

$$\mathcal{L} = \mathcal{L}_{\text{gauge}} + \mathcal{L}_{\text{chiral}}$$

$$-\sqrt{2}g\left[(\phi^* T^a \psi)\lambda^a + \lambda^{\dagger a}(\psi^\dagger T^a \phi)\right]$$
$$+g(\phi^* T^a \phi)D^a. \tag{3.72}$$

Here $\mathcal{L}_{\text{chiral}}$ means the chiral supermultiplet lagrangian found in section 3.2 [e.g., eq. (3.46) or (3.48)], but with ordinary derivatives replaced everywhere by gauge-covariant derivatives, and $\mathcal{L}_{\text{gauge}}$ was given in eq. (3.54). To prove that eq. (3.72) is invariant under the supersymmetry transformations, one must use the identity

$$W^i (T^a)_i^j \phi_j = 0, \tag{3.73}$$

which is precisely the condition that must be satisfied anyway in order for the superpotential (and thus $\mathcal{L}_{\text{chiral}}$) to be gauge invariant.

The last two lines in eq. (3.72) are interactions whose strengths are fixed to be gauge couplings by the requirements of supersymmetry, even though they are not gauge interactions from the point of view of an ordinary field theory. The second line is a direct coupling of gauginos to matter fields which is the "supersymmetrization" of the usual gauge boson coupling to matter fields. The last line combines with the $(1/2)D^a D^a$ term in $\mathcal{L}_{\text{gauge}}$ to provide an equation of motion

$$D^a = -g(\phi^* T^a \phi). \tag{3.74}$$

Like the auxiliary fields F_i and F^{*i}, the D^a are expressible purely algebraically in terms of the scalar fields. Replacing the auxiliary fields in eq. (3.72) using eq. (3.74), one finds that the complete scalar potential is (recall $\mathcal{L} \supset -V$):

$$V(\phi, \phi^*) = F^{*i} F_i + \frac{1}{2}\sum_a D^a D^a = W_i^* W^i + \frac{1}{2}\sum_a g_a^2 (\phi^* T^a \phi)^2. \tag{3.75}$$

The two types of terms in this expression are commonly called "F-term" and "D-term" contributions, respectively. In the second term in eq. (3.75), we have written an explicit sum $\sum_a$ to cover the case that the gauge group has several distinct factors with different gauge couplings g_a. [For instance, in the MSSM the three factors $SU(3)_C$, $SU(2)_L$ and $U(1)_Y$ have different gauge couplings g_3, g and g'.] Since $V(\phi, \phi^*)$ is a sum of squares, it is always greater than or equal to zero for every field configuration. It is a very interesting and unique feature of supersymmetric theories that the scalar potential is completely determined by the *other* interactions in the theory. The F-terms are fixed by Yukawa couplings and fermion mass terms, and the D-terms are fixed by the gauge interactions.

By using Noether's procedure [see eq. (3.19)], one finds the conserved supercurrent

$$J_\alpha^\mu = (\sigma^\nu \overline{\sigma}^\mu \psi_i)_\alpha \, D_\nu \phi^{*i} - i(\sigma^\mu \psi^{\dagger i})_\alpha \, W_i^*$$
$$- \frac{1}{2\sqrt{2}} (\sigma^\nu \overline{\sigma}^\rho \sigma^\mu \lambda^{\dagger a})_\alpha \, F_{\nu\rho}^a - \frac{i}{\sqrt{2}} g \phi^* T^a \phi \, (\sigma^\mu \lambda^{\dagger a})_\alpha, \tag{3.76}$$

generalizing the expression given in eq. (3.20) for the Wess-Zumino model. This expression will be useful when we discuss certain aspects of spontaneous supersymmetry breaking in section 6.2.

3.5 Summary: How to build a supersymmetric model

In a renormalizable supersymmetric field theory, the interactions and masses of all particles are determined just by their gauge transformation properties and by the superpotential W. By construction, we found that W had to be an analytic function of the complex scalar fields ϕ_i, which are always defined to transform under supersymmetry into *left*-handed Weyl fermions. We should mention that in an equivalent language, W is said to be a function of chiral superfields.[36] A superfield is a single object which contains as components all of the bosonic, fermionic, and auxiliary fields within the corresponding supermultiplet, e.g. $\Phi_i \supset (\phi_i, \psi_i, F_i)$. (This is analogous to the way in which one often describes a weak isospin doublet by a single field with more than one component.) The gauge quantum numbers and mass dimension of a chiral superfield are the same as that of its scalar component. In the superfield formulation, one writes instead of eq. (3.41)

$$W = \frac{1}{2} M^{ij} \Phi_i \Phi_j + \frac{1}{6} y^{ijk} \Phi_i \Phi_j \Phi_k \tag{3.77}$$

which means exactly the same thing. While this entails no difference in practical results, the fancier version eq. (3.77) at least serves to remind us that W determines not only the scalar interactions in the theory, but the fermion masses and Yukawa couplings as well. The derivation of all of our preceding results can be obtained somewhat more elegantly using superfield methods, which have the advantage of making invariance under supersymmetry transformations manifest. We have avoided this extra layer of notation on purpose, in favor of the more pedestrian but hopefully more familiar component field approach. The latter is at least more appropriate for making contact with phenomenology in a universe with supersymmetry breaking. The only (occasional) use we will make of superfield notation is the purely cosmetic one of following the common practice of specifying superpotentials like eq. (3.77)

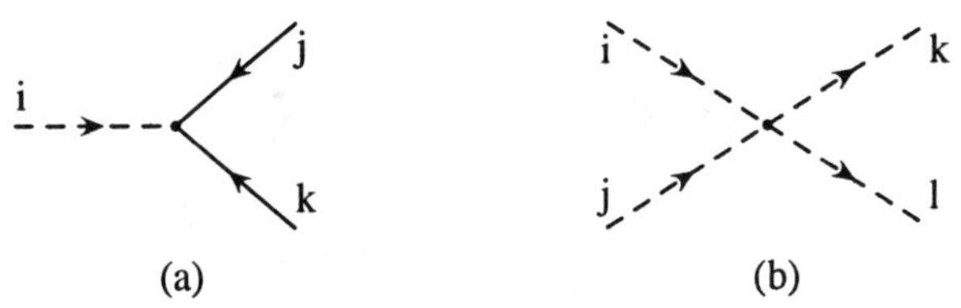

(a)   (b)

Figure 3: The dimensionless non-gauge interaction vertices in a supersymmetric theory: (a) scalar-fermion-fermion Yukawa interaction y^{ijk}, (b) quartic scalar interaction $y^{ijn}y^*_{kln}$.

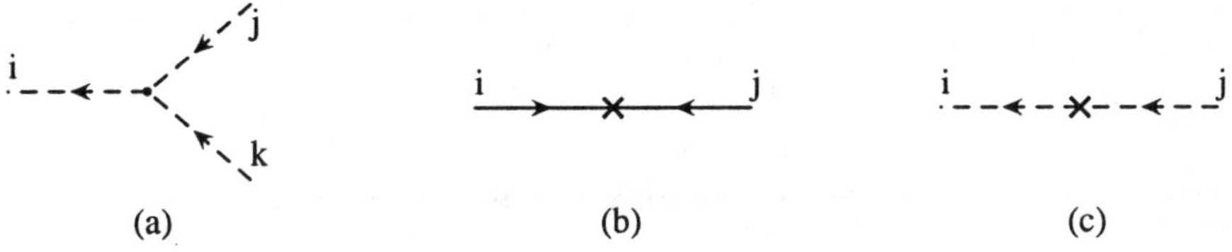

(a)   (b)   (c)

Figure 4: Supersymmetric dimensionful couplings: (a) $(\text{scalar})^3$ interaction vertex $M^*_{in}y^{jkn}$, (b) fermion mass term M^{ij}, (c) scalar $(\text{mass})^2$ term $M^*_{ik}M^{kj}$.

rather than (3.41). The specification of the superpotential is really a code for the terms that it implies in the lagrangian, so the reader may feel free to think of the superpotential either as a function of the scalar fields ϕ_i or as the same function of the superfields Φ_i which contain them.

Given the supermultiplet content of the theory, the form of the superpotential is restricted by gauge invariance. In any given theory, only a subset of the couplings M^{ij} and y^{ijk} will be allowed to be non-zero. The mass term M^{ij} can only be non-zero when the supermultiplets Φ_i and Φ_j transform under the gauge group in representations which are conjugates of each other. (In fact, in the MSSM there is only one such term, as we will see.) Likewise, the Yukawa couplings y^{ijk} can only be non-zero when Φ_i, Φ_j, and Φ_k transform in representations which can combine to form a singlet.

The interactions implied by the superpotential eq. (3.77) are shown [9] in Figs. 3 and 4. Those in Fig. 3 are all determined by the dimensionless parameters y^{ijk}. The Yukawa interaction in Fig. 3a corresponds to the next-to-last term in eq. (3.48). For each particular Yukawa coupling of $\phi_i\psi_j\psi_k$ with strength y^{ijk}, there must be equal couplings of $\phi_j\psi_i\psi_k$ and $\phi_k\psi_i\psi_j$, since y^{ijk}

---

[9]Here, the auxiliary fields have been eliminated using their equations of motion ("integrated out") as in eq. (3.48). It is quite possible instead to give Feynman rules which include the auxiliary fields, although this tends to be less useful in phenomenological applications.

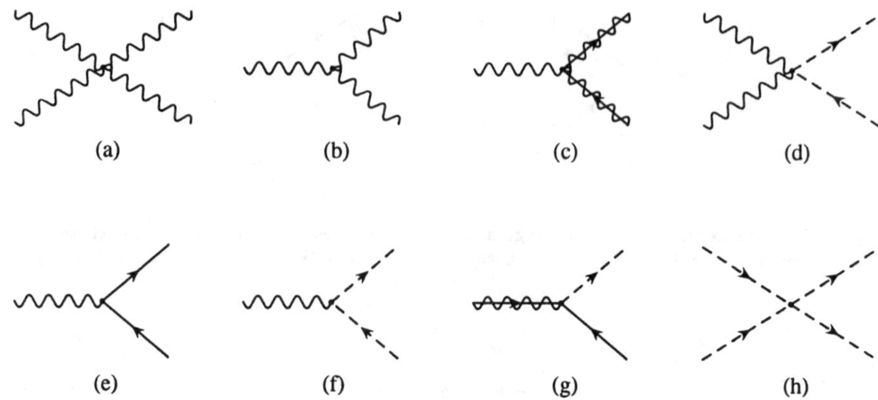

Figure 5: Supersymmetric gauge interaction vertices.

is completely symmetric under interchange of any two of its indices as shown in section 3.2. There is also a dimensionless coupling for $\phi_i \phi_j \phi^{*k} \phi^{*l}$, with strength $y^{ijn} y^*_{kln}$ as required by supersymmetry [see the last term in eq. (3.47)]. The arrows on both the fermion and scalar lines follow the chirality; i.e., one direction for propagation of ϕ and ψ and the other for the propagation of ϕ^* and $\psi^\dagger$. Thus there is a vertex corresponding to the one in Fig. 3a but with all arrows reversed, corresponding to the complex conjugate [the last term in eq. (3.48)]. The relationship between the interactions in Figs. 3a and 3b is exactly the type needed to cancel the quadratic divergences in quantum corrections to scalar masses, as discussed in the Introduction [compare Fig. 1].

In Fig. 4, we show the only interactions corresponding to renormalizable and supersymmetric vertices with dimensions of (mass) and (mass)^2. First, there are (scalar)^3 couplings which are entirely determined by the superpotential mass parameters M^{ij} and Yukawa couplings y^{ijk}, as indicated by the second and third terms in eq. (3.47). The propagators of the fermions and scalars in the theory are constructed in the usual way using the fermion mass M^{ij} and scalar (mass)^2 $M^*_{in} M^{nj}$. Of particular interest is the fact that the fermion mass term M^{ij} leads to a chirality-changing insertion in the fermion propagator; note the directions of the arrows in Fig. 4b. There is no such arrow-reversal for a scalar propagator in a theory with exact supersymmetry; as shown in Fig. 4c, if one treats the scalar (mass)^2 term as an insertion in the propagator, the arrow direction is preserved. Again, for each of Figures 4a and 4b there is an interaction with all arrows reversed.

In Fig. 5 we show in a similar manner the gauge interactions in a supersymmetric theory. Figures 5a,b,c occur only when the gauge group is non-abelian (e.g. for $SU(3)_C$ color and $SU(2)_L$ weak isospin in the MSSM). Figures 5a and 5b are the interactions of gauge bosons which derive from the first term in eq. (3.54). In the MSSM these are exactly the same as the well-known QCD gluon and electroweak gauge boson vertices of the Standard Model. (We do not show the interactions of ghost fields, which are necessary only for consistent loop amplitudes.) Figures 5c,d,e,f are just the standard interactions between gauge bosons and fermion and scalar fields which must occur in any gauge theory because of the form of the covariant derivative; they come from eqs. (3.56) and (3.65)-(3.67) inserted in the kinetic part of the lagrangian. Figure 5c shows the coupling of a gaugino to a gauge boson; the gaugino line in a Feynman diagram is traditionally drawn as a solid fermion line superimposed on a gauge boson squiggly line. In Fig. 5g we have the coupling of a gaugino to a chiral fermion and a complex scalar [the first term in the second line in eq. (3.72)]. One can think of this as the "supersymmetrization" of Figure 5e or 5f; any of these three vertices may be obtained from any other (up to a factor of $\sqrt{2}$) by replacing two of the particles by their supersymmetric partners. There is also an interaction like Fig. 5g but with all arrows reversed, corresponding to the complex conjugate term in the lagrangian [the second term in the second line in eq. (3.72)]. Finally in Fig. 5h we have a scalar quartic interaction vertex [the last term in eq. (3.75)] which is also determined by the gauge coupling.

The results of this section can be used as a recipe for constructing the supersymmetric interactions for any model. In the case of the MSSM, we already know the gauge group, particle content and the gauge transformation properties, so it only remains to decide on the superpotential. This we will do in section 5.1.

4 Soft supersymmetry breaking interactions

A realistic phenomenological model must contain supersymmetry breaking. From a theoretical perspective, we expect that supersymmetry, if it exists at all, should be an exact symmetry which is spontaneously broken. In other words, the ultimate model should have a lagrangian density which is invariant under supersymmetry, but a vacuum state which is not. In this way, supersymmetry is hidden at low energies in a manner exactly analogous to the fate of the electroweak symmetry in the ordinary Standard Model.

Many models of spontaneous symmetry breaking have indeed been proposed. These always involve extending the MSSM to include new particles and interactions at very high mass scales, and there is no consensus on exactly how this should be done. However, from a practical point of view, it is extremely useful to simply parameterize our ignorance of these issues by just introducing extra terms which break supersymmetry explicitly in the effective MSSM lagrangian. As was argued in the Introduction, the extra supersymmetry-breaking couplings should be soft (of positive mass dimension) in order to be able to naturally maintain a hierarchy between the electroweak scale and the Planck (or some other very large) mass scale. This means in particular that we should not consider any dimensionless supersymmetry-breaking couplings.

So, the possible soft supersymmetry-breaking terms in the lagrangian are

$$\mathcal{L}_{\text{soft}} = -\frac{1}{2}\left(M_\lambda\,\lambda^a\lambda^a + \text{c.c.}\right) - (m^2)^i_j\phi^{j*}\phi_i$$
$$-\left(\frac{1}{2}b^{ij}\phi_i\phi_j + \frac{1}{6}a^{ijk}\phi_i\phi_j\phi_k + \text{c.c.}\right), \tag{4.1}$$

$$\mathcal{L}_{\text{maybe soft}} = -\frac{1}{2}c^{jk}_i\phi^{*i}\phi_j\phi_k + \text{c.c.} \tag{4.2}$$

They consist of gaugino masses M_λ for each gauge group, scalar (mass)^2 terms $(m^2)^j_i$ and b^{ij}, and (scalar)^3 couplings a^{ijk} and c^{jk}_i. One might wonder why we have not included possible soft mass terms for the chiral supermultiplet fermions. The reason is that including such terms would be redundant; they can always be absorbed into a redefinition of the superpotential and the terms $(m^2)^j_i$ and c^{jk}_i. It has been shown rigorously that a softly-broken supersymmetric theory with $\mathcal{L}_{\text{soft}}$ as given by eq. (4.1) is indeed free of quadratic divergences in quantum corrections to scalar masses, to all orders in perturbation theory.[39] The situation is slightly more subtle if one tries to include the non-analytic (scalar)^3 couplings in $\mathcal{L}_{\text{maybe soft}}$. If any of the chiral supermultiplets in the theory are completely uncharged under all gauge symmetries, then non-zero c^{jk}_i terms can lead to quadratic divergences, despite the fact that they are formally soft. Now, this constraint does *not* apply to the MSSM, which does not have any gauge-singlet chiral supermultiplets. Nevertheless, the possibility of c^{jk}_i terms is nearly always neglected.[40] The real reason for this is that it is extremely difficult to construct any model of spontaneous supersymmetry breaking in which the c^{jk}_i are not utterly negligibly small. Equation (4.1) is therefore usually taken to be the most general soft supersymmetry-breaking lagrangian.

It should be clear that $\mathcal{L}_{\text{soft}}$ indeed breaks supersymmetry, since it involves only scalars and gauginos, and not their respective superpartners. In fact, the soft terms in $\mathcal{L}_{\text{soft}}$ are capable of giving masses to all of the scalars and gauginos in a theory, even if the gauge bosons and fermions in chiral supermultiplets are massless (or relatively light). The gaugino masses M_λ are always allowed by gauge symmetry. The $(m^2)^i_j$ terms are allowed for i, j such that ϕ_i, ϕ^{j*}

transform in complex conjugate representations of each other under all gauge symmetries; in particular this is true of course when $i = j$, so every scalar is eligible to get a mass in this way if supersymmetry is broken.

The remaining soft terms may or may not be allowed by the symmetries. In this regard it is useful to note that the b^{ij} and a^{ijk} terms have the same form as the M^{ij} and y^{ijk} terms in the superpotential [compare eq. (4.1) to eq. (3.41) or eq. (3.77)], so they will be allowed by gauge invariance if and only if a corresponding superpotential term is allowed.

5 The Minimal Supersymmetric Standard Model

In sections 3 and 4, we have found a general recipe for constructing lagrangians for softly broken supersymmetric theories. We are now ready to apply these general results to the MSSM. The particle content for the MSSM was described in the Introduction. In this section we will complete the model by specifying the superpotential and the soft-breaking terms.

5.1 The superpotential and supersymmetric interactions

The superpotential for the MSSM is given by

$$W_{\text{MSSM}} = \overline{u}\mathbf{y_u}QH_u - \overline{d}\mathbf{y_d}QH_d - \overline{e}\mathbf{y_e}LH_d + \mu H_u H_d \,. \tag{5.1}$$

The objects H_u, H_d, Q, L, $\overline{u}$, $\overline{d}$, $\overline{e}$ appearing in eq. (5.1) are chiral superfields corresponding to the chiral supermultiplets in Table 1. (Alternatively, they can be just thought of as the corresponding scalar fields, as was done in section 3, but we prefer not to put the tildes on Q, L, $\overline{u}$, $\overline{d}$, $\overline{e}$ in order to reduce clutter.) The dimensionless Yukawa coupling parameters $\mathbf{y_u}, \mathbf{y_d}, \mathbf{y_e}$ are $3{\times}3$ matrices in family space. Here we have suppressed all of the gauge [$SU(3)_C$ color and $SU(2)_L$ weak isospin] and family indices. The "μ term", as it is traditionally called, can be written out as $\mu(H_u)_\alpha (H_d)_\beta \epsilon^{\alpha\beta}$, where $\epsilon^{\alpha\beta}$ is used to tie together $SU(2)_L$ weak isospin indices $\alpha, \beta = 1, 2$ in a gauge-invariant way. Likewise, the term $\overline{u}\mathbf{y_u}QH_u$ can be written out as $\overline{u}^i_a (\mathbf{y_u})_i{}^j Q^a_{j\alpha} (H_u)_\beta \epsilon^{\alpha\beta}$, where $i = 1, 2, 3$ is a family index, and $a = 1, 2, 3$ is a color index which is raised (lowered) in the $\mathbf{3}$ $(\overline{\mathbf{3}})$ representation of $SU(3)_C$.

The μ term in eq. (5.1) is the supersymmetric version of the Higgs boson mass in the Standard Model. It is unique, because terms $H_u^* H_u$ or $H_d^* H_d$ are forbidden in the superpotential, since it must be analytic in the chiral superfields (or equivalently in the scalar fields) treated as complex variables as shown in section 3.2. We can also see from the form of eq. (5.1) why both H_u and H_d are needed in order to give Yukawa couplings, and thus masses, to all of the quarks and leptons. Since the superpotential must be analytic, the $\overline{u}QH_u$ Yukawa terms cannot be replaced by something like $\overline{u}QH_d^*$. Similarly, the $\overline{d}QH_d$ and $\overline{e}LH_d$ terms cannot be replaced by something like $\overline{d}QH_u^*$ and $\overline{e}LH_u^*$. The analogous Yukawa couplings would be allowed in a general two Higgs doublet model, but are forbidden by the structure of supersymmetry. So we need both H_u and H_d, even without the argument based on anomaly cancellation which was mentioned in the Introduction.

The Yukawa matrices determine the masses and CKM mixing angles of the ordinary quarks and leptons, after the neutral scalar components of H_u and H_d get VEVs. Since the top quark, bottom quark and tau lepton are the heaviest fermions in the Standard Model, it is often useful to make an approximation that only the $(3,3)$ family components of each of $\mathbf{y_u}$, $\mathbf{y_d}$ and $\mathbf{y_e}$ are important:

$$\mathbf{y_u} \approx \begin{pmatrix} 0 & 0 & 0 \\ 0 & 0 & 0 \\ 0 & 0 & y_t \end{pmatrix}; \quad \mathbf{y_d} \approx \begin{pmatrix} 0 & 0 & 0 \\ 0 & 0 & 0 \\ 0 & 0 & y_b \end{pmatrix}; \quad \mathbf{y_e} \approx \begin{pmatrix} 0 & 0 & 0 \\ 0 & 0 & 0 \\ 0 & 0 & y_\tau \end{pmatrix}. \quad (5.2)$$

In this limit, only the third family and Higgs fields contribute to the MSSM superpotential. It is instructive to write the superpotential in terms of the separate $SU(2)_L$ weak isospin components $[Q_3 = (t\,b); \ L_3 = (\nu_\tau\,\tau); \ H_u = (H_u^+\,H_u^0); \ H_d = (H_d^0\,H_d^-); \ \overline{u}_3 = \overline{t}; \ \overline{d}_3 = \overline{b}; \ \overline{e}_3 = \overline{\tau}]$, so:

$$W_{\mathrm{MSSM}} \approx y_t(\overline{t}tH_u^0 - \overline{t}bH_u^+) - y_b(\overline{b}tH_d^- - \overline{b}bH_d^0) - y_\tau(\overline{\tau}\nu_\tau H_d^- - \overline{\tau}\tau H_d^0)$$
$$+ \mu(H_u^+ H_d^- - H_u^0 H_d^0). \quad (5.3)$$

The minus signs inside the parentheses appear because of the antisymmetry of the $\epsilon^{\alpha\beta}$ symbol used to tie up the $SU(2)_L$ indices. The minus signs in eq. (5.1) were chosen so that the terms $y_t\overline{t}tH_u^0$, $y_b\overline{b}bH_d^0$, and $y_\tau\overline{\tau}\tau H_d^0$, which will become the top, bottom and tau masses when H_u^0 and H_d^0 get VEVs, have positive signs in eq. (5.3).

Since the Yukawa interactions y^{ijk} in a general supersymmetric theory must be completely symmetric under interchange of i, j, k, we know that $\mathbf{y_u}$, $\mathbf{y_d}$ and $\mathbf{y_e}$ imply not only Higgs-quark-quark and Higgs-lepton-lepton couplings as in the Standard Model, but also squark-Higgsino-quark and slepton-Higgsino-lepton interactions. To illustrate this, we show in Figs. 6a,b,c some of the interactions which involve the top-quark Yukawa coupling y_t. Figure 6a is the Standard Model-like coupling of the top quark to the neutral complex scalar Higgs boson, which follows from the first term in eq. (5.3). For variety, we have used t_L and $t_R^\dagger$ in place of their synonyms t and $\overline{t}$ in Fig. 6; see the discussion in the final paragraph in section 2. In Fig. 6b, we have the coupling of the left-handed top squark $\widetilde{t}_L$ to the neutral higgsino field $\widetilde{H}_u^0$ and right-handed

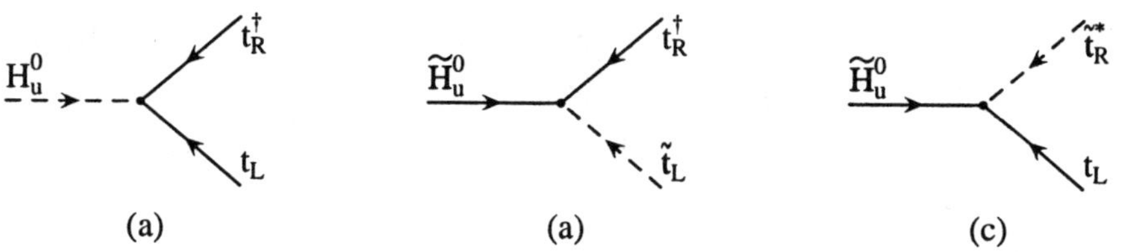

Figure 6: The top-quark Yukawa coupling (a) and its supersymmetrizations (b) and (c), all of strength y_t.

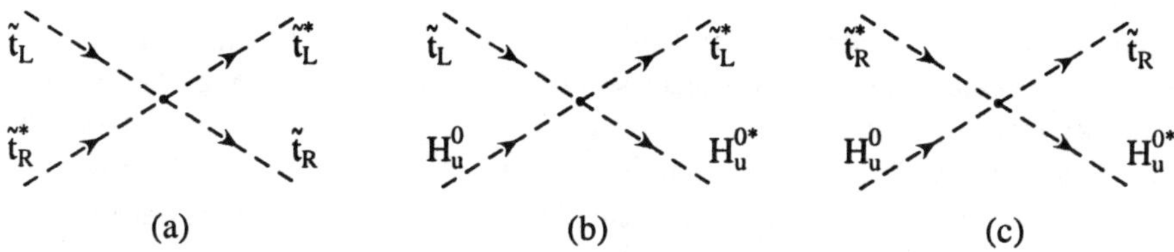

Figure 7: Some of the (scalar)^4 interactions with strength proportional to y_t^2.

top quark, while in Fig. 6c the right-handed top-squark field (known either as $\tilde{t}$ or $\tilde{t}_R^*$ depending on taste) couples to $\widetilde{H}_u^0$ and t_L. For each of the three interactions, there is another with $H_u^0 \to H_u^+$ and $t_L \to -b_L$ (with tildes where appropriate), corresponding to the second part of the first term in eq. (5.3). All of these interactions are required by supersymmetry to have the same strength y_t. This is also an incontrovertible prediction of softly-broken supersymmetry at tree-level, since these interactions are dimensionless and can be modified by the introduction of soft supersymmetry breaking only through finite (and small) radiative corrections. A useful mnemonic is that each of Fig. 6a,b,c can be obtained from any of the others by changing two of the particles into their superpartners.

There are also scalar quartic interactions with strength proportional to y_t^2, as can be seen from Fig. 3b or from the last term in eq. (3.47). Three of them are shown in Fig. 7. The reader is invited to check, using eq. (3.47) and eq. (5.3), that there are nine more, which can be obtained by replacing $\tilde{t}_L \to \tilde{b}_L$ and/or $H_u^0 \to H_u^+$ in each vertex. This illustrates the remarkable economy of supersymmetry; there are many interactions determined by only a single parameter! In a similar way, the existence of all the other quark and lepton Yukawa couplings in the superpotential eq. (5.1) leads not only to Higgs-quark-quark and Higgs-lepton-lepton lagrangian terms as in the ordinary

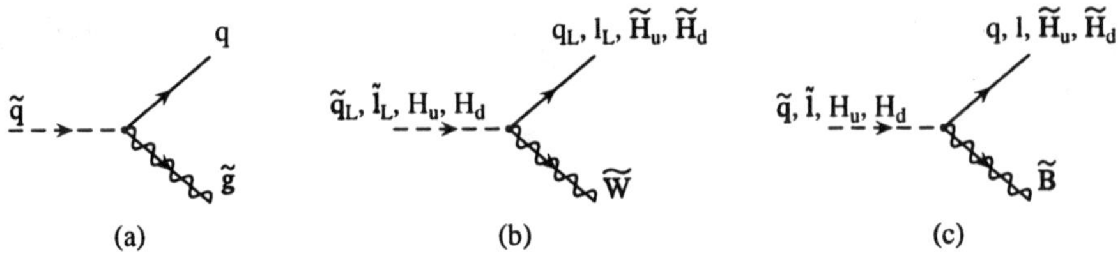

Figure 8: Couplings of the gluino, wino, and bino to MSSM (scalar, fermion) pairs.

Standard Model, but also to squark-higgsino-quark and slepton-higgsino-lepton terms, and scalar quartic couplings [(squark)^4, (slepton)^4, (squark)^2(slepton)^2, (squark)^2(Higgs)^2, and (slepton)^2(Higgs)^2]. If needed, these can all be obtained in terms of the Yukawa matrices $\mathbf{y_u}$, $\mathbf{y_d}$, and $\mathbf{y_e}$ as outlined above.

However, it is useful to note that the dimensionless interactions determined by the superpotential are often not the most important ones of direct interest for phenomenology. This is because the Yukawa couplings are already known to be very small, except for those of the third family (top, bottom, tau). Instead, decay and especially production processes for superpartners in the MSSM are typically dominated by the supersymmetric interactions of gauge-coupling strength. The couplings of the Standard Model gauge bosons (photon, $W^\pm$, Z^0 and gluons) to the MSSM particles are determined completely by the gauge invariance of the kinetic terms in the lagrangian. The gauginos also couple to (squark, quark) and (slepton, lepton) and (Higgs, higgsino) pairs as illustrated in the general case in Fig. 5g and the second line in eq. (3.72). For instance, each of the squark-quark-gluino couplings is given by $\sqrt{2}g_3(\tilde{q}T^a q\tilde{g} +$ c.c.) where T^a ($a = 1\ldots 8$) are the Gell-Mann matrices for $SU(3)_C$. The Feynman diagram for this interaction is shown in Fig. 8a. In Figs. 8b,c we show in a similar way the couplings of (squark, quark), (lepton, slepton) and (Higgs, higgsino) pairs to the winos and bino, with strengths proportional to the electroweak gauge couplings g and g' respectively. The winos and bino only couple to the left-handed squarks and sleptons, and the (lepton, slepton) and (Higgs, higgsino) pairs of course do not couple to the gluino. The bino couplings for each (scalar, fermion) pair are also proportional to the weak hypercharges Y as given in Table 1. The interactions shown in Fig. 8 provide for decays $\tilde{q} \to q\tilde{g}$ and $\tilde{q} \to \widetilde{W}q'$ and $\tilde{q} \to \tilde{B}q$ when the final states are kinematically allowed to be on-shell. However, a complication is that the $\widetilde{W}$ and $\tilde{B}$ states are not mass eigenstates, because of mixing due to electroweak symmetry breaking, as we will see in section 7.3.

There are also various scalar quartic interactions in the MSSM which are uniquely determined by gauge invariance and supersymmetry, according to the last term in eq. (3.75) illustrated in Fig. 5h. Among them are (Higgs)^4 terms proportional to g^2 and g'^2 in the scalar potential. These are the direct generalization of the last term in the Standard Model Higgs potential eq. (1.1) to the case of the MSSM. We will have occasion to identify them explicitly when we discuss the minimization of the MSSM Higgs potential in section 7.2.

The dimensionful terms in the supersymmetric part of the MSSM lagrangian are all dependent on μ. Following the general result of eq. (3.48), we find that μ provides for higgsino fermion mass terms

$$\mathcal{L} \supset -\mu(\tilde{H}_u^+ \tilde{H}_d^- - \tilde{H}_u^0 \tilde{H}_d^0) + \text{c.c.}, \tag{5.4}$$

as well as Higgs (mass)^2 terms in the scalar potential

$$-\mathcal{L} \supset V \supset |\mu|^2 (|H_u^0|^2 + |H_u^+|^2 + |H_d^0|^2 + |H_d^-|^2). \tag{5.5}$$

Since eq. (5.5) is positive-definite, it is clear that we cannot understand electroweak symmetry breaking without including supersymmetry-breaking (mass)^2 soft terms for the Higgs scalars, which can be negative. An explicit treatment of the Higgs scalar potential will therefore have to wait until we have introduced the soft terms for the MSSM. However, we can already see a puzzle: we expect that μ should be roughly of order 10^2 or 10^3 GeV in order to allow a Higgs VEV of order 174 GeV without too much miraculous cancellation between $|\mu|^2$ and the negative soft (mass)^2 terms that we have not written down yet. But why should μ be so small compared to, say, $M_{\rm P}$, and in particular why should it be roughly of the same order as $m_{\rm soft}$? The scalar potential of the MSSM seems to depend on two types of dimensionful parameters which are conceptually quite distinct, namely the supersymmetry-respecting mass μ and the supersymmetry-breaking soft mass terms. Yet the observed value for the electroweak breaking scale suggests that without miraculous cancellations, both of these apparently unrelated mass scales should be within an order of magnitude or so of 100 GeV. This puzzle is called "the μ problem". Several different solutions to the μ problem have been proposed, involving extensions of the MSSM of varying intricacy. They all work in roughly the same way; the parameter μ is completely absent at tree-level, and is to be replaced by the VEV(s) of some new field(s) which are determined by a potential which is in turn linked to the soft terms. In this way, the value of the effective parameter μ is no longer conceptually distinct from the mechanism of supersymmetry breaking; if we can explain why $m_{\rm soft} \ll M_{\rm P}$, we will also be able to understand why μ is of the same order. Some attractive solutions for the μ problem are proposed in Refs.[41,42,43,44] From the point of view of the MSSM, however, we can just treat μ as an independent parameter.

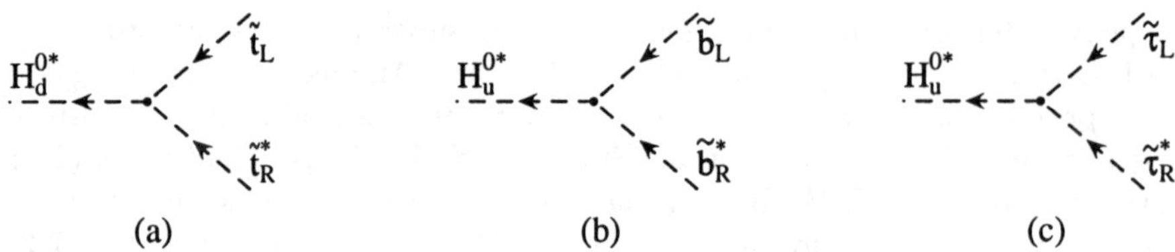

Figure 9: Some of the supersymmetric (scalar)^3 couplings proportional to $\mu^* y_t$, $\mu^* y_b$, and $\mu^* y_\tau$.

The μ-term and the Yukawa couplings in the superpotential eq. (5.1) combine to yield (scalar)^3 couplings [see the second and third terms on the right-hand side of eq. (3.47)] of the form

$$\mathcal{L} \supset \mu^*(\widetilde{\overline{u}}\mathbf{y_u}\tilde{u}H_d^{0*} + \widetilde{\overline{d}}\mathbf{y_d}\tilde{d}H_u^{0*} + \widetilde{\overline{e}}\mathbf{y_e}\tilde{e}H_u^{0*}$$
$$+ \widetilde{\overline{u}}\mathbf{y_u}\tilde{d}H_d^{-*} + \widetilde{\overline{d}}\mathbf{y_d}\tilde{u}H_u^{+*} + \widetilde{\overline{e}}\mathbf{y_e}\tilde{\nu}H_u^{+*}) + \text{c.c.} \tag{5.6}$$

In Fig. 9 we show some of these couplings which are proportional to $\mu^* y_t$, $\mu^* y_b$, and $\mu^* y_\tau$ respectively. These play an important role in determining the mixing of top squarks, bottom squarks, and tau sleptons, as we will see in section 7.5.

5.2 R-parity (also known as matter parity) and its consequences

The superpotential eq. (5.1) is minimal in the sense that it is sufficient to produce a phenomenologically viable model. However, there are other terms that one could write down which are gauge-invariant and analytic in the chiral superfields, but are not included in the MSSM because they violate either baryon number (B) or total lepton number (L). The most general gauge-invariant and renormalizable superpotential would include not only eq. (5.1), but also the terms

$$W_{\Delta L=1} = \frac{1}{2}\lambda^{ijk}L_iL_j\bar{e}_k + \lambda'^{ijk}L_iQ_j\bar{d}_k + \mu'^i L_i H_u \tag{5.7}$$

$$W_{\Delta B=1} = \frac{1}{2}\lambda''^{ijk}\bar{u}_i\bar{d}_j\bar{d}_k \tag{5.8}$$

where we have restored family indices $i = 1, 2, 3$. The chiral supermultiplets carry baryon number assignments B $= +1/3$ for Q_i; B $= -1/3$ for $\bar{u}_i, \bar{d}_i$; and B $= 0$ for all others. The total lepton number assignments are L $= +1$ for L_i, L $= -1$ for $\bar{e}_i$, and L $= 0$ for all others. Therefore, the terms in eq. (5.7) violate total lepton number by 1 unit (as well as the individual lepton flavors) and those in eq. (5.8) violate baryon number by 1 unit.

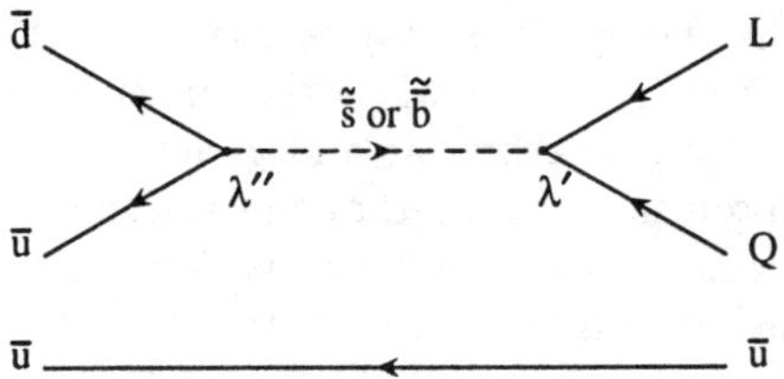

Figure 10: Squarks can mediate disastrously rapid proton decay if R-parity is violated.

The possible existence of such terms is rather disturbing, since B and L violating processes have never been seen experimentally. The most obvious experimental constraint comes from the non-observation of proton decay, which would violate both B and L by 1 unit. If both λ' and λ'' couplings are present and of order unity, then the lifetime of the proton would be measured in minutes or hours. For example, the Feynman graph in Fig. 10 would lead to $p^+ \to e^+\pi^0$ or e^+K^0 or $\mu^+\pi^0$ or μ^+K^0 or $\nu\pi^+$ or νK^+ etc. depending on which components of λ' are largest, and these processes would seem to be completely unsuppressed since the necessary couplings are all renormalizable. (The coupling λ'' must be antisymmetric in its last two flavor indices, since the color indices are contracted antisymmetrically. That is why the squark in Fig. 10 is $\tilde{\bar{s}}$ or $\tilde{\bar{b}}$ but not $\tilde{\bar{d}}$, for u, d quarks in the initial state.) In contrast, the decay time of the proton into these modes is measured to be in excess of 10^{32} years. Many other processes also give very significant constraints on the violation of lepton and baryon numbers; these are reviewed in Ref.[45]

One could simply try to take B and L conservation as a postulate in the MSSM. However, this is clearly a step backwards from the situation in the Standard Model, where the conservation of these quantum numbers is *not* assumed, but is rather a pleasantly "accidental" consequence of the fact that there are no possible renormalizable lagrangian terms which violate B or L. Furthermore, there is a quite general obstacle to treating B and L as fundamental symmetries of nature, since they are known to be violated by non-perturbative electroweak effects (even though those effects are negligible for experiments at ordinary energies). Therefore, in the MSSM one adds a new symmetry which has the effect of eliminating the possibility of B and L violating terms in the renormalizable superpotential, while allowing the good terms in eq. (5.1). This new symmetry is called "R-parity" [6] or equivalently "matter parity".[46]

Matter parity is a multiplicatively conserved quantum number defined as

$$P_M = (-1)^{3(\mathrm{B}-\mathrm{L})} \tag{5.9}$$

for each particle in the theory. It is easy to check that the quark and lepton supermultiplets all have $P_M = -1$, while the Higgs supermultiplets H_u and H_d have $P_M = +1$. The gauge bosons and gauginos of course do not carry baryon number or lepton number, so they are assigned matter parity $P_M = +1$. The symmetry principle to be enforced is that a term in the Lagrangian (or in the superpotential) is allowed only if the product of P_M for all of the fields in it is $+1$. It is easy to see that each of the terms in eq. (5.7) and (5.8) is thus forbidden, while the good terms in eq. (5.1) are allowed. This discrete symmetry commutes with supersymmetry, as all members of a given supermultiplet have the same matter parity. The advantage of matter parity is that it can in principle be an *exact* symmetry, which B and L themselves cannot, since they are known to be violated by non-perturbative electroweak effects. Thus even with exact matter parity conservation in the MSSM, one expects that small amounts of baryon number and total lepton number violation can occur, due to nonrenormalizable terms in the Lagrangian. However, the MSSM does not have renormalizable interactions that violate B or L, with the standard assumption of matter parity conservation.

It is sometimes useful to recast matter parity in terms of R-parity, defined for each particle as

$$P_R = (-1)^{3(\mathrm{B}-\mathrm{L})+2s} \tag{5.10}$$

where s is the spin of the particle. Now, matter parity conservation and R-parity conservation are exactly equivalent, since the product of $(-1)^{2s}$ is of course equal to $+1$ for the particles involved in any interaction vertex in a theory that conserves angular momentum. However, particles within the same supermultiplet do not have the same R-parity. In general, symmetries with the property that particles within the same multiplet have different charges are called R symmetries; they do not commute with supersymmetry. Continuous $U(1)$ R symmetries are often encountered in the model-building literature; they should not be confused with R-parity, which is a discrete Z_2 symmetry. In fact, the matter parity version of R-parity makes clear that there is really nothing intrinsically "R" about it; in other words it secretly does commute with supersymmetry, so its name is somewhat suboptimal. Nevertheless, the R-parity assignment is very useful for phenomenology because all of the Standard Model particles and the Higgs bosons have even R-parity ($P_R = +1$), while all of the squarks, sleptons, gauginos, and higgsinos have odd R-parity ($P_R = -1$).

The R-parity odd particles are known variously as "supersymmetric particles" or "sparticles" for short, and they are always distinguished with a tilde (see Tables 1 and 2). If R-parity is exactly conserved, then there can be no mixing between the sparticles and the $P_R = +1$ particles. Furthermore, every interaction vertex in the theory contains an even number of $P_R = -1$ sparticles. This has three extremely important phenomenological consequences:

- The lightest sparticle with $P_R = -1$, called the "lightest supersymmetric particle" or LSP, must be absolutely stable. If the LSP is electrically neutral, it interacts only weakly with ordinary matter, and so can make an attractive candidate [47] for the non-baryonic dark matter which seems to be required by cosmology.

- Each sparticle other than the LSP must eventually decay into a state which contains an odd number of LSPs (usually just one).

- In collider experiments, sparticles can only be produced in even numbers (usually two-at-a-time).

We *define* the MSSM to conserve R-parity or equivalently matter parity. While this decision seems to be well-motivated phenomenologically by proton decay constraints and the hope that the LSP will provide a good dark matter candidate, it might appear somewhat *ad hoc* from a theoretical point of view. After all, the MSSM would not suffer any internal inconsistency if we did not impose matter parity conservation. Furthermore, it is fair to ask why matter parity should be exactly conserved, given that the known discrete symmetries in the Standard Model (ordinary parity P, charge conjugation C, time reversal T, etc.) are all known to be inexact symmetries. Fortunately, it *is* sensible to formulate matter parity as a discrete symmetry which is exactly conserved. In general, exactly conserved, or "gauged" discrete symmetries [48] can exist provided that they satisfy certain anomaly cancellation conditions [49] (much like continuous gauged symmetries). One particularly attractive way this could occur is if B−L is a continuous $U(1)$ gauge symmetry which is spontaneously broken at some very high energy scale. From eq. (5.9), we observe that P_M is actually a discrete subgroup of the continuous $U(1)_{\text{B−L}}$ group. Therefore, if gauged $U(1)_{\text{B−L}}$ is broken by scalar VEVs (or other order parameters) which carry only even integer values of 3(B−L), then P_M will automatically survive as an exactly conserved remnant. A variety of extensions of the MSSM in which exact R-parity arises in just this way have been proposed.[50,51] It may also be possible to have gauged discrete symmetries which do not owe their exact conservation to an underlying continuous gauged symmetry, but rather to some other structure such as can occur in string theory. It is also possible that R-parity is broken, or is replaced by some alternative discrete symmetry.

5.3 Soft supersymmetry breaking in the MSSM

To complete the description of the MSSM, we need to specify the soft supersymmetry breaking terms. In section 4, we learned how to write down the
most general set of such terms in any supersymmetric theory. Applying this
recipe to the MSSM, we have:

$$
\begin{aligned}
\mathcal{L}_{\text{soft}}^{\text{MSSM}} = \ & -\frac{1}{2}\left(M_3\,\widetilde{g}\widetilde{g} + M_2\,\widetilde{W}\widetilde{W} + M_1\,\widetilde{B}\widetilde{B}\right) + \text{c.c.} \\
& -\left(\widetilde{\bar{u}}\,\mathbf{a_u}\,\widetilde{Q}H_u - \widetilde{\bar{d}}\,\mathbf{a_d}\,\widetilde{Q}H_d - \widetilde{\bar{e}}\,\mathbf{a_e}\,\widetilde{L}H_d\right) + \text{c.c.} \\
& -\widetilde{Q}^\dagger\,\mathbf{m_Q^2}\,\widetilde{Q} - \widetilde{L}^\dagger\,\mathbf{m_L^2}\,\widetilde{L} - \widetilde{\bar{u}}\,\mathbf{m_{\bar{u}}^2}\,\widetilde{\bar{u}}^\dagger - \widetilde{\bar{d}}\,\mathbf{m_{\bar{d}}^2}\,\widetilde{\bar{d}}^\dagger - \widetilde{\bar{e}}\,\mathbf{m_{\bar{e}}^2}\,\widetilde{\bar{e}}^\dagger \\
& - m_{H_u}^2\,H_u^*H_u - m_{H_d}^2\,H_d^*H_d - (bH_uH_d + \text{c.c.}).
\end{aligned}
\tag{5.11}
$$

In eq. (5.11), M_3, M_2, and M_1 are the gluino, wino, and bino mass terms. Here,
and from now on, we suppress the adjoint representation gauge indices on the
wino and gluino fields, and the gauge indices on all of the chiral supermultiplet
fields. The second line in eq. (5.11) contains the (scalar)^3 couplings [of the type
a^{ijk} in eq. (4.1)]. Each of $\mathbf{a_u}$, $\mathbf{a_d}$, $\mathbf{a_e}$ is a complex 3×3 matrix in family space,
with dimensions of (mass). They are in one-to-one correspondence with the
Yukawa coupling matrices in the superpotential. The third line of eq. (5.11)
consists of squark and slepton mass terms of the $(m^2)^j_i$ type in eq. (4.1). Each
of $\mathbf{m_Q^2}$, $\mathbf{m_{\bar{u}}^2}$, $\mathbf{m_{\bar{d}}^2}$, $\mathbf{m_L^2}$, $\mathbf{m_{\bar{e}}^2}$ is a 3×3 matrix in family space which can have
complex entries, but they must be hermitian so that the lagrangian is real. (To
avoid clutter, we do not put tildes on the $\mathbf{Q}$ in $\mathbf{m_Q^2}$, etc.) Finally, in the last
line of eq. (5.11) we have supersymmetry-breaking contributions to the Higgs
potential; $m_{H_u}^2$ and $m_{H_d}^2$ are (mass)^2 terms of the $(m^2)^j_i$ type, while b is the
only (mass)^2 term of the type b^{ij} in eq. (4.1) which can occur in the MSSM.[h]
Schematically, we can write

$$
M_1, \, M_2, \, M_3, \, \mathbf{a_u}, \, \mathbf{a_d}, \, \mathbf{a_e} \sim m_{\text{soft}};
\tag{5.12}
$$

$$
\mathbf{m_Q^2}, \, \mathbf{m_L^2}, \, \mathbf{m_{\bar{u}}^2}, \, \mathbf{m_{\bar{d}}^2}, \, \mathbf{m_{\bar{e}}^2}, \, m_{H_u}^2, \, m_{H_d}^2, \, b \sim m_{\text{soft}}^2
\tag{5.13}
$$

with a characteristic mass scale m_{soft} which is not much larger than 10^3 GeV,
as argued in the Introduction. The expression eq. (5.11) is the most general soft
supersymmetry-breaking Lagrangian of the form eq. (4.1) which is compatible
with gauge invariance and matter parity conservation.

Unlike the supersymmetry-preserving part of the lagrangian, $\mathcal{L}_{\text{soft}}^{\text{MSSM}}$ introduces many new parameters which were not present in the ordinary Standard

---

[h]The parameter we call b is often seen in the literature as m_{12}^2 or m_3^2 or $B\mu$.

Model. A careful count[52] reveals that there are 105 masses, phases and mixing angles in the MSSM lagrangian which cannot be rotated away by redefining the phases and flavor basis for the quark and lepton supermultiplets, and which have no counterpart in the ordinary Standard Model. Thus, in principle, supersymmetry (or more precisely, supersymmetry *breaking*) appears to introduce a tremendous arbitrariness in the lagrangian.

5.4 Hints of an Organizing Principle

Fortunately, there is already good experimental evidence that some sort of powerful "organizing principle" must govern the soft terms. This is because most of the new parameters in eq. (5.11) involve flavor mixing or CP violation of the type which is already severely restricted by experiment.[53] For example, suppose that $\mathbf{m}_{\tilde{e}}^2$ is not diagonal in a basis $(\tilde{e}_R, \tilde{\mu}_R, \tilde{\tau}_R)$ of sleptons whose superpartners are the right-handed pieces of the Standard Model mass eigenstates e, μ, τ. In that case slepton mixing occurs, and the individual lepton numbers will not be conserved. This is true even for processes which only involve the sleptons as virtual particles. A particularly strong limit on this possibility comes from the experimental constraint on $\mu \to e\gamma$ [54]. There are also important experimental constraints on the squark (mass)^2 matrices, for example from constraints on the neutral kaon system.

All of these potentially dangerous FCNC and CP-violating effects in the MSSM can be evaded if one assumes (or can explain!) that supersymmetry breaking should be suitably "universal". In particular, one can suppose that the squark and slepton (mass)^2 matrices are flavor-blind. This means that they are each proportional to the 3×3 identity matrix in family space:

$$\mathbf{m}_Q^2 = m_Q^2 \mathbf{1}; \quad \mathbf{m}_{\tilde{u}}^2 = m_{\tilde{u}}^2 \mathbf{1}; \quad \mathbf{m}_{\tilde{d}}^2 = m_{\tilde{d}}^2 \mathbf{1}; \quad \mathbf{m}_L^2 = m_L^2 \mathbf{1}; \quad \mathbf{m}_{\tilde{e}}^2 = m_{\tilde{e}}^2 \mathbf{1}. \quad (5.14)$$

If so, then all squark and slepton mixing angles are rendered trivial, because squarks and sleptons with the same electroweak quantum numbers will be degenerate in mass and can be rotated into each other at will. Supersymmetric contributions to FCNC processes will therefore vanish, modulo the mixing due to $\mathbf{a}_u, \mathbf{a}_d, \mathbf{a}_e$. One can make the further assumption that the (scalar)^3 couplings are each proportional to the corresponding Yukawa coupling matrix:

$$\mathbf{a}_u = A_{u0}\, \mathbf{y}_u; \qquad \mathbf{a}_d = A_{d0}\, \mathbf{y}_d; \qquad \mathbf{a}_e = A_{e0}\, \mathbf{y}_e. \qquad (5.15)$$

This ensures that only the squarks and sleptons of the third family can have large (scalar)^3 couplings. Finally, one can avoid disastrously large CP-violating effects with the assumption that the soft parameters do not introduce new complex phases. This is automatic for $m_{H_u}^2$ and $m_{H_d}^2$, and for m_Q^2, $m_{\tilde{u}}^2$ etc. if eq. (5.14) is assumed; if they were not real numbers, the lagrangian would not be real. One can also fix μ in the superpotential and b in eq. (5.11) to be real, by an appropriate phase rotation of H_u and H_d. If one then assumes that

$$\arg(M_1),\ \arg(M_2),\ \arg(M_3),\ \arg(A_{u0}),\ \arg(A_{d0}),\ \arg(A_{e0}) = 0 \text{ or } \pi, \qquad (5.16)$$

then the only CP-violating phase in the theory will be the ordinary CKM phase found in the ordinary Yukawa couplings. Together, the conditions eqs. (5.14)-(5.16) make up a rather weak version of what is often called the assumption of *soft-breaking universality*.

The soft-breaking universality relations eqs. (5.14)-(5.16) (or stronger versions of them) are presumed to be the result of some specific model for the origin of supersymmetry breaking, even though there is considerable disagreement among theorists as to what the specific model should actually be. In any case, they are indicative of an underlying simplicity or symmetry of the lagrangian at some very high energy scale Q_0, which we will call the "input scale". If we use this lagrangian to compute masses and cross-sections and decay rates for experiments at ordinary energies near the electroweak scale, the results will involve large logarithms of order $\ln(Q_0/m_Z)$. The large logarithms can be conveniently resummed using renormalization group (RG) equations. Therefore, eqs. (5.14)-(5.16) should be imposed as boundary conditions on the running soft parameters at the RG scale Q_0, and we must then RG-evolve the soft parameters, the superpotential parameters, and the gauge couplings down to the electroweak scale where humans perform experiments. At the electroweak scale, eqs. (5.14) and (5.15) will no longer hold. However, RG corrections due to gauge interactions will respect eqs. (5.14) and (5.15), while RG corrections due to Yukawa interactions are quite small except for couplings involving the top squarks (stops) and possibly the bottom squarks (sbottoms) and tau sleptons (staus). In particular, the (scalar)^3 couplings should be quite negligible for the squarks and sleptons of the first two families. Furthermore, RG evolution does not introduce new CP-violating phases. Therefore, if universality can be arranged to hold at the input scale, supersymmetric contributions to FCNC and CP-violating observables can be acceptably small in comparison to present limits (although quite possibly measurable in future experiments).

One good reason to be optimistic that such a program can succeed is the celebrated apparent unification of gauge couplings in the MSSM.[55] The 1-loop RG equations for the Standard Model gauge couplings g_1, g_2, g_3 are given by

$$\frac{d}{dt}g_a = \frac{1}{16\pi^2}b_a g_a^3 \qquad \Rightarrow \qquad \frac{d}{dt}\alpha_a^{-1} = -\frac{b_a}{2\pi} \qquad (a = 1, 2, 3) \qquad (5.17)$$

where $t = \ln(Q/Q_0)$ with Q the RG scale. In the Standard Model, $b_a^{\text{SM}} =$

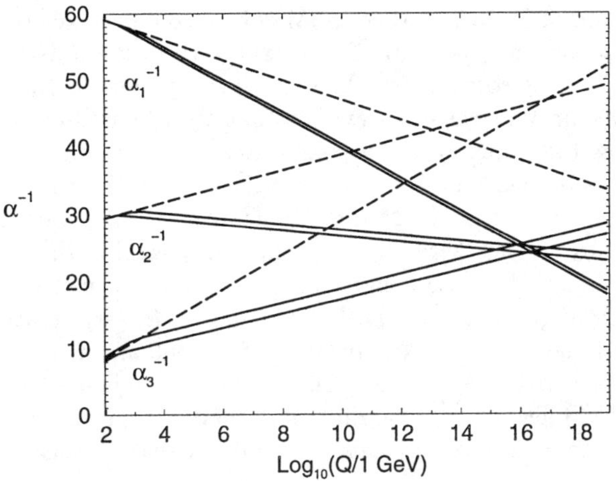

Figure 11: RG evolution of the inverse gauge couplings $\alpha_a^{-1}(Q)$ in the Standard Model (dashed lines) and the MSSM (solid lines). In the MSSM case, $\alpha_3(m_Z)$ is varied between 0.113 and 0.123, and the sparticle mass thresholds between 250 GeV and 1 TeV.

$(41/10,\ -19/6,\ -7)$, and $b_a^{\mathrm{MSSM}} = (33/5,\ 1,\ -3)$ in the MSSM. The latter coefficients are larger because of the virtual effects of the extra MSSM particles in loops. The normalization for g_1 here is chosen to agree with the canonical covariant derivative for grand unification of the gauge group $SU(3)_C \times SU(2)_L \times U(1)_Y$ into $SU(5)$ or $SO(10)$. Thus in terms of the conventional electroweak gauge couplings g and g' with $e = g\sin\theta_W = g'\cos\theta_W$, one has $g_2 = g$ and $g_1 = \sqrt{5/3}g'$. The quantities $\alpha_a = g_a^2/4\pi$ have the nice property that their reciprocals run linearly with RG scale at one-loop order. In Fig. 11 we compare the RG evolution of the α_a^{-1}, including two-loop effects, in the Standard Model (dashed lines) and the MSSM (solid lines). Unlike the Standard Model, the MSSM includes just the right particle content to ensure that the gauge couplings can unify, at a scale $M_U \sim 2 \times 10^{16}$ GeV. While the apparent unification of gauge couplings at M_U could be just an accident, it may also be taken as a strong hint in favor of a grand unified theory (GUT) or superstring models, both of which indeed predict gauge coupling unification below M_P. Furthermore, if we take this hint seriously, then it means that we can reasonably expect to apply a similar RG analysis to the other MSSM couplings and soft masses as well.

We must mention that there are two other possible types of explanations for the suppression of FCNCs in the MSSM, which could replace the universality hypothesis of eqs. (5.14)-(5.16). One might refer to them as "irrelevancy" and "alignment" of the soft masses. The "irrelevancy" idea is that the sparticles masses are simply *extremely* heavy, so that their contributions to FCNC and CP-violating diagrams are highly suppressed. In practice, however, the degree of suppression needed often requires $m_{\text{soft}} \gg 1$ TeV for at least some of the scalar masses; this seems to go directly against the motivation for supersymmetry as a cure for the hierarchy problem as discussed in the Introduction. Nevertheless, it is possible to arrange a scheme where this can work in a sensible way.[56] The "alignment" idea is that the squark (mass)^2 matrices do not have the flavor-blindness indicated in eq. (5.14), but are arranged in flavor space to be aligned with the relevant Yukawa matrices in just such a way as to avoid large FCNC effects.[40,57] The alignment models typically require rather special flavor symmetries. In any case, we will not discuss these possibilities further.

In practice, a given model for the origin of supersymmetry breaking may make predictions for the MSSM soft terms that are even stronger than eqs. (5.14)-(5.16). In the next section we will discuss the ideas that go into making such predictions, before turning to their implications for the MSSM spectrum in section 7.

6   Origins of supersymmetry breaking

6.1   *General considerations for supersymmetry breaking*

In the MSSM, supersymmetry breaking is simply introduced explicitly. However, we have seen that the soft parameters cannot be arbitrary. In order to understand how patterns like eqs. (5.14), (5.15) and (5.16) can emerge, it is necessary to consider models in which supersymmetry is spontaneously broken. By definition, this means that the vacuum state $|0\rangle$ is not invariant under supersymmetry transformations, so $Q_\alpha|0\rangle \neq 0$ and $Q^\dagger_{\dot\alpha}|0\rangle \neq 0$. Now, in global supersymmetry, the Hamiltonian operator H can be related to the supersymmetry generators through the algebra eq. (3.30):

$$H = P^0 = \frac{1}{4}(Q_1 Q_1^\dagger + Q_1^\dagger Q_1 + Q_2 Q_2^\dagger + Q_2^\dagger Q_2). \qquad (6.1)$$

If supersymmetry is unbroken in the vacuum state, it follows that $H|0\rangle = 0$ and the vacuum has zero energy. Conversely, if supersymmetry is spontaneously broken in the vacuum state, then the vacuum must have positive energy, since

$$\langle 0|H|0\rangle = \frac{1}{4}\left(||Q_1|0\rangle||^2 + ||Q_1^\dagger|0\rangle||^2 + ||Q_2|0\rangle||^2 + ||Q_2^\dagger|0\rangle||^2\right) > 0 \qquad (6.2)$$

if the Hilbert space is to have positive norm. If spacetime-dependent effects and fermion condensates can be neglected, then $\langle 0|H|0\rangle = \langle 0|V|0\rangle$, where V is the scalar potential in eq. (3.75). Therefore supersymmetry will be spontaneously broken if F_i and/or D^a does not vanish in the ground state. Note that if any state exists in which all F_i and D^a vanish, then it will have zero energy, implying that supersymmetry cannot be spontaneously broken in the true ground state. Therefore the way to achieve spontaneous supersymmetry breaking is to look for models in which the equations $F_i = 0$ and $D^a = 0$ cannot be simultaneously satisfied for *any* values of the fields.

Supersymmetry breaking with non-zero D-terms can be achieved through the Fayet-Iliopoulos mechanism.[58] If the gauge symmetry includes a $U(1)$ factor, then one can introduce a term linear in the corresponding auxiliary field of the gauge supermultiplet:

$$\mathcal{L}_{\text{Fayet}-\text{Iliopoulos}} = \kappa D \tag{6.3}$$

where κ is a constant parameter with dimensions of $(\text{mass})^2$. This term is gauge-invariant and supersymmetric by itself. [Note that the supersymmetry transformation δD in eq. (3.59) is a total derivative for a $U(1)$ gauge symmetry.] If we include it in the lagrangian, then D may get a non-zero VEV, depending on the other interactions of the scalar fields that are charged under the $U(1)$. To see this, we can write the relevant part of the scalar potential using eqs. (3.54) and (3.72) as

$$V = \frac{1}{2}D^2 - \kappa D + gD \sum_i q_i \phi^{*i} \phi_i \tag{6.4}$$

where the q_i are the charges of the scalar fields ϕ_i under the $U(1)$ gauge group in question. The presence of the Fayet-Iliopoulos term modifies the equation of motion eq. (3.74) to

$$D = \kappa - g \sum_i q_i \phi^{*i} \phi_i. \tag{6.5}$$

Now if the scalar fields ϕ_i have other interactions (such as large superpotential mass terms) which prevent them from getting VEVs, then the auxiliary field D will be forced to get a VEV equal to κ, and supersymmetry will be broken. This mechanism cannot work for non-abelian gauge groups, however, since eq. (6.3) would not be gauge-invariant.

In the MSSM, one can imagine that the D term for $U(1)_Y$ has a Fayet-Iliopoulos term which is the source of supersymmetry breaking. Unfortunately, this would be an immediate disaster, because at least some of the squarks and sleptons would just get non-zero VEVs (breaking color, electromagnetism, and/or lepton number, but not supersymmetry) in order to satisfy eq. (6.5), because they do not have superpotential mass terms. This means that a Fayet-Iliopoulos term for $U(1)_Y$ must be subdominant compared to other sources of supersymmetry breaking in the MSSM, if not absent altogether. One could also attempt to trigger supersymmetry breaking with a Fayet-Iliopoulos term for some other $U(1)$ gauge symmetry which is as yet unknown because it is spontaneously broken at a very high mass scale or because it does not couple to the Standard Model particles. However, if this is the ultimate source for supersymmetry breaking, it proves difficult to give appropriate masses to all of the MSSM particles, especially the gauginos. In any case, we will not discuss D-term breaking as the origin of supersymmetry violation any further, although it may not be ruled out.[59]

Models where supersymmetry breaking is due to non-zero F-terms, called O'Raifeartaigh models,[60] may have brighter phenomenological prospects. The idea is to pick a set of chiral supermultiplets $\Phi_i \supset (\phi_i, \psi_i, F_i)$ and a superpotential W in such a way that the equations $F_i = -\delta W^*/\delta\phi^{*i} = 0$ have no simultaneous solution. Then $V = \sum_i |F_i|^2$ will have to be positive at its minimum, ensuring that supersymmetry is broken. The simplest example which does this has three chiral supermultiplets with

$$W = -k\Phi_1 + m\Phi_2\Phi_3 + \frac{y}{2}\Phi_1\Phi_3^2. \tag{6.6}$$

Note that W contains a linear term, with k having dimensions of $(\text{mass})^2$. This is only possible if Φ_1 is a gauge singlet. In section 3 we cheated and did not mention such a term, because we knew that the MSSM contains no such singlet chiral supermultiplet. Nevertheless, it should be clear from retracing the derivation in section 3.2 that such a term is allowed if a gauge-singlet chiral supermultiplet is added to the theory. In fact, a linear term is absolutely necessary to achieve F-term breaking, since otherwise setting all $\phi_i = 0$ will always give a supersymmetric global minimum with all $F_i = 0$. Without loss of generality, we can choose k, m, and y to be real and positive (by a phase rotation of the fields). The scalar potential following from eq. (6.6) is

$$V = |F_1|^2 + |F_2|^2 + |F_3|^2; \tag{6.7}$$

$$F_1 = k - \frac{y}{2}\phi_3^{*2}; \qquad F_2 = -m\phi_3^*; \qquad F_3 = -m\phi_2^* - y\phi_1^*\phi_3^*. \tag{6.8}$$

Clearly, $F_1 = 0$ and $F_2 = 0$ are not compatible, so supersymmetry must indeed be broken. If $m^2 > yk$ (which we assume from now on), then it is easy to show that the absolute minimum of the potential is at $\phi_2 = \phi_3 = 0$ with ϕ_1

undetermined, so $F_1 = k$ and $V = k^2$ at the minimum of the potential. The fact that ϕ_1 is undetermined is an example of a "flat direction" in the scalar potential; this is a common feature of supersymmetric models.[i]

If we presciently choose to expand V around $\phi_1 = 0$, the mass spectrum of the theory consists of 6 real scalars with tree-level squared masses

$$0, \; 0, \; m^2, \; m^2, \; m^2 - yk, \; m^2 + yk. \tag{6.9}$$

Meanwhile, there are 3 Weyl fermions with masses

$$0, \; m, \; m. \tag{6.10}$$

The non-degeneracy of scalars and fermions is a clear sign that supersymmetry has been spontaneously broken. The 0 eigenvalues in eqs. (6.9) and (6.10) correspond to the complex scalar ϕ_1 and its fermionic partner ψ_1. However, ϕ_1 and ψ_1 have different reasons for being massless. The masslessness of ϕ_1 corresponds to the existence of the flat direction, since any value of ϕ_1 gives the same energy at tree-level. This flat direction is an accidental feature of the classical scalar potential, and in this case it is removed ("lifted") by quantum corrections. This can be seen by computing the Coleman-Weinberg one-loop effective potential.[61] After some calculation, one finds the result that the global minimum is indeed fixed at $\phi_1 = \phi_2 = \phi_3 = 0$, with the complex scalar ϕ_1 receiving a small positive-definite (mass)^2 equal to

$$m_{\phi_1}^2 = \frac{1}{32\pi^2} \left[\left(\frac{ym^4}{k} + y^3 k \right) \ln\left(\frac{m^2 + yk}{m^2 - yk} \right) + 2y^2 m^2 \left(\ln[1 - \frac{y^2 k^2}{m^4}] - 1 \right) \right]. \tag{6.11}$$

[In the limit $yk \ll m^2$, one has $m_{\phi_1}^2 = y^4 k^2 / (48\pi^2 m^2)$.] In contrast, the Weyl fermion ψ_1 remains exactly massless because of a general feature of all models with spontaneously broken supersymmetry. To understand this, recall that the spontaneous breaking of any global symmetry always gives rise to a massless Nambu-Goldstone mode with the same quantum numbers as the broken symmetry generator. In the case of supersymmetry, the broken generator is the fermionic charge Q_α, so the Nambu-Goldstone particle must be a massless neutral Weyl fermion called the *goldstino*. In the O'Raifeartaigh model example, ψ_1 is the goldstino because it is the fermionic partner of the auxiliary field F_1 which got a VEV. We will prove these statements in a more general context in section 6.2.

---

[i] More generally, "flat directions" are non-compact lines and surfaces in the space of scalar fields along which the scalar potential vanishes. The classical scalar potential of the MSSM would have many flat directions if supersymmetry were not broken.

The O'Raifeartaigh superpotential determines the mass scale of supersymmetry breaking $\sqrt{F_1}$ in terms of a dimensionful parameter k which is put in by hand. This is somewhat *ad hoc* since $\sqrt{k}$ will have to be much less than $M_{\rm P}$ in order to give the right order of magnitude for the MSSM soft terms. We would like to have a mechanism which can instead generate such scales naturally. This can be done in models of dynamical supersymmetry breaking. In such theories, the small (compared to $M_{\rm P}$) mass scales associated with supersymmetry breaking arise by dimensional transmutation. They generally feature a new asymptotically-free non-Abelian gauge symmetry with a gauge coupling g which is perturbative at $M_{\rm P}$ and which gets strong in the infrared at some smaller scale $\Lambda \sim e^{-8\pi^2/|b|g_0^2} M_{\rm P}$, where g_0 is the running gauge coupling at $M_{\rm P}$ with beta function $-|b|g^3/16\pi^2$. Just as in QCD, it is perfectly natural for Λ to be many orders of magnitude below the Planck scale. Supersymmetry breaking may then be best described in terms of the effective dynamics of the strongly coupled theory. One possibility is that the auxiliary F field for a composite chiral supermultiplet (built out of the fundamental fields which transform under the new strongly-coupled gauge group) obtains a VEV. Constructing models which actually break supersymmetry in an acceptable way is a highly non-trivial business; for more information we refer the reader to Ref.[62]

The one thing that is now clear about spontaneous supersymmetry breaking (dynamical or not) is that it requires us to extend the MSSM. The ultimate supersymmetry-breaking order parameter cannot belong to any of the supermultiplets of the MSSM. Therefore one must ask how supersymmetry breakdown is "communicated" to the MSSM particles. It is very difficult to achieve this in a phenomenologically viable way with renormalizable interactions at tree-level. First, it is problematic to give masses to the MSSM gauginos, because supersymmetry does not allow (scalar)-(gaugino)-(gaugino) couplings which could turn into gaugino mass terms when the scalar gets a VEV. Second, at least some of the MSSM squarks and sleptons would have to be unacceptably light, and should have been discovered already. This can be understood in a general way from the existence of a sum rule which governs the tree-level squared masses of scalars and chiral fermions in theories with spontaneous supersymmetry breaking:

$$\mathrm{Tr}[M^2_{\text{real scalars}}] = 2\mathrm{Tr}[M^2_{\text{chiral fermions}}]. \tag{6.12}$$

If supersymmetry were not broken, then eq. (6.12) would follow immediately from the degeneracy of complex scalars [with two real scalar components, hence the factor of 2] and their Weyl fermion superpartners. However, eq. (6.12) still holds when supersymmetry is broken spontaneously by F-terms and D-terms, as one can verify in general by explicitly computing the (mass)^2 matrices for

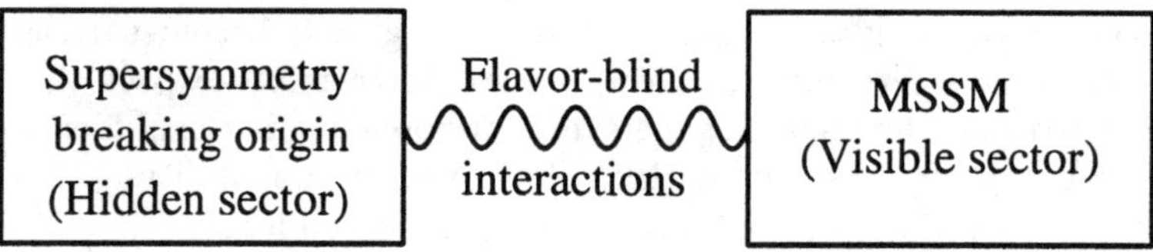

Figure 12: The presumed schematic structure for supersymmetry breaking.

arbitrary values of the fields.[j] One can easily see, for example, that with the O'Raifeartaigh spectrum of eqs. (6.9) and (6.10), the sum rule eq. (6.12) is indeed satisfied. This sum rule seems to be bad news for a phenomenologically viable model, because the masses of all of the MSSM chiral fermions are already known to be small (except for the top quark and the higgsinos). Even if one could succeed in doing so, there is no reason why the resulting MSSM soft terms in such a model should satisfy conditions like eqs. (5.14) or (5.15).

For these reasons, we expect that the MSSM soft terms arise radiatively, rather than from tree-level renormalizable couplings to the supersymmetry-breaking order parameters. Supersymmetry breaking evidently occurs in a "hidden sector" of particles which have no (or only very small) direct couplings to the "visible sector" chiral supermultiplets of the MSSM. However, the two sectors do share some interactions which are responsible for mediating supersymmetry breaking from the hidden sector to the visible sector, where they appear as calculable soft terms. (See Fig. 12.) In this scenario, the tree-level sum rule eq. (6.12) need not hold for the visible sector fields, so that a phenomenologically viable superpartner mass spectrum is in principle achievable. As a bonus, if the mediating interactions are flavor-blind, then the soft terms appearing in the MSSM will automatically obey conditions like eqs. (5.14), (5.15) and (5.16). There are two main competing proposals for what the mediating interactions might be. The first (and historically the more popular) is that they are gravitational. In this *gravity-mediated supersymmetry breaking* scenario, if supersymmetry is broken in the hidden sector by a VEV $\langle F \rangle$, then the soft terms in the visible sector should be roughly of order

$$m_{\text{soft}} \sim \frac{\langle F \rangle}{M_{\text{P}}},\tag{6.13}$$

by dimensional analysis. This is because we know that m_{soft} must vanish in the limit $\langle F \rangle \to 0$ where supersymmetry is unbroken, and also in the limit

---

[j] This assumes only that the $U(1)$ generators in the theory satisfy $\text{Tr}[T^a] = 0$, which does hold for $U(1)_Y$ in the MSSM and more generally for any non-anomalous gauge symmetry.

$M_{\rm P} \to \infty$ (corresponding to $G_{\rm Newton} \to 0$) in which gravity becomes irrelevant. For $m_{\rm soft}$ of order a few hundred GeV, one would therefore expect that the scale associated with the origin of supersymmetry breaking in the hidden sector should be roughly $\sqrt{\langle F \rangle} \sim 10^{10}$ or 10^{11} GeV. Another possibility is that the supersymmetry breaking order parameter is a gaugino condensate $\langle 0 | \lambda^a \lambda^b | 0 \rangle = \delta^{ab} \Lambda^3 \neq 0$. If the composite field $\lambda^a \lambda^b$ is part of an auxiliary field F for some (perhaps composite) chiral superfield, then by dimensional analysis we expect supersymmetry breaking soft terms of order

$$m_{\rm soft} \sim \frac{\Lambda^3}{M_{\rm P}^2}. \tag{6.14}$$

In that case, the scale associated with dynamical supersymmetry breaking should be more like $\Lambda \sim 10^{13}$ GeV.

The second main possibility is that the flavor-blind mediating interactions for supersymmetry breaking are the ordinary electroweak and QCD gauge interactions. In this *gauge-mediated supersymmetry breaking* scenario, the MSSM soft terms arise from loop diagrams involving some *messenger* particles in the hidden sector which have $SU(3)_C \times SU(2)_L \times U(1)_Y$ interactions and which couple to a supersymmetry-breaking VEV $\langle F \rangle$. Then one estimates for the MSSM soft terms

$$m_{\rm soft} \sim \frac{\alpha_a}{4\pi} \frac{\langle F \rangle}{M_{\rm mess}} \tag{6.15}$$

where the $\alpha_a / 4\pi$ is a loop factor for Feynman diagrams involving gauge interactions, and $M_{\rm mess}$ is a characteristic scale of the masses of the messenger fields. So if $M_{\rm mess} \sim \sqrt{\langle F \rangle}$, then the scale of supersymmetry breaking can be as low as about $\sqrt{\langle F \rangle} \sim 10^4$ or 10^5 GeV (much lower than in the gravity-mediated case!) to give $m_{\rm soft}$ of the right order of magnitude.

6.2 *The goldstino and the gravitino*

As explained in the previous section, the spontaneous breaking of global supersymmetry implies the existence of a massless Weyl fermion, the goldstino. In the particular case of the O'Raifeartaigh model, the goldstino was identified to be ψ_1. More generally, we might expect that in the case of F-term or D-term breaking, the goldstino is the fermionic component of the supermultiplet whose auxiliary field obtains a VEV.

Let us make this more precise by actually proving that the goldstino exists and, in the process, identifying it. This is actually rather easy. Consider a general supersymmetric model with both gauge and chiral supermultiplets as in section 3. The fermionic degrees of freedom consist of gauginos (λ^a) and chiral fermions (ψ_i). After some of the scalar fields in the theory obtain VEVs, the fermion mass matrix will have the form:

$$M_{\text{fermion}} = \begin{pmatrix} 0 & \sqrt{2}g_a(\langle\phi^*\rangle T^a)^i \\ \sqrt{2}g_a(\langle\phi^*\rangle T^a)^j & \langle W^{ij}\rangle \end{pmatrix} \tag{6.16}$$

in the (λ^a, ψ_i) basis. [The off-diagonal entries in this matrix come from the second line in eq. (3.72), and the lower right entry can be seen in eq. (3.46).] Now we simply note that M_{fermion} annihilates the vector

$$\widetilde{G} = \left(\frac{\langle D^a \rangle}{\sqrt{2}}, \, \langle F_i \rangle \right). \tag{6.17}$$

The first row of M_{fermion} annihilates $\widetilde{G}$ by virtue of the requirement eq. (3.73) that the superpotential is gauge invariant, and the second row annihilates $\widetilde{G}$ because of the condition $\langle \partial V / \partial \phi_i \rangle = 0$ which must be satisfied at the minimum of the scalar potential. Eq. (6.17) is proportional to the goldstino wavefunction; it is non-trivial if and only if at least one of the auxiliary fields has a VEV, breaking supersymmetry. So we have proven that if global supersymmetry is spontaneously broken, then the goldstino exists and has zero mass, and that its components among the various fermions in the theory are just proportional to the corresponding auxiliary field VEVs.

We can derive another very important property of the goldstino by considering the form of the conserved supercurrent eq. (3.76). Suppose for simplicity that the non-vanishing auxiliary field VEV is $\langle F \rangle$ and that its goldstino superpartner is $\widetilde{G}$. Then the supercurrent conservation equation tells us that

$$0 = \partial_\mu J^\mu_\alpha = i\langle F \rangle (\sigma^\mu \partial_\mu \widetilde{G}^\dagger)_\alpha + \partial_\mu j^\mu_\alpha + \ldots \tag{6.18}$$

where j^μ_α is the part of the supercurrent which involves all of the other supermultiplets, and the ellipses represent other contributions of the goldstino supermultiplet to $\partial_\mu J^\mu_\alpha$ which we can ignore. [The first term in eq. (6.18) comes from the second term in eq. (3.76), using the equation of motion $F_i = -W_i^*$ for the goldstino's auxiliary field.] This equation of motion for the goldstino field allows us to write an effective lagrangian

$$\mathcal{L}_{\text{goldstino}} = -i\widetilde{G}^\dagger \overline{\sigma}^\mu \partial_\mu \widetilde{G} - \frac{1}{\langle F \rangle}(\widetilde{G} \partial_\mu j^\mu + \text{c.c.}) \tag{6.19}$$

which describes the interactions of the goldstino with all of the other fermion-boson pairs.[63] In particular, since $j^\mu_\alpha = (\sigma^\nu \overline{\sigma}^\mu \psi_i)_\alpha \partial_\nu \phi^{*i} - (1/2\sqrt{2})\sigma^\nu \overline{\sigma}^\rho \sigma^\mu \lambda^{\dagger a} F^a_{\nu\rho} +$

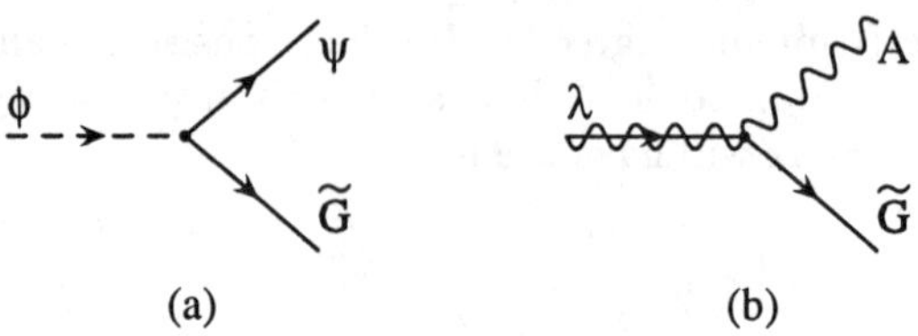

Figure 13: Goldstino/gravitino interactions with superpartner pairs (ϕ, ψ) and (λ^a, A^a).

..., there are goldstino-scalar-chiral fermion and goldstino-gaugino-gauge boson vertices as shown in Fig. 13. Since this derivation depends only on supercurrent conservation, eq. (6.19) holds independently of the details of how supersymmetry breaking is communicated from $\langle F \rangle$ to the MSSM sector fields (ϕ_i, ψ_i) and (λ^a, A^a). It may appear strange at first that the interaction terms in eq. (6.19) get larger as $\langle F \rangle$ goes to zero. However, the interaction term $\widetilde{G} \partial_\mu j^\mu$ contains two derivatives which turn out to always give a kinematical factor proportional to the (mass)^2 difference of the superpartners when they are on-shell, i.e. $m_{\phi_i}^2 - m_{\psi_i}^2$ and $m_\lambda^2 - m_A^2$ for Figs. 13a and 13b respectively. These can be non-zero only by virtue of supersymmetry breaking, so they must also vanish as $\langle F \rangle \to 0$, and the interaction is well-defined in that limit. Nevertheless, for fixed values of $m_{\phi_i}^2 - m_{\psi_i}^2$ and $m_\lambda^2 - m_A^2$, the interaction term in eq. (6.19) can be phenomenologically important if $\langle F \rangle$ is not too large.[63,64,65,66]

The above remarks apply to the breaking of global supersymmetry. However, when one takes into account gravity, supersymmetry must be a local symmetry. This means that the spinor parameter ϵ^α which first appeared in section 3.1 is no longer a constant, but can vary from point to point in spacetime. The resulting locally supersymmetric theory is called *supergravity*.[67,68] It necessarily unifies the spacetime symmetries of ordinary general relativity with local supersymmetry transformations. In supergravity, the spin-2 graviton has a spin-3/2 fermion superpartner called the gravitino, which we will denote $\widetilde{\Psi}_\mu^\alpha$. The gravitino has odd R-parity ($P_R = -1$), as can be seen from the definition eq. (5.10). It carries both a vector and a spinor index, and transforms inhomogeneously under local supersymmetry transformations:

$$\delta \widetilde{\Psi}_\mu^\alpha = -\partial_\mu \epsilon^\alpha + \dots \tag{6.20}$$

Thus the gravitino is the "gauge" particle of local supersymmetry transformations [compare eq. (3.52)]. As long as supersymmetry is unbroken, the graviton and the gravitino are both massless, each with two spin helicity states. Once supersymmetry is spontaneously broken, the gravitino acquires a mass by absorbing ("eating") the goldstino, which becomes its longitudinal (helicity $\pm 1/2$) components. This is called the *super-Higgs* mechanism. It is entirely analogous to the ordinary Higgs mechanism for gauge theories, by which the $W^{\pm}$ and Z^0 gauge bosons in the Standard Model gain mass by absorbing the Nambu-Goldstone bosons associated with the spontaneously broken electroweak gauge invariance. The counting works, because the massive spin-3/2 gravitino now has four helicity states, of which two were originally assigned to the would-be goldstino. The gravitino mass is traditionally called $m_{3/2}$, and in the case of F-term breaking it is given by [69]

$$m_{3/2} = \frac{\langle F \rangle}{\sqrt{3}M_{\mathrm{P}}}, \tag{6.21}$$

if the goldstino supermultiplet has canonically-normalized kinetic terms and the cosmological constant is required to vanish. This means that one has very different expectations for the mass of the gravitino in gravity-mediated and in gauge-mediated models, because they usually make very different predictions for $\langle F \rangle$.

In the gravity-mediated supersymmetry breaking case, the gravitino mass is usually comparable to the masses of the MSSM sparticles [compare eqs. (6.13) and (6.21)]. Therefore $m_{3/2}$ is expected to be at least 100 GeV or so. Its interactions will be of gravitational strength, so the gravitino will not play any role in collider physics, but it can be a very important consideration in cosmology.[70] If it is the LSP, then it is stable and its primordial density could easily exceed the critical density, causing the universe to become matter-dominated too early. Even if it is not the LSP, the gravitino can cause problems unless its density is diluted by inflation at late times, or it decays sufficiently rapidly.

In gauge-mediated supersymmetry breaking models, the gravitino is probably much lighter than the MSSM sparticles if $M_{\mathrm{mess}} \ll M_{\mathrm{P}}$, as can be seen by comparing eqs. (6.15) and (6.21) The gravitino is almost certainly the LSP in this case, and all of the MSSM sparticles will eventually decay into final states that include it. Naively, one might expect that these decays are extremely slow. However, this is not necessarily true, because the gravitino inherits the *non*-gravitational interactions of the goldstino it has absorbed. This means that the gravitino, or more precisely its longitudinal (goldstino) components, can play an important role in collider physics experiments. The mass of the gravitino can generally be ignored for kinematical purposes, as can its helicity $\pm 3/2$ components which really do have only gravitational interactions. So, by using eq. (6.19), one can compute that the decay rate of any sparticle $\widetilde{X}$ into its Standard Model partner X plus a gravitino/goldstino $\widetilde{G}$ is given by

$$\Gamma(\widetilde{X} \to X\widetilde{G}) = \frac{1}{16\pi} m_{\widetilde{X}} \frac{(m_{\widetilde{X}}^2 - m_X^2)^4}{\langle F \rangle^2} = \frac{1}{48\pi} \frac{m_{\widetilde{X}}^5}{M_{\mathrm{P}}^2 m_{3/2}^2} \left(1 - \frac{m_X^2}{m_{\widetilde{X}}^2}\right)^4 \qquad (6.22)$$

corresponding to either Fig. 13a or 13b, with $(\widetilde{X}, X) = (\phi, \psi)$ or (λ, A) respectively. A factor $(m_{\widetilde{X}}^2 - m_X^2)^2$ came from the derivatives in the interaction term in eq. (6.19) evaluated for on-shell final states, and another factor $(m_{\widetilde{X}}^2 - m_X^2)^2$ comes from the kinematic phase space integral with $m_{3/2} \ll m_{\widetilde{X}}, m_X$. The second equality in eq. (6.22) relies on eq. (6.21); it actually may be more general in that it can apply even when supersymmetry breaking is not due to an F-term getting a VEV. Note that the decay width is larger for smaller $\langle F \rangle$, or equivalently for smaller $m_{3/2}$, if the other masses are fixed. If $\widetilde{X}$ is a mixture of superpartners of different Standard Model particles X, then eq. (6.22) should be multiplied by a suppression factor equal to the square of the cosine of the appropriate mixing angle. If $m_{\widetilde{X}}$ is of order 100 GeV or more, and $\sqrt{\langle F \rangle} \lesssim$ few$\times 10^6$ GeV [corresponding to $m_{3/2}$ less than roughly 1 keV according to eq. (6.21)], then the decay $\widetilde{X} \to X\widetilde{G}$ can occur quickly enough to be observed in a modern collider detector. This gives rise to some very interesting phenomenological signatures.

We now turn to a slightly more systematic analysis of the way in which the MSSM soft terms arise, considering in turn the gravity-mediated and gauge-mediated scenarios.

6.3 Gravity-mediated supersymmetry breaking models

The defining feature of these models is that the hidden sector of the theory communicates with our MSSM only (or dominantly) through gravitational interactions. In an effective field theory format, this means that the supergravity lagrangian contains nonrenormalizable terms which communicate between the two sectors and which are suppressed by powers of the Planck mass, since the gravitational coupling is proportional to $1/M_{\mathrm{P}}$. These will include

$$\begin{aligned}
\mathcal{L}_{\mathrm{NR}} = \;& -\frac{1}{M_{\mathrm{P}}} F_S \sum_a \frac{1}{2} f_a \lambda^a \lambda^a + \text{c.c.} \\
& -\frac{1}{M_{\mathrm{P}}^2} F_S F_S^* \, x_j^i \phi_i \phi^{*j} \\
& -\frac{1}{M_{\mathrm{P}}} F_S \left(\frac{1}{6} y'^{ijk} \phi_i \phi_j \phi_k + \frac{1}{2} \mu'^{ij} \phi_i \phi_j\right) + \text{c.c.} \qquad (6.23)
\end{aligned}$$

where F_S is the auxiliary field for a chiral supermultiplet S in the hidden sector, and ϕ_i and λ^a are the scalar and gaugino fields in the MSSM. By themselves, the terms in eq. (6.23) are not supersymmetric, but it is possible to show (see Appendix) that they are part of a nonrenormalizable supersymmetric lagrangian which contains other terms that we can ignore. Now if one assumes that $\langle F_S \rangle \sim 10^{10}$ or 10^{11} GeV, then $\mathcal{L}_{\mathrm{NR}}$ will give us nothing other than a lagrangian of the form $\mathcal{L}_{\mathrm{soft}}$ in eq. (4.1), with MSSM soft terms of order a few hundred GeV. [Note that terms of the form $\mathcal{L}_{\mathrm{maybe\ soft}}$ in eq. (4.2) do not arise.]

The dimensionless parameters f_a, x^i_j, y'^{ijk} and μ'^{ij} in $\mathcal{L}_{\mathrm{NR}}$ are to be determined by the underlying theory. This is a difficult enterprise in general, but a dramatic simplification occurs if one assumes a "minimal" form for the normalization of kinetic terms and gauge interactions in the full, nonrenormalizable supergravity lagrangian (see Appendix). In that case, one finds that there is a common $f_a = f$ for the three gauginos; $x^i_j = x\delta^i_j$ is the same for all scalars; and the other couplings are proportional to the corresponding superpotential parameters, so that $y'^{ijk} = \alpha y^{ijk}$ and $\mu'^{ij} = \beta\mu^{ij}$ with universal dimensionless constants α and β. Then one finds that the soft terms in $\mathcal{L}_{\mathrm{soft}}^{\mathrm{MSSM}}$ can all be written in terms of just four parameters:

$$m_{1/2} = f\frac{\langle F_S \rangle}{M_\mathrm{P}}; \qquad m_0^2 = x\frac{|\langle F_S \rangle|^2}{M_\mathrm{P}^2}; \qquad A_0 = \alpha\frac{\langle F_S \rangle}{M_\mathrm{P}}; \qquad B_0 = \beta\frac{\langle F_S \rangle}{M_\mathrm{P}}. \tag{6.24}$$

In terms of these, one can write for the parameters appearing in eq. (5.11):

$$M_3 = M_2 = M_1 = m_{1/2}; \tag{6.25}$$

$$\mathbf{m}_Q^2 = \mathbf{m}_{\bar{\mathbf{u}}}^2 = \mathbf{m}_{\bar{\mathbf{d}}}^2 = \mathbf{m}_L^2 = \mathbf{m}_{\bar{\mathbf{e}}}^2 = m_0^2\,\mathbf{1}; \quad m_{H_u}^2 = m_{H_d}^2 = m_0^2; \tag{6.26}$$

$$\mathbf{a_u} = A_0\mathbf{y_u}; \qquad \mathbf{a_d} = A_0\mathbf{y_d}; \qquad \mathbf{a_e} = A_0\mathbf{y_e}; \tag{6.27}$$

$$b = B_0\mu. \tag{6.28}$$

It is a matter of some controversy whether the assumptions going into this parameterization are completely well-motivated on purely theoretical grounds,[k] but from a phenomenological perspective they are clearly very nice. This framework successfully evades the most dangerous types of FCNC and CP-violation as discussed in section 5.4. In particular, eqs. (6.26) and (6.27) are just stronger versions of eqs. (5.14) and (5.15), respectively. If $m_{1/2}$, A_0 and B_0 all have the same complex phase, then eq. (5.16) will also be satisfied.

Equations (6.25)-(6.28) also have the virtue of being highly predictive. [Of course, eq. (6.28) is content-free unless one can relate B_0 to the other

---

[k] The familiar flavor-blindness of gravitational interactions due to Einstein's equivalence principle does not, by itself, tell us anything about the form of eq. (6.23).

parameters in some non-trivial way.] They should be applied as RG boundary conditions at the scale M_P. The RG evolution of the soft parameters down to the electroweak scale will then allow us to predict the entire MSSM spectrum in terms of just five parameters $m_{1/2}$, m_0^2, A_0, B_0, and μ (plus the already-measured gauge and Yukawa couplings of the MSSM). In practice, the approximation is usually made of starting this RG running from the unification scale $M_U \approx 2 \times 10^{16}$ GeV instead of M_P. The reason for this is that the apparent unification of gauge couplings gives us a strong hint that we know something about how the RG equations are behaving up to M_U, but gives us little guidance about what to expect at scales between M_U and M_P. The error made in neglecting these effects is proportional to a loop suppression factor times $\ln(M_P/M_U)$ and can be partially absorbed into a redefinition of m_0^2, $m_{1/2}$, A_0 and B_0, but in some cases can lead to important effects.[71] The framework described in the above few paragraphs has been the subject of the bulk of phenomenological studies of supersymmetry. It is sometimes referred to as the *minimal supergravity* or *supergravity-inspired* scenario for the soft terms. A few examples of the many useful numerical RG studies of the MSSM spectrum which have been performed in this framework can be found in Ref.[72]

Particular models of gravity-mediated supersymmetry breaking can be even more predictive, relating some of the parameters $m_{1/2}$, m_0^2, A_0 and B_0 to each other and to the mass of the gravitino $m_{3/2}$. For example, three popular kinds of models for the soft terms are:

- Dilaton-dominated: [73] $\quad m_0^2 = m_{3/2}^2; \quad m_{1/2} = -A_0 = \sqrt{3}m_{3/2}.$

- Polonyi: [74] $\quad m_0^2 = m_{3/2}^2; \quad A_0 = (3 - \sqrt{3})m_{3/2}; \quad m_{1/2} = \mathcal{O}(m_{3/2}).$

- "No-scale": [75] $\quad m_{1/2} \gg m_0, A_0, m_{3/2}.$

The dilaton-dominated scenario arises in a particular limit of superstring theory. While it appears to be highly predictive, it can easily be generalized in other limits.[76] The Polonyi model has the advantage of being the simplest possible model for supersymmetry breaking in the hidden sector, but it is rather *ad hoc* and does not seem to have a special place in grander schemes like superstrings. The "no-scale" limit may arise in a low-energy limit of superstrings in which the gravitino mass scale is undetermined at tree-level (hence the name). It implies that only the gaugino masses are appreciable at M_P. As we will see in section 7.1, RG evolution feeds $m_{1/2}$ into the squark, slepton and Higgs (mass)^2 parameters with sufficient magnitude to give acceptable phenomenology at the electroweak scale. More recent versions of the no-scale scenario, however, also can give significant A_0 and m_0^2 at M_P. In many cases B_0 can also be predicted in terms of the other parameters, but this is quite sensitive to model assumptions. For phenomenological studies, $m_{1/2}$, m_0^2, A_0 and B_0 are usually just taken to be convenient independent parameters of our ignorance of the supersymmetry breaking mechanism.

6.4   Gauge-mediated supersymmetry breaking models

A strong alternative to the scenario described in the previous section is provided by the gauge-mediated supersymmetry breaking proposal.[77,78] The basic idea is to introduce some new chiral supermultiplets, called messengers, which couple to the ultimate source of supersymmetry-breaking, and which couple indirectly to the fields of the MSSM through the ordinary $SU(3)_C \times SU(2)_L \times U(1)_Y$ gauge interactions. In this way, the ordinary gauge interactions, rather than gravity, are responsible for the appearance of soft terms in the MSSM. There is still gravitational communication between the MSSM and the source of supersymmetry breaking, of course, but that effect is now relatively unimportant compared to the gauge interaction effects.

In the simplest such model, the messenger fields are a set of chiral supermultiplets q, $\bar{q}$, ℓ, $\bar{\ell}$ which transform under $SU(3)_C \times SU(2)_L \times U(1)_Y$ like

$$q \sim (3, 1, -\frac{1}{3}); \qquad \bar{q} \sim (\bar{3}, 1, \frac{1}{3}); \qquad \ell \sim (1, 2, \frac{1}{2}); \qquad \bar{\ell} \sim (1, 2, -\frac{1}{2}). \qquad (6.29)$$

These particles, called messenger (s)quarks and (s)leptons, must get very large masses in order not to have been discovered already. They manage to do so by coupling to a gauge singlet chiral supermultiplet S through a superpotential

$$W = y_2 S \ell \bar{\ell} + y_3 S q \bar{q}. \qquad (6.30)$$

The scalar component and the auxiliary component of S, denoted S (also) and F_S respectively, are both assumed to acquire VEVs.[l] It is not too hard to show that the fermionic components of $q, \bar{q}$ then get a mass $y_3 \langle S \rangle$ while their scalar partners get squared masses $|y_3 \langle S \rangle|^2 \pm |y_3 \langle F_S \rangle|$. Likewise, the fermionic components of $\ell, \bar{\ell}$ get a mass $y_2 \langle S \rangle$ while the scalars get squared masses $|y_2 \langle S \rangle|^2 \pm |y_2 \langle F_S \rangle|$.

The supersymmetry violation apparent in this messenger spectrum is communicated to the MSSM sparticles through radiative corrections with gauge-interaction strength, as follows. The MSSM gauginos obtain masses from the 1-loop graph shown in Fig. 14. The scalar and fermion lines in the loop are messenger fields, with $q, \bar{q}$ loops giving mass to the gluino and the bino, and $\ell, \bar{\ell}$ loops giving mass to the wino and bino fields. If $\langle F_S \rangle$ were 0, then the

---

[l]This can be accomplished either by putting S into an O'Raifeartaigh-type model,[77] or by a dynamical mechanism.[78]

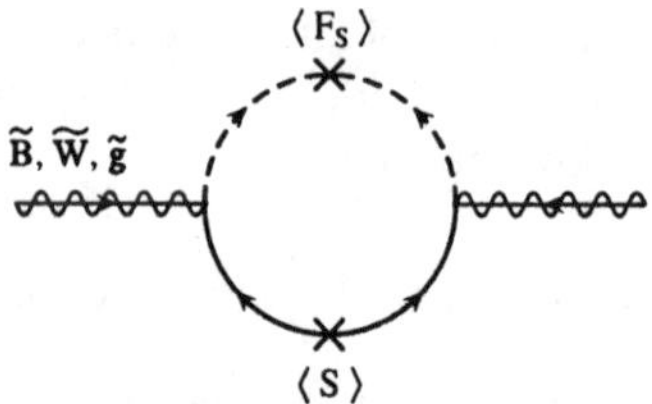

Figure 14: Contributions to the MSSM gaugino masses in gauge-mediated supersymmetry breaking models arise from one-loop graphs involving virtual messenger particles.

messenger scalars would be degenerate with their fermionic superpartners and there would be no contribution to the MSSM gaugino masses. In the limit that $\langle F_S \rangle \ll y_i \langle S \rangle^2$, one finds[78] that the MSSM gaugino masses are given in terms of

$$\Lambda \equiv \langle F_S \rangle / \langle S \rangle \tag{6.31}$$

by

$$M_a = \Lambda \frac{\alpha_a}{4\pi} \qquad (a = 1, 2, 3). \tag{6.32}$$

(Recall that the interaction vertices in Fig. 14 are of gauge interaction strength.) The corresponding MSSM gauge bosons do not get a corresponding mass shift, since they are protected by gauge invariance from obtaining masses. So supersymmetry breaking has been successfully communicated to the visible sector. To a good approximation, eq. (6.32) holds for the running gaugino masses at an RG scale Q_0 corresponding to the average characteristic mass of the heavy messenger particles. (If there are significant hierarchies within the spectrum of messenger particles, then one must be a little more careful.[79]) The running masses can then be RG-evolved down to the electroweak scale to make predictions.

The scalars of the MSSM do not get any radiative corrections to their masses at one-loop order. The leading contribution to their masses comes from the two-loop graphs shown in Fig. 15, with the messenger fermions (heavy solid lines) and messenger scalars (heavy dashed lines) and ordinary gauge bosons and gauginos running around the loops. In the same limit $\langle F_S \rangle \ll y_i \langle S \rangle^2$, one finds that each MSSM scalar ϕ gets a (mass)^2 given by the formula:

$$m_\phi^2 = 2\Lambda^2 \left[\left(\frac{\alpha_3}{4\pi} \right)^2 C_3^\phi + \left(\frac{\alpha_2}{4\pi} \right)^2 C_2^\phi + \left(\frac{\alpha_1}{4\pi} \right)^2 C_1^\phi \right]. \tag{6.33}$$

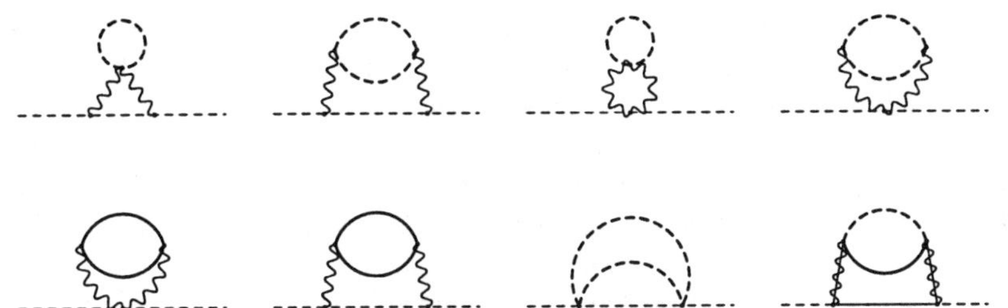

Figure 15: Contributions to MSSM scalar squared masses in gauge-mediated supersymmetry breaking models arise in leading order from these two-loop Feynman graphs.

Here C_a^ϕ are the quadratic Casimir group theory invariants for the scalar ϕ for each gauge group. They are defined by $C_a^\phi \delta_i^j = (T^a T^a)_i^j$ where the T^a are the group generators which act on the scalar ϕ, or explicitly:

$$C_3^\phi = \begin{cases} 4/3 & \text{for } \phi = \widetilde{Q}_i, \widetilde{\overline{u}}_i, \widetilde{\overline{d}}_i; \\ 0 & \text{for } \phi = \widetilde{L}_i, \widetilde{\overline{e}}_i, H_u, H_d \end{cases} \tag{6.34}$$

$$C_2^\phi = \begin{cases} 3/4 & \text{for } \phi = \widetilde{Q}_i, \widetilde{L}_i, H_u, H_d; \\ 0 & \text{for } \phi = \widetilde{\overline{u}}_i, \widetilde{\overline{d}}_i, \widetilde{\overline{e}}_i \end{cases} \tag{6.35}$$

$$C_1^\phi = 3Y_\phi^2/5 \quad \text{for each } \phi \text{ with weak hypercharge } Y_\phi. \tag{6.36}$$

The squared masses in eq. (6.33) are positive (fortunately!). Because the gaugino masses arise at *one*-loop order and the scalar (mass)^2 contributions appear at *two*-loop order, both eq. (6.32) and (6.33) correspond to the estimate eq. (6.15) for m_{soft}, with $M_{\text{mess}} \sim y_i \langle S \rangle$. The terms $\mathbf{a_u}$, $\mathbf{a_d}$, $\mathbf{a_e}$ arise first at two-loop order, and are suppressed by an extra factor of $\alpha_a/(4\pi)$ compared to m_{soft}, so at the messenger mass scale one has, to a very good approximation,

$$\mathbf{a_u} = \mathbf{a_d} = \mathbf{a_e} = 0, \tag{6.37}$$

a significantly stronger condition than eq. (5.15). Again, eqs. (6.33) and (6.37) should be applied at an RG scale equal to the average mass of the messenger fields running in the loops. However, after evolving the RG equations down to the electroweak scale, non-zero $\mathbf{a_u}$, $\mathbf{a_d}$ and $\mathbf{a_e}$ are generated proportional to the corresponding Yukawa matrices and the non-zero gaugino masses, as we will see in section 7.1. These will only be large for the third family squarks and sleptons, in the approximation of eq. (5.2). The parameter b may also be taken to vanish near the messenger scale, but this is quite model-dependent, and in any case b will be non-zero when it is RG-evolved to the electroweak scale. In the following, we will refer to the boundary conditions eqs. (6.32)-(6.37) as the "gauge-mediated" scenario for the soft terms.

With the particle content of eq. (6.29), the approximate unification of gauge couplings will still occur when they are extrapolated up to the same scale M_U (but with a different unified value for the gauge couplings), if all of the messenger masses are not very different from each other. This nice property continues to hold if there are n_{mess} copies of $q, \overline{q}, \ell, \overline{\ell}$. (There are certain other combinations of messenger supermultiplets which will do this same trick, but such models usually give very similar results.) If n_{mess} is too large, however, then the running gauge couplings will blow up before they can reach the unification scale. For messenger masses of order 10^6 GeV or less, for example, one needs $n_{\mathrm{mess}} \leq 4$. The case $n_{\mathrm{mess}} = 1$ is often called the minimal model of gauge-mediated supersymmetry breaking. For non-minimal models, the effect on the MSSM soft terms is to multiply each of eqs. (6.32) and (6.33) by n_{mess}, since now there are n_{mess} messengers running around the loops. Therefore, in non-minimal models the gauginos tend to have relatively larger masses than in the minimal model, because they scale like n_{mess}, while the scalar masses scale only like $\sqrt{n_{\mathrm{mess}}}$. There are many other possible generalizations of the basic gauge-mediated supersymmetry breaking scheme as described above. The common feature which makes all of these models very attractive is that the masses of the squarks and sleptons depend only on their gauge quantum numbers, leading automatically to the degeneracy of squark and slepton masses needed for suppression of FCNC effects. But the most distinctive phenomenological prediction of gauge-mediated models may be the fact that the gravitino is the LSP. This can have crucial consequences for both cosmology and collider physics.

7　The mass spectrum of the MSSM

In this section, we will study the mass spectrum of the MSSM. We begin by looking at the RG equations for the soft breaking parameters. These play an important role in determining the lagrangian at the electroweak scale, given a set of boundary conditions on the soft parameters at the (much higher) input scale. Of course, the boundary conditions on soft parameters are model-dependent, but there are some quite general lessons to be learned from the form of the RG equations. In section 7.2 we will discuss electroweak symmetry breaking and the Higgs scalars. Sections 7.3, 7.4, 7.5 are devoted to the sparticle masses and mixings. Finally in section 7.6 we will summarize some of the general features and expectations for the MSSM spectrum.

7.1 Renormalization Group Equations

In order to translate a set of predictions at the input scale into physically meaningful quantities which describe physics at the electroweak scale, it is necessary to evolve the gauge couplings, superpotential parameters, and soft terms using the RG equations.

The one-loop RG equations for the gauge couplings g_1, g_2, g_3 have already been listed in eq. (5.17). The one-loop RG equations for the three gaugino mass parameters in the MSSM are determined by the same quantities b_a^{MSSM} which appear in the gauge coupling RG eqs. (5.17):

$$\frac{d}{dt} M_a = \frac{1}{8\pi^2} b_a g_a^2 M_a \qquad (b_a = 33/5, 1, -3) \tag{7.1}$$

for $a = 1, 2, 3$. It is therefore easy to show that the three ratios M_a/g_a^2 are each constant (RG-scale independent) up to small two-loop corrections. In minimal supergravity models, we can therefore write

$$M_a(Q) = \frac{g_a^2(Q)}{g_a^2(Q_0)} m_{1/2} \qquad (a = 1, 2, 3) \tag{7.2}$$

at any RG scale $Q < Q_0$, where Q_0 is the input scale which is presumably nearly equal to M_P. Since the gauge couplings are observed to unify at $M_U \sim 0.01 M_P$, one expects [m] that $g_1^2(M_P) \approx g_2^2(M_P) \approx g_3^2(M_P)$. Therefore, one finds that

$$\frac{M_1}{g_1^2} = \frac{M_2}{g_2^2} = \frac{M_3}{g_3^2} \tag{7.3}$$

at any RG scale, up to small two-loop effects and possibly larger threshold effects near M_U and M_P. The common value in eq. (7.3) is also equal to $m_{1/2}/g_U^2$ in minimal supergravity models, where g_U is the unified gauge coupling at the scale where $m_{1/2}$ is the common gaugino mass. Interestingly, eq. (7.3) is *also* the solution to the one-loop RG equations in the case of gauge-mediated boundary conditions applied at the messenger mass scale. This is true even though there is no such thing as a unified gaugino mass $m_{1/2}$ in the gauge-mediated case, because of the fact that the gaugino masses are proportional to the g_a^2 times a constant. So eq. (7.3) is theoretically well-motivated (but certainly not inevitable) in both frameworks. The prediction eq. (7.3) is particularly useful

---

[m] In a GUT model, it is automatic that the gauge couplings and gaugino masses are unified at all scales $Q > M_U$ and in particular at $Q \approx M_P$, because in the unified theory the gauginos all live in the same representation of the unified gauge group. In many superstring models, this is also known to be a good approximation.

since the gauge couplings g_1^2, g_2^2, and g_3^2 are already quite well known at the electroweak scale from experiment. Therefore they can be extrapolated up to at least M_U, assuming that the apparent unification of gauge couplings is not a fake. The gaugino mass parameters feed into the RG equations for all of the other soft terms.

Next we consider the 1-loop RG equations for the analytic soft parameters $\mathbf{a_u}$, $\mathbf{a_d}$, $\mathbf{a_e}$. In models obeying eq. (5.15), these matrices start off proportional to the corresponding Yukawa couplings at the input scale, and the RG evolution respects this property. With the approximation of eq. (5.2), one can therefore also write, at any RG scale,

$$\mathbf{a_u} \approx \begin{pmatrix} 0 & 0 & 0 \\ 0 & 0 & 0 \\ 0 & 0 & a_t \end{pmatrix}; \quad \mathbf{a_d} \approx \begin{pmatrix} 0 & 0 & 0 \\ 0 & 0 & 0 \\ 0 & 0 & a_b \end{pmatrix}; \quad \mathbf{a_e} \approx \begin{pmatrix} 0 & 0 & 0 \\ 0 & 0 & 0 \\ 0 & 0 & a_\tau \end{pmatrix}, \quad (7.4)$$

which defines[n] running parameters a_t, a_b, and a_τ. The RG equations for these parameters and b are given by

$$16\pi^2 \frac{d}{dt} a_t = a_t \left[18|y_t|^2 + |y_b|^2 - \frac{16}{3}g_3^2 - 3g_2^2 - \frac{13}{15}g_1^2 \right]$$
$$+ 2a_b y_b^* y_t + y_t \left[\frac{32}{3}g_3^2 M_3 + 6g_2^2 M_2 + \frac{26}{15}g_1^2 M_1 \right]; \qquad (7.5)$$

$$16\pi^2 \frac{d}{dt} a_b = a_b \left[18|y_b|^2 + |y_t|^2 + |y_\tau|^2 - \frac{16}{3}g_3^2 - 3g_2^2 - \frac{7}{15}g_1^2 \right]$$
$$+ 2a_t y_t^* y_b + 2a_\tau y_\tau^* y_b + y_b \left[\frac{32}{3}g_3^2 M_3 + 6g_2^2 M_2 + \frac{14}{15}g_1^2 M_1 \right]; \quad (7.6)$$

$$16\pi^2 \frac{d}{dt} a_\tau = a_\tau \left[12|y_\tau|^2 + 3|y_b|^2 - 3g_2^2 - \frac{9}{5}g_1^2 \right]$$
$$+ 6a_b y_b^* y_\tau + y_\tau \left[6g_2^2 M_2 + \frac{18}{5}g_1^2 M_1 \right]; \qquad (7.7)$$

$$16\pi^2 \frac{d}{dt} b = b \left[3|y_t|^2 + 3|y_b|^2 + |y_\tau|^2 - 3g_2^2 - \frac{3}{5}g_1^2 \right]$$
$$+ \mu \left[6a_t y_t^* + 6a_b y_b^* + 2a_\tau y_\tau^* + 6g_2^2 M_2 + \frac{6}{5}g_1^2 M_1 \right] \qquad (7.8)$$

in this approximation. Note that even if A_0 and B_0 appearing in eqs. (6.27) and (6.28) vanish at the input scale, the RG corrections proportional to gaugino

---

[n] We must warn the reader that rescaled soft parameters $A_t = a_t/y_t$, $A_b = a_b/y_b$, and $A_\tau = a_\tau/y_\tau$ are commonly used in the literature. We do not follow this notation, because it cannot be generalized beyond the approximation of eqs. (5.2), (7.4) without introducing horrible complications such as non-polynomial RG equations, and because a_t, a_b and a_τ are the parameters that actually appear in the lagrangian anyway.

masses appearing in eqs. (7.5)-(7.8) ensure that a_t, a_b, a_τ and b will still be non-zero at the electroweak scale.

Next let us consider the RG equations for the scalar masses in the MSSM. In the approximation of eq. (5.2), the squarks and sleptons of the first two families have only gauge interactions. This means that if the scalar masses satisfy a boundary condition like eq. (5.14) at an input RG scale, then when renormalized to any other RG scale, they will be diagonal, with the approximate form

$$\mathbf{m_Q^2} \approx \begin{pmatrix} m_{Q_1}^2 & 0 & 0 \\ 0 & m_{Q_1}^2 & 0 \\ 0 & 0 & m_{Q_3}^2 \end{pmatrix} ; \qquad \mathbf{m_{\overline{u}}^2} \approx \begin{pmatrix} m_{\overline{u}_1}^2 & 0 & 0 \\ 0 & m_{\overline{u}_1}^2 & 0 \\ 0 & 0 & m_{\overline{u}_3}^2 \end{pmatrix} ; \qquad (7.9)$$

etc. The first and second family squarks and sleptons with given gauge quantum numbers remain very nearly degenerate, but the third family squarks and sleptons feel the effects of the Yukawa couplings and get renormalized differently. The one-loop RG equations for the first and second family squark and slepton squared masses can be written as [o]

$$16\pi^2 \frac{d}{dt} m_\phi^2 = - \sum_{a=1,2,3} 8 g_a^2 C_a^\phi |M_a|^2 \qquad (7.10)$$

for each scalar ϕ, where the $\sum_a$ is over the three gauge groups $U(1)_Y$, $SU(2)_L$ and $SU(3)_C$; M_a are the corresponding running gaugino mass parameters which are known from eq. (7.3); and the constants C_a^ϕ are the same quadratic Casimir group theory invariants which appeared in eqs. (6.34)-(6.36). An important feature of eq. (7.10) is that the right-hand sides are strictly negative, so that the scalar (mass)^2 parameters *grow* as they are RG-evolved from the input scale down to the electroweak scale. Even if the scalars have zero or very small masses at the input scale, as in the "no-scale" boundary condition limit $m_0^2 = 0$, they will obtain large positive squared masses at the electroweak scale, thanks to the gaugino masses.

The RG equations for the (mass)^2 parameters of the Higgs scalars and third family squarks and sleptons get the same gauge contributions as in eq. (7.10),

---

[o]There are also terms in the scalar (mass)^2 RG equations which are proportional to $\mathrm{Tr}[Y m^2]$ (the sum of the weak hypercharge times the soft (mass)^2 for all scalars in the theory). However, these contributions vanish in both the cases of minimal supergravity and gauge-mediated boundary conditions for the soft terms, as one can see by explicitly calculating $\mathrm{Tr}[Y m^2]$ in each case. If $\mathrm{Tr}[Y m^2]$ is zero at the input scale, then it will remain zero under RG evolution. Therefore we neglect such terms in our discussion, although they can have an important effect in more general situations.

but they also have contributions due to the large Yukawa $(y_{t,b,\tau})$ and soft $(a_{t,b,\tau})$ couplings. At one-loop order, these only appear in three combinations:

$$X_t = 2|y_t|^2(m_{H_u}^2 + m_{Q_3}^2 + m_{\bar{u}_3}^2) + 2|a_t|^2, \tag{7.11}$$

$$X_b = 2|y_b|^2(m_{H_d}^2 + m_{Q_3}^2 + m_{\bar{d}_3}^2) + 2|a_b|^2, \tag{7.12}$$

$$X_\tau = 2|y_\tau|^2(m_{H_d}^2 + m_{L_3}^2 + m_{\bar{e}_3}^2) + 2|a_\tau|^2. \tag{7.13}$$

In terms of these quantities the RG equations for the soft Higgs (mass)^2 parameters $m_{H_u}^2$ and $m_{H_d}^2$ are

$$16\pi^2 \frac{d}{dt} m_{H_u}^2 = 3X_t - 6g_2^2|M_2|^2 - \frac{6}{5}g_1^2|M_1|^2, \tag{7.14}$$

$$16\pi^2 \frac{d}{dt} m_{H_d}^2 = 3X_b + X_\tau - 6g_2^2|M_2|^2 - \frac{6}{5}g_1^2|M_1|^2. \tag{7.15}$$

Note that X_t, X_b, and X_τ are positive, so their effect is always to *decrease* the Higgs masses as one evolves the RG equations downward from the input scale to the electroweak scale. Since y_t is the largest of the Yukawa couplings because of the experimental fact that the top quark is heavy, X_t is typically expected to be larger than X_b and X_τ. This can cause the RG-evolved $m_{H_u}^2$ to run negative near the electroweak scale, helping to destabilize the point $H_u = 0$ and so provoking a Higgs VEV which is just what we want.^{P} Thus a large top Yukawa coupling favors the breakdown of the electroweak symmetry breaking because it induces negative radiative corrections to the Higgs (mass)^2.

The third family squark and slepton (mass)^2 parameters also get contributions which depend on X_t, X_b and X_τ. Their RG equations are given by

$$16\pi^2 \frac{d}{dt} m_{Q_3}^2 = X_t + X_b - \frac{32}{3}g_3^2|M_3|^2 - 6g_2^2|M_2|^2 - \frac{2}{15}g_1^2|M_1|^2 \tag{7.16}$$

$$16\pi^2 \frac{d}{dt} m_{\bar{u}_3}^2 = 2X_t - \frac{32}{3}g_3^2|M_3|^2 - \frac{32}{15}g_1^2|M_1|^2 \tag{7.17}$$

$$16\pi^2 \frac{d}{dt} m_{\bar{d}_3}^2 = 2X_b - \frac{32}{3}g_3^2|M_3|^2 - \frac{8}{15}g_1^2|M_1|^2 \tag{7.18}$$

$$16\pi^2 \frac{d}{dt} m_{L_3}^2 = X_\tau - 6g_2^2|M_2|^2 - \frac{3}{5}g_1^2|M_1|^2 \tag{7.19}$$

$$16\pi^2 \frac{d}{dt} m_{\bar{e}_3}^2 = 2X_\tau - \frac{24}{5}g_1^2|M_1|^2. \tag{7.20}$$

---

^{P}One should think of "$m_{H_u}^2$" as a parameter unto itself, and not as the square of some mythical real number m_{H_u}. Thus there is nothing strange about having $m_{H_u}^2 < 0$. However, strictly speaking $m_{H_u}^2 < 0$ is neither necessary nor sufficient for electroweak symmetry breaking; see section 7.2.

In eqs. (7.14)-(7.20), the terms proportional to $|M_3|^2$, $|M_2|^2$ and $|M_1|^2$ are just the same ones indicated by eq. (7.10). Note that the terms proportional to X_t appear with smaller numerical coefficients in the $m^2_{Q_3}$ and $m^2_{\bar{u}_3}$ RG equations than they did for the Higgs scalars, and they do not appear at all in the $m^2_{\bar{d}_3}$, $m^2_{L_3}$ and $m^2_{\bar{e}_3}$ RG equations. Furthermore, the third-family squark (mass)^2 get a large positive contribution proportional to $|M_3|^2$ from the RG evolution, which the Higgs scalars do not get. These facts make it easy to understand why the Higgs scalars in the MSSM can get VEVs, but the squarks and sleptons, having large positive (mass)^2, do not. An examination of the RG equations, (7.10) and (7.14)-(7.20) reveals that if the gaugino mass parameters M_1, M_2, and M_3 are non-zero at the input scale, then all of the other soft terms will be generated. This is why the "no-scale" limit with $m_{1/2} \gg m_0, A_0, B_0$ can be phenomenologically viable even though the squarks and sleptons are massless at tree-level. On the other hand, if the gaugino masses were to vanish at tree-level, then they would not get any contributions to their masses at one-loop order; in that case M_1, M_2, and M_3 would be extremely small.

7.2 Electroweak symmetry breaking and the Higgs bosons

In the MSSM, the description of electroweak symmetry breaking is slightly complicated by the fact that there are two complex Higgs doublets $H_u = (H_u^+, H_u^0)$ and $H_d = (H_d^0, H_d^-)$ rather than just one in the ordinary Standard Model. The classical scalar potential for the Higgs scalar fields in the MSSM is given by

$$
\begin{aligned}
V \;=\; & (|\mu|^2 + m_{H_u}^2)(|H_u^0|^2 + |H_u^+|^2) + (|\mu|^2 + m_{H_d}^2)(|H_d^0|^2 + |H_d^-|^2) \\
& + b\,(H_u^+ H_d^- - H_u^0 H_d^0) + \text{c.c.} \\
& + \frac{1}{8}(g^2 + g'^2)(|H_u^0|^2 + |H_u^+|^2 - |H_d^0|^2 - |H_d^-|^2)^2 \\
& + \frac{1}{2}g^2 |H_u^+ H_d^{0*} + H_u^0 H_d^{-*}|^2.
\end{aligned}
\tag{7.21}
$$

The terms proportional to $|\mu|^2$ come from F-terms [see the first term on the right-hand side of eq. (5.5)]. The terms proportional to $m_{H_u}^2$, $m_{H_d}^2$ and b are nothing but a rewriting of the last three terms of eq. (5.11). Finally, the terms proportional to g^2 and g'^2 are the D-term contributions which may be derived from the general formula eq. (3.75), after some rearranging. The full scalar potential of the theory will also include many terms involving the squark and slepton fields which do not get VEVs because they have large positive (mass)^2.

We now have to demand that the minimum of this potential should break electroweak symmetry down to electromagnetism $SU(2)_L \times U(1)_Y \to U(1)_{\text{EM}}$, in accord with experiment. We can use the freedom to make gauge transformations to simplify this analysis. First, the freedom to make $SU(2)_L$ gauge transformations allows us to rotate away a possible VEV for one of the weak isospin components of one of the scalar fields; so without loss of generality we can take $H_u^+ = 0$ at the minimum of the potential. Then one finds that a minimum of the potential satisfying $\partial V/\partial H_u^+ = 0$ must also have $H_d^- = 0$. This is good, because it means that at the minimum of the potential electromagnetism is necessarily unbroken, since the charged components of the Higgs scalars cannot get VEVs. So after setting $H_u^+ = H_d^- = 0$ we are left to consider the scalar potential

$$
\begin{aligned}
V &= (|\mu|^2 + m_{H_u}^2)|H_u^0|^2 + (|\mu|^2 + m_{H_d}^2)|H_d^0|^2 - (b\, H_u^0 H_d^0 + \text{c.c.}) \\
&\quad + \frac{1}{8}(g^2 + g'^2)(|H_u^0|^2 - |H_d^0|^2)^2.
\end{aligned}
\tag{7.22}
$$

The only term in this potential which depends on the phases of the fields is the b-term. Therefore a redefinition of the phases of H_u and H_d can absorb any phase in b, so we can take b to be real and positive. Then it is clear that a minimum of the potential V requires that $H_u^0 H_d^0$ is also real and positive, so $\langle H_u^0 \rangle$ and $\langle H_d^0 \rangle$ must have opposite phases. We can therefore use a $U(1)_Y$ gauge transformation to make them both be real and positive without loss of generality, since H_u and H_d have opposite weak hypercharges ($\pm 1/2$). It follows that CP cannot be spontaneously broken by the Higgs scalar potential, since all of the VEVs and couplings can be simultaneously chosen to be real. This means that the Higgs scalar mass eigenstates can be assigned well-defined eigenvalues of CP.

Note that the b-term always favors electroweak symmetry breaking. The combination of the b term and the terms $m_{H_u}^2$ and $m_{H_d}^2$ can allow for one linear combination of H_u^0 and H_d^0 to have a negative (mass)^2 near $H_u^0 = H_d^0 = 0$. This requires that

$$
b^2 > (|\mu|^2 + m_{H_u}^2)(|\mu|^2 + m_{H_d}^2).
\tag{7.23}
$$

If this inequality is not satisfied, then $H_u^0 = H_d^0 = 0$ will be a stable minimum of the potential, and electroweak symmetry breaking will not occur. Note that a negative value for $|\mu|^2 + m_{H_u}^2$ will help eq. (7.23) to be satisfied, but it is not necessary. Furthermore, even if $m_{H_u}^2 < 0$, there may be no electroweak symmetry breaking if $|\mu|$ is too large or if b is too small.

In order for the MSSM scalar potential to be viable, it is not enough that the point $H_u^0 = H_d^0 = 0$ is destabilized by a negative (mass)^2 direction; we must also make sure that the potential is bounded from below for arbitrarily large values of the scalar fields, so that V will really have a minimum. (Recall from the discussion in section 3.2 and 3.4 that scalar potentials in purely supersymmetric theories are automatically positive and so clearly bounded from below, but now that we have introduced supersymmetry breaking we must be careful.) The scalar quartic interactions in V will stabilize the potential for almost all arbitrarily large values of H_u^0 and H_d^0. However, there are special directions in field space with $|H_u^0| = |H_d^0|$, along which the quartic contributions to V [the second line in eq. (7.22)] are identically zero. Such directions in field space are called D-flat directions, because along them the part of the scalar potential coming from D-terms vanishes. In order for the potential to be bounded from below, we need the quadratic part of the scalar potential to be positive along the D-flat directions. This requirement amounts to

$$2b < 2|\mu|^2 + m_{H_u}^2 + m_{H_d}^2. \tag{7.24}$$

Interestingly, if $m_{H_u}^2 = m_{H_d}^2$, the constraints eqs. (7.23) and (7.24) cannot both be satisfied. In models derived from the minimal supergravity or gauge-mediated boundary conditions, $m_{H_u}^2 = m_{H_d}^2$ holds at tree-level at the input scale, but the X_t contribution to the RG equation for $m_{H_u}^2$ naturally pushes it to negative or small values $m_{H_u}^2 < m_{H_d}^2$ at the electroweak scale, as we saw in section 7.1. Unless this effect is large, the parameter space in which the electroweak symmetry is broken would be quite small. So in these models electroweak symmetry breaking is actually driven purely by quantum corrections; this mechanism is therefore known as *radiative electroweak symmetry breaking*. The realization that this works most naturally with a large top-quark Yukawa coupling provides additional motivation for these models.[80,81]

Having established the conditions necessary for H_u^0 and H_d^0 to get non-zero VEVs, we can now require that they are compatible with the observed phenomenology of electroweak symmetry breaking $SU(2)_L \times U(1)_Y \to U(1)_{\rm EM}$. Let us write $\langle H_u^0 \rangle = v_u$ and $\langle H_d^0 \rangle = v_d$ for the VEVs at the minimum of the potential. These VEVs can be connected to the known mass of the Z^0 boson and the electroweak gauge couplings:

$$v_u^2 + v_d^2 = v^2 = 2m_Z^2/(g^2 + g'^2) \approx 174\,{\rm GeV}. \tag{7.25}$$

The ratio of the two VEVs is traditionally written as

$$\tan\beta \equiv v_u/v_d. \tag{7.26}$$

The value of $\tan\beta$ is not fixed by present experiments, but it depends on the lagrangian parameters of the MSSM in a calculable way. Since $v_u = v\sin\beta$ and $v_d = v\cos\beta$ were taken to be real and positive, we have $0 < \beta < \pi/2$, a requirement that will be sharpened below. Now one can write down the conditions $\partial V/\partial H_u^0 = \partial V/\partial H_d^0 = 0$ under which the potential eq. (7.22) will have a minimum satisfying eqs. (7.25) and (7.26):

$$|\mu|^2 + m_{H_d}^2 = b\tan\beta - (m_Z^2/2)\cos 2\beta; \qquad (7.27)$$

$$|\mu|^2 + m_{H_u}^2 = b\cot\beta + (m_Z^2/2)\cos 2\beta. \qquad (7.28)$$

It is easy to check that these equations indeed satisfy the necessary conditions eqs. (7.23) and (7.24). They allow us to eliminate two of the lagrangian parameters b and $|\mu|$ in favor of $\tan\beta$, but do not determine the phase of μ.

As an aside, we note that eqs. (7.27) and (7.28) highlight the "μ problem" already mentioned in section 5.1. If we view $|\mu|^2$, b, $m_{H_u}^2$ and $m_{H_d}^2$ as input parameters, and m_Z and $\tan\beta$ as output parameters obtained by solving these two equations, then without miraculous cancellations we expect that all of the input parameters ought to be within an order of magnitude or two of m_Z^2. However, in the MSSM, μ is a supersymmetry-respecting parameter appearing in the superpotential, while b, $m_{H_u}^2$, $m_{H_d}^2$ are supersymmetry-breaking parameters. This has lead to a widespread belief that the MSSM must be extended at very high energies to include a mechanism which relates the effective value of μ to the supersymmetry-breaking mechanism in some way; see Refs.[41,42,43,44] for examples.

The Higgs scalar fields in the MSSM consist of two complex $SU(2)_L$-doublet, or eight real, scalar degrees of freedom. When the electroweak symmetry is broken, three of them are the would-be Nambu-Goldstone bosons G^0, $G^\pm$ which become the longitudinal modes of the Z^0 and $W^\pm$ massive vector bosons. The remaining five Higgs scalar mass eigenstates consist of one CP-odd neutral scalar A^0, a charge $+1$ scalar H^+ and its conjugate charge -1 scalar H^-, and two CP-even neutral scalars h^0 and H^0. In terms of the original gauge-eigenstate fields, the mass eigenstates and would-be Nambu-Goldstone bosons are given by

$$\begin{pmatrix} G^0 \\ A^0 \end{pmatrix} = \sqrt{2}\begin{pmatrix} \sin\beta & -\cos\beta \\ \cos\beta & \sin\beta \end{pmatrix}\begin{pmatrix} \mathrm{Im}[H_u^0] \\ \mathrm{Im}[H_d^0] \end{pmatrix}, \qquad (7.29)$$

$$\begin{pmatrix} G^+ \\ H^+ \end{pmatrix} = \begin{pmatrix} \sin\beta & -\cos\beta \\ \cos\beta & \sin\beta \end{pmatrix}\begin{pmatrix} H_u^+ \\ H_d^{-*} \end{pmatrix}, \qquad (7.30)$$

with $G^- = G^{+*}$ and $H^- = H^{+*}$, and

$$\begin{pmatrix} h^0 \\ H^0 \end{pmatrix} = \sqrt{2}\begin{pmatrix} \cos\alpha & -\sin\alpha \\ \sin\alpha & \cos\alpha \end{pmatrix}\begin{pmatrix} \mathrm{Re}[H_u^0] - v_u \\ \mathrm{Re}[H_d^0] - v_d \end{pmatrix}, \qquad (7.31)$$

which defines a mixing angle α. The tree-level masses of these fields can be found by expanding the scalar potential around the minimum. One obtains

$$m_{A^0}^2 = 2b/\sin 2\beta \tag{7.32}$$

$$m_{H^\pm}^2 = m_{A^0}^2 + m_W^2 \tag{7.33}$$

$$m_{h^0,H^0}^2 = \frac{1}{2}\left(m_{A^0}^2 + m_Z^2 \mp \sqrt{(m_{A^0}^2 + m_Z^2)^2 - 4m_Z^2 m_{A^0}^2 \cos^2 2\beta}\right). \tag{7.34}$$

In terms of these masses, the mixing angle α appearing in eq. (7.31) is determined at tree-level by

$$\frac{\sin 2\alpha}{\sin 2\beta} = -\frac{m_{A^0}^2 + m_Z^2}{m_{H^0}^2 - m_{h^0}^2}; \qquad \frac{\cos 2\alpha}{\cos 2\beta} = -\frac{m_{A^0}^2 - m_Z^2}{m_{H^0}^2 - m_{h^0}^2}. \tag{7.35}$$

The Feynman rules for couplings of the mass eigenstate Higgs scalars to the Standard Model quarks and leptons and the electroweak vector bosons, as well as to the various sparticles, have been worked out in detail in Ref.[82,83]

The masses of A^0, H^0 and $H^\pm$ can in principle be arbitrarily large since they all grow with $b/\sin 2\beta$. In contrast, the mass of h^0 is bounded from above. It is not hard to show from eq. (7.34) that

$$m_{h^0} < |\cos 2\beta| m_Z \tag{7.36}$$

at tree-level.[84] If this inequality were robust, it would guarantee that the lightest Higgs boson of the MSSM would be kinematically accessible to LEP2, with large regions of parameter space already ruled out. However, the tree-level mass formulas given above for the Higgs mass eigenstates are subject to quite significant quantum corrections which are especially important to take into account in the case of h^0. The largest such contributions typically come from top-stop loop corrections to the terms in the scalar potential. In the limit of stop squark masses $m_{\tilde{t}_1}$, $m_{\tilde{t}_2}$ much greater than the top quark mass m_t, one finds a one-loop radiative correction to eq. (7.34):

$$\Delta(m_{h^0}^2) = \frac{3}{4\pi^2} v^2 y_t^4 \sin^4\beta \, \ln\left(\frac{m_{\tilde{t}_1} m_{\tilde{t}_2}}{m_t^2}\right). \tag{7.37}$$

Including this and other corrections,[85,86] one can obtain only a considerably weaker, but still very interesting, bound

$$m_{h^0} \lesssim 130\,\mathrm{GeV} \tag{7.38}$$

in the MSSM. This assumes that all of the sparticles that can contribute to $\Delta(m_{h^0}^2)$ in loops have masses that do not exceed 1 TeV. By adding extra supermultiplets to the MSSM, this bound can be made even weaker. However, assuming that none of the MSSM sparticles have masses exceeding 1 TeV and that all of the couplings in the theory remain perturbative up to the unification scale, one still finds [87]

$$m_{h^0} \lesssim 150\,\text{GeV}. \tag{7.39}$$

This bound is also weakened if, for example, the top squarks are heavier than 1 TeV, but the upper bound rises only logarithmically with the soft masses, as can be seen from eq. (7.37). Thus it is a fairly robust prediction of supersymmetry at the electroweak scale that at least one of the Higgs scalar bosons must be light.

An interesting limit occurs when $m_{A^0} \gg m_Z$. In that case, m_{h^0} can saturate the upper bound just mentioned with $m_{h^0} \approx m_Z|\cos 2\beta|$ at tree-level, but subject to large positive quantum corrections. The particles A^0, H^0, and $H^\pm$ are much heavier and nearly degenerate, forming an isospin doublet which decouples from sufficiently low-energy experiments. The angle α is fixed to be approximately $\beta - \pi/2$. In this limit, h^0 has the same couplings to quarks and leptons and electroweak gauge bosons as would the physical Higgs boson of the ordinary Standard Model without supersymmetry. Indeed, model-building experiences have shown that it is quite common for h^0 to behave in a way nearly indistinguishable from a Standard Model-like Higgs boson, even if m_{A^0} is not too huge. On the other hand, it is important to keep in mind that the couplings of h^0 might turn out to deviate in important ways from those of a Standard Model Higgs boson. For a given set of model parameters, it is very important to take into account the complete set of one-loop corrections and even the dominant two-loop effects in a leading logarithm approximation in order to get accurate predictions for the Higgs masses and mixing angles.[85,86]

In the MSSM, the masses and CKM mixing angles of the quarks and leptons are determined by the Yukawa couplings of the superpotential and the parameter $\tan\beta$. This is because the top, charm and up quarks get masses proportional to $v_u = v\sin\beta$ and the bottom, strange, and down quarks and the charge leptons get masses proportional to $v_d = v\cos\beta$. Therefore one finds at tree-level

$$y_t = \frac{gm_t}{\sqrt{2}m_W\sin\beta}; \qquad y_b = \frac{gm_b}{\sqrt{2}m_W\cos\beta}; \qquad y_\tau = \frac{gm_\tau}{\sqrt{2}m_W\cos\beta}. \tag{7.40}$$

These relations hold for the running masses of t, b, τ rather than the physical pole masses which are significantly larger.[88] Including those corrections, one can relate the Yukawa couplings to $\tan\beta$ and the known fermion masses and CKM mixing angles. It is now clear why we have not neglected y_b and y_τ, even though $m_b, m_\tau \ll m_t$. To a first approximation, $y_b/y_t = (m_b/m_t)\tan\beta$ and $y_\tau/y_t = (m_\tau/m_t)\tan\beta$, so that y_b and y_τ cannot be neglected if $\tan\beta$ is much larger than 1. In fact, there are good theoretical motivations for considering models with large $\tan\beta$. For example, models based on the GUT gauge group $SO(10)$ (or certain of its subgroups) can unify the running top, bottom and tau Yukawa couplings at the unification scale; this requires $\tan\beta$ to be very roughly of order m_t/m_b.[89,90]

Note that if one tries to make $\sin\beta$ too small, y_t will become nonperturbatively large. Requiring that y_t does not blow up above the electroweak scale, one finds that $\tan\beta \gtrsim 1.2$ or so, depending on the mass of the top quark, the QCD coupling, and other fine details. In principle, one can also determine a lower bound on $\cos\beta$ and thus an upper bound on $\tan\beta$ by requiring that y_b and y_τ are not nonperturbatively large. This gives a rough upper bound of $\tan\beta \lesssim 65$. However, this is complicated slightly by the fact that the bottom quark mass gets significant one-loop corrections in the large $\tan\beta$ limit.[90] One can obtain a slightly stronger upper bound on $\tan\beta$ in models where $m_{H_u}^2 = m_{H_d}^2$ at the input scale, by requiring that y_b does not significantly exceed y_t. [Otherwise, X_b would be larger than X_t in eqs. (7.14) and (7.15), so one would find $m_{H_d}^2 < m_{H_u}^2$ at the electroweak scale, and the minimum of the potential would have to be at $\langle H_d^0 \rangle > \langle H_u^0 \rangle$ which would be a contradiction with the supposition that $\tan\beta$ is large.] In the following, we will see that the parameter $\tan\beta$ has an important effect on the masses and mixings of the MSSM sparticles.

7.3 Neutralinos and charginos

The higgsinos and electroweak gauginos mix with each other because of the effects of electroweak symmetry breaking. The neutral higgsinos ($\widetilde{H}_u^0$ and $\widetilde{H}_d^0$) and the neutral gauginos ($\widetilde{B}$, $\widetilde{W}^0$) combine to form four neutral mass eigenstates called *neutralinos*. The charged higgsinos ($\widetilde{H}_u^+$ and $\widetilde{H}_d^-$) and winos ($\widetilde{W}^+$ and $\widetilde{W}^-$) mix to form two mass eigenstates with charge ± 1 called *charginos*. We will denote [q] the neutralino and chargino mass eigenstates by $\widetilde{N}_i$ ($i = 1, 2, 3, 4$) and $\widetilde{C}_i^\pm$ ($i = 1, 2$) with $m_{\widetilde{N}_1} < m_{\widetilde{N}_2} < m_{\widetilde{N}_3} < m_{\widetilde{N}_4}$ and $m_{\widetilde{C}_1} < m_{\widetilde{C}_2}$. The lightest neutralino, $\widetilde{N}_1$, is usually assumed to be the LSP, unless there is a lighter gravitino or unless R-parity is not conserved, because it is the only MSSM particle which can make a good cold dark matter candidate. In this subsection, we will describe the mass spectrum and mixing of the neutralinos and charginos in the MSSM.

---

[q]Other common notations use $\widetilde{\chi}_i^0$ or $\widetilde{Z}_i$ for neutralinos, and $\widetilde{\chi}_i^\pm$ or $\widetilde{W}_i^\pm$ for charginos.

78

In the gauge-eigenstate basis $\psi^0 = (\widetilde{B}, \widetilde{W}^0, \widetilde{H}_d^0, \widetilde{H}_u^0)$, the neutralino mass terms in the lagrangian are

$$\mathcal{L} \supset -\frac{1}{2}(\psi^0)^T \mathbf{M}_{\widetilde{N}} \psi^0 + \text{c.c.} \tag{7.41}$$

where

$$\mathbf{M}_{\widetilde{N}} = \begin{pmatrix} M_1 & 0 & -c_\beta\, s_W\, m_Z & s_\beta\, s_W\, m_Z \\ 0 & M_2 & c_\beta\, c_W\, m_Z & -s_\beta\, c_W\, m_Z \\ -c_\beta\, s_W\, m_Z & c_\beta\, c_W\, m_Z & 0 & -\mu \\ s_\beta\, s_W\, m_Z & -s_\beta\, c_W\, m_Z & -\mu & 0 \end{pmatrix}. \tag{7.42}$$

Here we have introduced abbreviations $s_\beta = \sin\beta$, $c_\beta = \cos\beta$, $s_W = \sin\theta_W$, and $c_W = \cos\theta_W$. The entries M_1 and M_2 in this matrix come directly from the MSSM soft Lagrangian [see eq. (5.11)] while the entries $-\mu$ are the supersymmetric higgsino mass terms [see eq. (5.4)]. The terms proportional to m_Z are the result of Higgs-higgsino-gaugino couplings [see eq. (3.72)], with the Higgs scalars getting their VEVs [eqs. (7.25),(7.26)]. The mass matrix $\mathbf{M}_{\widetilde{N}}$ can be diagonalized by a unitary matrix $\mathbf{N}$ with $\widetilde{N}_i = \mathbf{N}_{ij}\psi_j^0$, so that

$$\mathbf{M}_{\widetilde{N}}^{\text{diag}} = \mathbf{N}^* \mathbf{M}_{\widetilde{N}} \mathbf{N}^{-1} \tag{7.43}$$

has positive real entries $m_{\widetilde{N}_1}$, $m_{\widetilde{N}_2}$, $m_{\widetilde{N}_3}$, $m_{\widetilde{N}_4}$ on the diagonal. These are the absolute values of the eigenvalues of $\mathbf{M}_{\widetilde{N}}$, or equivalently the square roots of the eigenvalues of $\mathbf{M}_{\widetilde{N}}^\dagger \mathbf{M}_{\widetilde{N}}$. The indices (i, j) on $\mathbf{N}_{ij}$ are (mass, gauge) eigenstate labels. The mass eigenvalues and the mixing matrix $\mathbf{N}_{ij}$ can be given in closed form in terms of the parameters M_1, M_2, μ and $\tan\beta$, but the results are very complicated and not very illuminating.

In general, the parameters M_1, M_2, and μ could have arbitrary complex phases. In the broad class of minimal supergravity or gauge-mediated models satisfying the gaugino mass unification conditions $M_1 = M_2 = M_3$ at some scale, M_2 and M_1 will have the same complex phase which is preserved by their RG evolution eq. (7.1). In that case, a redefinition of the phases of $\widetilde{B}$ and $\widetilde{W}$ allows us to make M_1 and M_2 both real and positive. The phase of μ is then really a physical parameter which cannot be rotated away. [We have already used up the freedom to redefine the phases of the Higgs fields, since we have picked b and $\langle H_u^0 \rangle$ and $\langle H_d^0 \rangle$ to be real and positive, which guarantees that the off-diagonal entries in eq. (7.42) proportional to m_Z are real.] However, if μ is not real, then there can be potentially disastrous CP-violating effects in low-energy physics, including electric dipole moments for both the electron and the neutron. Therefore, it is usual (although not absolutely mandatory) to assume that μ is real in the same set of phase conventions which make M_1, M_2, b, $\langle H_u^0 \rangle$ and $\langle H_d^0 \rangle$ real and positive. The sign of μ is still undetermined by this constraint.

In models which satisfy eq. (7.3), one has the nice prediction

$$M_1 \approx \frac{5}{3} \tan^2 \theta_W\, M_2 \approx 0.5 M_2 \qquad (7.44)$$

at the electroweak scale. If so, then the neutralino masses and mixing angles depend on only three unknown parameters. This assumption is sufficiently theoretically compelling that it has been made in almost all phenomenological studies; nevertheless it should be recognized as an assumption, to be tested someday by experiment.

Specializing further, there is an interesting and not unlikely limit in which electroweak symmetry breaking effects can be viewed as a small perturbation on the neutralino mass matrix. If

$$m_Z \ll |\mu \pm M_1|, |\mu \pm M_2| \qquad (7.45)$$

then the neutralino mass eigenstates are very nearly $\tilde{N}_1 \approx \tilde{B}$; $\tilde{N}_2 \approx \widetilde{W}^0$; $\tilde{N}_3, \tilde{N}_4 \approx (\tilde{H}_u^0 \pm \tilde{H}_d^0)/\sqrt{2}$, with mass eigenvalues:

$$m_{\tilde{N}_1} = M_1 - \frac{m_Z^2 s_W^2 (M_1 + \mu \sin 2\beta)}{\mu^2 - M_1^2} + \ldots \qquad (7.46)$$

$$m_{\tilde{N}_2} = M_2 - \frac{m_W^2 (M_2 + \mu \sin 2\beta)}{\mu^2 - M_2^2} + \ldots \qquad (7.47)$$

$$m_{\tilde{N}_3}, m_{\tilde{N}_4} = |\mu| + \frac{m_Z^2 (1 - \epsilon \sin 2\beta)(|\mu| + M_1 c_W^2 + M_2 s_W^2)}{2(|\mu| + M_1)(|\mu| + M_2)} + \ldots, \qquad (7.48)$$

$$|\mu| + \frac{m_Z^2 (1 + \epsilon \sin 2\beta)(|\mu| - M_1 c_W^2 - M_2 s_W^2)}{2(|\mu| - M_1)(|\mu| - M_2)} + \ldots \qquad (7.49)$$

where ϵ is the sign of μ. The labeling of the mass eigenstates $\tilde{N}_1$ and $\tilde{N}_2$ assumes $M_1 < M_2 < |\mu|$; otherwise the subscripts may need to be rearranged. It turns out that a "bino-like" LSP $\tilde{N}_1$ can very easily have the right cosmological abundance to make a good dark matter candidate, so the large $|\mu|$ limit may be preferred from that point of view. In addition, this limit tends to emerge from minimal supergravity boundary conditions on the soft parameters, which often require $|\mu|$ to be significantly larger than M_2 in order to get correct electroweak symmetry breaking.

The chargino spectrum can be analyzed in a similar way. In the gauge-eigenstate basis $\psi^\pm = (\widetilde{W}^+, \widetilde{H}_u^+, \widetilde{W}^-, \widetilde{H}_d^-)$, the chargino mass terms in the lagrangian are

$$\mathcal{L} \supset -\frac{1}{2}(\psi^\pm)^T \mathbf{M}_{\widetilde{C}} \psi^\pm + \text{c.c.} \tag{7.50}$$

where, in 2×2 block form,

$$\mathbf{M}_{\widetilde{C}} = \begin{pmatrix} \mathbf{0} & \mathbf{X}^T \\ \mathbf{X} & \mathbf{0} \end{pmatrix} ; \qquad \mathbf{X} = \begin{pmatrix} M_2 & \sqrt{2}s_\beta\, m_W \\ \sqrt{2}c_\beta\, m_W & \mu \end{pmatrix} . \tag{7.51}$$

The mass eigenstates are related to the gauge eigenstates by two unitary 2×2 matrices $\mathbf{U}$ and $\mathbf{V}$ according to

$$\begin{pmatrix} \widetilde{C}_1^+ \\ \widetilde{C}_2^+ \end{pmatrix} = \mathbf{V} \begin{pmatrix} \widetilde{W}^+ \\ \widetilde{H}_u^+ \end{pmatrix} ; \qquad \begin{pmatrix} \widetilde{C}_1^- \\ \widetilde{C}_2^- \end{pmatrix} = \mathbf{U} \begin{pmatrix} \widetilde{W}^- \\ \widetilde{H}_d^- \end{pmatrix} . \tag{7.52}$$

Note that there are different mixing matrices for the positively charged states and for the negatively charged states. They are to be chosen so that

$$\mathbf{U}^T \mathbf{X} \mathbf{V}^{-1} = \begin{pmatrix} m_{\widetilde{C}_1} & 0 \\ 0 & m_{\widetilde{C}_2} \end{pmatrix} . \tag{7.53}$$

Because these are only 2×2 matrices, it is not hard to solve for the masses explicitly:

$$m_{\widetilde{C}_1}^2, m_{\widetilde{C}_2}^2 = \frac{1}{2}\Big[(M_2^2 + \mu^2 + 2m_W^2) \tag{7.54}$$

$$\mp \sqrt{(M_2^2 + \mu^2 + 2m_W^2)^2 - 4(\mu M_2 - m_W^2 \sin 2\beta)^2}\Big]. \tag{7.55}$$

It should be noted that these are the (doubly degenerate) eigenvalues of the 4×4 matrix $\mathbf{M}_{\widetilde{C}}^\dagger \mathbf{M}_{\widetilde{C}}$, or equivalently the eigenvalues of $\mathbf{X}^T\mathbf{X}$, but they are *not* the squares of the eigenvalues of $\mathbf{X}$. In the limit of eq. (7.45), one finds that the charginos mass eigenstates consist of a wino-like $\widetilde{C}_1^\pm$ and and a higgsino-like $\widetilde{C}_2^\pm$, with masses

$$m_{\widetilde{C}_1} = M_2 - \frac{m_W^2(M_2 + \mu \sin 2\beta)}{\mu^2 - M_2^2} + \dots \tag{7.56}$$

$$m_{\widetilde{C}_2} = |\mu| + \frac{m_W^2(|\mu| + \epsilon M_2 \sin 2\beta)}{\mu^2 - M_2^2} + \dots . \tag{7.57}$$

Here again the labeling assumes $M_2 < |\mu|$. Amusingly, the lighter chargino $\widetilde{C}_1$ is nearly degenerate with the second lightest neutralino $\widetilde{N}_2$ in this limit, but this is not an exact result. Their higgsino-like colleagues $\widetilde{N}_3$, $\widetilde{N}_4$ and $\widetilde{C}_2$ have masses of order $|\mu|$. The case of $M_1 \approx 0.5 M_2 \ll |\mu|$ is not uncommonly found in viable models following from the boundary conditions in section 6, and it has been elevated to the status of a benchmark scenario in many recent phenomenological studies. However it cannot be overemphasized that such expectations are not mandatory.

In practice, the masses and mixing angles for the neutralinos and charginos are best computed numerically. The corresponding Feynman rules may be inferred in terms of $\mathbf{N}$, $\mathbf{U}$ and $\mathbf{V}$ from the MSSM lagrangian as discussed above; they are collected in Refs.[19,82]

7.4   The gluino

The gluino is a color octet fermion, so it cannot mix with any other particle in the MSSM, even if R-parity is violated. In this regard, it is unique among all of the MSSM sparticles. In the models following from minimal supergravity or gauge-mediated boundary conditions, the gluino mass parameter M_3 is related to the bino and wino mass parameters M_1 and M_2 by eq. (7.3)

$$M_3 = \frac{\alpha_S}{\alpha} \sin^2 \theta_W \, M_2 = \frac{3}{5} \frac{\alpha_S}{\alpha} \cos^2 \theta_W \, M_1 \tag{7.58}$$

at any RG scale, up to small two-loop corrections. If we use values $\alpha_S = 0.118$, $\alpha = 1/128$, $\sin^2 \theta_W = 0.23$, then one finds the rough prediction

$$M_3 : M_2 : M_1 \approx 7 : 2 : 1 \tag{7.59}$$

at the electroweak scale. Therefore we expect that the gluino should be much heavier than the lighter neutralinos and charginos.

For more precise estimates, one must take into account the fact that the parameter M_3 is really a running mass which has an implicit dependence on the RG scale Q. Because the gluino is a strongly interacting particle, M_3 runs rather quickly with Q [see eq. (7.1)]. A more useful quantity physically is the RG scale-independent mass $m_{\widetilde{g}}$ at which the renormalized gluino propagator has a pole. Including one-loop corrections to the gluino propagator due to gluon exchange and quark-squark loops, one finds that the pole mass is given in terms of the running mass in the $\overline{\mathrm{DR}}$ renormalization scheme by [91]

$$m_{\widetilde{g}} = M_3(Q)\left(1 + \frac{\alpha_S}{4\pi}\left[15 + 6\ln(Q/M_3) + \sum A_{\widetilde{q}}\right]\right) \tag{7.60}$$

where

$$A_{\tilde{q}} = \int_0^1 dx\, x \ln\left[xm_{\tilde{q}}^2/M_3^2 + (1-x)m_q^2/M_3^2 - x(1-x)\right]. \tag{7.61}$$

The sum in eq. (7.60) is over all 12 squark-quark supermultiplets, and we have neglected small effects due to squark mixing. It is easy to check that requiring $m_{\tilde{g}}$ to be independent of Q in eq. (7.60) reproduces the one-loop RG equation for $M_3(Q)$ in eq. (7.1). The correction terms proportional to α_S in eq. (7.60) can be quite significant, so that $m_{\tilde{g}}/M_3(M_3)$ can exceed unity by 25% or more. The reasons for this are that the gluino is strongly interacting, with a large group theory factor [the 15 in eq. (7.60)] due to its color octet nature, and that it couples to all the squark-quark pairs. Of course, there are similar corrections which relate the running masses of all the other MSSM particles to their physical masses. These have been systematically evaluated at one-loop order in Ref.[92] They are more complicated in form and usually numerically smaller than for the gluino, but in some cases they could be quite important in future efforts to connect a given candidate model for the soft terms to experimentally measured masses and mixing angles of the MSSM particles.

7.5 The squark and slepton mass spectrum

In principle, any scalars with the same electric charge, R-parity, and color quantum numbers can mix with each other. This means that with completely arbitrary soft terms, the mass eigenstates of the squarks and sleptons of the MSSM should be obtained by diagonalizing three 6×6 (mass)^2 matrices for up-type squarks ($\tilde{u}_L$, $\tilde{c}_L$, $\tilde{t}_L$, $\tilde{u}_R$, $\tilde{c}_R$, $\tilde{t}_R$), down-type squarks ($\tilde{d}_L$, $\tilde{s}_L$, $\tilde{b}_L$, $\tilde{d}_R$, $\tilde{s}_R$, $\tilde{b}_R$), and charged sleptons ($\tilde{e}_L$, $\tilde{\mu}_L$, $\tilde{\tau}_L$, $\tilde{e}_R$, $\tilde{\mu}_R$, $\tilde{\tau}_R$), and one 3×3 matrix for sneutrinos ($\tilde{\nu}_e$, $\tilde{\nu}_\mu$, $\tilde{\nu}_\tau$). Fortunately, the general hypothesis of flavor-blind soft parameters eqs. (5.14) and (5.15) predicts that most of these mixing angles are very small. The third-family squarks and sleptons can have very different masses compared to their first- and second-family counterparts, because of the effects of large Yukawa (y_t, y_b, y_τ) and soft (a_t, a_b, a_τ) couplings in the RG equations. Furthermore, they can have substantial mixing in pairs ($\tilde{t}_L$, $\tilde{t}_R$), ($\tilde{b}_L$, $\tilde{b}_R$) and ($\tilde{\tau}_L$, $\tilde{\tau}_R$). In contrast, the first- and second-family squarks and sleptons have negligible Yukawa couplings, so they end up in 7 very nearly degenerate, unmixed pairs ($\tilde{e}_R, \tilde{\mu}_R$), ($\tilde{\nu}_e, \tilde{\nu}_\mu$), ($\tilde{e}_L, \tilde{\mu}_L$), ($\tilde{u}_R, \tilde{c}_R$), ($\tilde{d}_R, \tilde{s}_R$), ($\tilde{u}_L, \tilde{c}_L$), ($\tilde{d}_L, \tilde{s}_L$). As we have already discussed in section 5.4, this avoids the problem of disastrously large virtual sparticle contributions to FCNC processes.

Let us first consider the spectrum of first- and second-family squarks and sleptons. In models fitting into both of the broad categories of minimal supergravity or gauge-mediated boundary conditions, their running masses can be conveniently parameterized in the following way:

$$m_{Q_1}^2 = m_{Q_2}^2 = m_0^2 + K_3 + K_2 + \frac{1}{36}K_1, \tag{7.62}$$

$$m_{\bar{u}_1}^2 = m_{\bar{u}_2}^2 = m_0^2 + K_3 \qquad + \frac{4}{9}K_1, \tag{7.63}$$

$$m_{\bar{d}_1}^2 = m_{\bar{d}_2}^2 = m_0^2 + K_3 \qquad + \frac{1}{9}K_1, \tag{7.64}$$

$$m_{L_1}^2 = m_{L_2}^2 = m_0^2 \qquad + K_2 + \frac{1}{4}K_1, \tag{7.65}$$

$$m_{\bar{e}_1}^2 = m_{\bar{e}_2}^2 = m_0^2 \qquad + K_1. \tag{7.66}$$

In minimal supergravity models, m_0^2 is the common scalar (mass)^2 which appears in eq. (6.26). It can be 0 in the "no-scale" limit, but it could also be the dominant source of the scalar masses. The contributions K_3, K_2 and K_1 are strictly positive and are due to the RG running proportional to the gaugino masses; see eq. (7.10). The key point is that the same K_3, K_2 and K_1 appear everywhere in eqs. (7.62)-(7.66), since all of the chiral supermultiplets couple to the same gauginos with the same gauge couplings. The different coefficients in front of K_1 just correspond to the various values of weak hypercharge squared for each scalar. The quantities K_1, K_2, K_3 depend on the RG scale Q at which they are evaluated. Explicitly, they are found by solving eq. (7.10):

$$K_a(Q) = \left\{\begin{matrix} 3/5 \\ 3/4 \\ 4/3 \end{matrix}\right\} \times \frac{1}{2\pi^2} \int_{\ln Q}^{\ln Q_0} dt \; g_a^2(t) \, |M_a(t)|^2 \qquad (a = 1, 2, 3). \tag{7.67}$$

Here Q_0 is the input RG scale and Q should be taken to be evaluated near the squark and slepton mass under consideration, presumably less than about 1 TeV or so. The values of the running parameters $g_a(Q)$ and $M_a(Q)$ can be found using eqs. (5.17) and (7.3). If the input scale is approximated by the apparent scale of gauge coupling unification $Q_0 = M_U \approx 2 \times 10^{16}$ GeV, one finds that numerically

$$K_1 \approx 0.15 m_{1/2}^2; \qquad K_2 \approx 0.5 m_{1/2}^2; \qquad K_3 \approx (4.5 \text{ to } 6.5) m_{1/2}^2. \tag{7.68}$$

for Q near 1 TeV. Here $m_{1/2}$ is the common gaugino mass parameter at the unification scale. Note that $K_3 \gg K_2 \gg K_1$; this is a direct consequence of the relative sizes of the gauge couplings g_3, g_2, and g_1. The large uncertainty

84

in K_3 is due in part to the experimental uncertainty in the QCD coupling constant, and in part to the uncertainty in where to choose Q, since K_3 runs rather quickly below 1 TeV. If the gauge couplings and gaugino masses are unified between M_U and M_P, as would occur in a GUT model, then the effect of RG running for $M_U < Q < M_P$ can be absorbed into a redefinition of m_0^2. Otherwise, it adds a further uncertainty which is roughly proportional to $\ln(M_P/M_U)$, compared to the larger contributions in eq. (7.67) which go roughly like $\ln(M_U/1 \text{ TeV})$.

In gauge-mediated models, the same parameterization eqs. (7.62)-(7.66) holds, but m_0^2 is always 0. At the input scale Q_0, each MSSM scalar gets contributions to its (mass)^2 which depend only on its gauge interactions, as in eq. (6.33). It is not hard to see that in general these contribute in exactly the same pattern as K_1, K_2, and K_3 in eq. (7.62)-(7.66). The subsequent evolution of the scalar squared masses down to the electroweak scale again just yields more contributions to the K_1, K_2, and K_3 parameters. It is somewhat more difficult to give meaningful numerical estimates for these parameters in gauge-mediated models than in the minimal supergravity models, because of uncertainties in the messenger mass scale(s) and in the multiplicities of the messenger fields. However, in the gauge-mediated case one quite generally expects that the ratios K_3/K_2, K_3/K_1 and K_2/K_1 should be even larger than in eq. (7.68). There are two reasons for this. First, the running squark squared masses start off larger than slepton squared masses already at the input scale in gauge-mediated models, rather than having a common value m_0^2. Furthermore, in the gauge-mediated case, the input scale Q_0 is typically much lower than M_P or M_U, so that the RG evolution gives relatively more weight to smaller RG scales where the hierarchies $g_3 > g_2 > g_1$ and $M_3 > M_2 > M_1$ are already in effect. In general, one therefore expects that the squarks should be considerably heavier than the sleptons, with the effect being more pronounced in gauge-mediated supersymmetry breaking models than in minimal supergravity models. For any specific choice of model, this effect can be easily quantified with an RG analysis. The hierarchy $m_{\text{squark}} > m_{\text{slepton}}$ tends to hold in great generality, because the RG contributions to squark masses from the gluino are always present and usually quite large, since QCD has a larger gauge coupling than the electroweak interactions.

There is also a "hyperfine" splitting in the squark and slepton mass spectrum produced by electroweak symmetry breaking. Each squark and slepton ϕ will get a contribution Δ_ϕ to its (mass)^2 from the $SU(2)_L$ and $U(1)_Y$ D-term quartic interactions [see the last term in eq. (3.75)] of the form (squark)^2(Higgs)^2 and (slepton)^2(Higgs)^2, when the neutral Higgs scalars H_u^0 and H_d^0 get VEVs.

They are model-independent for a given value of $\tan\beta$, and are given by

$$\Delta_\phi = (T_3^\phi - Q_{\rm EM}^\phi \sin^2\theta_W)\cos 2\beta\, m_Z^2, \qquad (7.69)$$

where T_3^ϕ and $Q_{\rm EM}^\phi$ are the third component of weak isospin and the electric charge of the chiral supermultiplet to which ϕ belongs. [For example, $\Delta_u = (\frac{1}{2} - \frac{2}{3}\sin^2\theta_W)\cos 2\beta\, m_Z^2$ and $\Delta_{\overline{u}} = (\frac{2}{3}\sin^2\theta_W)\cos 2\beta\, m_Z^2$]. These D-term contributions are typically smaller than the m_0^2 and K_1, K_2, K_3 contributions, but should not be neglected. They split apart the components of the $SU(2)_L$-doublet sleptons and squarks $L_1 = (\widetilde{\nu}_e, \widetilde{e}_L)$, etc. Including them, the first-family squark and slepton masses are given by:

$$m_{\widetilde{d}_L}^2 = m_0^2 + K_3 + K_2 + \frac{1}{36}K_1 + \Delta_d, \qquad (7.70)$$

$$m_{\widetilde{u}_L}^2 = m_0^2 + K_3 + K_2 + \frac{1}{36}K_1 + \Delta_u, \qquad (7.71)$$

$$m_{\widetilde{u}_R}^2 = m_0^2 + K_3 \qquad\quad + \frac{4}{9}K_1 + \Delta_{\overline{u}}, \qquad (7.72)$$

$$m_{\widetilde{d}_R}^2 = m_0^2 + K_3 \qquad\quad + \frac{1}{9}K_1 + \Delta_{\overline{d}}, \qquad (7.73)$$

$$m_{\widetilde{e}_L}^2 = m_0^2 \qquad\quad + K_2 + \frac{1}{4}K_1 + \Delta_e, \qquad (7.74)$$

$$m_{\widetilde{\nu}}^2 = m_0^2 \qquad\quad + K_2 + \frac{1}{4}K_1 + \Delta_\nu, \qquad (7.75)$$

$$m_{\widetilde{e}_R}^2 = m_0^2 \qquad\qquad\quad + K_1 + \Delta_{\overline{e}}, \qquad (7.76)$$

with identical formulas for the second-family squarks and sleptons. The mass splittings for the left-handed squarks and sleptons are governed by model-independent sum rules

$$m_{\widetilde{e}_L}^2 - m_{\widetilde{\nu}_e}^2 = m_{\widetilde{d}_L}^2 - m_{\widetilde{u}_L}^2 = -\cos 2\beta\, m_W^2. \qquad (7.77)$$

Since $\cos 2\beta < 0$ in the allowed range $\tan\beta > 1$, it follows that $m_{\widetilde{e}_L} > m_{\widetilde{\nu}_e}$ and $m_{\widetilde{d}_L} > m_{\widetilde{u}_L}$, with the magnitude of the splittings constrained by electroweak symmetry breaking.

Let us next consider the masses of the top squarks, for which there are several non-negligible contributions. First, there are (mass)^2 terms for $\widetilde{t}_L^* \widetilde{t}_L$ and $\widetilde{t}_R^* \widetilde{t}_R$ which are just equal to $m_{Q_3}^2 + \Delta_u$ and $m_{\overline{u}_3}^2 + \Delta_{\overline{u}}$, respectively, as for the first- and second-family squarks. Second, there are contributions equal to m_t^2 for each of $\widetilde{t}_L^* \widetilde{t}_L$ and $\widetilde{t}_R^* \widetilde{t}_R$. These come from F-terms in the scalar potential of the form $y_t^2 H_u^{0*} H_u^0 \widetilde{t}_L^* \widetilde{t}_L$ and $y_t^2 H_u^{0*} H_u^0 \widetilde{t}_R^* \widetilde{t}_R$ (see Figs. 7b and 7c), with the

Higgs fields replaced by their VEVs. These contributions are of course present for all of the squarks and sleptons, but they are much too small to worry about except in the case of the top squarks. Third, there are contributions to the scalar potential from F-terms of the form $-\mu y_t \widetilde{\bar{t}} \tilde{t} H_d^{0*} + \text{c.c.}$; see eqs. (5.6) and Fig. 9a. These become $-\mu v y_t \cos\beta\, \tilde{t}_R^* \tilde{t}_L + \text{c.c.}$ when H_d^0 is replaced by its VEV. Finally, there are contributions to the scalar potential from the soft (scalar)^3 couplings $a_t \widetilde{\bar{t}} \widetilde{Q}_3 H_u^0 + \text{c.c.}$ [see the first term of the second line of eq. (5.11), which become $a_t v \sin\beta\, \tilde{t}_L \tilde{t}_R^* + \text{c.c.}$ when H_u^0 is replaced by its VEV. Putting these all together, we have a (mass)^2 matrix for the top squarks, which in the gauge-eigenstate basis $(\tilde{t}_L, \tilde{t}_R)$ is given by

$$-\mathcal{L} \supset (\, \tilde{t}_L^* \quad \tilde{t}_R^* \,)\, \mathbf{m}_{\tilde{t}}^2 \begin{pmatrix} \tilde{t}_L \\ \tilde{t}_R \end{pmatrix} \tag{7.78}$$

where

$$\mathbf{m}_{\tilde{t}}^2 = \begin{pmatrix} m_{Q_3}^2 + m_t^2 + \Delta_u & v(a_t \sin\beta - \mu y_t \cos\beta) \\ v(a_t \sin\beta - \mu y_t \cos\beta) & m_{\bar{u}_3}^2 + m_t^2 + \Delta_{\bar{u}} \end{pmatrix}. \tag{7.79}$$

This matrix can be diagonalized to give mass eigenstates

$$\begin{pmatrix} \tilde{t}_1 \\ \tilde{t}_2 \end{pmatrix} = \begin{pmatrix} \cos\theta_{\tilde{t}} & \sin\theta_{\tilde{t}} \\ -\sin\theta_{\tilde{t}} & \cos\theta_{\tilde{t}} \end{pmatrix} \begin{pmatrix} \tilde{t}_L \\ \tilde{t}_R \end{pmatrix} \tag{7.80}$$

with $m_{\tilde{t}_1}^2 < m_{\tilde{t}_2}^2$ being the eigenvalues of eq. (7.79) and $0 \le \theta_{\tilde{t}} \le \pi$. Because of the large RG effects proportional to X_t in eq. (7.16) and eq. (7.17), at the electroweak scale one finds that $m_{\bar{u}_3}^2 < m_{Q_3}^2$, and both of these quantities are usually significantly smaller than the squark squared masses for the first two families. The diagonal terms m_t^2 in eq. (7.79) tend to mitigate this effect somewhat, but the off-diagonal entries will typically induce a significant mixing which always reduces the lighter top-quark (mass)^2 eigenvalue. For this reason, it is often found in models that $\tilde{t}_1$ is the lightest squark of all.

A very similar analysis can be performed for the bottom squarks and charged tau sleptons, which in their respective gauge-eigenstate bases $(\tilde{b}_L, \tilde{b}_R)$ and $(\tilde{\tau}_L, \tilde{\tau}_R)$ have (mass)^2 matrices:

$$\mathbf{m}_{\tilde{b}}^2 = \begin{pmatrix} m_{Q_3}^2 + \Delta_d & v(a_b \cos\beta - \mu y_b \sin\beta) \\ v(a_b \cos\beta - \mu y_b \sin\beta) & m_{\bar{d}_3}^2 + \Delta_{\bar{d}} \end{pmatrix}; \tag{7.81}$$

$$\mathbf{m}_{\tilde{\tau}}^2 = \begin{pmatrix} m_{L_3}^2 + \Delta_e & v(a_\tau \cos\beta - \mu y_\tau \sin\beta) \\ v(a_\tau \cos\beta - \mu y_\tau \sin\beta) & m_{\bar{e}_3}^2 + \Delta_{\bar{e}} \end{pmatrix}. \tag{7.82}$$

These can be diagonalized to give mass eigenstates $\tilde{b}_1, \tilde{b}_2$ and $\tilde{\tau}_1, \tilde{\tau}_2$ in exact analogy with eq. (7.80).

The magnitude and importance of mixing in the sbottom and stau sectors depends on how large $\tan\beta$ is. If $\tan\beta$ is not too large (in practice, this usually means less than about 10 or so, depending on the situation under study), the sbottoms and staus do not get a very large effect from the mixing terms and the RG effects due to X_b and X_τ, because $y_b, y_\tau \ll y_t$ from eq. (7.40). In that case the mass eigenstates are very nearly the same as the gauge eigenstates $\tilde{b}_L$, $\tilde{b}_R$, $\tilde{\tau}_L$ and $\tilde{\tau}_R$. The latter three, and $\tilde{\nu}_\tau$, will be nearly degenerate with their first- and second-family counterparts with the same $SU(3)_C \times SU(2)_L \times U(1)_Y$ quantum numbers. However, even in the case of small $\tan\beta$, $\tilde{b}_L$ will feel the effects of the large top Yukawa coupling because it is part of the doublet $\tilde{Q}_3$ which contains $\tilde{t}_L$. In particular, from eq. (7.16) we see that X_t acts to decrease $m^2_{\tilde{Q}_3}$ as it is RG-evolved down from the input scale to the electroweak scale. So the mass of $\tilde{b}_L$ can be significantly less than the masses of $\tilde{d}_L$ and $\tilde{s}_L$.

For larger values of $\tan\beta$, the mixing in eqs. (7.81) and (7.82) can be quite significant, because y_b, y_τ and a_b, a_τ are non-negligible. Just as in the case of the top squarks, the lighter sbottom and stau mass eigenstates (denoted $\tilde{b}_1$ and $\tilde{\tau}_1$) can be significantly lighter than their first- and second-family counterparts. Furthermore, $\tilde{\nu}_\tau$ can be much lighter than the nearly degenerate $\tilde{\nu}_e$, $\tilde{\nu}_\mu$.

The requirement that the third-family squarks and sleptons should all have positive (mass)^2 implies limits on the sizes of $a_t \sin\beta - \mu y_t \cos\beta$, $a_b \cos\beta - \mu y_b \sin\beta$, and $a_\tau \cos\beta - \mu y_\tau \sin\beta$. If they are too large, the smaller eigenvalue of eq. (7.79), (7.81) or (7.82) will be driven negative, implying that a squark or charged slepton gets a VEV, breaking $SU(3)_C$ or electromagnetism. Since this is clearly unacceptable, one can put bounds on the (scalar)^3 couplings, or equivalently on the parameter A_0 in minimal supergravity models. Even if all of the (mass)^2 eigenvalues are positive, the presence of large (scalar)^3 couplings can yield global minima of the scalar potential with non-zero squark and/or charged slepton VEVs which are disconnected from the vacuum which conserves $SU(3)_C$ and electromagnetism.[93] However, it is not always clear whether the non-existence of such disconnected global minima should really be taken as a constraint, because the tunneling rate from our "good" vacuum to the "bad" vacua can easily be much longer than the age of the universe.[94]

7.6 Summary: the MSSM sparticle spectrum

In the MSSM there are 32 distinct masses corresponding to undiscovered particles, not including the gravitino. In this section we have explained how the masses and mixing angles for these particles can be computed, given an un-

Table 3: Undiscovered particles in the Minimal Supersymmetric Standard Model

Names	Spin	P_R	Mass Eigenstates	Gauge Eigenstates
Higgs bosons	0	$+1$	$h^0 \; H^0 \; A^0 \; H^\pm$	$H_u^0 \; H_d^0 \; H_u^+ \; H_d^-$
squarks	0	-1	$\tilde{u}_L \; \tilde{u}_R \; \tilde{d}_L \; \tilde{d}_R$ $\tilde{s}_L \; \tilde{s}_R \; \tilde{c}_L \; \tilde{c}_R$ $\tilde{t}_1 \; \tilde{t}_2 \; \tilde{b}_1 \; \tilde{b}_2$	" " " " $\tilde{t}_L \; \tilde{t}_R \; \tilde{b}_L \; \tilde{b}_R$
sleptons	0	-1	$\tilde{e}_L \; \tilde{e}_R \; \tilde{\nu}_e$ $\tilde{\mu}_L \; \tilde{\mu}_R \; \tilde{\nu}_\mu$ $\tilde{\tau}_1 \; \tilde{\tau}_2 \; \tilde{\nu}_\tau$	" " " " $\tilde{\tau}_L \; \tilde{\tau}_R \; \tilde{\nu}_\tau$
neutralinos	1/2	-1	$\tilde{N}_1 \; \tilde{N}_2 \; \tilde{N}_3 \; \tilde{N}_4$	$\tilde{B}^0 \; \widetilde{W}^0 \; \tilde{H}_u^0 \; \tilde{H}_d^0$
charginos	1/2	-1	$\tilde{C}_1^\pm \; \tilde{C}_2^\pm$	$\widetilde{W}^\pm \; \tilde{H}_u^+ \; \tilde{H}_d^-$
gluino	1/2	-1	$\tilde{g}$	" "
gravitino/ goldstino	3/2	-1	$\tilde{G}$	" "

derlying model for the soft terms at some input scale. Assuming only that the mixing of first- and second-family squarks and sleptons is negligible, the mass eigenstates of the MSSM are listed in Table 3. A complete set of Feynman rules for the interactions of these particles with each other and with the Standard Model quarks, leptons, and gauge bosons can be found in Refs.[19,82] Specific models for the soft terms typically predict the masses and the mixing angles angles for the MSSM in terms of far fewer parameters. For example, in the minimal supergravity models, one has only five parameters μ, m_0^2, $m_{1/2}$, A_0, and B_0 which are not already measured by experiment. After RG evolving the soft terms down to the electroweak scale, one can impose that the scalar potential gives correct electroweak symmetry breaking. This allows us to trade $|\mu|$ and B_0 for one parameter $\tan\beta$, as in eqs. (7.27)-(7.28). So, to a reasonable approximation, the entire mass spectrum is determined by only four unknown parameters and one choice of sign: m_0^2, $m_{1/2}$, A_0, $\tan\beta$, and $\text{sign}(\mu)$. On the other hand, in gauge-mediated models, the free parameters are the integer numbers of messenger fields with given $SU(3)_C \times SU(2)_L \times U(1)_Y$ quantum numbers, the individual messenger particle masses, and the mass parameter Λ, in addition to μ and perhaps b. Both frameworks are highly predictive. Of course, it is easy to imagine that the essential physics of supersymmetry breaking is not captured by either of these two scenarios, at least in their minimal forms.

While it would be a mistake to underestimate the uncertainties in the MSSM mass and mixing spectrum, it is also useful to keep in mind some general lessons that recur in various different scenarios. Indeed, there has emerged a sort of folklore concerning likely features of the MSSM spectrum, which is partly based on theoretical bias and partly on the constraints inherent in any supersymmetric theory. We remark on these features in part because they represent the prevailing prejudice among supersymmetry theorists, which is certainly a useful thing for the reader to know even if he or she makes the wise decision to remain sceptical. For example, it is perhaps not unlikely that:

- The LSP is the lightest neutralino $\widetilde{N}_1$, unless the gravitino is lighter or R-parity is not conserved. If $\mu > M_1, M_2$, then $\widetilde{N}_1$ is likely to be bino-like, with a mass roughly 0.5 times the masses of $\widetilde{N}_2$ and $\widetilde{C}_1$. In the opposite case $\mu < M_1, M_2$, then $\widetilde{N}_1$ has a large higgsino content and $\widetilde{N}_2$ and $\widetilde{C}_1$ are not much heavier.

- The gluino will be much heavier than the lighter neutralinos and charginos. This is certainly true in the case of the "standard" gaugino mass relation; more generally, the running gluino mass parameter grows relatively quickly as it is RG-evolved into the infrared because the QCD coupling is larger than the electroweak gauge couplings. So even if there are big corrections to the gaugino mass boundary conditions, the gluino mass parameter M_3 is likely to come out larger than M_1 and M_2.

- The squarks of the first and second families are nearly degenerate and much heavier than the sleptons. This is because each squark mass gets the same large positive-definite radiative corrections from loops involving the gluino. The left-handed squarks $\widetilde{u}_L$, $\widetilde{d}_L$, $\widetilde{s}_L$ and $\widetilde{c}_L$ are likely to be heavier than their right-handed counterparts $\widetilde{u}_R$, $\widetilde{d}_R$, $\widetilde{s}_R$ and $\widetilde{c}_R$, because of the effect of K_2 in eqs. (7.70)-(7.76).

- The squarks of the first two families cannot be lighter than about 0.8 times the mass of the gluino in minimal supergravity models, and about 0.6 times the mass of the gluino in the simplest gauge-mediated models if the number of messenger squark pairs is $n_{\text{mess}} \leq 4$. In the minimal supergravity case this is because the gluino mass feeds into the squark masses through RG evolution; in the gauge-mediated case it is because the gluino and squark masses are connected by the supersymmetry breaking mechanism.

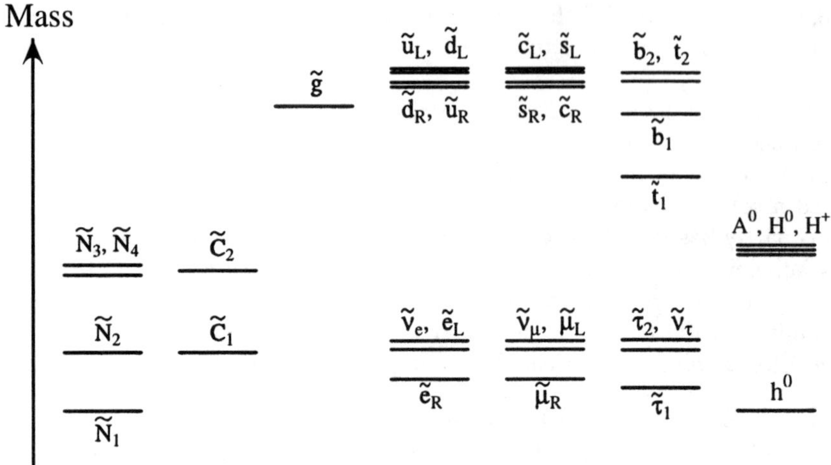

Figure 16: A schematic sample spectrum for the undiscovered particles in the MSSM. This spectrum is presented for entertainment purposes only. No warranty, expressed or implied, guarantees that this spectrum looks anything like the real world.

- The lighter stop $\tilde{t}_1$ and the lighter sbottom $\tilde{b}_1$ are probably the lightest squarks. This is because stop and sbottom mixing effects and the effects of X_t and X_b in eqs. (7.16)-(7.18) both tend to decrease the lighter stop and sbottom masses.

- The lightest charged slepton is probably a stau $\tilde{\tau}_1$. The mass difference $m_{\tilde{e}_R} - m_{\tilde{\tau}_1}$ is likely to be significant if $\tan\beta$ is large, because of the effects of a large tau Yukawa coupling. For smaller $\tan\beta$, $\tilde{\tau}_1$ is predominantly $\tilde{\tau}_R$ and it is not so much lighter than $\tilde{e}_R$, $\tilde{\mu}_R$.

- The left-handed charged sleptons $\tilde{e}_L$ and $\tilde{\mu}_L$ are likely to be heavier than their right-handed counterparts $\tilde{e}_R$ and $\tilde{\mu}_R$. This is because of the effect of K_2 in eq. (7.74), and the fact that $\Delta_e - \Delta_{\bar{e}}$ is positive (and very small because of the numerical accident $\sin^2\theta_W \approx 1/4$).

- The lightest neutral Higgs boson h^0 should be lighter than about 150 GeV, and may be much lighter than the other Higgs scalar mass eigenstates A^0, $H^\pm$, H^0.

In Figure 16 we show a qualitative sketch of a sample MSSM mass spectrum which illustrates these features. Variations in the model parameters can have important and predictable effects. For example, taking larger (smaller) m_0^2 in minimal supergravity models will tend to move the entire spectrum of squarks, sleptons and the Higgs scalars A^0, $H^\pm$, H^0 higher (lower) compared to the neutralinos, charginos and gluino; taking larger values of $\tan\beta$ with other model parameters held fixed will usually tend to lower $\tilde{b}_1$ and $\tilde{\tau}_1$ masses compared to those of the other sparticles, etc. The important point is that by measuring the masses and mixing angles of the MSSM particles we will be able to gain a great deal of information which can rule out or bolster evidence for competing proposals for the origin of supersymmetry breaking. Testing the various possible organizing principles will provide the high-energy physicists of the next millennium with an exciting challenge.

8 Concluding remarks

In this primer, I have attempted to convey some of the more essential features of supersymmetry as it is known so far. One of the most amazing features of supersymmetry is that so much is known about it already, despite the present lack of direct experimental data. Even the terms and stakes of many of the important outstanding questions, especially the paramount issue "How is supersymmetry broken?", are already rather clear. That this can be so is a testament to the unreasonably predictive quality of the symmetry itself.

The experimental verification of supersymmetry will not be an end, but rather a revolution in high energy physics. It seems likely to present us with questions and challenges which at present we can only guess at. The measurement of sparticle masses, production cross-sections, and decays modes will rule out some models for supersymmetry breaking and lend credence to others. We will be able to test the principle of R-parity conservation, the idea that supersymmetry has something to do with the dark matter, and possibly make connections to other aspects of cosmology including baryogenesis and inflation. Other fundamental questions, like the origin of the μ parameter and the rather peculiar hierarchical structure of the Yukawa couplings may be brought into sharper focus with the discovery of the MSSM spectrum. Understanding the precise connection of supersymmetry to the electroweak scale will surely open the window to even deeper levels of fundamental physics.

Acknowledgments

I am grateful to G. Kane and J. Wells for many helpful comments on this primer. I am also indebted to my collaborators on supersymmetric matters, S. Ambrosanio, D. Castaño, T. Gherghetta, C. Kolda, G. Kribs, S. Mrenna, M. Vaughn, and especially P. Ramond, for countless illuminating and inspiring conversations on the subjects discussed here. I thank the Aspen Center for Physics and the Stanford Linear Accelerator Center for their hospitality. This work was supported in part by the U.S. Department of Energy.

References

1. S. Weinberg, *Phys. Rev.* D **13**, 974 (1976); *Phys. Rev.* D **19**, 1277 (1979); L. Susskind, *Phys. Rev.* D **20**, 2619 (1979); G. 't Hooft, in *Recent developments in gauge theories*, Proceedings of the NATO Advanced Summer Institute, Cargese 1979, ed. G. 't Hooft *et al.* (Plenum, New York 1980).
2. E. Witten, *Nucl. Phys.* **B188**, 513 (1981); N. Sakai, *Z. Phys.* C **11**, 153 (1981); S. Dimopoulos and H. Georgi, *Nucl. Phys.* **B193**, 150 (1981); R.K. Kaul and P. Majumdar, *Nucl. Phys.* **B199**, 36 (1982).
3. S. Coleman and J. Mandula, *Phys. Rev.* **159**, 1251 (1967); R. Haag, J. Lopuszanski, and M. Sohnius, *Nucl. Phys.* **B88**, 257 (1975).
4. P. Fayet, *Phys. Lett.* B **64**, 159 (1976).
5. P. Fayet, *Phys. Lett.* B **69**, 489 (1977); *Phys. Lett.* B **84**, 416 (1979).
6. G.R. Farrar and P. Fayet, *Phys. Lett.* B **76**, 575 (1978).
7. P. Ramond, *Phys. Rev.* D **3**, 2415 (1971); A. Neveu and J.H. Schwarz, *Nucl. Phys.* **B31**, 86 (1971); J.L. Gervais and B. Sakita, *Nucl. Phys.* **B34**, 632 (1971).
8. Yu. A. Gol'fand and E. P. Likhtman, *JETP Lett.* **13**, 323 (1971).
9. J. Wess and B. Zumino, *Nucl. Phys.* **B70**, 39 (1974).
10. D.V. Volkov and V.P. Akulov, *Phys. Lett.* B **46**, 109 (1973).
11. J. Wess and J. Bagger, *Introduction to Supersymmetry*, (Princeton University Press, Princeton NJ, 1992).
12. G.G. Ross, *Grand Unified Theories*, (Addison-Wesley, Redwood City CA, 1985).
13. P.P. Srivastava, *Supersymmetry and superfields*, (Adam-Hilger, Bristol England, 1986).
14. P.G.O. Freund, *Introduction to Supersymmetry*, (Cambridge University Press, Cambridge England, 1986).
15. P.C. West, *Introduction to Supersymmetry and Supergravity*, (World Scientific, Singapore, 1990).
16. R.N. Mohapatra, *Unification and Supersymmetry: The Frontiers of Quark-Lepton Physics*, Springer-Verlag, New York 1992.
17. D. Bailin and A. Love, *Supersymmetric Gauge Field Theory and String Theory*, (Institute of Physics Publishing, Bristol England, 1994).
18. P. Ramond, *Beyond the Standard Model*, Frontiers in Physics series, Addison-Wesley, to appear.

19. H.E. Haber and G.L. Kane, *Phys. Rep.* **117**, 75 (1985).

20. H.P. Nilles, *Phys. Rep.* **110**, 1 (1984).

21. *Superspace or One Thousand and One Lessons in Supersymmetry*, S.J. Gates, M.T. Grisaru, M. Rocek and W. Siegel, Benjamin/Cummings 1983.

22. R. Arnowitt, A. Chamseddine and P. Nath, *N=1 Supergravity*, World Scientific, Singapore, 1984.

23. D.R.T. Jones, "Supersymmetric gauge theories", in *TASI Lectures in Elementary Particle Physics 1984*, ed. D.N. Williams, TASI publications, Ann Arbor 1984.

24. H.E. Haber, "Introductory low-energy supersymmetry", TASI-92 lectures, in *Recent Directions in Particle Theory*, eds. J. Harvey and J. Polchinski, World Scientific, 1993.

25. P. Ramond, "Introductory lectures on low-energy supersymmetry", TASI-94 lectures, hep-th/9412234.

26. J.A. Bagger, "Weak-scale supersymmetry: theory and practice", TASI-95 lectures, hep-ph/9604232.

27. H. Baer et al., "Low energy supersymmetry phenomenology", hep-ph/9503479, in *Report of the Working Group on Electroweak Symmetry Breaking and New Physics* of the 1995 study of the future of particle physics in the USA, to be published by World Scientific.

28. M. Drees and S.P. Martin, "Implications of SUSY model building", as in Ref.[27]

29. J.D. Lykken, "Introduction to Supersymmetry", TASI-96 lectures, hep-th/9612114

30. S. Dawson, "SUSY and such", Lectures given at NATO Advanced Study Institute on Techniques and Concepts of High-energy Physics, hep-ph/9612229.

31. M. Dine, "Supersymmetry Phenomenology (With a Broad Brush)", hep-ph/9612389.

32. J.F. Gunion, "A simplified summary of supersymmetry", to appear in *Future High Energy Colliders*, AIP Press, hep-ph/9704349.

33. M. Shifman, "Non-perturbative dynamics in supersymmetric gauge theories", hep-ph/9704114.

34. X. Tata, "What is supersymmetry and how do we find it?", Lectures presented at the IX Jorge A. Swieca Summer School, Campos do Jordão, Brazil, hep-ph/9706307.

35. S. Ferrara, editor, *Supersymmetry*, (World Scientific, Singapore, 1987).

36. A. Salam and J. Strathdee, *Nucl. Phys.* **B76**, 477 (1974); S. Ferrara, J. Wess and B. Zumino, *Phys. Lett.* B **51**, 239 (1974). See Ref.[11] for a

pedagogical introduction to the superfield formalism.

37. J. Wess and B. Zumino, *Phys. Lett.* B **49**, 52 (1974); J. Iliopoulos and B. Zumino, *Nucl. Phys.* **B76**, 310 (1974).

38. J. Wess and B. Zumino, *Nucl. Phys.* **B78**, 1 (1974).

39. L. Girardello and M.T. Grisaru *Nucl. Phys.* **B194**, 65 (1982).

40. See, however, L.J. Hall and L. Randall, *Phys. Rev. Lett.* **65**, 2939 (1990).

41. H.P. Nilles, M. Srednicki and D. Wyler, *Phys. Lett.* B **120**, 346 (1983); J. Ellis et al., *Phys. Rev.* D **39**, 844 (1989). U. Ellwanger, M. Rausch de Traubenberg, and C.A. Savoy, *Phys. Lett.* B **315**, 331 (1993), *Nucl. Phys.* **B492**, 21 (1997); S.A. Abel, S. Sarkar and I.B. Whittingham, *Nucl. Phys.* **B392**, 83 (1993); F. Franke, H. Fraas and A. Bartl, *Phys. Lett.* B **336**, 415 (1994); S.F. King and P.L. White, *Phys. Rev.* D **52**, 4183 (1995);

42. J.E. Kim and H. P. Nilles, *Phys. Lett.* B **138**, 150 (1984) J.E. Kim and H. P. Nilles, *Phys. Lett.* B **263**, 79 (1991); E.J. Chun, J.E. Kim and H.P. Nilles, *Nucl. Phys.* **B370**, 105 (1992).

43. G.F. Giudice and A. Masiero, *Phys. Lett.* B **206**, 480 (1988); J.A. Casas and C. Muñoz, *Phys. Lett.* B **306**, 288 (1993).

44. G. Dvali, G.F. Giudice and A. Pomarol, *Nucl. Phys.* **B478**, 31 (1996).

45. See, for example, F. Zwirner, *Phys. Lett.* B **132**, 103 (1983); R. Barbieri and A. Masiero, *Nucl. Phys.* **B267**, 679 (1986); S. Dimopoulos and L. Hall, *Phys. Lett.* B **207**, 210 (1987); V. Barger, G. Giudice, and T. Han, *Phys. Rev.* D **40**, 2987 (1989); R. Godbole, P. Roy and X. Tata, *Nucl. Phys.* **B401**, 67 (1992); G. Bhattacharya and D. Choudhury, *Mod. Phys. Lett.* **A10**, 1699 (1995); G. Bhattacharya, "*R*-parity-violating supersymmetric Yukawa couplings: a Mini-review", hep-ph/9608415, in *Supersymmetry '96*, ed. R.N. Mohapatra and A. Rasin; H. Dreiner, "An Introduction to explicit *R*-parity violation", hep-ph/9707435, this volume.

46. S. Dimopoulos and H. Georgi, *Nucl. Phys.* **B193**, 150 (1981); S. Weinberg, *Phys. Rev.* D **26**, 2878 (1982); N. Sakai and T. Yanagida, *Phys. Rev.* D **B197**, 533 (1982); S. Dimopoulos, S. Raby and F. Wilczek, *Phys. Lett.* B **112**, 133 (1982).

47. H. Goldberg, *Phys. Rev. Lett.* **50**, 1419 (1983); J. Ellis et al, *Nucl. Phys.* **B238**, 453 (1984).

48. L. Krauss and F. Wilczek, *Phys. Rev. Lett.* **62**, 1221 (1989).

49. L.E. Ibáñez and G. Ross, *Phys. Lett.* B **260**, 291 (1991); T. Banks and M. Dine, *Phys. Rev.* D **45**, 1424 (1995); L.E. Ibáñez, *Nucl. Phys.* **B398**, 301 (1993).

50. R.N. Mohapatra, *Phys. Rev.* D **34**, 3457 (1986); A. Font, L.E. Ibáñez and F. Quevedo, *Phys. Lett.* B **228**, 79 (1989); S.P. Martin *Phys. Rev.* D

46, 2769 (1992); *Phys. Rev.* D **54**, 2340 (1996).

51. R. Kuchimanchi and R.N. Mohapatra, *Phys. Rev.* D **48**, 4352 (1993); *Phys. Rev. Lett.* **75**, 3989 (1995); C.S. Aulakh, K. Benakli and G. Senjanović, hep-ph/9703434; C.S. Aulakh, A. Melfo and G. Senjanović.

52. S. Dimopoulos and D. Sutter, *Nucl. Phys.* **B452**, 496 (1995).

53. For a comprehensive analysis, see F. Gabbiani, E. Gabrielli, A. Masiero and L. Silvestrini, *Nucl. Phys.* **B477**, 321 (1996) and references therein.

54. L.J. Hall, V.A. Kostalecky and S. Raby, *Nucl. Phys.* **B267**, 415 (1986); F. Gabbiani and A. Masiero, *Phys. Lett.* B **209**, 289 (1988); R. Barbieri and L.J. Hall, *Phys. Lett.* B **338**, 212 (1994).

55. P. Langacker, in Proceedings of the PASCOS90 Symposium, Eds. P. Nath and S. Reucroft, (World Scientific, Singapore 1990) J. Ellis, S. Kelley, and D. Nanopoulos, *Phys. Lett.* B **260**, 131 (1991); U. Amaldi, W. de Boer, and H. Furstenau, *Phys. Lett.* B **260**, 447 (1991); P. Langacker and M. Luo, *Phys. Rev.* D **44**, 817 (1991); C. Giunti, C.W. Kim and U. W. Lee, *Mod. Phys. Lett.* **A6**, 1745 (1991).

56. A.G. Cohen, D.B. Kaplan and A.E. Nelson, *Phys. Lett.* B **388**, 588 (1996).

57. Y. Nir and N. Seiberg, *Phys. Lett.* B **309**, 337 (1993).

58. P. Fayet and J. Iliopoulos, *Phys. Lett.* B **51**, 461 (1974); P. Fayet, *Nucl. Phys.* **B90**, 104 (1975).

59. However, see for example P. Binétruy and E. Dudas, *Phys. Lett.* B **389**, 503 (1996); G. Dvali and A. Pomarol, *Phys. Rev. Lett.* **77**, 3728 (1996); R.N. Mohapatra and A. Riotto, *Phys. Rev.* D **55**, 4262 (1997). A non-zero Fayet-Iliopoulos term for an anomalous $U(1)$ symmetry is commonly found in superstring models: M. Green and J. Schwarz, *Phys. Lett.* B **149**, 117 (1984); M. Dine, N. Seiberg and E. Witten, *Nucl. Phys.* **B289**, 589 (1987); J. Atick, L. Dixon and A. Sen, *Nucl. Phys.* **B292**, 109 (1987); This may even help to explain the observed structure of the Yukawa couplings: L.E. Ibáñez, *Phys. Lett.* B **303**, 55 (1993); L.E. Ibáñez and G.G. Ross, *Phys. Lett.* B **332**, 100 (1994); P. Binétruy, S. Lavignac, P. Ramond, *Nucl. Phys.* **B477**, 353 (1996). P. Binétruy, N.Irges, S. Lavignac and P. Ramond, *Phys. Lett.* B **403**, 38 (1997).

60. L. O'Raifeartaigh, *Nucl. Phys.* **B96**, 331 (1975).

61. S. Coleman and E. Weinberg, *Phys. Rev.* D **7**, 1888 (1973).

62. E. Witten, *Nucl. Phys.* **B202**, 253 (1982); I. Affleck, M. Dine and N. Seiberg, *Nucl. Phys.* **B241**, 493 (1981); *Nucl. Phys.* **B256**, 557 (1986). For reviews, see L. Randall, "Models of Dynamical Supersymmetry Breaking", hep-ph/9706474; A. Nelson, "Dynamical Supersymmetry Breaking", hep-ph/9707442; G.F. Giudice and R. Rattazzi, "Theories

with Gauge-Mediated Supersymmetry Breaking", hep-ph/9801271.

63. P. Fayet, *Phys. Lett.* B **70**, 461 (1977); *Phys. Lett.* B **86**, 272 (1979); and in *Unification of the fundamental particle interactions* (Plenum, New York, 1980).

64. N. Cabibbo, G.R. Farrar and L. Maiani, *Phys. Lett.* B **105**, 155 (1981); M.K. Gaillard, L. Hall and I. Hinchliffe, *Phys. Lett.* B **116**, 279 (1982); J. Ellis and J.S. Hagelin, *Phys. Lett.* B **122**, 303 (1983), D.A. Dicus, S. Nandi and J. Woodside, *Phys. Lett.* B **258**, 231 (1991); D.R. Stump, M. Wiest, and C.P. Yuan, *Phys. Rev.* D **54**, 1936 (1996); S. Ambrosanio, G. Kribs, and S.P. Martin, *Phys. Rev.* D **56**, 1761 (1997).

65. S. Dimopoulos, M. Dine, S. Raby and S.Thomas, *Phys. Rev. Lett.* **76**, 3494 (1996); S. Dimopoulos, S. Thomas and J. D. Wells, *Phys. Rev.* D **54**, 3283 (1996).

66. S. Ambrosanio et al., *Phys. Rev. Lett.* **76**, 3498 (1996); *Phys. Rev.* D **54**, 5395 (1996).

67. S. Ferrara, D.Z. Freedman and P. van Nieuwenhuizen, *Phys. Rev.* D **13**, 3214 (1976); S. Deser and B. Zumino, *Phys. Lett.* B **62**, 335 (1976); D.Z. Freedman and P. van Nieuwenhuizen, *Phys. Rev.* D **14**, 912 (1976); E. Cremmer et al., *Nucl. Phys.* **B147**, 105 (1979); J. Bagger, *Nucl. Phys.* **B211**, 302 (1983).

68. E. Cremmer, S. Ferrara, L. Girardello, and A. van Proeyen, *Nucl. Phys.* **B212**, 413 (1983).

69. S. Deser and B. Zumino, *Phys. Rev. Lett.* **38**, 1433 (1977); E. Cremmer et al., *Phys. Lett.* B **79**, 1978 (231).

70. H. Pagels, J.R. Primack, *Phys. Rev. Lett.* **48**, 1982 (223); T. Moroi, H. Murayama, M. Yamaguchi, *Phys. Lett.* B **303**, 1993 (289).

71. P. Moxhay and K. Yamamoto, *Nucl. Phys.* **B256**, 130 (1985); K. Grassie, *Phys. Lett.* B **159**, 32 (1985); B. Gato, *Nucl. Phys.* **B278**, 189 (1986); N. Polonsky and A. Pomarol, *Phys. Rev. Lett.* **73**, 2292 (1994).

72. G.G. Ross and R.G. Roberts, *Nucl. Phys.* **B377**, 571 (1992). H. Arason et al., *Phys. Rev. Lett.* **67**, 2933 (1991); R. Arnowitt and P. Nath, *Phys. Rev. Lett.* **69**, 725 (1992) and *Phys. Rev.* D **46**, 3981 (1992); D.J. Castaño, E.J. Piard, P. Ramond, *Phys. Rev.* D **49**, 4882 (1994); V. Barger, M. Berger and P. Ohmann, *Phys. Rev.* D **49**, 4908 (1994); G.L. Kane, C. Kolda, L. Roszkowski, J.D. Wells, *Phys. Rev.* D **49**, 6173 (1994); B. Ananthanarayan, K.S. Babu and Q. Shafi, *Nucl. Phys.* **B428**, 19 (1994); M. Carena, M. Olechowski, S. Pokorski, C.E.M. Wagner, *Nucl. Phys.* **B419**, 213 (1994); W. de Boer, R. Ehret and D. Kazakov, *Z. Phys.* C **67**, 647 (1995); M. Carena, P. Chankowski, M. Olechowski, S. Pokorski, C.E.M. Wagner, *Nucl. Phys.* **B491**, 103 (1997).

73. V. Kaplunovsky and J. Louis, *Phys. Lett.* B **306**, 269 (1993); R. Barbieri, J. Louis and M. Moretti, *Phys. Lett.* B **312**, 451 (1993); A. Brignole, L.E. Ibáñez and C. Muñoz, *Nucl. Phys.* **B422**, 125 (1994), erratum *Nucl. Phys.* **B436**, 747 (1995).

74. J. Polonyi, Hungary Central Research Institute report KFKI-77-93 (1977) (unpublished). See Ref.[17] for a pedagogical account.

75. For a review, see A.B. Lahanas and D.V. Nanopoulos, *Phys. Rep.* **145**, 1 (1987).

76. For a review, see A. Brignole, L.E. Ibáñez and C. Muñoz, "Soft supersymmetry-breaking terms from supergravity and supersring models", hep-ph/9707209, this volume.

77. M. Dine and W. Fischler, *Phys. Lett.* B **110**, 227 (1982); C.R. Nappi and B.A. Ovrut, *Phys. Lett.* B **113**, 175 (1982); L. Alvarez-Gaumé, M. Claudson, and M. B. Wise, *Nucl. Phys.* **B207**, 96 (1982).

78. M. Dine, A. E. Nelson, *Phys. Rev.* D **48**, 1277 (1993); M. Dine, A.E. Nelson, Y. Shirman, *Phys. Rev.* D **51**, 1362 (1995); M. Dine, A.E. Nelson, Y. Nir, Y. Shirman, *Phys. Rev.* D **53**, 2658 (1996).

79. S. Dimopoulos, G.F. Giudice and A. Pomarol, *Phys. Lett.* B **389**, 37 (1996); S.P. Martin *Phys. Rev.* D **55**, 3177 (1997); E. Poppitz and S.P. Trivedi, *Phys. Lett.* B **401**, 38 (1997).

80. K. Inoue, A. Kakuto, H. Komatsu and H. Takeshita, *Prog. Theor. Phys.* **68**, 927 (1982) and **71**, 413 (1984); N. K. Falck *Z. Phys.* C **30**, 247 (1986).

81. L.E. Ibáñez and G.G. Ross, *Phys. Lett.* B **110**, 215 (1982); L.E. Ibáñez, *Phys. Lett.* B **118**, 73 (1982); J. Ellis, D.V. Nanopoulos and K. Tamvakis, *Phys. Lett.* B **121**, 123 (1983); L. Alvarez-Gaumé, J. Polchinski, and M. Wise, *Nucl. Phys.* **B221**, 495 (1983).

82. J.F. Gunion and H.E. Haber, *Nucl. Phys.* **B272**, 1 (1986); *Nucl. Phys.* **B278**, 449 (1986); *Nucl. Phys.* **B307**, 445 (1988). (E: *Nucl. Phys.* **B402**, 567 (1993)).

83. J.F. Gunion, H.E. Haber, G.L. Kane and S. Dawson, *The Higgs Hunter's Guide* Addison-Wesley 1991, errata: hep-ph/9302272.

84. K. Inoue, A. Kakuto, H. Komatsu and S. Takeshita, *Prog. Theor. Phys.* **67**, 1889 (1982); R.A.Flores and M. Sher *Ann. Phys.* (NY) **148**, 95, 1983.

85. H.E. Haber and R. Hempfling, *Phys. Rev. Lett.* **66**, 1815 (1991); Y. Okada, M. Yamaguchi and T. Yanagida, *Prog. Theor. Phys.* **85**, 1 (1991), *Phys. Lett.* B **262**, 54 (1991); J. Ellis, G. Ridolfi and F. Zwirner, *Phys. Lett.* B **257**, 83 (1991), *Phys. Lett.* B **262**, 477 (1991).

86. G. Gamberini, G. Ridolfi and F. Zwirner, *Nucl. Phys.* **B331**, 331 (1990); R. Barbieri, M.Frigeni and F. Caravaglio, *Phys. Lett.* B **258**, 167 (1991);

A. Yamada, *Phys. Lett.* B **263**, 233 (1991) and *Z. Phys.* C **61**, 247 (1994); J.R. Espinosa and M. Quiros, *Phys. Lett.* B **266**, 389 (1991); A. Brignole, *Phys. Lett.* B **281**, 284 (1992); M. Drees and M.M. Nojiri, *Phys. Rev.* D **45**, 2482 (1992) and *Nucl. Phys.* **B369**, 54 (1992); H.E. Haber and R. Hempfling, *Phys. Rev.* D **48**, 4280 (1993); P.H. Chankowski, S. Pokorski and J. Rosiek, *Phys. Lett.* B **274**, 191 (1992) and *Nucl. Phys.* **B423**, 437 (1994); R. Hempfling and A.H. Hoang, *Phys. Lett.* B **331**, 99 (1994); J. Kodaira, Y. Yasui and K. Sasaki, *Phys. Rev.* D **50**, 7035 (1994); J.A. Casas, J.R. Espinosa, M. Quiros and A. Riotto, *Nucl. Phys.* **B436**, 3 (1995) [E: *Nucl. Phys.* **B439**, 466 (1995)]; M. Carena, M. Quirós and C. Wagner, *Nucl. Phys.* **B461**, 407 (1996).

87. See G.L. Kane, C. Kolda and J.D. Wells, *Phys. Rev. Lett.* **70**, 2686 (1993); J.R. Espinosa and M. Quirós, *Phys. Lett.* B **302**, 51 (1993), and references therein.

88. R. Tarrach, *Nucl. Phys.* **B183**, 384 (1980); H. Gray, D.J. Broadhurst, W. Grafe and K. Schilcher, *Z. Phys.* C **48**, 673 (1990); H. Arason et al., *Phys. Rev.* D **46**, 3945 (1992).

89. L.E. Ibáñez and C. Lopez, *Phys. Lett.* B **126**, 54 (1983); H. Arason et al., *Phys. Rev. Lett.* **67**, 2933 (1991); V. Barger, M.S. Berger and P. Ohmann, *Phys. Rev.* D **47**, 1093 (1993); P. Langacker and N. Polonsky, *Phys. Rev.* D **49**, 1454 (1994); P. Ramond, R.G. Roberts, G.G. Ross, *Nucl. Phys.* **B406**, 19 (1993); M. Carena, S. Pokorski and C. Wagner, *Nucl. Phys.* **B406**, 59 (1993); G. Anderson et al, *Phys. Rev.* D **49**, 3660 (1994).

90. L.J. Hall, R. Rattazzi and U. Sarid, *Phys. Rev.* D **50**, 7048 (1994); M. Carena, M. Olechowski, S. Pokorski and C.E.M. Wagner, *Nucl. Phys.* **B426**, 269 (1994); R. Hempfling, *Phys. Rev.* D **49**, 6168 (1994); R. Rattazzi and U. Sarid, *Phys. Rev.* D **53**, 1553 (1996).

91. S.P. Martin and M.T. Vaughn, *Phys. Lett.* B **318**, 331 (1993).

92. D. Pierce and A. Papadopoulos, *Phys. Rev.* D **50**, 565 (1994); *Nucl. Phys.* **B430**, 278 (1994); D. Pierce, J.A. Bagger, K. Matchev, and R.-J. Zhang, *Nucl. Phys.* **B491**, 3 (1997).

93. J. Frère, D.R.T. Jones and S. Raby, *Nucl. Phys.* **B222**, 11 (1983); M. Claudson, L.J. Hall, and I. Hinchliffe, *Nucl. Phys.* **B228**, 501 (1983); J.A. Casas, A. Lleyda, C. Muñoz, *Nucl. Phys.* **B471**, 3 (1996). T. Falk, K.A. Olive, L. Roszkowski, and M. Srednicki *Phys. Lett.* B **367**, 183 (1996). H. Baer, M. Brhlik and D.J. Castaño, *Phys. Rev.* D **54**, 6944 (1996). For a review, see J.A. Casas, hep-ph/9707475, this volume.

94. A. Kusenko, P. Langacker and G. Segre, *Phys. Rev.* D **54**, 5824 (1996).

SUPERSYMMETRY AND PARTICLE PHYSICS: A ROADMAP OF FUTURE DIRECTIONS [a]

KEITH R. DIENES[1,2] AND CHRISTOPHER KOLDA[1]

[1] *School of Natural Sciences, Institute for Advanced Study*
Olden Lane, Princeton, NJ 08540 USA
[2] *CERN Theory Division, CH-1211 Geneva 23, Switzerland*

We give a brief survey of several open theoretical questions in supersymmetric particle physics. The questions we have chosen range from the GeV scale to the Planck scale, and include issues pertaining to the Minimal Supersymmetric Standard Model and its extensions, SUSY-breaking, cosmology, grand unified theories, and string theory. Throughout, our goal is to address those topics in which supersymmetry plays a fundamental role, and which are areas of active research in the field. A more complete discussion of these and other open questions can be found in the e-print **hep-ph/9712322**.

At first glance, supersymmetry appears to be a theoretical success story writ large. With one simple idea, such diverse issues as extending the Lorentz group, solving the gauge hierarchy problem, coupling gauge theories to gravity, and generating gauge coupling unification all seem to fall into place. The lack of direct experimental evidence for supersymmetry (SUSY) dampens our enthusiasm somewhat, but only now are experiments really beginning to probe the domain where SUSY should be expected to be manifest.

Nonetheless, there is a price to be paid for the successes of SUSY, and we mean more than simply the doubling of the Standard Model (SM) particle spectrum. SUSY introduces into physics a host of new questions which must be addressed, and hopefully, answered. Most of these questions were once real problems — problems which seemed to detract from, or perhaps even invalidate, SUSY as a viable fundamental symmetry of nature. But as competing answers to these questions have been found, these questions became opportunities not only to simply discover SUSY, but also to probe physics well beyond the scale of SUSY.

The directions that we will discuss in this work have been chosen because they satisfy two criteria. First and foremost, each defines an area of active research in the field. Second, these are questions which are intrinsically supersymmetric and may not even arise in the SM alone. Thus, generic questions in

---

[a]This work was supported in part by DOE Grant #DE-FG02-90ER40542, the generosity of Helen and Martin Chooljian, and the hospitality of the Aspen Center for Physics.

the SM (such as the cosmological-constant, inflation, baryogenesis, and fermion mass hierarchy questions, to name a few) are not included here.

Many of the topics that we shall discuss here will be covered in much greater detail in the topical chapters of this book, and we will try to indicate the relevant chapters as we go along. Note that a more complete discussion of these and other open issues can be found in Ref. [1].

Section I: Open Questions in the MSSM

The minimal extension of the SM which exhibits unbroken SUSY is a simple model with fewer free parameters than the SM itself, despite the large number of new fields. It is only in the breaking of SUSY that the number of free parameters becomes large. In the SM, the field content along with the gauge symmetries serve to provide a number of accidental symmetries at the renormalizable level. These include baryon number B, and lepton number L. Flavor-changing neutral currents (FCNC's) are also forbidden up to small corrections arising from Yukawa couplings in loops. The Minimal Supersymmetric Standard Model (MSSM) shares neither of these properties. As we will discuss, a generic MSSM would have the proton decay with a weak-interaction lifetime and large FCNC's. The reason is this same proliferation of fields and free parameters. Even after imposing symmetries to forbid the fast proton decay (see below), one finds [2] that the MSSM contains 106 new, independent (real) parameters above and beyond those of the SM. These consist of 26 masses (resulting from 12 squark, 9 slepton, and 3 gaugino masses, plus μ- and B_μ-terms), 37 mixing angles, and 43 CP-violating phases. Understanding, constraining, and ultimately measuring these parameters is one of the primary goals of the SUSY program in particle physics.

Many of the "problems" in the MSSM are really questions of naturalness: what are the natural values that these 106 new parameters should take, and why in many cases don't they? In SUSY, the usual issue of 't Hooft (or technical) naturalness rarely arises thanks to the so-called non-renormalization theorem for the superpotential. However, issues of Dirac naturalness are common; that is, we often have to explain why some parameter is small when it need not be from symmetry considerations alone. Being supersymmetric, small parameters are stable against radiative corrections, but that does not explain why they are small in the first place. Since answers to these questions of Dirac naturalness should be provided by a complete theory, we suspect that even the MSSM is an incomplete description of the world.

The first such naturalness question arises immediately when one attempts to write down the MSSM in full generality. As discussed in the article of

S. Martin in this volume, the most general, renormalizable superpotential for the Minimal Supersymmetric Standard Model (MSSM) can be organized into three pieces, $W = W_0 + W_{\not B} + W_{\not L}$, where

$$W_0 = y_{ij}^U Q^i u^j H_U + y_{ij}^D Q^i d^j H_D + y_{ij}^E L^i e^j H_D + \mu H_U H_D \tag{1}$$

$$W_{\not L} = \lambda_{ijk} Q^i d^j L^k + \lambda'_{ijk} e^i L^j L^k + \mu'_i H_U L^i \tag{2}$$

$$W_{\not B} = \lambda''_{ijk} u^i d^j d^k . \tag{3}$$

The first piece preserves the global B and L quantum numbers of the SM, while the other two each violate one of either B or L. This is unlike the SM in which one can obtain B-violating operators only by going to dimension-six, or L-violating operators only by going to dimension-five. Simultaneous B- and L-violation would be a disaster. For example, given non-zero λ and λ'', one can form a four-fermion operator $QudL$ which can mediate proton decay and which is suppressed by only squark masses [3,4], not M_{GUT} or M_{Pl}. Resulting proton lifetimes would be on the order of 10^{-17} years!

There is a simple remedy for this: one can set $W_{\not B} = W_{\not L} = 0$ by hand. Because of the non-renormalization theorem, this is technically natural, but in order to have Dirac naturalness we must have some symmetry which forbids the unwanted terms. The obvious candidates, global B and L symmetries, are probably not suitable to play this role since we expect from a number of arguments that they will be broken by non-renormalizable or non-perturbative terms. Fortunately it is easy to invent a new discrete symmetry which simultaneously forbids all the unwanted terms but allows those that are phenomenologically necessary. This is "matter parity" (or R-parity), a $\mathbb{Z}_2$ symmetry under which all the matter fields (L, e, Q, u, d) are odd and the Higgs fields are even [5]. An added benefit of such a symmetry is the exact stability of the lightest supersymmetric particle (LSP), even against higher-order interactions, thereby providing a candidate for the (cold) dark matter.

Of course, the original argument against $W_{\not B}$ and $W_{\not L}$ was that both could not be allowed without leading to rapid nucleon decay. This argument suffices to forbid one or the other, but not both. In fact, much work has been done considering extensions of the MSSM with so-called "R-parity violation" in which either $W_{\not B}$ or $W_{\not L}$ is non-zero, but not both. The price one pays is that the LSP is no longer a dark-matter candidate. The phenomenology of and motivations for R-parity violation are rather complicated, and are discussed in the contribution of H. Dreiner to this volume.

Thus far, we have said nothing about proton decay from higher-dimension operators. In particular, there are many terms of dimension five *which are invariant under matter parity* and which can mediate proton decay. For example, consider the term $W = (\frac{\eta}{M})QQQL$. Even if $M = M_{\rm Pl}$, current proton lifetime bounds require η to be unnaturally small: $\eta \lesssim 10^{-7}$.

Is matter parity conserved in nature, or must it be augmented/replaced in order to fix the dimension-five proton decay? Will this lead to observable signals at colliders? Will it provide a source of dark matter? And can matter parity arise naturally from some larger symmetry [like $SO(10)$] so that it need not be imposed by hand on the theory? These are a few of the unresolved issues that await both experimental and theoretical inputs.

While we can find symmetries which forbid rapid proton decay, there are other naturalness questions in the MSSM which seem to have no simple symmetry solutions. The SUSY flavor and CP problems are two such questions.

We know that in the SM, quark and lepton mass matrices are not diagonal in the interaction eigenbasis (since $\theta_c \neq 0$). However this non-diagonality does not lead to large flavor-changing neutral currents (FCNC's) because of the GIM mechanism. In unbroken SUSY, the mass and interaction eigenstates of the scalars should also be rotated, but the same rotations that diagonalize the fermions should also diagonalize their scalar partners. Again, the GIM mechanism suppresses any FCNC's.

But after SUSY-breaking, one rotates fermions by some unitary matrix U, while their partners are rotated by an *independent* $\widetilde{U}$. Since gauginos couple to both a particle *and* its superpartner simultaneously, their couplings have the form $\widetilde{U}^\dagger U$, which need not be diagonal. This can generate FCNC's[3,6].

The tightest bounds on FCNC's presently come from $K^0 - \overline{K^0}$ mixing. Requiring that the MSSM prediction for the $K^0 - \overline{K^0}$ mass difference not exceed the measured value yields limits of the form[9]:

$$\frac{\mathcal{A}^2}{m_{\tilde{Q}}^2} \left(\frac{\delta m^2}{m_{\tilde{Q}}^2}\right)^2 \leq 5 \times 10^{-9} \text{ GeV}^{-2} \tag{4}$$

where $\mathcal{A}^2$ is a product of angles which rotate from the quark mass basis to the squark mass basis, and where δm^2 is the mass difference between the $\widetilde{d_L}$ and $\widetilde{s_L}$ squarks. There are corresponding limits on $\widetilde{d_R} - \widetilde{s_R}$ splittings as well as mixed left-right limits. Given the above constraint, it is clear that if the mass splittings are $\mathcal{O}(1)$ and the angles take average values ($\mathcal{A}^2 \sim 1/20$), the squark mass scale must be $\gtrsim 3\,\text{TeV}$.

There are three primary proposals for solving the SUSY flavor problem: degeneracy, alignment, and decoupling. The last of these probably does not work in detail[7], so we will not discuss it here. Degeneracy attempts to solve the flavor problem by positing that squarks and sleptons of a given flavor are mass-degenerate[8,9], *i.e.*, $\delta m^2 = 0$. (One way to see that this would forbid FCNC's is to note that if, for example, the $\tilde{d}$ and $\tilde{s}$ were to have equal mass, they would exactly cancel each other's contributions to any flavor-changing process.) This extension of the GIM mechanism to the SUSY sector is often referred to as "super-GIM": $m_{\tilde{u}}^2 = m_{\tilde{c}}^2 \simeq m_{\tilde{t}}^2$ for L,R squarks separately, likewise $m_{\tilde{e}}^2 = m_{\tilde{\mu}}^2 \simeq m_{\tilde{\tau}}^2$. Given the weaker bounds on FCNC's involving the third generation, the final equalities need only be approximate.

Alignment solves the flavor problem not by setting $\delta\tilde{m}^2 = 0$ as with degeneracy, but by enforcing $U = \tilde{U}$ (or equivalently $\mathcal{A} = 0$) to a very high accuracy [10]. Here the flavor physics is typically generated by one or more $U(1)$ gauge interactions, and ties the generation of the scalar soft masses to that of the fermion mass/Yukawa matrices.

Once SUSY is discovered, it may not take long to discern which of these paths nature has chosen to follow. Observation of scalar partners should quickly tell us whether or not they are degenerate (*i.e.*, super-GIM or aligned). However, it will take much more experimental and theoretical work to determine just how each of these choices is concretely realized.

Similar issues arise when one considers sources of CP violation in the MSSM. In the MSSM there are 43 physical CP-violating phases above and beyond that of the SM (ignoring θ_{QCD}), and these phases can appear in large, easily observed, and easily constrained experimental processes. The tightest constraints come from CP-violating in the kaon system and the electric dipole moment of the neutron. The former receives SUSY contributions only if there are also flavor-changing SUSY contributions such as those discussed above; the latter exists even if SUSY is flavor-preserving. (For a review, see Ref. [11] and the contribution of A. Masiero and L. Silvestrini to this volume.)

Let us consider the second case first, with all SUSY flavor-changing effects put to zero through universal soft masses. Assuming that the trilinear couplings are all proportional to the Yukawas, and that the gaugino masses are universal at some scale, then only two physical phases beyond the SM remain: $\phi_A = \arg(A^* M_3)$ and $\phi_B = \arg(B_\mu^* \mu M_3)$. At one-loop order, gluinos and squarks can contribute to the electric dipole moment (EDM) of quarks, and then in turn nuclei. The EDM of the neutron is calculated to be [12] $d_N \simeq 2(100\,\mathrm{GeV}/\tilde{m})^2 \sin(\phi_A - \phi_B) \times 10^{-23}\,e\,\mathrm{cm}$ (for $M_3 \simeq m_{\tilde{q}} \equiv \tilde{m}$), where experimentally $d_N < 1.1 \times 10^{-25}\,e\,\mathrm{cm}$. Clearly, if the phases $\phi_{A,B}$ take values of $\mathcal{O}(1)$, then the sparticles must be heavier than $1\,\mathrm{TeV}$; or for sparticles around $100\,\mathrm{GeV}$, the phases must be $\lesssim 10^{-2}$. In either case some degree of unnaturalness is introduced.

In the most general case where flavor violations are also allowed, not only does the number of physical CP-violating phases increase, but the bounds

104

from observables become much stronger. For example, gluinos and squarks can appear in the internal lines of box diagrams contributing to ϵ_K. One finds for $M_3 \simeq m_{\widetilde{Q}} \simeq m_{\widetilde{d}} \equiv \widetilde{m}$ that [9]:

$$\epsilon_K \simeq \left(\frac{10^{10}\,\text{GeV}^2}{\widetilde{m}^2} \right) \text{Im} \left(\frac{\delta m_{\widetilde{Q}}^2}{m_{\widetilde{Q}}^2} \frac{\delta m_{\widetilde{d}}^2}{m_{\widetilde{d}}^2} \right) . \tag{5}$$

Experimentally, $\epsilon_K \simeq 2.3 \times 10^{-3}$. Thus, given $\mathcal{O}(1)$ mass splittings and phases, the squarks and sleptons must be $\gtrsim 1000\,\text{TeV}$! Alternatively, either the mass splittings or the CP-violating angles must be made unnaturally small.

The CP problem is solved in ways that are similar to those used for the flavor problem. For the ϵ_K problem, degeneracy is the most attractive solution. The EDM problem is not so simply solved, but then it is not quite as serious.

The μ-term of the MSSM prompts yet another question of Dirac naturalness: why would a parameter take a value far below its "natural" scale? For the μ-parameter of the MSSM (*i.e.*, $W = \mu H_U H_D$), the natural value is the UV cutoff of the MSSM. Yet phenomenologically we know that $\mu \sim m_Z$ by minimization of the MSSM Higgs potential. So why would a SUSY-invariant and G_{SM}-invariant mass parameter have a value typical of SUSY-breaking masses? Within the context of a GUT model, the problem is exacerbated. In $SU(5)$ parlance, the μ-term provides the mass for the complete $\mathbf{5}$ and $\overline{\mathbf{5}}$ of Higgs, thereby becoming entangled in the doublet-triplet splitting problem.

There is an almost-default solution to the μ-problem which goes by the acronym NMSSM (Next-to-Minimal...). In this model, a gauge singlet N is introduced whose role is to produce a μ-term through its VEV. The superpotential would have the form $W = \lambda N H_U H_D + \lambda' N^3$ while N itself is presumed to have a soft $(\text{mass})^2$ term which is negative. Obviously a μ-term arises: $\mu = \lambda \langle N \rangle$. However, the singlet solution to the μ-problem is not without problems, including possible generation of cosmological domain walls at the electroweak scale [13], and tadpole contributions to the Lagrangian that can potentially destabilize the gauge hierarchy [14].

Within SUGRA-mediated models of SUSY-breaking (see Section II) there is actually a more attractive solution, known as the Giudice-Masiero mechanism [15]. If the μ-term is forbidden by some symmetry (*e.g.*, a discrete symmetry) which is violated in the hidden sector, then a μ-term can arise through a non-minimal Kähler potential: $K = \cdots + (S/M_{\text{Pl}}) H_U H_D$, where S is a hidden sector field with F-term $\langle F_S \rangle = m_Z M_{\text{Pl}}$. Then $\int d^4\theta S H_U H_D / M_{\text{Pl}} = \int d^2\theta F_S H_U H_D / M_{\text{Pl}} \equiv \int d^2\theta\, \mu H_U H_D$. In this way the μ-term, which is SUSY-preserving, is actually tied to the breaking of SUSY and is thus naturally $\sim m_Z$. The corresponding value of B_μ will then also be $\sim m_Z^2$.

Within non-SUGRA models, there are no such simple solutions. In particular, GMSB models struggle with a severe μ-problem. In generic models one typically finds $B_\mu/\mu^2 \gg 1$ while one needs $B_\mu \sim \mu^2$ phenomenologically. In some special cases one finds $B_\mu/\mu^2 \ll 1$. Here one would expect an axion, but one-loop corrections to B_μ pull the axion mass above m_Z. The hallmark of such a scenario[16] is a very large value for $\tan\beta$ (~ 50).

Section II: Open Questions on SUSY-Breaking

All of the questions we have discussed thus far have begun with the structure of the MSSM. Implicit in that construction is the fact that as a symmetry of nature, SUSY makes some profoundly (and obviously) wrong predictions. Specifically, SUSY requires a spectrum in which every fermionic degree of freedom has a bosonic counterpart with identical mass and quantum numbers. Therefore, in order for SUSY to play any role in low-energy physics, it must be broken (or hidden), just as is the $SU(2)$ gauge symmetry of weak interactions.

Once it is clear that SUSY must be broken, the next logical step is to find some way in which to break SUSY spontaneously. We now know that breaking SUSY spontaneously entails adding to the MSSM new fields and interactions to act as a SUSY-breaking sector, just as the Higgs sector is added to the SM, and that none of the MSSM fields can themselves play this role.

By putting in a set of new fields, it is easy to break SUSY (*e.g.*, through an O'Raifeartaigh-type superpotential). But how, if at all, can these fields couple to the usual MSSM fields? One of the properties of SUSY which is preserved after spontaneous breaking[3] is the supertrace formula $\mathrm{STr}\, M^2 = 0$, applied separately to each supermultiplet. This formula is phenomenologically untenable, for it predicts that the masses of scalars are distributed evenly above and below the fermion masses. But $\mathrm{STr}\, M^2 = 0$ only at tree-level. Even so, this constraint requires that a SUSY-breaking sector must not have renormalizable tree-level couplings to ordinary matter. Because the SUSY-breaking sector must be kept at some distance from the SM sector, it has been called "hidden" or "secluded", while the SM sector is called "visible".

Because the actual breaking of SUSY is far removed from the SM, the question of how SUSY is broken is also far removed from experimental probes. Thus, this question becomes subsidiary to the next issue we will consider, namely that of communicating SUSY-breaking to the SM. This will be discussed in the chapter of M. Peskin.

If SUSY-breaking is generated spontaneously in some sector far from the MSSM, then how does the MSSM "learn" that SUSY has been broken? Remember that *at tree-level*, $\mathrm{STr}\, M^2 = 0$. Two routes therefore seem open to

106

us for communicating SUSY-breaking to the visible sector: loops and non-renormalizable interactions. Both have found application in realistic models.

The "default" mechanism is through the non-renormalizable interactions found in supergravity theories. (A full introduction to supergravity can be found in the contribution of R. Arnowitt and P. Nath to this volume, or in the standard references [18].) The minimal supergravity-mediation model relies on the following assumptions: *(i)* The superpotential can be written in the form $W = W_H(X_i) + W_V(\varphi_i)$ where $W_H(X_i)$ is the superpotential for the hidden sector fields X_i, and $W_V(\varphi_i)$ is the superpotential for the visible sector fields φ_i; *(ii)* The Kähler potential is minimal; *(iii)* The gauge kinetic function is given as $f = cX/M_{\text{Pl}}$ for some constant $c \sim \mathcal{O}(1)$; and *(iv)* In the hidden sector, SUSY breaks such that $\langle F_X \rangle \neq 0$, $\langle W_H \rangle \sim \langle F_X \rangle M_{\text{Pl}}$, and $\langle V \rangle = 0$. After SUSY-breaking, the visible sector potential of supergravity reduces to

$$V(\varphi) = e^{\langle K \rangle / M_{\text{Pl}}^2} \frac{|\langle W_H \rangle|^2}{M_{\text{Pl}}^4} \sum_i |\varphi_i|^2 , \tag{6}$$

giving masses to all visible sector fields $\sim F_X/M_{\text{Pl}}$. One of the more well-known features of this mechanism is that all the visible-sector scalars receive exactly the same mass, leading to mass universalities which might solve the FCNC problem discussed earlier. Similar analyses give universal bilinear, trilinear, and gaugino mass terms, also of sizes $\sim F_X/M_{\text{Pl}}$.

If supergravity is the *dominant* mediator of SUSY-breaking, then the weak scale can be *defined* to be F_X/M_{Pl}, *i.e.*, $F_X \sim m_Z M_{\text{Pl}}$. (In fact, even if supergravity is *not* the dominant mediator, it will still communicate SUSY-breaking to the visible sector and the masses induced will always be $\sim F_X/M_{\text{Pl}}$.) The output of minimal supergravity consists of four universal mass parameters which describe all the soft masses of the MSSM: a scalar mass m_0, a gaugino mass $M_{1/2}$, a A-term A_0, and a B-term B_0. These four parameters can then be evolved from the GUT or Planck scale down to the weak scale in order to form the basis for realistic phenomenology studies [19].

It is by now well-known that, beyond the lowest order, there are many effects in supergravity models that can significantly disrupt the mass universality at the weak scale. For example, there is no way to forbid all terms of the form $y_{ij}(X^\dagger, X)\varphi_i^\dagger \varphi_j$ from appearing in the Kähler potential. Though such terms are suppressed by powers of $1/M_{\text{Pl}}$, the hidden-sector fields receive VEV's $\sim m_Z M_{\text{Pl}}$ so that terms of this type contribute to the scalar masses with size $\sim m_Z$. Furthermore, there is no reason for y_{ij} to be diagonal in the same basis as the fermion masses. There are also other effects, including RGE running in the third generation, which can spoil universality.

One might hope for some way of communicating SUSY-breaking that yields mass universality more robustly. In this vein, there has been much recent interest in so-called "gauge-mediated" models [20]. The basic principle for these models is rather simple: If the soft scalar masses are functions of only sparticle gauge charges, universality is automatic. (Remember that universality in this context refers only to sparticles with identical quantum numbers, such as the $\tilde{d}_L, \tilde{s}_L, \tilde{b}_L$-squarks.) Furthermore, if the communication of SUSY-breaking takes place at a scale well below the $M_{\rm Pl}$, then the Planckian "corrections" discussed above cannot disrupt the universality ($F_X/M_{\rm Pl} \ll m_Z$).

Can experiments differentiate this type of mediation from supergravity mediation? Perhaps most significantly for phenomenology, these models predict that the lightest SUSY particle will be the gravitino and that other SUSY particles can decay into it with observable lifetimes. Thus, for an interesting portion of the parameter space of these models, the missing-energy signal typical of SUSY models is augmented by two hard photons. However, the rest of their phenomenology is very similar to that of the supergravity models. Readers interested in the details of gauge-mediation should see the contributions of M. Peskin and S. Dimopoulos to this volume.

Section III: Open Questions on SUSY Cosmology

One of the successes of modern physics has been the transformation of cosmology from metaphysics into real physics. We now turn to some issues in which cosmology and particle physics have a particularly strong interplay.

In supersymmetric models with matter-parity conservation, sparticles can interact only in pairs, thereby guaranteeing that the lightest SUSY particle (LSP) is absolutely stable. This provides both an important constraint and the exciting possibility that SUSY may produce stable cosmological relics. It is known that non-luminous matter is needed to explain the rotation curves of galaxies (and galactic clusters). And even larger densities of "dark matter" are needed in order to place the universe in a stable evolutionary trajectory such that its current age and density are not fine-tuned. Such densities are also needed in order to reproduce the otherwise successful predictions of inflation.

However, a number of constraints place strong limitations on the form of the dark matter: nucleosynthesis does not allow very much of the dark matter to be baryons; heavy isotope searches constrain the ability of strongly or electromagnetically interacting dark matter; and structure formation simulations generally rule out neutrinos and other types of "hot" dark matter. Of all the classes of particles, those which remain as good candidates for the dark matter are the so-called WIMP's — Weakly Interacting Massive Particles.

The MSSM provides two ideal candidates for the dark matter, both of them WIMP's: the sneutrino and the neutralino. A detailed discussion of how well each of these particles serves as a dark-matter candidate, as well as a complete list of references, can be found in the contribution of J. Wells to this volume; for now let us simply summarize the results for the neutralino.

The neutralino can be either a good candidate or a bad one, because the neutralino is itself an admixture of the bino, wino, and the higgsinos. Bino-like neutralinos are the most common in realistic models and also provide the best source of dark matter [21]. They can provide reasonable relic densities throughout a broad mass range from tens to hundreds of GeV. Higgsino-like neutralinos, on the other hand, are generally poor dark-matter candidates. They tend to interact too strongly and therefore stay in thermal equilibrium until their densities are depleted. Even if they could somehow provide the galactic dark matter, they are very easily detected by a variety of searches.

What is most inspiring about the possibility of SUSY dark matter is that models of particle physics devised solely to satisfy particle-physics constraints and prejudices nevertheless simultaneously provide a candidate for the long-sought-after dark matter. Furthermore, it may be possible to study the dark-matter candidate both at accelerators and in dark-matter detectors, hopefully verifying that the properties observed in one match those seen in the other. This would close the dark-matter question once and for all.

Going beyond the MSSM, on encounters several more general issues. When global SUSY is broken (at a scale $\sqrt{F}$), there is always a spin-1/2 goldstino state $\widetilde{G}_\alpha$ in the massless spectrum. When SUSY is promoted to a local symmetry (supergravity), the goldstino is eaten by the massless spin-3/2 gravitino. The resulting fermion has mass $m_{3/2} \sim F/M_{\rm Pl}$, "transverse" components which interact with matter gravitationally, and "longitudinal" components which couple derivatively to the SUSY current. As is typical for a Goldstone field [22], $\mathcal{L} = (1/F)\widetilde{G}_\alpha \partial_\mu J^{\alpha\mu}$, where $J^{\alpha\mu}$ is the SUSY current. A lower bound on the gravitino mass is provided by the requirement that $F \gtrsim m_Z^2$ which means $m_{3/2} \gtrsim 10^{-5}\,{\rm eV}$; an upper bound comes from demanding that $F \lesssim m_Z M_{\rm Pl}$, or $m_{3/2} \lesssim 1\,{\rm TeV}$.

In the early universe, gravitinos were in thermal equilibrium with a plasma of visible-sector fields. As the universe cooled, the gravitinos decoupled, locking in their relic density. Calculating the relevant cross-sections and solving the Boltzmann equation allows one to put an upper bound on the mass of a stable gravitino in order to avoid overclosing the universe. If there exists no mechanism for diluting the gravitino densities, one finds [23] that $m_{3/2} \lesssim 2h^2\,{\rm keV}$ where h is the Hubble constant in units of $100\,{\rm km/sec/Mpc}$. On the other hand, if the gravitino is very, very light, then its interactions are quite strong and it can stay in equilibrium below the QCD phase transition. From standard Big Bang nucleosynthesis, we know that the number of neutrino species allowed is $\lesssim 3$; a coupled gravitino behaves as an additional species, violating the bounds. Thus there are stringent limits on the mass of a gravitino.

As with any unwanted relic, the gravitino excess can be diluted by a period of inflation. However, reheating must not produce a new population of gravitinos. This places upper bounds on the reheating temperature T_R.

The detailed limits placed on the gravitino mass and/or cosmological processes near the weak scale have been the source of much recent activity. We will not follow this complicated discussion here but instead refer readers to recent papers which summarize our current understanding[24].

Closely related to the questions of SUSY-breaking implicit in the discussion of the gravitino are those that arise in the "moduli problem." The moduli problem is actually a large class of problems corresponding to the physics of the different moduli which occur in the MSSM and its extensions. The original example (the "Polonyi problem") is provided by that hidden-sector field (the "Polonyi field") which is a gauge singlet, has a nearly flat potential and no renormalizable couplings to other matter, and whose F-VEV breaks SUSY in supergravity models. After receiving a scalar VEV $\sim M_{\rm Pl}$ and F-VEV $\sim m_Z M_{\rm Pl}$, the physical Polonyi (scalar) field Φ emerges as the scalar partner of the goldstino/gravitino and thus has a mass $\sim m_Z$. The VEV of Φ, which sets the natural scale of its oscillations, is much larger than its mass, and so at temperatures far above the weak scale, Φ feels only the potential induced by the (SUSY-breaking) temperature and vacuum energy. In particular, the minimum in which Φ finds itself at high temperature T and large Hubble constant H need not correspond to the minimum of the $T = H = 0$ potential.

Once T and H fall below $\langle\Phi\rangle$, Φ finds itself far from its true minimum and begins to oscillate with amplitude $\sim \sqrt{m_Z M_{\rm Pl}}$. Since it is only weakly coupled to matter, there is little friction to damp the oscillations (*i.e.*, Γ_Φ is small), allowing the oscillations to continue for times approaching minutes. During this time, Φ will dominate the energy density of the universe; if Φ is too long-lived, it will in fact overclose the universe. But even if Φ is not so long-lived, it is the decays of Φ, occurring long after baryogenesis and perhaps even nucleosynthesis, that are the principal concern. After several minutes of slow Φ oscillations, the temperature of the universe is far below the weak scale, but Φ decays are still occurring, dumping entropy into the universe in quantities which are more than sufficient to overdilute the baryon density (and/or dissociate light nuclei) without increasing the temperature enough to restart the original B-violating processes that could replenish it.

The above argument can be generalized and refined by identifying Φ with other moduli of the theory. These include (but are not limited to): string moduli whose VEV's $\sim M_{\rm Pl}$ parametrize the size of compact dimensions; the string dilaton whose VEV $\sim M_{\rm Pl}$ parametrizes the string coupling constant; D- and F-flat directions of the MSSM which have no potential before SUSY-breaking; and Higgs fields responsible for breaking GUT's to the MSSM.

For stringy moduli, the conditions to avoid washing out the baryon density are particularly hard to fulfill. When string moduli begin oscillating, their amplitudes are generally $\sim M_{\rm Pl}$; and their couplings to matter are always suppressed by powers of $m_{3/2}/M_{\rm Pl}$, so that the oscillations are long-lived. Thus, more generally than the case above, one can expect a moduli problem to arise whenever the finite-T, H minimum is displaced from the $T = H = 0$ minimum.

Finally, it would seem natural to use the moduli themselves as the inflatons. We shall not comment on the successes or difficulties in doing so, but leave that discussion for the contribution of L. Randall to this volume.

Section IV: Open Questions on SUSY Grand Unification

One particularly attractive idea for physics at high energy scales concerns the possible appearance of a grand unified theory, or GUT, with a single symmetry group that is large enough to incorporate the $SU(3) \times SU(2) \times U(1)$ gauge group of the Standard Model as a subgroup [25]. The idea of grand unification has a long history independent of SUSY, but once SUSY is included in the picture, a number of new results and predictions arise. We will consider some of these issues in this section.

There are several profound attractions to the idea of grand unification. Perhaps the most obvious is that GUT's have the potential to unify the diverse set of particle representations and parameters found in the MSSM into a single, comprehensive, and hopefully predictive framework. For example, through the GUT symmetry one might hope to explain the quantum numbers of the fermion spectrum, or even the origins of fermion mass. Moreover, by unifying all $U(1)$ generators within a non-abelian theory, GUT's would also provide an explanation for the quantization of electric charge; note that this is a puzzle in the Standard Model due to the abelian $U(1)_Y$ hypercharge group factor whose allowed eigenvalues are arbitrary. Furthermore, because they generally lead to baryon-number violation, GUT's have the potential to explain the cosmological baryon/anti-baryon asymmetry. By combining GUT's with supersymmetry in the context of SUSY GUT's [26], we then hope to realize the attractive features of GUT's simultaneously with those of supersymmetry in a single theory.

There is also one compelling piece of *experimental* evidence for the existence of *supersymmetric* GUT's. It is a straightforward matter to extrapolate the strong, electroweak, and hypercharge gauge couplings of either the Standard Model or the MSSM to higher energies by using their one-loop renormalization group equations[27]. One finds that if this extrapolation is performed within the non-supersymmetric Standard Model, these couplings fail to unify at any scale. However, performing this extrapolation within the context of the supersymmetric MSSM with superpartners near the Z scale, one obtains an apparent gauge coupling unification[28] of the form

$$\frac{5}{3}\,\alpha_Y(M_{\rm GUT}) \;=\; \alpha_2(M_{\rm GUT}) \;=\; \alpha_3(M_{\rm GUT}) \;\approx\; \frac{1}{25} \tag{7}$$

at the scale $M_{\rm GUT} \approx 2 \times 10^{16}$ GeV. This single unified gauge coupling is then easy to interpret as that of a single GUT group $G_{\rm GUT}$. Note that it is the introduction of *supersymmetry* which enables this gauge coupling unification to take place without any further intermediate-scale structure.

While there are *a priori* many choices for such possible groups $G_{\rm GUT}$, the list can be narrowed down by requiring groups of rank ≥ 4 that have complex representations. The smallest possibilities are then $SU(5)$, $SU(6)$, $SO(10)$, and E_6. Amongst these choices, $SO(10)$ is particularly attractive because $SO(10)$ is the smallest simple Lie group for which a single anomaly-free irreducible representation (namely the spinor **16** representation) can accommodate the entire MSSM fermion content of each generation, including a right-handed neutrino. Furthermore, such $SU(5)$-based unification scenarios provide a natural explanation for the normalization factor 5/3 which appears in Eq. (7): this is simply the group-theoretic factor by which the Standard-Model hypercharge generator must be rescaled in order to join with the non-abelian generators into a single $SU(5)$ non-abelian multiplet.

The apparent gauge coupling unification of the MSSM is strong circumstantial evidence in favor of the emergence of a SUSY GUT near 10^{16} GeV. However, GUT theories naturally lead to a variety of outstanding questions whose solutions provide a window into high-scale physics.

Perhaps the most important problem that SUSY GUT's must address is the proton-lifetime problem. In general, GUT's lead to a number of processes that can mediate proton decay. For example, proton decay can be mediated via the off-diagonal $SU(5)$ X gauge bosons that connect quarks to quarks and/or leptons. Such gauge bosons arise, along with the Standard-Model gauge bosons, in the decomposition of the $SU(5)$ adjoint representation, and have fractional electric charges $-4/3$ and $-1/3$ respectively. Because interactions via these X bosons violate baryon number (B) and lepton number (L), these bosons can mediate proton decay via processes such as $p \to \pi^0 e^+$. However, in supersymmetric GUT's this is not the dominant source of proton decay because the X boson must have a mass $M_X \approx M_{\mathrm{GUT}} \approx 2 \times 10^{16}$ GeV. This is two orders of magnitude higher than the expected "unification" scale of non-supersymmetric GUT's. Furthermore, since this decay is gauge-mediated, the contribution to the branching ratio for proton decay via this process goes as $\Gamma \sim g^4 m_p^5 / M_X^4$ where $g \approx 0.7$ is the unified gauge coupling and m_p is the proton mass. The factor M_X^{-4} is a typical suppression for a dimension-six operator, and results in an expected lifetime $\tau(p \to \pi^0 e^+) \approx 10^{36}$ years.

A much more problematic dimension-five contribution arises in supersymmetric GUT's via mediation by colored Higgsino triplets [29]. Because the MSSM requires two electroweak Higgs doublets, and because a minimal $SU(5)$ GUT gauge structure forces these doublets to be part of a larger (*e.g.*, five-dimensional) Higgs representation, the electroweak doublet Higgs will necessarily have a colored triplet Higgs counterpart which contains a fermionic colored Higgsino component. *A priori*, a given $SU(5)$-invariant mass term for this Higgs multiplet will tend to give the same mass to the doublet Higgs as to the triplet Higgs(ino). Therefore, since the electroweak doublet Higgs is expected to have a mass ~ 100 GeV, it is generally quite difficult to give the color triplet Higgs(ino) a large mass. However, a large mass is precisely what we need if we are to avoid rapid proton decay, for this fermionic Higgsino component of the color Higgs triplet can mediate decay processes such as $p \to K^+ \bar{\nu}$. In this case, the branching ratios go as $\Gamma \sim h^4 m_p^5 / M_{\tilde{H}}^2 M_{\mathrm{SUSY}}^2$ where $h \approx 10^{-5}$ is the Higgs(ino) Yukawa coupling to the light generation and M_{SUSY} is the scale of SUSY-breaking. Despite the fact that this branching ratio is Yukawa-suppressed (by the factor of h^4) relative to the dimension-six case, we have only a factor of $M_{\tilde{H}}^{-2}$ mass suppression because the Higgsino mediator is a fermion. Thus, in order to protect against proton decay (and also to preserve gauge coupling unification), the color triplet Higgs must be substantially heavier than the electroweak doublet Higgs. Indeed, in order not to violate current experimental bounds, we must ensure that $\tau(p \to K^+ \bar{\nu}) \gtrsim 10^{32}$ years. This is the problem of "doublet-triplet splitting". Once the doublets and triplets are somehow split, supersymmetric non-renormalization theorems should protect this splitting against radiative corrections.

It is striking that the dominant proton decay mode depends so crucially on whether or not supersymmetry is present. Discovery of the $p \to K^+ \bar{\nu}$ decay mode can thus serve as a clear signal for supersymmetry.

There are a number of potential solutions to the doublet-triplet splitting problem. Proposals include the so-called "sliding singlet" [30] and "missing partner" [31] mechanisms which apply in the case of $SU(5)$, and also a "Higgs-as-pseudo-Goldstone" mechanism [32] which applies in the case of $SU(6)$. Perhaps the most attractive proposal, however, is the "missing VEV" solution for $SO(10)$, originally proposed by Dimopoulos and Wilczek [33]. Making use of the fact that the **45** representation contains two Standard-Model singlets, the Dimopoulos-Wilczek mechanism entails giving a VEV to only one of these singlets, and keeping the other VEV fixed at zero. By carefully selecting which singlet receives the VEV, an effective superpotential term such as $\Phi_{10}\Phi_{45}\Phi'_{10}$ can naturally be used to give a mass to the triplet Higgs but not the doublet Higgs. Constructing a fully consistent $SO(10)$ model in which this mechanism is implemented in a natural way remains an active area of research.

Fermion masses are another generic issue that SUSY GUT's must address. Specifically, by extrapolating the observed fermion Yukawa couplings λ_i up to the unification scale $M_{\rm GUT}$, one finds the approximate relations

$$\lambda_d(M_{\rm GUT}) \approx 3\lambda_e(M_{\rm GUT}) , \quad \lambda_s(M_{\rm GUT}) \approx \frac{1}{3}\lambda_\mu(M_{\rm GUT}) ,$$
$$\lambda_b(M_{\rm GUT}) \approx \lambda_\tau(M_{\rm GUT}) . \tag{8}$$

The issue, then, is to "explain" these relations within the context of a consistent GUT model. We would also like to explain additional features of the fermion mass spectrum, such as the inter-generation mass hierarchy, the masses of the *up*-type quarks, and the (near?)-masslessness of the neutrinos. Reviews of the fermion mass problem within GUT scenarios can be found in Ref. [34].

Certain features are easy to explain. For example, the factors of three that appear in Eq. (8) can be understood, as first suggested by Georgi and Jarlskog [35], as Clebsch-Gordon coefficients of the GUT gauge group (which in turn ultimately stem from the fact that there are three quarks for every lepton). This requires the appearance of certain *textures* (*i.e.*, patterns of zero and non-zero entries) in the fermion mass matrices. Likewise, the inter-generation mass hierarchy might be explained if the first-generation mass terms are of higher dimension in the effective superpotential than those of the second and third generations. Indeed, within $SO(10)$, even small neutrino masses can be accommodated via the see-saw mechanism [36].

The goal, however, is to realize all of these mechanisms simultaneously within the context of a self-consistent supersymmetric GUT model. There are many ways in which such mechanisms can be implemented. For example, judicious use of a **126** representation of $SO(10)$ can give rise to a heavy Majorana right-handed neutrino mass, the proper Georgi-Jarlskog factors of three in the light quark/lepton mass ratios, and GUT symmetry-breaking with automatic R-parity conservation. Use of the **120** and **144** representations can also accomplish some (but not all) of the same goals.

Recently, much attention has focused on deriving GUT models that are consistent with the additional constraints that come from string theory. String theory, in particular, tends to severely restrict not only the GUT representations that might be available for model-building, but also their couplings[37,62]. It turns out that large representations are often entirely excluded, and only very minimal sets of representations and couplings are allowed.

Section V: Open Questions on SUSY and String Theory

In this final section, we will discuss some of the phenomenological connections between supersymmetry and string theory. We will only focus on general themes and basic introductory ideas, since many details will be provided in the subsequent chapters. For an overall introduction to string theory, we recommend Ref.[38]. There are also a number of review articles that deal with more specific aspects of string theory. For example, recent discussions concerning string phenomenology and string model-building can be found in Ref.[39]. Likewise, reviews of methods of supersymmetry-breaking in string theory can be found in Ref.[40], and a review of gauge coupling unification in string theory can be found in Ref.[41]. Note that in these sections we will *not* be discussing some of the more formal aspects of string theory such as string duality; reviews of this topic can be found in Ref.[42].

The fundamental tenet underlying string theory is that all of the known elementary particles and gauge bosons can be realized as the different excitation modes of a single fundamental closed string of size $\sim 10^{-33}$ cm. Thus, within string theory, the physics of zero-dimensional points is replaced by the physics of one-dimensional strings, and likewise one-dimensional worldlines are replaced by two-dimensional worldsheets. There are several profound attractions to this idea. First, one finds that among the excitations of such a string there exists a spin-two massless excitation that is naturally identified as the graviton. Thus string theory is a theory of quantized gravity, which sets the fundamental scale for string theory to be the Planck scale. Second, it turns out that string theory enjoys a measure of finiteness that is not found in ordinary point-particle field theories, and therefore many of the divergences associated with field theory are absent in string theory. Third, it is found that string theory in some sense *requires* gauge symmetry for its internal consistency, and moreover predicts gauge coupling unification. But the most intriguing aspect of string theory may be that it seems to predict supersymmetry. Specifically, the only known method of constructing a string model which has a stable vacuum is to demand that the string model give rise to a supersymmetric spectrum.[b]

---

[b]We emphasize, however, that this does not constitute a proof that the only consistent string theories are those that are supersymmetric. In particular, we shall see that non-supersymmetric string models also have a remarkable degree of finiteness and stability. A

Given the unique role of supersymmetry in string theory, and given that our low-energy world is non-supersymmetric, the next issue that arises is the means by which supersymmetry can be *broken* in string theory. Although there are many different proposals, these can be grouped into essentially three methods: one can break SUSY within perturbative string theory itself (so that one obtains a non-supersymmetric string); one can break SUSY within the low-energy effective field theory derived from a supersymmetric string; and one can break SUSY via a new scenario (the Hořava-Witten scenario) that makes use of certain features of non-perturbative string theory.

Perhaps the most direct way of breaking supersymmetry in string theory is within the full string theory itself. Thus, one obtains a string that has no spacetime supersymmetry at *any* scale, not even the Planck scale. As we stated above, such strings are generally not stable, but it is not known whether there might exist a special subset of non-supersymmetric strings which *are* stable. In many ways, this question is the stringy analogue of the cosmological constant problem: how can one find a non-supersymmetric ground state which preserves a near-exact (if not absolutely exact) cancellation of the cosmological constant? Indeed, in string theory these two questions are actually related in a deep way, and various proposals exist for solving this problem [46].

Breaking supersymmetry within the string theory itself can be done in a variety of ways. In all cases, however, the basic idea is to implement a carefully chosen "twist" when compactifying the string so that all superpartner states (including the gravitinos) suffer so-called "GSO projections" and are removed from the string spectrum. In many (but not all) cases, this method is equivalent to the well-known Scherk-Schwarz mechanism [47] in which supersymmetry is broken through the special dependence that compactified fields have on the coordinates of compactified dimensions. This procedure was introduced into string theory in Ref. [48], and has since been pursued in a number of contexts [49,50,51,52].

Breaking supersymmetry this way offers several distinct advantages. The most important may be that it *preserves the string itself*. Specifically, because this method results in another *string* theory, it preserves the string symmetries (such as modular invariance) that underlie many of the properties of string theory (such as finiteness) that we would like to preserve even after SUSY-breaking. For example, it has been shown that even though spacetime supersymmetry is broken in such scenarios, there is a hidden "misaligned supersymmetry" [49] that remains in the string spectrum. This misaligned supersymmetry tightly constrains the distribution of bosonic and fermionic states throughout the string spectrum in such a way that even though SUSY is bro-

---

recent study of this question, including non-perturbative effects, can be found in Ref. [43].

ken, bosons and fermions nevertheless provide cancelling contributions to string amplitudes, and certain mass supertraces continue to vanish [49,50]. Indeed, the phenomenology of misaligned supersymmetry ensures that these supertraces cancel not in the usual scale-by-scale manner, multiplet-by-multiplet, but rather through subtle simultaneous conspiracies between physics at different energy scales. This may have important phenomenological consequences.

Finally, we remark that even though breaking SUSY through the string itself does not provide supersymmetry at any scale below the Planck scale, this need not be in conflict with gauge coupling unification. Discussions of this point can be found in Refs. [41,51].

The second way of breaking SUSY in string theory is to start with a supersymmetric string at the Planck scale, and then break SUSY within the low-energy effective field theory that is derived from the massless (observable) modes of the string. Since this method is essentially field-theoretic in nature, occurring purely within the language of the effective field theory, it does not necessarily result in a particle spectrum that can be interpreted as the low-energy limit of a non-supersymmetric string. This method therefore presumably breaks some or all of the consistency constraints that underlie string theory, and destroys the fundamental finiteness properties of the string. However, it offers the advantage that a purely field-theoretic treatment of SUSY-breaking will suffice.

Because this method of SUSY-breaking is field-theoretic, all of the SUSY-breaking mechanisms we have outlined in Section II apply to this case as well. The most commonly assumed scenario is that the dynamics of extra "hidden" string sectors will break supersymmetry through some mechanism (*e.g.*, gaugino condensation) which is then communicated to the observable sector through either gravitational or gauge interactions. The ensuing phenomenologies are then analyzed in purely field-theoretic terms.

Finally, there also exists a third scenario for SUSY-breaking within string theory. At strong coupling, it has been proposed [53] that the ten-dimensional $E_8 \times E_8$ heterotic string can be described as the compactification of an *eleven*-dimensional theory known as 'M-theory' on a line segment of finite length ρ. The two E_8 gauge factors are presumed to exist at opposite endpoints of this line segment. In order to incorporate GUT-scale gauge coupling unification within this scenario, it turns out [54] that the length ρ of the eleventh dimension must be substantially larger than the eleven-dimensional Planck length. Thus, one has a situation in which the two E_8 gauge factors communicate primarily with their own ten-dimensional worlds located at opposite ends of an eleven-dimensional bulk, and only gravitational interactions connect these two "ends of the world" with each other.

If we now imagine further compactifying this picture to four dimensions, we obtain a scenario in which one four-dimensional world is the "observable" world that descends from one of the E_8 gauge factors, while the other four-dimensional world represents the "hidden" sector that descends from the other E_8 gauge factor. Since the radii of the additional six-dimensional compactification must be considerably smaller than the length of the eleventh dimension, one obtains an effective situation in which two four-dimensional worlds are connected through a five-dimensional bulk. Most interestingly, if strong-coupling dynamics in the hidden E_8 gauge factor causes SUSY-breaking to occur in that sector (such as via gaugino condensation), the effects of such SUSY-breaking will be communicated gravitationally to the observable world through the five-dimensional interior bulk. This scenario thereby places the question of SUSY-breaking in an entirely new geometric context. For example, in some circumstances the resulting gravitino mass can be identified with the radius ρ of the fifth dimension. Various phenomenological consequences of this picture of SUSY-breaking are currently being explored, in the context of both gaugino condensation and Scherk-Schwarz compactification.

We now turn to another important issue connected with strings and supersymmetry, namely gauge coupling unification. As we have seen, the strong, electroweak, and hypercharge gauge couplings appear to unify at approximately $M_{\rm MSSM} \approx 2 \times 10^{16}$ GeV when extrapolated within the framework of the MSSM. From the point of view of string theory, this is quite compelling, for it turns out that *independently of the existence of a unified gauge symmetry*, heterotic string theories also predict a natural unification of gauge couplings. Indeed, in heterotic string theory, the gauge and gravitational couplings automatically unify[55] to form a single coupling constant $g_{\rm string}$:

$$8\pi \frac{G_N}{\alpha'} \;=\; g_i^2\, k_i \;=\; g_{\rm string}^2 \,. \tag{9}$$

Here G_N is the gravitational (Newton) coupling; α' is the Regge slope (which sets the mass scale for string theory); g_i are the gauge couplings; and the normalization constants k_i are the *affine levels* (also sometimes called *Kač-Moody* levels) at which the different group factors are realized. For non-abelian group factors we have $k_i \in \mathbb{Z}^+$, while for $U(1)$ gauge factors the k_i are arbitrary. Thus, string theory appears to give us precisely the features we want.

There is an important catch, however. The unification scale expected within heterotic string theory is set by α' (which in turn is set by Planck scale), and at the one-loop level one finds[56]:

$$M_{\rm string} \;\approx\; g_{\rm string} \;\times\; 5 \times 10^{17} \text{ GeV} \,. \tag{10}$$

Since extrapolation of low-energy data suggests that $g_{\text{string}} \approx \mathcal{O}(1)$, we thus find that $M_{\text{string}} \approx 5 \times 10^{17}$ GeV — a factor of 20 discrepancy relative to the MSSM prediction! This is a major problem for string phenomenology. Indeed, if we were to start our MSSM running of gauge couplings down from M_{string} rather than from M_{MSSM}, we find that string theory predicts values for these quantities which differ from their experimentally observed values by many standard deviations.

Over the past decade, a number of solutions to this problem have been proposed. We shall here outline only six possible classes of solutions, and summarize their current status. The reader should consult Ref.[41] for a more complete discussion of these and other solutions.

The first solution reconciles M_{MSSM} and M_{string} by assuming that the three low-energy gauge couplings indeed unify at M_{MSSM} because of the presence of a unifying gauge symmetry group G at that scale, whereupon the new unified gauge coupling g_G runs upwards to M_{string} where it unifies with the gravitational coupling. Thus, at the string scale, we are essentially realizing the GUT group G as our gauge symmetry: these are "string GUT models". Because GUT groups such as $SU(5)$ or $SO(10)$ require adjoint or larger representations in order to serve as Higgses for GUT symmetry breaking, it turns out that string GUT models must be realized at high affine levels $k_{\text{GUT}} \geq 2$, and historically it has proven to be a highly non-trivial task to construct such a higher-level string GUT model with three generations[57,58,59]. At present, the three-generation problem has been solved at level two only in the case of $SU(5)$[58,59,60]. At level three, however, there currently exist three-generation models for $SU(5)$, $SU(6)$, $SO(10)$, and E_6[61]. However, much phenomenological analysis of these models still remains to be done. In some cases, these models tend to have extra chiral matter, or unsuitable couplings. Doublet-triplet splitting also remains a problem, and appears to require fine-tuning. There are also rather tight constraints[62] concerning the allowed representations and couplings for these models which restrict their phenomenologies significantly.

The second possible solution to the unification problem makes use of the affine levels k_i that appear in the string unification relation in Eq. (9). Indeed, in string theory, these levels k_i need not take the values $(k_Y, k_2, k_3) = (5/3, 1, 1)$ that we naïvely expect them to have in the MSSM. It is then possible that non-standard values for these levels could alter the runnings in such a way as to reconcile the string unification scale with the MSSM unification scale. This would clearly be a stringy effect. A straightforward analysis[63,64] shows that in order to do the job, the required levels would be:

$$k_2 = k_3 = 1, 2 \; ; \qquad k_Y/k_2 \approx 1.45 - 1.5 \; . \tag{11}$$

Thus, restricting our attention the simpler models with $k_2 = k_3 = 1$, the question arises: can one even realize realistic string models with k_Y in this range? Note that one must always have $k_Y \geq 1$ in any string model containing at least the MSSM spectrum[65].

The current status of this approach is as follows. In general, it is very difficult to arrange to have $k_Y < 5/3$ in string theory[64,59]. However, some self-consistent string models with $k_Y < 5/3$ have been constructed[59]. Unfortunately, all of these models have unwanted fractionally charged states that could survive in their light spectra. This is to be expected, since there is a general result[65] that if a string model is to completely avoid fractionally-charged color-neutral string states, then its affine levels must obey the relation

$$3\,k_Y + 3\,k_2 + 4\,k_3 = 0 \quad (\text{mod } 12)\ . \tag{12}$$

For $k_2 = k_3 = 1, 2$, this implies $k_Y/k_2 \geq 5/3$. Of course, it is possible that fractionally charged states appear but are extremely massive, or that they might bind together into color-neutral objects under the influence of extra hidden-sector interactions. A general classification of the binding scenarios that can eliminate such fractionally charged states has been performed[64], but no string model has yet been constructed which realizes these scenarios.

The third solution to the string unification problem supposes that there can be large "heavy string threshold corrections" at the string scale. These corrections represent the contributions from the infinite towers of massive (Planck-scale) string states that are otherwise neglected in an analysis of the purely massless string spectrum. This would also be an intrinsically stringy effect. In order to reconcile the values of the three low-energy gauge couplings g_i with string-scale unification, it turns out that such corrections Δ_i must have the relative sizes

$$\Delta_{\hat{Y}} - \Delta_2 \approx -28\ , \quad \Delta_{\hat{Y}} - \Delta_3 \approx -58\ , \quad \Delta_2 - \Delta_3 \approx -30\ . \tag{13}$$

where $\hat{Y} \equiv Y/\sqrt{5/3}$ is the renormalized hypercharge. These corrections are quite sizable, and the fundamental question is then how to obtain corrections of this size.

The formalism for calculating these corrections was first derived in Ref.[56] and more recently refined in Ref.[66]. From these results, a number of theoretical mechanisms were identified for making these corrections sufficiently large. Perhaps the most obvious mechanism[67] is to construct a string model with a large modulus (such as a large compactified dimension), for as the size of such a radius is increased, various momentum states become lighter and lighter. The contributions of such states to the threshold corrections therefore become more substantial, ultimately leading to a decompactification of the theory. Unfortunately, it is not known why a given string model should be expected to have such a large modulus. Indeed, the general expectation is that moduli should settle at or near the self-dual point, for which moduli are of order one[68].

Explicit calculations of these threshold corrections have been carried out within several realistic string models[69]. Such models all have $N = 1$ spacetime SUSY; appropriate gauge groups [such as $SU(3) \times SU(2) \times U(1)$, Pati-Salam $SO(6) \times SO(4)$, or flipped $SU(5) \times U(1)$]; the proper massless observable spectrum (including three complete chiral MSSM generations with correct quantum numbers, hypercharges, and Higgs scalar representations); and anomaly cancellation. Unfortunately, the results found within these models are not encouraging: in each of the realistic string models of Ref. [69], it is found[70] that threshold corrections are unexpectedly small, and moreover they have the wrong sign. Thus it seems that threshold corrections by themselves are not able to resolve the discrepancy with the low-energy couplings in these realistic string models. Indeed, despite some interesting proposals[71], there do not presently exist any realistic string models with small moduli for which the threshold corrections are sufficiently large.

A fourth solution to the string unification problem involves "light SUSY thresholds" — the corrections that arise due to the breaking of supersymmetry — and are typically analyzed in field theory. Similar contributions may also come, in principle, from the breaking of intermediate-scale gauge symmetries. Within the context of the low-energy effective theories derived from the realistic string models in Ref. [69], detailed calculations of both of these effects have been performed[70]. The results indicate that the light SUSY thresholds are generally insufficient to resolve the discrepancies, and that the effects of intermediate gauge structure in the realistic string models only *enlarge* the disagreement with experiment! These results serve to illustrate the rather tight (and predictive) constraints that a given string model provides.

A fifth solution involves extra matter beyond the MSSM at intermediate mass scales. While introducing such matter may seem *ad hoc*, it turns out that certain exotic non-MSSM states appear in, and are actually *required for the self-consistency of*, many realistic string models. Such states usually appear in vector-like representations, and ultimately have masses determined by cubic and higher-order terms in the superpotential. In one string model, for example, it has been estimated[72] that such extra states will naturally sit at an intermediate scale $\approx 10^{11}$ GeV. In the realistic string models[69] with $SU(3) \times SU(2) \times U(1)$ gauge groups, such matter typically arises in rather specific $SU(3) \times SU(2) \times U(1)_Y$ representations such as $(\mathbf{3}, \mathbf{2})_{1/6}$, $(\overline{\mathbf{3}}, \mathbf{1})_{1/3}$, $(\overline{\mathbf{3}}, \mathbf{1})_{1/6}$, and $(\mathbf{1}, \mathbf{2})_0$. While the first two representations can be fit into standard $SO(10)$ mutliplets, the remaining two cannot, and are truly exotic.

What is remarkable, however, is that this extra matter is just what is needed: because of their unusual hypercharge assignments, these representations have the potential to modify the running of the $SU(2)$ and $SU(3)$ couplings without seriously affecting the $U(1)$ coupling. Moreover, in some string models, these extra non-MSSM matter representations also appear in precisely the right *combinations* to do the job and resolve the string unification problem. Details can be found in Ref.[70]. Thus, on the basis of this evidence, it appears that extra intermediate-scale matter beyond the MSSM may turn out to be the string-preferred route to string-scale unification. It is remarkable that string theory, which predicts an unexpectedly high unification scale, often also simultaneously predicts precisely the extra exotic matter necessary to reconcile this higher scale with the observed low-energy couplings.

Finally, a sixth solution[54] involves possible effects due to non-perturbative string physics. For example, recent developments in string duality suggest that at strong coupling, the behavior of heterotic strings can be modelled by other theories for which the heterotic string prediction in Eq. (9) is no longer valid. This then loosens the tight constraints between the gauge couplings and the gravitational coupling, which in turn enables one to separate the gauge-coupling unification scale from the gravitationally-determined string scale. An outstanding question in this approach, however, is the development of a mechanism which fixes the separation of these two scales. At present, this issue is not understood, and presumably requires some new non-perturbative dynamics.

References

1. K.R. Dienes and C. Kolda, `hep-ph/9712322`.
2. S. Dimopoulos and D. Sutter, *Nucl. Phys.* **B452** (1995) 496.
3. S. Ferrara, L. Girardello, and F. Palumbo, *Phys. Rev.* **D20** (1979) 403; S. Dimopoulos and H. Georgi, *Nucl. Phys.* **B193** (1981) 150.
4. S. Weinberg, *Phys. Rev.* **D26** (1982) 287.
5. G. Farrar and P. Fayet, *Phys. Lett.* **B76** (1978) 575; S. Dimopoulos, S. Raby, and F. Wilczek, *Phys. Lett.* **B112** (1982) 133.
6. L. Hall, V. Kostelecky, and S. Raby, *Nucl. Phys.* **B267** (1986) 415; H. Georgi, *Phys. Lett.* **B169** (1986) 231.
7. N. Arkani-Hamed and H. Murayama, `hep-ph/9703259`.
8. J. Donoghue, H.P. Nilles, and D. Wyler, *Phys. Lett.* **B128** (1983) 55.
9. F. Gabbiani, E. Gabrielli, A. Masiero, and L. Silvestrini, *Nucl. Phys.* **B477** (1996) 321.
10. Y. Nir and N. Seiberg, *Phys. Lett.* **B309** (1993) 337.

11. Y. Grossman, Y. Nir, and R. Rattazzi, hep-ph/9701231.

12. W. Fischler, S. Paban, and S. Thomas, *Phys. Lett.* **B289** (1992) 373.

13. For a recent work, see: S. Abel, S. Sarkar, and P. White, *Nucl. Phys.* **B454** (1995) 663.

14. J. Bagger, E. Poppitz, and L. Randall, *Nucl. Phys.* **B455** (1995) 59.

15. G. Guidice and A. Masiero, *Phys. Lett.* **B206** (1988) 480.

16. K.S. Babu, C. Kolda, and F. Wilczek, *Phys. Rev. Lett.* **77** (1996) 3070; M. Dine, Y. Nir, and Y. Shirman, *Phys. Rev.* **D55** (1997) 1501; J. Bagger *et al.*, *Phys. Rev.* **D55** (1997) 3188.

17. J. Casas, A. Lleyda, and C. Muñoz, *Nucl. Phys.* **B471** (1996) 3.

18. J. Wess and J. Bagger, *Introduction to Supersymmetry, 2nd Ed.* (Princeton University Press, 1992); H.P. Nilles, *Phys. Rep.* **110** (1984) 1; H. Haber and G. Kane, *Phys. Rep.* **117** (1985) 75.

19. G. Kane, L. Roszkowski, C. Kolda, and J. Wells, *Phys. Rev.* **D49** (1994) 6173 and references therein.

20. For a review, see: C. Kolda, hep-ph/9707450.

21. For a review, see: G. Jungman, M. Kamionkowski, and K. Griest, *Phys. Rep.* **267** (1996) 195.

22. P. Fayet, *Phys. Rep.* **105** (1984) 21.

23. H. Pagels and J. Primack, *Phys. Rev. Lett.* **48** (1982) 223.

24. T. Gherghetta, hep-ph/9712343; F. Zwirner, hep-th/9801003.

25. J. Pati and A. Salam, *Phys. Rev.* **D8** (1973) 1240; H. Georgi and S. Glashow, *Phys. Rev. Lett.* **32** (1974) 438.

26. S. Dimopoulos, S. Raby, and F. Wilczek, *Phys. Rev.* **D24** (1981) 1681; S. Dimopoulos and H. Georgi, *Nucl. Phys.* **B193** (1981) 150; N. Sakai, *Zeit. Phys.* **C11** (1981) 153; L. Ibáñez and G. Ross, *Phys. Lett.* **B105** (1981) 439.

27. H. Georgi, H. Quinn, and S. Weinberg, *Phys. Rev. Lett.* **33** (1974) 451.

28. M. Einhorn and D. Jones, *Nucl. Phys.* **B196** (1982) 475; W. Marciano and G. Senjanovic, *Phys. Rev.* **D25** (1982) 3092.

29. N. Sakai and T. Yanagida, *Nucl. Phys.* **B197** (1982) 533; S. Weinberg, *Phys. Rev.* **D26** (1982) 287.

30. E. Witten, *Phys. Lett.* **B105** (1981) 267; D. Nanopoulos and K. Tamvakis, *Phys. Lett.* **B113** (1982) 151; S. Dimopoulos and H. Georgi, *Phys. Lett.* **B117** (1982) 287; L. Ibáñez and G. Ross, *Phys. Lett.* **B110** (1982) 215.

31. A. Buras *et al.*, *Nucl. Phys.* **B135** (1985) 66; H. Georgi, *Phys. Lett.* **B108** (1982) 283; A. Masiero *et al.*, *Phys. Lett.* **B115** (1982) 380; B. Grinstein, *Nucl. Phys.* **B206** (1982) 387.

32. K. Inoue, A. Kakuto, and T. Takano, *Prog. Theor. Phys.* **75** (1984) 664;

A. Anselm and A. Johansen, *Phys. Lett.* **B200** (1988) 331; Z. Berezhiani and G. Dvali, *Sov. Phys. Lebedeev Inst. Reps.* **5** (1989) 55; R. Barbieri, G. Dvali, and M. Moretti, *Phys. Lett.* **B312** (1993) 137.

33. S. Dimopoulos and F. Wilczek, ITP preprint NSF-ITP-82-07 (1982). For a recent discussion, see: K. Babu and S. Barr, *Phys. Rev.* **D48** (1993) 5354; *Phys. Rev.* **D50** (1994) 3529.
34. S. Raby, hep-ph/9501349; Z. Berezhiani, hep-ph/9602325.
35. H. Georgi and C. Jarlskog, *Phys. Lett.* **B86** (1979) 297.
36. M. Gell-Mann, P. Ramond, and R. Slansky, in *Supergravity* (edited by P. van Nieuwenhuizen and D. Freedman), North-Holland, Amsterdam (1979); T. Yanagida, in *Proceedings of the Workshop on Unified Theories and Baryon Number in the Universe* (edited by A. Sawada and A. Sugamoto), K.E.K. preprint 79-18 (1979); R. Mohapatra and G. Senjanovic, *Phys. Rev. Lett.* **44** (1980) 912.
37. J. Ellis, J. Lopez, and D. Nanopoulos, *Phys. Lett.* **B245** (1990) 375.
38. M. Green, J. Schwarz, and E. Witten, *Superstring Theory, Vols. 1 & 2* (Cambridge University Press, Cambridge, 1987); M. Dine, ed., *String Theory in Four Dimensions* (North-Holland, Amsterdam, 1988); J. Polchinski, hep-th/9411028.
39. L.E. Ibáñez, hep-th/9112050; hep-th/9505098; I. Antoniadis, hep-th/9307002; M. Dine, hep-ph/9309319; J.L. Lopez, *Surveys H.E. Phys.* **8** (1995) 135; J. Lykken, hep-ph/9511456; hep-th/9607144; F. Quevedo, hep-th/9603074; hep-ph/9707434; A. Faraggi, hep-ph/9707311; Z. Kakushadze *et al.*, hep-th/9710149.
40. T. Taylor, hep-ph/9510281; F. Quevedo, hep-th/9511131.
41. K.R. Dienes, *Phys. Reports* **287** (1997) 447 [hep-th/9602045].
42. J. Polchinski, *Prog. Theor. Phys. Suppl.* **123** (1996) 9; *Rev. Mod. Phys.* **68** (1996) 1245; J. Schwarz, hep-th/9607201; A. Sen, hep-th/9609176.
43. J. Blum and K. Dienes, *Phys. Lett.* **B414** (1997) 260; hep-th/9707160.
44. M. Green and J. Schwarz, *Phys. Lett.* **B149** (1984) 117.
45. M. Dine, N. Seiberg, and E. Witten, *Nucl. Phys.* **B289** (1987) 589.
46. See, *e.g.*: G. Moore, *Nucl. Phys.* **B293** (1987) 139; Erratum: *ibid.* **B299** (1988) 847; K.R. Dienes, *Phys. Rev.* **D42** (1990) 2004; *Phys. Rev. Lett.* **65** (1990) 1979; T. Gannon and C. Lam, *Phys. Rev.* **D46** (1992) 1710.
47. J. Scherk and J. Schwarz, *Phys. Lett.* **B82** (1979) 60.
48. R. Rohm, *Nucl. Phys.* **B237** (1984) 553.
49. K.R. Dienes, *Nucl. Phys.* **B429** (1994) 533; hep-th/9409114; hep-th/9505194.
50. K.R. Dienes, M. Moshe, and R. Myers, *Phys. Rev. Lett.* **74** (1995) 4767; hep-th/9506001.

124

51. I. Antoniadis, *Phys. Lett.* **B246** (1990) 377; I. Antoniadis, C. Muñoz, and M. Quirós, *Nucl. Phys.* **B397** (1993) 515.

52. S. Ferrara, C. Kounnas, and M. Porrati, *Nucl. Phys.* **B304** (1988) 500; S. Ferrara *et al.*, *Nucl. Phys.* **B318** (1989) 75.

53. P. Hořava and E. Witten, *Nucl. Phys.* **B460** (1996) 506.

54. E. Witten, *Nucl. Phys.* **B471** (1996) 135; P. Hořava and E. Witten, *Nucl. Phys.* **B475** (1996) 94.

55. P. Ginsparg, *Phys. Lett.* **B197** (1987) 139.

56. V. Kaplunovsky, *Nucl. Phys.* **B307** (1988) 145; Erratum: *ibid.* **B382** (1992) 436.

57. D. Lewellen, *Nucl. Phys.* **B337** (1990) 61; A. Font, L. Ibáñez, and F. Quevedo, *Nucl. Phys.* **B345** (1990) 389; G. Aldazabal *et al.*, *Nucl. Phys.* **B452** (1995) 3; S. Chaudhuri *et al.*, *Nucl. Phys.* **B456** (1995) 89; J. Erler, hep-th/9602032; G. Cleaver, hep-th/9604183.

58. G. Aldazabal *et al.*, *Nucl. Phys.* **B465** (1996) 34.

59. S. Chaudhuri, G. Hockney, and J. Lykken, *Nucl. Phys.* **B469** (1996) 357.

60. D. Finnell, *Phys. Rev.* **D53** (1996) 5781.

61. Z. Kakushadze and S.-H.H. Tye, *Phys. Rev. Lett.* **77** (1996) 2612; *Phys. Lett.* **B392** (1997) 335; hep-ph/9705202.

62. K.R. Dienes and J. March-Russell, *Nucl. Phys.* **B479** (1996) 113; K.R. Dienes, *Nucl. Phys.* **B488** (1997) 141.

63. L. Ibáñez, *Phys. Lett.* **B318** (1993) 73.

64. K.R. Dienes, A. Faraggi, and J. March-Russell, *Nucl. Phys.* **B467** (1996) 44.

65. A. Schellekens, *Phys. Lett.* **B237** (1990) 363.

66. E. Kiritsis and C. Kounnas, *Nucl. Phys.* **B442** (1995) 472.

67. L. Dixon, V. Kaplunovsky, and J. Louis, *Nucl. Phys.* **B355** (1991) 649.

68. A. Font *et al.*, *Phys. Lett.* **B245** (1990) 401; M. Cvetič *et al.*, *Nucl. Phys.* **B361** (1991) 194.

69. I. Antoniadis, J. Ellis, J. Hagelin, and D. Nanopoulos, *Phys. Lett.* **B231** (1989) 65; I. Antoniadis, G. Leontaris, and J. Rizos, *Phys. Lett.* **B245** (1990) 161; A. Faraggi, *Phys. Lett.* **B278** (1992) 131; *Nucl. Phys.* **B387** (1992) 239; *Phys. Lett.* **B302** (1993) 202.

70. K.R. Dienes and A. Faraggi, *Phys. Rev. Lett.* **75** (1995) 2646; *Nucl. Phys.* **B457** (1995) 409.

71. H.-P. Nilles and S. Stieberger, *Phys. Lett.* **B367** (1996) 126.

72. A. Faraggi, *Phys. Rev.* **D46** (1992) 3204.

SOFT SUPERSYMMETRY–BREAKING TERMS FROM SUPERGRAVITY AND SUPERSTRING MODELS

A. BRIGNOLE

Theory Division, CERN, CH-1211 Geneva 23, Switzerland

L.E. IBÁÑEZ and C. MUÑOZ

Departamento de Física Teórica, Universidad Autónoma de Madrid
Cantoblanco, 28049 Madrid, Spain

We review the origin of soft supersymmetry–breaking terms in $N = 1$ supergravity models of particle physics. We first consider general formulae for those terms in general models with a hidden sector breaking supersymmetry at an intermediate energy scale. The results for some simple models are given. We then consider the results obtained in some simple superstring models in which particular assumptions about the origin of supersymmetry breaking are made. These are models in which the seed of supersymmetry breaking is assumed to be originated in the dilaton/moduli sector of the theory.

1 Introduction

A phenomenological implementation of the idea of supersymmetry (SUSY) in the standard model requires the presence of SUSY breaking. There are essentially two large families of models in this context, depending on whether the scale of spontaneous SUSY breaking is high (of order 10^{10}–10^{13} GeV) or low (of order 1–10^2 TeV). We will focus on the former possibility. The latter possibility, considered in other chapters of this book, has only recently received sufficient attention, since it was realized from the very first days of SUSY phenomenology that the existence of certain supertrace constraints in spontaneously broken SUSY theories made the building of realistic models quite complicated. Possible solutions to these early difficulties are discussed elsewhere and we are not going to discuss them further here.

A more pragmatic attitude to the issue of SUSY breaking is the addition of explicit soft SUSY-breaking terms of the appropriate size (of order 10^2–10^3 GeV) in the Lagrangian and with appropriate flavour symmetries to avoid dangerous flavour-changing neutral currents (FCNC) transitions. The problem with this pragmatic attitude is that, taken by itself, lacks any theoretical explanation. Supergravity theories provide an attractive context that can justify such a procedure. Indeed, if one considers the SUSY standard model and couples it to $N = 1$ supergravity, the spontaneous breaking of local SUSY in a hidden sector generates explicit soft SUSY-breaking terms of the required form in the effective low-energy Lagrangian[1,2]. If SUSY is broken at a scale Λ_S, the soft terms have a scale of order Λ_S^2/M_{Planck}. Thus one obtains the required size if SUSY is broken at an intermediate scale $\Lambda_S \sim 10^{10}$ GeV, as mentioned above. Large classes of supergravity models, as we discuss in section 2, give rise to universal soft SUSY-breaking terms, providing for an understanding of FCNC supression. In the last few years it has often been stated in the literature that this class of supergravity models have a flavour-changing problem. We think more appropriate to say that some particular models get interesting *constraints* from FCNC bounds. A generic statement like that seems unjustified, since it is usually based on a strong assumption, i.e. the existence of a region in between the grand unified theory (GUT) scale and the Planck (or superstring) scale in which important flavour non-diagonal renormalization effects take place.

Recently there have been studies of supergravity models obtained in particularly simple classes of superstring compactifications[3]. Such heterotic models have a natural hidden sector built-in: the complex dilaton field S and the complex moduli fields T_i. These gauge singlet fields are generically present in four-dimensional models: the dilaton arises from the gravitational sector of the theory and the moduli parametrize the size and shape of the compactified variety. Assuming that the auxiliary fields of those multiplets are the seed of SUSY breaking, interesting predictions for this simple class of models are obtained. These are reviewed in section 3. The analysis does not assume any specific SUSY-breaking mechanism. We leave section 4 for some final comments and additional references to recent work.

2 Soft terms from supergravity

2.1 General computation of soft terms

The full N=1 supergravity Lagrangian[1] (up to two derivatives) is specified in terms of two functions which depend on the chiral superfields ϕ_M of the theory (denoted by the same symbol as their scalar components): the analytic gauge kinetic function $f_a(\phi_M)$ and the real gauge-invariant Kähler function $G(\phi_M, \phi_M^*)$. f_a determines the kinetic terms for the fields in the vector multiplets and in particular the gauge coupling constant, $Ref_a = 1/g_a^2$. The subindex a is associated with the different gauge groups of the theory since in general $\mathcal{G} = \prod_a \mathcal{G}_a$. For example, in the case of the pure SUSY standard model coupled to supergravity, a would correspond to $SU(3)_c$, $SU(2)_L$, $U(1)_Y$. G is a combination of two functions

$$G(\phi_M, \phi_M^*) = K(\phi_M, \phi_M^*) + \log |W(\phi_M)|^2 \, , \tag{1}$$

where K is the Kähler potential, W is the complete analytic superpotential, and we use from now on the standard supergravity mass units where $M_P \equiv M_{Planck}/\sqrt{8\pi} = 1$. W is related with the Yukawa couplings (which eventually determine the fermion masses) and also includes possibly non-perturbative effects

$$W = \hat{W}(h_m) + \frac{1}{2}\mu_{\alpha\beta}(h_m)C^\alpha C^\beta + \frac{1}{6}Y_{\alpha\beta\gamma}(h_m)C^\alpha C^\beta C^\gamma + \dots , \qquad (2)$$

where we assume two different types of scalar fields $\phi_M = h_m, C^\alpha$: C^α correspond to the observable sector and in particular include the SUSY standard model fields, while h_m correspond to a hidden sector. The latter fields may develop large ($\gg M_W$) vacuum expectation values (VEVs) and are responsible for SUSY breaking if some auxiliary components F^m (see below) develop nonvanishing VEVs. The ellipsis indicates terms of higher order in C^α whose coefficients are suppressed by negative powers of M_P. The second derivative of K determines the kinetic terms for the fields in the chiral supermultiplets and is thus important for obtaining the proper normalization of the fields. Expanding in powers of C^α and $C^{*\overline{\alpha}}$ we have

$$\begin{aligned}
K &= \hat{K}(h_m, h_m^*) + \tilde{K}_{\overline{\alpha}\beta}(h_m, h_m^*)C^{*\overline{\alpha}}C^\beta \\
&\quad + \left[\frac{1}{2}Z_{\alpha\beta}(h_m, h_m^*)C^\alpha C^\beta + h.c.\right] + \dots ,
\end{aligned} \qquad (3)$$

where the ellipsis indicates terms of higher order in C^α and $C^{*\overline{\alpha}}$. Notice that the coefficients $\tilde{K}_{\overline{\alpha}\beta}$, $Y_{\alpha\beta\gamma}$, $\mu_{\alpha\beta}$, and $Z_{\alpha\beta}$ which appear in (2) and (3) may depend on the hidden sector fields in general. The bilinear terms associated with $\mu_{\alpha\beta}$ and $Z_{\alpha\beta}$ are often forbidden by gauge invariance in specific models, but they may be *relevant* in order to solve the so-called μ problem in the context of the minimal supersymmetric standard model (MSSM), as we will discuss below. In this case the two Higgs doublets, which are necessary to break the electroweak symmetry, have opposite hypercharges. Therefore those terms are allowed and may generate both the μ parameter and the corresponding soft bilinear term.

The (F part of the) tree-level supergravity scalar potential, which is crucial to analyze the breaking of SUSY, is given by

$$V(\phi_M, \phi_M^*) = e^G\left(G_M K^{M\bar{N}} G_{\bar{N}} - 3\right) = \left(\bar{F}^{\bar{N}} K_{\bar{N}M} F^M - 3e^G\right) , \qquad (4)$$

where $G_M \equiv \partial_M G \equiv \partial G/\partial\phi_M$ and the matrix $K^{M\bar{N}}$ is the inverse of the Kähler metric $K_{\bar{N}M} \equiv \partial_{\bar{N}}\partial_M K$. We have also written V as a function of the

ϕ_M auxiliary fields, $F^M = e^{G/2}K^{M\bar{P}}G_{\bar{P}}$. When, at the minimum of the scalar potential, some of the hidden sector fields h_m acquire VEVs in such a way that at least one of their auxiliary fields ($\hat{K}^{m\bar{n}}$ is the inverse of the hidden field metric $\hat{K}_{\bar{n}m}$)

$$F^m = e^{G/2}\hat{K}^{m\bar{n}}G_{\bar{n}} \tag{5}$$

is non-vanishing, then SUSY is spontaneously broken and soft SUSY-breaking terms are generated in the observable sector. Let us remark that, for simplicity, we are assuming vanishing D-term contributions to SUSY breaking. When this is not the case, their effects on soft terms can be found e.g. in [4]. The goldstino, which is a combination of the fermionic partners of the above fields, is swallowed by the gravitino via the superHiggs effect. The gravitino becomes massive and its mass

$$m_{3/2} = e^{G/2} \tag{6}$$

sets the overall scale of the soft parameters.

General results

Using the above information, the soft SUSY-breaking terms in the observable sector can be computed. They are obtained by replacing h_m and their auxiliary fields F^m by their VEVs in the supergravity Lagrangian and taking the so-called flat limit where $M_P \to \infty$ but $m_{3/2}$ is kept fixed. Then the non-renormalizable gravity corrections are formally eliminated and one is left with a global SUSY Lagrangian plus a set of soft SUSY-breaking terms. On the one hand, from the fermionic part of the supergravity Lagrangian, soft gaugino masses for the canonically *normalized* gaugino fields can be obtained

$$M_a = \frac{1}{2}\left(Ref_a\right)^{-1}F^m\partial_m f_a , \tag{7}$$

as well as the *un-normalized* Yukawa couplings of the observable sector fermions and the SUSY *un-normalized* masses of some of them (those with bilinear terms either in the superpotential or in the Kähler potential, e.g. the Higgsinos in the case of the MSSM)

$$Y'_{\alpha\beta\gamma} = \frac{\hat{W}^*}{|\hat{W}|}e^{\hat{K}/2}Y_{\alpha\beta\gamma} , \tag{8}$$

$$\mu'_{\alpha\beta} = \frac{\hat{W}^*}{|\hat{W}|}e^{\hat{K}/2}\mu_{\alpha\beta} + m_{3/2}Z_{\alpha\beta} - \overline{F}^{\overline{m}}\partial_{\overline{m}}Z_{\alpha\beta} . \tag{9}$$

On the other hand, scalar soft terms arise from the expansion of the supergravity scalar potential (4)

$$V_{soft} = m'^2_{\overline{\alpha}\beta} C^{*\overline{\alpha}} C^{\beta} + \left(\frac{1}{6} A'_{\alpha\beta\gamma} C^{\alpha} C^{\beta} C^{\gamma} + \frac{1}{2} B'_{\alpha\beta} C^{\alpha} C^{\beta} + h.c. \right) . \tag{10}$$

In the most general case, when hidden and observable sector matter metrics are not diagonal, the *un-normalized* soft scalar masses, trilinear and bilinear parameters are given respectively by[5]

$$m'^2_{\overline{\alpha}\beta} = \left(m^2_{3/2} + V_0 \right) \tilde{K}_{\overline{\alpha}\beta}$$
$$- \overline{F}^{\overline{m}} \left(\partial_{\overline{m}} \partial_n \tilde{K}_{\overline{\alpha}\beta} - \partial_{\overline{m}} \tilde{K}_{\overline{\alpha}\gamma} \tilde{K}^{\gamma\overline{\delta}} \partial_n \tilde{K}_{\overline{\delta}\beta} \right) F^n , \tag{11}$$

$$A'_{\alpha\beta\gamma} = \frac{\hat{W}^*}{|\hat{W}|} e^{\hat{K}/2} F^m \left[\hat{K}_m Y_{\alpha\beta\gamma} + \partial_m Y_{\alpha\beta\gamma} \right.$$
$$\left. - \left(\tilde{K}^{\delta\overline{\rho}} \partial_m \tilde{K}_{\overline{\rho}\alpha} Y_{\delta\beta\gamma} + (\alpha \leftrightarrow \beta) + (\alpha \leftrightarrow \gamma) \right) \right] , \tag{12}$$

$$B'_{\alpha\beta} = \frac{\hat{W}^*}{|\hat{W}|} e^{\hat{K}/2} \left\{ F'^m \left[\hat{K}_m \mu_{\alpha\beta} + \partial_m \mu_{\alpha\beta} \right.\right.$$
$$\left. - \left(\tilde{K}^{\delta\overline{\rho}} \partial_m \tilde{K}_{\overline{\rho}\alpha} \mu_{\delta\beta} + (\alpha \leftrightarrow \beta) \right) \right] - m_{3/2} \mu_{\alpha\beta} \Big\}$$
$$+ \left(2m^2_{3/2} + V_0 \right) Z_{\alpha\beta} - m_{3/2} \overline{F}^{\overline{m}} \partial_{\overline{m}} Z_{\alpha\beta}$$
$$+ m_{3/2} F^m \left[\partial_m Z_{\alpha\beta} - \left(\tilde{K}^{\delta\overline{\rho}} \partial_m \tilde{K}_{\overline{\rho}\alpha} Z_{\delta\beta} + (\alpha \leftrightarrow \beta) \right) \right]$$
$$- \overline{F}^{\overline{m}} F^n \left[\partial_{\overline{m}} \partial_n Z_{\alpha\beta} - \left(\tilde{K}^{\delta\overline{\rho}} \partial_n \tilde{K}_{\overline{\rho}\alpha} \partial_{\overline{m}} Z_{\delta\beta} + (\alpha \leftrightarrow \beta) \right) \right] , \tag{13}$$

where $\tilde{K}^{\alpha\overline{\beta}}$ is the inverse of the observable matter metric $\tilde{K}_{\overline{\beta}\gamma}$. V_0 is the VEV of the scalar potential (4), i.e. the tree-level cosmological constant

$$V_0 = \overline{F}^{\overline{m}} \hat{K}_{\overline{m}n} F^n - 3m^2_{3/2} . \tag{14}$$

It has a bearing on measurable quantities like scalar masses and therefore it is preferable to leave it undetermined (see[6,7,8] for a discussion on this point). Notice that, after normalizing the fields to get canonical kinetic terms, the first piece in (11) will lead to universal diagonal soft masses but the second piece will generically induce *off-diagonal* contributions. Actually, universality is a desirable property not only to reduce the number of independent parameters in SUSY models, but also for phenomenological reasons, particularly to avoid FCNC. The latter is a *low-energy* phenomenological constraint that must be satisfied by any supergravity model. It is worth mentioning in this context that one–loop corrections to the soft parameters (7), (11), (12) and (13) have recently been computed in [9]. They may induce FCNC phenomena even when the tree-level computation gives a universal soft mass. Also a discussion about the loop effects on the contribution of the cosmological constant to the soft terms can be found there. Concerning the A and B parameters, notice that we have not factored out the Yukawa couplings and mass terms respectively as usual, since proportionality is not guaranteed in (12) and (13). Finally, from (5) and (6), one can easily see that the (tree-level) soft parameters in (7), (11) and (12) are generically $\mathcal{O}(m_{3/2})$, as mentioned above. A departure from this result is only possible when some of them vanish. We will mention below some examples of this type, i.e. the case of no-scale supergravity models and special limits of superstring models.

The μ problem

The set of mass parameters in the MSSM Higgs potential includes, besides $\mathcal{O}(m_{3/2}^2)$ soft masses, the square of $\mu'_{\alpha\beta}$ (9) and the B parameter (13), which is $\mathcal{O}(m_{3/2}\mu'_{\alpha\beta})$. In order to have correct electroweak symmetry breaking, the SUSY mass term $\mu'_{\alpha\beta}$ should also be $\mathcal{O}(m_{3/2})$. This is the so-called μ problem[2]. In this respect, notice that $\mu'_{\alpha\beta} = \mathcal{O}(m_{3/2})$ is *naturally* achieved in the presence of a non-vanishing $Z_{\alpha\beta}$ in the Kähler potential[10,11]. The other possible source of the mass $\mu'_{\alpha\beta}$ is the SUSY mass $\mu_{\alpha\beta}$ in the superpotential[12]. This case is more involved since in principle the natural scale of $\mu_{\alpha\beta}$ would be M_P. However, a possible solution can be obtained if the superpotential contains e.g. a non-renormalizable term[11,12]

$$\lambda(h_m)\hat{W}(h_m)H_1 H_2 \,, \tag{15}$$

characterized by the coupling λ, which mixes the observable sector with the hidden sector. Since $m_{3/2} = e^{G/2} = e^{\hat{K}/2}|\hat{W}|$, if that term exists then an effective μ parameter $\mathcal{O}(m_{3/2})$ is generated *dynamically* when h_m acquire VEVs:

$$\mu = \lambda(h_m)\hat{W}(h_m) \,. \tag{16}$$

We should add that both mechanisms to generate $\mu'_{\alpha\beta}$, a bilinear term in the Kähler potential or in the superpotential, could be present simultaneously. Notice also that the two mechanisms are equivalent if Z depends only on h_m (not on h_m^*). Indeed, in that case the supergravity theory is equivalent to the one with a Kähler potential K without the terms $ZH_1 H_2 + h.c.$ and a superpotential $We^{ZH_1 H_2}$, since the G function (1) is the same for both. After expanding the exponential, the superpotential will have a contribution

$Z\hat{W}H_1H_2$, i.e. a term of the type (15). Finally, let us mention that several new sources of the μ term due to loop effects on (9), which are naturally of order the weak scale, have recently been computed in [9].

We recall that the solutions mentioned here in order to solve the μ problem are *naturally* present in superstring models. For instance, in large classes of superstring models the Kähler potential does contain bilinear terms analytic in the observable fields as in (3), with specific coefficients $Z_{\alpha\beta}$ [13,14,15], so that a μ parameter may be naturally generated. Concerning superpotential contributions, we recall that a 'direct' $\mu H_1 H_2$ term in W (2) is naturally absent (otherwise the natural scale for μ would be M_P), since in supergravity models deriving from superstring theory mass terms for light fields are forbidden in the superpotential by scale invariance of the theory. However, the superpotential (2) may well contain an 'effective' $\mu H_1 H_2$ term, e.g. a term of the type (15) [11,15] induced by non-perturbative SUSY-breaking mechanisms like gaugino-squark condensation in the hidden sector.

The low-energy spectrum

The results (7), (8), (9), (11), (12) and (13) should be understood as being valid at some high scale $\mathcal{O}(M_P)$ and the standard RGEs must be used to obtain the low-energy values. Although the SUSY spectrum will depend in general on the details of $SU(2)_L \times U(1)_Y$ breaking, there are several particles whose mass is rather independent of those details and is mostly given by the boundary conditions and the renormalization group running. In particular, neglecting all Yukawa couplings except the one of the top, that is the case of the gluino g, all the squarks (except stops and left sbottom) $Q_L = (u_L, d_L), u_L^c, d_L^c$ and all the sleptons $L_L = (v_L, e_L), e_L^c$. For all these particles one can write explicit expressions for the masses in terms of the soft parameters (after normalizing the fields to get canonical kinetic terms). For instance, assuming that gauginos have a common initial mass (e.g. due to a universal f function) and that there is nothing but the MSSM in between the weak scale and the Planck scale, one obtains the approximate numerical expressions:

$$
\begin{aligned}
M_g^2(M_Z) &\simeq 9.8\, M^2\,, \\
m_{Q_L}^2(M_Z) &\simeq m_{Q_L}^2 + 8.3\, M^2\,, \\
m_{u_L^c, d_L^c}^2(M_Z) &\simeq m_{u_L^c, d_L^c}^2 + 8\, M^2\,, \\
m_{L_L}^2(M_Z) &\simeq m_{L_L}^2 + 0.7\, M^2\,, \\
m_{e_L^c}^2(M_Z) &\simeq m_{e_L^c}^2 + 0.23\, M^2\,,
\end{aligned}
\tag{17}
$$

where the second term in the expression of the scalar masses is the effect of gaugino loop contributions. In the above formulae we have neglected the scalar potential D-term contributions, which are normally small compared to the terms above, and the contribution of the $U(1)_Y$ D-term in the RGEs of scalar masses. These may be found e.g. in [16].

2.2 Supergravity models

We now specialize the above general discussion to the case of supergravity models where the observable (here MSSM) matter fields have diagonal metric:

$$\tilde{K}_{\overline{\alpha}\beta}(h_m, h_m^*) = \delta_{\overline{\alpha}\beta}\tilde{K}_\alpha(h_m, h_m^*) \; . \tag{18}$$

This possibility is particularly interesting due to its simplicity and also for phenomenological reasons related to the absence of FCNC in the effective low-energy theory (see [6,17,18,19,20,9] for a discussion on this point). Besides, the supergravity models that will be studied below correspond to this situation. Then the Kähler potential (3), to lowest order in the observable fields C^α, and the superpotential (2) have the form

$$
\begin{aligned}
K &= \hat{K}(h_m, h_m^*) + \tilde{K}_\alpha(h_m, h_m^*)C^{*\overline{\alpha}}C^\alpha + [Z(h_m, h_m^*)H_1 H_2 + h.c.] & (19) \\
W &= \hat{W}(h_m) + \mu(h_m)H_1 H_2 + \sum_{generations} [Y_u(h_m)Q_L H_2 u_L^c \\
&\quad + Y_d(h_m)Q_L H_1 d_L^c + Y_e(h_m)L_L H_1 e_L^c] \; , & (20)
\end{aligned}
$$

where $C^\alpha = Q_L, u_L^c, d_L^c, L_L, e_L^c, H_1, H_2$, and we have taken for simplicity diagonal Yukawa couplings ($Y_{\alpha\beta\gamma} = Y_u, Y_d, Y_e$, in a self-explanatory notation). Now the form of the effective soft Lagrangian obtained from (7) and (10) is given by

$$
\begin{aligned}
\mathcal{L}_{soft} &= \frac{1}{2}(M_a \hat{\lambda}^a \hat{\lambda}^a + h.c.) - m_\alpha^2 \hat{C}^{*\overline{\alpha}}\hat{C}^\alpha \\
&\quad - \left(\frac{1}{6}A_{\alpha\beta\gamma}\hat{Y}_{\alpha\beta\gamma}\hat{C}^\alpha\hat{C}^\beta\hat{C}^\gamma + B\hat{\mu}\hat{H}_1\hat{H}_2 + h.c. \right) \; , & (21)
\end{aligned}
$$

with

$$m_\alpha^2 = \left(m_{3/2}^2 + V_0\right) - \overline{F}^{\overline{m}}F^n \partial_{\overline{m}}\partial_n \log \tilde{K}_\alpha \; , \tag{22}$$

$$A_{\alpha\beta\gamma} = F^m \left[\hat{K}_m + \partial_m \log Y_{\alpha\beta\gamma} - \partial_m \log(\tilde{K}_\alpha \tilde{K}_\beta \tilde{K}_\gamma) \right] \; , \tag{23}$$

$$B = \hat{\mu}^{-1}(\tilde{K}_{H_1}\tilde{K}_{H_2})^{-1/2} \left\{ \frac{\hat{W}^*}{|\hat{W}|}e^{\hat{K}/2}\mu \left(F^m \left[\hat{K}_m + \partial_m \log \mu \right. \right. \right.$$

$$\left. - \partial_m \log(\tilde{K}_{H_1}\tilde{K}_{H_2})\right] - m_{3/2}\right)$$
$$+ \left(2m_{3/2}^2 + V_0\right) Z - m_{3/2}\overline{F^m}\partial_{\overline{m}}Z$$
$$+ m_{3/2}F^m \left[\partial_m Z - Z\partial_m \log(\tilde{K}_{H_1}\tilde{K}_{H_2})\right]$$
$$\left. - \overline{F^m}F^n \left[\partial_{\overline{m}}\partial_n Z - \partial_{\overline{m}}Z\partial_n \log(\tilde{K}_{H_1}\tilde{K}_{H_2})\right]\right\} , \tag{24}$$

where $\widehat{C}^\alpha$ and $\widehat{\lambda}^a$ are the scalar and gaugino canonically *normalized* fields respectively

$$\widehat{C}^\alpha = \tilde{K}_\alpha^{1/2}C^\alpha , \tag{25}$$
$$\widehat{\lambda}^a = (Ref_a)^{1/2}\lambda^a , \tag{26}$$

and the rescaled Yukawa couplings and μ parameter

$$\widehat{Y}_{\alpha\beta\gamma} = Y_{\alpha\beta\gamma}\frac{\hat{W}^*}{|\hat{W}|}e^{\hat{K}/2}(\tilde{K}_\alpha\tilde{K}_\beta\tilde{K}_\gamma)^{-1/2} , \tag{27}$$
$$\widehat{\mu} = \left(\frac{\hat{W}^*}{|\hat{W}|}e^{\hat{K}/2}\mu + m_{3/2}Z - \overline{F^m}\partial_{\overline{m}}Z\right)(\tilde{K}_{H_1}\tilde{K}_{H_2})^{-1/2} , \tag{28}$$

have been factored out in the A and B terms as usual.

Now we are ready to study specific supergravity models. As follows from the above discussion, the particular values of the soft parameters depend on the type of supergravity theory from which the MSSM derives and, in general, on the mechanism of SUSY breaking (through the presence of $\hat{W}(h_m)$ in $m_{3/2}$ and F terms). However, it is still possible to learn things about soft parameters without knowing all the details of SUSY breaking. In order to show this, let us consider two simple and interesting supergravity models studied extensively in the literature: minimal supergravity and no-scale supergravity.

i) *Minimal supergravity*

This model corresponds to use the form of K that leads to minimal (canonical) kinetic terms in the supergravity Lagrangian, namely

$$\tilde{K}_\alpha(h_m, h_m^*) = 1 \tag{29}$$

in (19). Then, irrespective of the SUSY-breaking mechanism, the scalar masses and the A, B parameters can be straightforwardly computed using (22), (23) and (24)

$$m_\alpha^2 = m_{3/2}^2 + V_0 , \tag{30}$$

$$A_{\alpha\beta\gamma} = F^m \left(\hat{K}_m + \partial_m \log Y_{\alpha\beta\gamma} \right) , \tag{31}$$

$$
\begin{aligned}
B = \hat{\mu}^{-1} \Bigg\{ &\frac{\hat{W}^*}{|\hat{W}|} e^{\hat{K}/2} \mu \left[F^m \left(\hat{K}_m + \partial_m \log \mu \right) - m_{3/2} \right] \\
&+ \left(2m_{3/2}^2 + V_0 \right) Z + m_{3/2} \left(F^m \partial_m Z - \overline{F^m} \partial_{\overline{m}} Z \right) \\
&- \overline{F^m} F^n \partial_{\overline{m}} \partial_n Z \Bigg\} ,
\end{aligned}
\tag{32}
$$

where

$$\hat{\mu} = \frac{\hat{W}^*}{|\hat{W}|} e^{\hat{K}/2} \mu + m_{3/2} Z - \overline{F^m} \partial_{\overline{m}} Z . \tag{33}$$

Notice that the scalar masses are automatically *universal* in this model. Further simplifications can be obtained if $Z = 0$ and if the superpotential parameters $Y_{\alpha\beta\gamma}$ and μ do not depend on the hidden sector fields. Under such assumptions, which are common in the literature, (31) and (32) generate universal A parameters, as well as the relation

$$B = A - m_{3/2} . \tag{34}$$

Furthermore, if we assume $V_0 = 0$, then $m \equiv m_\alpha = m_{3/2}$ and the well known result for the B parameter, $B = A - m$, is recovered. This supergravity model is attractive for its simplicity and for the natural explanation that it offers to the universality of the soft scalar masses.

We remark that although minimal (canonical) kinetic terms for hidden matter, $\hat{K}(h_m, h_m^*) = \sum_m h_m h_m^*$, are also usually assumed, we have seen that it is not a necessary condition in order to obtain the above results. Concerning the kinetic terms for vector multiplets, it can be seen from (7) that the minimal (canonical) choice $f_a = const.$ is not phenomenologically interesting, since it implies $M_a = 0$. Nonvanishing and universal gaugino masses can be obtained if all the f_a have the same dependence on the hidden sector fields, i.e. $f_a(h_m) = c_a f(h_m)$ for the different gauge group factors of the theory. This is in fact what happens, at tree level, in supergravity models deriving from superstring theory, as we will see in the next section. As an additional comment, we stress that relation (34) depends on the particular mechanism that is used to generate the μ parameter. As a counter-example, notice that if one takes e.g. an h_m-dependent μ as in (16) with $\lambda = const.$, instead of taking $\mu = const.$, then (32) gives

$$B = 2m_{3/2} + \frac{V_0}{m_{3/2}} , \tag{35}$$

with $\hat{\mu} = m_{3/2}\lambda$ from (33). Thus the relation (34) does not hold. The above result (35) can be obtained also if one takes $\mu = 0$ in the superpotential (20) and $Z = const.$ in the Kähler potential (19). This also follows from our discussion above about the equivalence between the two mechanisms when Z is an analytic function.

ii) *No-scale supergravity*

In no-scale supergravity models [21], after the spontaneous breaking of SUSY, the tree-level potential vanishes identically along some directions. A simple example of this type of models has just one hidden field h, a Kähler potential (19) with

$$\hat{K} = -3\,\log(h + h^*)\,, \quad \tilde{K}_\alpha = (h + h^*)^{-1}\,, \tag{36}$$

and a superpotential (20) with a hidden field independent $\hat{W}$, i.e. $\hat{W} = const.$ Then, the attractive result of a *vanishing* (flat) tree-level effective potential for the hidden sector (14) is obtained

$$V_0 = 0\,, \tag{37}$$

for all VEVs of h. On the other hand, the soft parameters, using (22), (23) and (24), are given by

$$m_\alpha^2 \;=\; 0\,, \tag{38}$$

$$A_{\alpha\beta\gamma} \;=\; -m_{3/2}(h + h^*)\partial_h \log Y_{\alpha\beta\gamma}\,, \tag{39}$$

$$B \;=\; -\hat{\mu}^{-1} m_{3/2}(h + h^*)^2 \left\{ \frac{\hat{W}^*}{|\hat{W}|}(h + h^*)^{-3/2}\partial_h \mu \right.$$
$$\left. + m_{3/2}\left[\partial_{h^*} Z + \partial_h Z + (h + h^*)\partial_h \partial_{h^*} Z\right] \right\}\,, \tag{40}$$

where

$$\hat{\mu} \;=\; (h + h^*)\left[\frac{\hat{W}^*}{|\hat{W}|}(h + h^*)^{-3/2}\mu + m_{3/2}Z + m_{3/2}(h + h^*)\partial_{h^*} Z\right]\,. \tag{41}$$

Assuming now that the μ and Z coefficients and the Yukawa couplings are hidden field independent, the well known result for the soft parameters is recovered:

$$m_\alpha = A_{\alpha\beta\gamma} = B = 0\,. \tag{42}$$

Although the above parameters are vanishing at the high scale, gaugino masses (7) can induce non-vanishing values at the electroweak scale due to radiative corrections.

In conclusion, both supergravity models considered in this section are interesting and give rise to concrete predictions for the soft parameters. However, one can think of many possible supergravity models (with different K, W and f) leading to *different* results for the soft terms. This arbitrariness, as we will see in the next section, can be ameliorated in supergravity models deriving from superstring theory, where K, f, and the hidden sector are more constrained. We can already anticipate, however, that in such a context the kinetic terms are generically *not* canonical. Besides, although Kähler potentials of the no-scale type may appear at tree-level, the superpotentials are in general hidden field *dependent*. Moreover, the Yukawa couplings $Y_{\alpha\beta\gamma}$ and the bilinear coefficients μ and Z are also generically hidden field *dependent*.

Finally, we remark that further constraints on the soft parameter space of the MSSM can be obtained if one wishes to avoid low-energy charge and color breaking minima deeper than the standard vacuum[22]. On these grounds, and assuming also radiative symmetry breaking with nothing but the MSSM in between the weak scale and the Planck scale, e.g. large regions in the parameter space (m, M, A, B) of the minimal supergravity model i) are forbidden. In the limiting case $m = 0$ the whole parameter space turns out to be excluded. This has obvious implications, e.g. for the no-scale supergravity model ii). If the same kind of analysis is applied to the soft parameters of superstring models, again strong constraints can be obtained, as we will comment below.

3 Soft terms from superstring theory

3.1 General parametrization of SUSY breaking

We are going to consider N=1 four-dimensional superstrings where the rôle of hidden sector fields is effectively played by r moduli fields T_i, $i = 1, ..., r$ and the dilaton field S, i.e. $h_m = S, T_i$ following the notation of the previous section. We recall that we are denoting the T- and U-type (Kähler class and complex structure in the Calabi-Yau language) moduli collectively by T_i. The associated effective N=1 supergravity Kähler potentials (3), to lowest order in the matter fields, are of the type:

$$K = \hat{K}(S, S^*, T_i, T_i^*) + \tilde{K}_{\overline{\alpha}\beta}(T_i, T_i^*)C^{*\overline{\alpha}}C^\beta$$
$$+ \left[\frac{1}{2}Z_{\alpha\beta}(T_i, T_i^*)C^\alpha C^\beta + h.c.\right], \tag{43}$$

where at the superstring tree level

$$\hat{K}(S, S^*, T_i, T_i^*) = -\log(S + S^*) + \hat{K}(T_i, T_i^*). \tag{44}$$

The first piece in (44) is the usual term corresponding to the complex dilaton S that is present for any compactification. The second piece is the Kähler potential of the moduli fields, which in general depends on the compactification scheme and can be a complicated function. For the moment we leave it generic, but in the next subsection we will analyze some specific classes of superstring models where it has been computed. The same comment applies to $\tilde{K}_{\bar{\alpha}\beta}(T_i, T_i^*)$ and $Z_{\alpha\beta}(T_i, T_i^*)$. In the case of the superpotential (2), $Y_{\alpha\beta\gamma}(T_i)$ is also independent of S, but the non-perturbative contributions $\hat{W}(S, T_i)$ and $\mu_{\alpha\beta}(S, T_i)$ may depend in general on both S and T_i. Finally, for any four-dimensional superstring the tree-level gauge kinetic function is independent of the moduli sector and is simply given by

$$f_a = k_a S , \tag{45}$$

where k_a is the Kac–Moody level of the gauge factor. Usually (level one case) one takes $k_3 = k_2 = \frac{3}{5}k_1 = 1$, but this is irrelevant for our tree-level computation since k_a will not contribute to the soft parameters.

As we will show below, it is important to know what fields, either S or T_i, play the predominant role in the process of SUSY breaking. This will have relevant consequences in determining the pattern of soft parameters, and therefore the spectrum of physical particles[6]. That is why it is very useful to introduce the following parametrization, consistent with (14), for the VEVs of dilaton and moduli auxiliary fields

$$F^S = \sqrt{3}C m_{3/2} K_{\bar{S}S}^{-1/2} \sin\theta\, e^{-i\gamma_S} ,$$
$$F^i = \sqrt{3}C m_{3/2} \cos\theta\, P^{i\bar{j}}\Theta_{\bar{j}} , \tag{46}$$

where the constant C is defined as follows

$$C^2 = 1 + \frac{V_0}{3m_{3/2}^2} . \tag{47}$$

This parametrization is valid for the general case of off-diagonal moduli metric, since P is a matrix canonically normalizing the moduli fields, i.e. $P^\dagger \hat{K} P = 1$ where $\hat{K} \equiv \hat{K}_{\bar{i}j}$ and 1 stands for the unit matrix. The angle θ and the complex parameters $\Theta_{\bar{j}}$ just parametrize the direction of the goldstino in the S, T_i field space (see below (5)) and $\sum_j \Theta_j^* \Theta_{\bar{j}} = 1$. We have also allowed for the possibility of some complex phases which could be relevant for the CP structure of the theory (see [6,23,17,18,20,24] for a discussion on this point). Notice that if the tree-level cosmological constant V_0 is assumed to vanish, one has $C = 1$, but we prefer for the moment to leave it undetermined as we did in the previous section (see below (14)).

138

Notice that such a phenomenological approach allows us to 'reabsorb' (or circumvent) our ignorance about the (nonperturbative) S- and T_i- dependent part of the superpotential (2), $\hat{W}(S, T_i)$, which is responsible for SUSY breaking.

It is now a straightforward exercise, plugging (43), (44), (45) and (46) into (7), (11), (12) and (13), to compute the soft SUSY-breaking parameters as functions of θ and $\Theta_{\bar{j}}$. On the one hand, since the tree-level gauge kinetic function is given for any four-dimensional superstring by (45), the tree-level gaugino masses are universal, independent of the moduli sector, and simply given by:

$$M_a \;=\; \sqrt{3}Cm_{3/2}\sin\theta e^{-i\gamma s} \;. \tag{48}$$

On the other hand, the bosonic soft parameters depend in general on the moduli sector (i.e. on the functions $\tilde{K}_{\bar{\alpha}\beta}, Z_{\alpha\beta}(T_i, T_i^*), \ldots$ and on the parameters $\cos\theta$ and $\Theta_{\bar{j}}$) and therefore they should be studied in the context of specific classes of superstring models. However, we will first focus on the very interesting limit $\cos\theta = 0$, which corresponds to the case where the dilaton sector is the source of all the SUSY breaking (see (46)) and the results are compactification independent.

Dilaton SUSY breaking

Since the dilaton couples in a universal manner to all particles, this limit is quite model *independent*[13,6]. Indeed, the expressions for all the soft parameters (except B) are quite simple and independent of the four-dimensional superstring considered. After canonically normalizing the fields, one obtains:

$$m_\alpha^2 \;=\; m_{3/2}^2 + V_0 \;, \tag{49}$$

$$M_a \;=\; \sqrt{3}Cm_{3/2}e^{-i\gamma s} \;, \tag{50}$$

$$A_{\alpha\beta\gamma} \;=\; -M_a \;, \tag{51}$$

$$B \;=\; \hat{\mu}^{-1}(\tilde{K}_{H_1}\tilde{K}_{H_2})^{-1/2}\left\{\frac{\hat{W}^*}{|\hat{W}|}e^{\hat{K}/2}\mu m_{3/2}\,(-1\right.$$

$$\left. -\sqrt{3}Ce^{-i\gamma s}\,[1-(S+S^*)\,\partial_S\log\mu]) + Z\left(2m_{3/2}^2 + V_0\right)\right\} \tag{52}$$

where

$$\hat{\mu} \;=\; \left(\frac{\hat{W}^*}{|\hat{W}|}e^{\hat{K}/2}\mu + m_{3/2}Z\right)(\tilde{K}_{H_1}\tilde{K}_{H_2})^{-1/2} \;. \tag{53}$$

Although the general expression for B is more involved than the ones of the other soft parameters, a considerable simplification occurs if Z is the only source of the μ term. In this case B reduces to

$$B \;=\; 2m_{3/2} \;+\; \frac{V_0}{m_{3/2}} \,. \tag{54}$$

and thus becomes independent of the four-dimensional superstring considered, as the other parameters. It is easy to check that the same result (54) is also obtained if $Z = 0$ and the superpotential contains a μ coefficient of the form (16), where now $\mu = \lambda(T_i)\hat{W}(S,T_i)$. Notice that the expressions for m_α (49) and B (54) coincide with the corresponding ones obtained in the minimal supergravity model i), (30) and (35) respectively. Furthermore, if $Z = 0$ and $\partial_S \mu = 0$, the expression for B obtained from (52) coincides with the corresponding one (34) of the minimal supergravity model.

This dilaton-dominated scenario is attractive for its simplicity and for the natural explanation that it offers to the *universality* of the soft terms. For possible explicit SUSY–breaking mechanisms where this limit might be obtained see [25]. Because of the simplicity of this scenario, the low-energy predictions are quite precise [26,6,27,28]. Assuming a vanishing cosmological constant and imposing, e.g. from the limits on the electric dipole moment of the neutron, $\gamma_S = 0 \bmod \pi$ (49), (50) and (51) give[a]

$$m_\alpha \;=\; m_{3/2} \,, \quad M_a \;=\; \pm\sqrt{3}\,m_{3/2} \,, \quad A_{\alpha\beta\gamma} \;=\; -M_a \,. \tag{55}$$

Since scalars are lighter than gauginos at the high scale and all mass ratios are fixed, at low-energy ($\sim M_Z$) one finds the following mass ratios for the gluino, slepton and squark (except stops and left sbottom) masses

$$M_g : m_{Q_L} : m_{u_L^c} : m_{d_L^c} : m_{L_L} : m_{e_L^c} \simeq 1 : 0.94 : 0.92 : 0.92 : 0.32 : 0.24 \,, \tag{56}$$

as can be computed e.g. from (17). Although squarks and sleptons have the same soft mass at the high scale, at low-energy the former are much heavier than the latter because of the gluino contribution to the renormalization of their masses. The rest of the spectrum is very dependent on the details of $SU(2)_L \times U(1)_Y$ breaking and therefore on the values of B and $\hat{\mu}$. For $B = 2m_{3/2}$ (see(54)) and μ treated as a free parameter this analysis can be found in [26]. Modifications to this scenario due to the effect of possible superstring non–perturbative corrections to the Kähler potential can be found in [30].

It is worth noticing here that, although the value of $\hat{\mu}$ (53) is compactification dependent even in this dilaton-dominated scenario, the simple result

---

[a]It is worth remarking that these particular boundary conditions have also interesting finiteness properties. In particular, they preserve one-loop finiteness of $N = 1$ finite theories [29].

140

$\hat{\mu} = m_{3/2}$ can be obtained in any compactification scheme where the source of $\hat{\mu}$ is a Z term in the Kähler potential fulfilling the property $Z = (\tilde{K}_{H_1}\tilde{K}_{H_2})^{1/2}$. In fact, we will see in the next subsection that this is the case of some classes of orbifold models. Notice that, when such a property holds, the whole SUSY spectrum depends only on one parameter, $m_{3/2}$, since

$$m_\alpha = m_{3/2} \ , \ M_a = \pm\sqrt{3}m_{3/2} \ , \ A_{\alpha\beta\gamma} = -M_a \ , \ B = 2m_{3/2} \ , \ \hat{\mu} = m_{3/2} \ . \quad (57)$$

Besides, this parameter can be fixed from the phenomenological requirement of correct electroweak breaking $2M_W^2/g_2^2 = \langle|H_1|^2\rangle + \langle|H_2|^2\rangle$. Thus at the end of the day we are left essentially with no free parameters. In [31] the consistency of the above boundary conditions with the appropriate radiative electroweak symmetry breaking was explored. Unfortunately, it was found that they are not consistent with the measured value of the top-quark mass, namely the mass obtained in this scheme turns out to be too small. A possible way-out to this situation is to assume that also the moduli fields contribute to SUSY breaking, since the soft terms are then modified. Of course, this amounts to a departure of the pure dilaton-dominated scenario. This possibility will be discussed in the context of orbifold models in the next subsection.

Finally, we recall that the phenomenological problem of the pure dilaton-dominated limit mentioned above is also obtained in a different context, namely from requiring the absence of low-energy charge and color breaking minima deeper than the standard vacuum [32]. In fact, on these grounds, the dilaton-dominated limit is excluded not only for a μ term generated through the Kähler potential but for any possible mechanism solving the μ problem. The results indicate that the whole free parameter space $(m_{3/2}, B, \mu)$ is excluded after imposing the present experimental data on the top mass. Again this rests on the assumption of radiative symmetry breaking with nothing but the MSSM in between the weak scale and the superstring scale.

Dilaton/Moduli SUSY breaking

In general the moduli fields T_i may also contribute to SUSY breaking, i.e. $F^i \neq 0$ in (46), and therefore their effects on soft parameters must also be included [6,8,33,19,34]. In this sense it is interesting to note that explicit possible scenarios of SUSY breaking by gaugino condensation in superstrings, when analyzed at the one–loop level, lead to the mandatory inclusion of the moduli in the game (in fact the moduli are the main source of SUSY breaking in these cases) [35]. Since different compactification schemes give rise to different expressions for the moduli-dependent part of the Kähler potential (43), the computation of the bosonic soft parameters will be model *dependent*. The results are discussed below in the context of some specific superstring models.

3.2 Superstring models

To illustrate the main features of mixed dilaton/moduli SUSY breaking, we will concentrate mainly on the case of diagonal moduli and matter metrics. For instance, under this assumption the parametrization (46) is simplified to

$$
\begin{aligned}
F^S &= \sqrt{3}Cm_{3/2}\hat{K}_{\overline{S}S}^{-1/2}\sin\theta e^{-i\gamma_S} , \\
F^i &= \sqrt{3}Cm_{3/2}\hat{K}_{\overline{i}i}^{-1/2}\cos\theta\,\Theta_i e^{-i\gamma_i} ,
\end{aligned}
\tag{58}
$$

where $\sum_i \Theta_i^2 = 1$. Although this is the generic case e.g. in most orbifolds, off–diagonal metrics are present in general in Calabi–Yau compactifications. This may lead to FCNC effects in the low–energy effective N=1 softly broken Lagrangian. The analysis of soft SUSY-breaking parameters in Calabi–Yau compactifications is therefore more involved and can be found in [36] using parametrization (46). A similar analysis for the few orbifolds with off–diagonal metrics was carried out in [34]. Some comments about the "off-diagonal" results will be made below. Also in the case of orbifold compactifications with continuous Wilson lines off-diagonal moduli metrics arise, due to the moduli–Wilson line mixing. However, this analysis turns out to be simple [37] and the results are similar to the ones studied below in the diagonal case.

Since the moduli part of the Kähler potential (43) has been computed for $(0,2)$ symmetric Abelian orbifolds, we will concentrate here on these models. They contain generically three T-type moduli (the exceptions are the orbifolds Z_3, Z_4 and Z_6', which have 9, 5 and 5 respectively, and are precisely the ones with off-diagonal metrics) and, at most, three U-type moduli. We will denote them collectively by T_i, where e.g. $T_i = U_{i-3}$; $i = 4, 5, 6$. For this class of models the Kähler potential has the form

$$
K = -\log(S + S^*) - \sum_i \log(T_i + T_i^*) + \sum_\alpha |C^\alpha|^2 \Pi_i (T_i + T_i^*)^{n_\alpha^i} .
\tag{59}
$$

Here n_α^i are (zero or negative) fractional numbers usually called "modular weights" of the matter fields C^α. For each given Abelian orbifold, independently of the gauge group or particle content, the possible values of the modular weights are very restricted. For a classification of modular weights for all Abelian orbifolds see [38]. The piece proportional to $Z_{\alpha\beta}$ in (43) has been shown to be present in Calabi–Yau compactifications and orbifolds. In particular, in the case of orbifolds, such a term arises when the untwisted sector has at least one complex–structure field U and has been explicitly computed. We will analyze separately this case below, as well as the associated μ and B parameters, whereas we will concentrate here on the other bosonic soft parameters. Plugging the particular form (59) of the Kähler potential and the parametrization

(58) in (22) and (23) we obtain the following results for the scalar masses and trilinear parameters [34,33,19]:

$$m_\alpha^2 = m_{3/2}^2 \left(1 + 3C^2 \cos^2\theta \; \vec{n_\alpha}.\vec{\Theta^2}\right) + V_0 \,, \tag{60}$$

$$A_{\alpha\beta\gamma} = -\sqrt{3}Cm_{3/2}\left(\sin\theta e^{-i\gamma_S} + \cos\theta \sum_{i=1}^{6} e^{-i\gamma_i}\Theta_i \,[1 \right.$$
$$\left. + n_\alpha^i + n_\beta^i + n_\gamma^i - (T_i + T_i^*)\partial_i \log Y_{\alpha\beta\gamma}]\right) \,. \tag{61}$$

It is easy to check that the results (49) and (51) are recovered in the limit where $\cos\theta \to 0$. Notice that neither the scalar (60) nor the gaugino masses (48) have any explicit dependence on S or T_i: they only depend on the gravitino mass and the goldstino angles. This is one of the advantages of a parametrization in terms of such angles. Although in the case of the A-parameter an explicit T_i-dependence may appear in the term proportional to $\partial_i \log Y_{\alpha\beta\gamma}$, it disappears in several interesting cases [34]. Using the above information, we can now analyze the structure of soft parameters available in Abelian orbifolds.

In the dilaton-dominated case ($\cos\theta = 0$) the soft parameters are universal, as already studied in the previous section. However, in general, they show a lack of universality due to the modular weight dependence (see (60) and (61)). So, even with diagonal matter metrics, FCNC effects may appear. However, we recall that the low-energy running of the scalar masses has to be taken into account. In particular, in the squark case, for gluino masses heavier than (or of the same order as) the scalar masses at the boundary scale, there are large flavour-independent gluino loop contributions which are the dominant source of scalar masses (see (17)). We will show below that this situation is very common in orbifold models. The above effect can therefore help in fulfilling the FCNC constraints.

Another feature of the case under study is that, depending on the goldstino direction, tachyons may appear. For $\cos^2\theta \geq 1/3$, the goldstino direction cannot be chosen arbitrarily if one is interested in avoiding tachyons (see (60)). Nevertheless, having a tachyonic sector is not necessarily a problem, it may even be an advantage [34]. In the case of superstring GUTs (or the standard model with extra U(1) interactions), the negative squared mass may just induce gauge symmetry breaking by forcing a VEV for a particular scalar, GUT-Higgs field, in the model. The latter possibility provides us with interesting phenomenological consequences: the breaking of SUSY could directly induce further gauge symmetry breaking.

Finally, let us consider three particles C^α, C^β and C^γ, coupled through a Yukawa $Y_{\alpha\beta\gamma}$. They may belong both to the untwisted (U) sector or to a twisted (**T**) sector, i.e. we consider couplings of the type **UUU, UTT, TTT**. Then, using the above formulae (60) and (48), with negligible V_0, one finds [34] that in general for *any* choice of goldstino direction

$$m_\alpha^2 + m_\beta^2 + m_\gamma^2 \leq |M_a|^2 = 3m_{3/2}^2 \sin^2\theta \ . \tag{62}$$

Remarkably, the same sum rule is fulfilled even in the presence of off-diagonal metrics, as it is the case of the orbifolds Z_3, Z_4 and Z_6'. The three scalar mass eigenvalues will be in general non-degenerate, which in turn may induce FCNC. This can be automatically avoided in the dilaton dominated limit or under special conditions (for instance, when $\hat{W}$ does not depend on the moduli, a no-scale scenario arises and the mass eigenvalues vanish). The same problem is present in Calabi-Yau compactifications, where again the mass eigenvalues are typically non-degenerate. Besides, the sum rule (62) is violated in general [36]. Coming back to the orbifold case, notice that the above sum rule implies that on average scalars are lighter than gauginos. For small $\sin\theta$, some particular scalar mass may become bigger than the gaugino mass, but in that case at least one of the other scalars involved in the sum rule would be forced to have a *negative* squared mass. This situation is quite dangerous in the context of standard model four-dimensional superstrings, since some observable particles, like Higgses, squarks or sleptons, could be forced to acquire large VEVs (of order the superstring scale). If the above sum rule is applied and squared soft masses are (conservatively) required to be non-negative in order to avoid instabilities of the scalar potential, then the tree level soft masses of observable scalars are constrained to be always smaller than gaugino masses at the boundary scale:

$$m_\alpha \ < \ M_a \ . \tag{63}$$

In turn, this implies that, at low-energy ($\sim M_Z$), the masses of gluinos, sleptons and squarks (except stops and left sbottom) are ordered as

$$m_l \ < \ m_q \ \simeq M_g \ , \tag{64}$$

where gluinos are slightly heavier than scalars. Therefore, in spite of the different set of (non-universal) soft scalar masses, the low-energy phenomenological predictions of the mixed dilaton/moduli SUSY breaking become qualitatively similar to those of the pure dilaton-dominated SUSY breaking. This holds especially for the squark masses, as follows e.g. from (63) and (17). In the case of sleptons, which do not feel gluino loop effects, the boundary values of the soft masses (63) can be relatively more important at low-energy, and larger deviations from the numerical results of (56) can be obtained. Analyses of the low-energy predictions of the dilaton/moduli scenario taking account of the radiative symmetry breaking can be found in [6,28,39].

Before concluding, we recall that exceptions to the above pattern (63), (64) can arise in several situations [34]. For instance, since the total squared Higgs masses receive a positive contribution μ^2, the corresponding soft masses may be allowed to be negative: in this case the restrictions from the sum rule would be relaxed. Another example concerns MSSM Yukawa couplings that arise effectively from higher dimension operators: in this case the three-particle sum rule itself may not hold. Finally, a departure from relations (63) and (64) can also arise when both scalar and gaugino masses vanish at tree level. Such a vanishing can happen in the fully moduli-dominated SUSY breaking, e.g. if SUSY breaking is equally shared among T_1, T_2, T_3 and one consider untwisted particles: then superstring loop effects become important and tend to make scalars heavier than gauginos [6]. In any event, we stress again that potential violations of the pattern in (63) and (64) can occur only when SUSY breaking is mainly moduli dominated (specifically, $\cos^2 \theta \geq 2/3$), since only in this case the gaugino masses can decrease below $m_{3/2}$ and possibly become lighter than some scalar mass.

The B parameter and the μ problem

As already discussed in section 2.1, the two mechanisms to solve the μ problem in the context of supergravity are *naturally* present in superstring models. We will concentrate here on the case in which μ arises from a bilinear term in the Kähler potential (3). The alternative mechanism which generates μ from the superpotential, as in (16), may also be present in orbifolds, but the results are more model dependent. They can be found in [34]. We recall that, in any orbifold with at least one complex-structure field U, the Kähler potential of the untwisted sector possesses the structure $Z(T_i, T_i^*)C_1 C_2 + h.c.$ [14,15] and can therefore generate a μ term. Specifically, the Z_N orbifolds based on Z_4, Z_6, Z_8, Z_{12}' and the $Z_N \times Z_M$ orbifolds based on $Z_2 \times Z_4$ and $Z_2 \times Z_6$ do all have a U-type field in (say) the third complex plane. In addition, the $Z_2 \times Z_2$ orbifold has U fields in the three complex planes. In all these models the piece of the Kähler potential involving the moduli and the untwisted matter fields $C_{1,2}$ in the third complex plane has the form

$$K_3 = -\log[(T_3 + T_3^*)(U_3 + U_3^*) - (C_1 + C_2^*)(C_1^* + C_2)] \tag{65}$$

$$\simeq -\log(T_3 + T_3^*) - \log(U_3 + U_3^*) + \frac{(C_1 + C_2^*)(C_1^* + C_2)}{(T_3 + T_3^*)(U_3 + U_3^*)} . \tag{66}$$

Now, from the expansion shown in (66), one can easily read off the functions

Z, $\tilde{K}_1$, $\tilde{K}_2$ associated to C_1 and C_2:

$$Z = \tilde{K}_1 = \tilde{K}_2 = \frac{1}{(T_3 + T_3^*)(U_3 + U_3^*)} . \tag{67}$$

Let us assume that the MSSM can be obtained from a superstring model of the kind mentioned above and let us identify the fields C_1 and C_2 with the electroweak Higgs fields H_1 and H_2. Plugging back the expressions (67) in (24) and (28) with $\mu = 0$, and using the parametrization (58), one can compute μ and B for this interesting class of models [34]:

$$\hat{\mu} = m_{3/2} \left[1 + \sqrt{3} C \cos\theta (e^{i\gamma_3}\Theta_3 + e^{i\gamma_6}\Theta_6) \right] , \tag{68}$$

$$\begin{aligned} B\hat{\mu} = {} & 2m_{3/2}^2 \Big[1 + \sqrt{3} C \cos\theta (\cos\gamma_3 \Theta_3 + \cos\gamma_6 \Theta_6) \\ & + 3C^2 \cos^2\theta \cos(\gamma_3 - \gamma_6)\Theta_3\Theta_6 \Big] + V_0 . \end{aligned} \tag{69}$$

Notice that, in the limit where $\cos\theta \to 0$, the results in (57) are recovered. In addition, we recall from (60) that the soft masses are

$$m_{H_1}^2 = m_{H_2}^2 = m_{3/2}^2 \left[1 - 3C^2 \cos^2\theta(\Theta_3^2 + \Theta_6^2) \right] + V_0 . \tag{70}$$

In general, the quadratic part of the Higgs potential after SUSY breaking has the form (see(21))

$$V_2 = (m_{H_1}^2 + |\hat{\mu}|^2)|\hat{H}_1|^2 + (m_{H_2}^2 + |\hat{\mu}|^2)|\hat{H}_2|^2 + (B\hat{\mu}\hat{H}_1\hat{H}_2 + h.c.) , \tag{71}$$

where we recall that $\hat{H}_1$ and $\hat{H}_2$ are the canonically normalized Higgs fields. In the specific case under consideration, from (68), (69) and (70) we find the remarkable result that the three coefficients in V_2 are equal, i.e.

$$m_{H_1}^2 + |\hat{\mu}|^2 = m_{H_2}^2 + |\hat{\mu}|^2 = B\hat{\mu} . \tag{72}$$

so that V_2 has the simple form

$$V_2 = B\hat{\mu} \, (\hat{H}_1 + \hat{H}_2^*)(\hat{H}_1^* + \hat{H}_2) . \tag{73}$$

and therefore $\tan\beta = \frac{\langle \hat{H}_2 \rangle}{\langle \hat{H}_1 \rangle} = -1$. Of course, this corresponds to the boundary condition on the scalar potential at the superstring scale: at lower energies the renormalization group equations should be used. Although the common value of the three coefficients in (72) depends on the Goldstino direction via the parameters $\cos\theta$, Θ_3, $\Theta_6,\ldots$ (see e.g. the expression of $B\hat{\mu}$ in (69)), we stress that the equality itself and the form of V_2 hold *independently* of the Goldstino direction.

Starting from such 'superstringy' boundary conditions for the MSSM parameters, one can explore their consistency with radiative electroweak-symmetry breaking[31] (see also[40]). One finds that consistency with the measured value of the top-quark mass cannot be achieved in the dilaton-dominated scenario (as already mentioned in section 3.1). The only SUSY-breaking scenario compatible with such constraints requires a suppressed dilaton contribution and important (often dominant) contributions from the T_3, U_3 moduli.

4 Final Comments and Outlook

It is worth remarking that most of the above results for soft terms in superstring models refer to certain simple *perturbative* heterotic compactifications. In addition, it is assumed that the goldstino is a fermionic partner of some combination of the dilaton and/or the moduli fields. Recently some information about the non-perturbative regime in superstring theory has been obtained in terms of the S-dualities[41] of the theory. All superstring theories seem to correspond to some points in the parameter space of a unique eleven-dimensional underlying theory, M-theory[42]. Although the structure of this theory is largely unknown, some preliminary attempts have been made to extract some information of phenomenological interest. A scenario to understand the difference between the GUT scale and the superstring scale has been put forward[43]. Supersymmetry breaking and other phenomenological issues have also been explored within this context in[44]. Studies in progress concerning *non-perturbative* superstring vacua with $N = 1$ SUSY will certainly bring us new surprises.

References

1. For a review, see: H.P. Nilles, *Phys. Rep.* **110** (1984) 1, and references therein.
2. For a recent review, see: C. Muñoz, *hep-th/9507108*, and references therein.
3. For a review, see: C. Muñoz, *hep-ph/9601325*, and references therein.
4. Y. Kawamura, T. Kobayashi and T. Komatsu, *hep-ph/9609462*, and references therein.
5. S.K. Soni and H.A. Weldon, *Phys. Lett.* **B126** (1983) 215.
6. A. Brignole, L.E. Ibáñez and C. Muñoz, *Nucl. Phys.* **B422** (1994) 125 [Erratum: **B436** (1995) 747].
7. K. Choi, J.E. Kim and H.P. Nilles, *Phys. Rev. Lett.* **73** (1994) 1758; K. Choi, J.E. Kim and G.T. Park, *Nucl. Phys.* **B442** (1995) 3.

8. S. Ferrara, C. Kounnas and F. Zwirner, *Nucl. Phys.* **B429** (1994) 589 [Erratum: **B433** (1995) 255].

9. K. Choi, J.S. Lee and C. Muñoz, *hep-ph/9709250*.

10. G.F. Giudice and A. Masiero, *Phys. Lett.* **B206** (1988) 480.

11. J.A. Casas and C. Muñoz, *Phys. Lett.* **B306** (1993) 288.

12. J.E. Kim and H.P. Nilles, *Phys. Lett.* **B138** (1984) 150, *Phys. Lett.* **B263** (1991) 79; E.J. Chun, J.E. Kim and H.P. Nilles, *Nucl. Phys.* **B370** (1992) 105.

13. V.S. Kaplunovsky and J. Louis *Phys. Lett.* **B306** (1993) 269.

14. G. Lopes-Cardoso, D. Lüst and T. Mohaupt, *Nucl. Phys.* **B432** (1994) 68.

15. I. Antoniadis, E. Gava, K.S. Narain and T.R. Taylor, *Nucl. Phys.* **B432** (1994) 187.

16. A. Lleyda and C. Muñoz, *Phys. Lett.* **B317** (1993) 82.

17. D. Choudhury, F. Eberlein, A. Köning, J. Louis and S. Pokorski, *Phys. Lett.* **B342** (1995) 180.

18. J. Louis and Y. Nir, *Nucl. Phys.* **B447** (1995) 18.

19. P. Brax and M. Chemtob, *Phys.Rev.* **D51** (1995) 6550.

20. P. Brax and C.A. Savoy, *Nucl. Phys.* **B447** (1995) 227.

21. For a review, see: A.B. Lahanas and D.V. Nanopoulos, *Phys. Rep.* **145** (1987) 1, and references therein.

22. J.A. Casas, A. Lleyda and C. Muñoz, *Nucl. Phys.* **B471** (1996) 3, *Phys. Lett.* **B389** (1996) 305.

23. K. Choi, *Phys. Rev. Lett.* **72** (1994) 1592.

24. B. Acharya, D. Bailin, A. Love, W.A. Sabra and S. Thomas, *Phys. Lett.* **B357** (1995) 387; D. Bailin, G.V. Kraniotis and A. Love, *hep-th/9705244, hep-th/9707105*.

25. A. de la Macorra and G.G. Ross, *Nucl. Phys.* **B404** (1993) 321; V. Halyo and E. Halyo, *Phys. Lett.* **B382** (1996) 89.

26. R. Barbieri, J. Louis and M. Moretti, *Phys. Lett.* **B312** (1993) 451 [Erratum: **B316** (1993) 632].

27. J.L. Lopez, D.V. Nanopoulos and A. Zichichi, *Phys. Lett.* **B319** (1993) 451.

28. S. Khalil, A. Masiero and F. Vissani *Phys. Lett.* **B375** (1996) 154.

29. See e.g. L.E. Ibáñez, *hep-th/9505098*, Proc. of Strings 95, World Scientific (1995), and references therein.

30. J.A. Casas, *Phys. Lett.* **B384** (1996) 103.

31. A. Brignole, L.E. Ibáñez and C. Muñoz, *Phys. Lett.* **B387** (1996) 305.

32. J.A. Casas, A. Lleyda and C. Muñoz, *Phys. Lett.* **B380** (1996) 59.

33. T. Kobayashi, D. Suematsu, K. Yamada and Y. Yamagishi, *Phys. Lett.*

B348 (1995) 402.

34. A. Brignole, L.E. Ibáñez, C. Muñoz and C. Scheich, *Z. Phys.* **C74** (1997) 157.

35. A. Font, L.E. Ibañez, D. Lüst and F. Quevedo, *Phys. Lett.* **B245** (1990) 401; S. Ferrara, N. Magnoli, T.R. Taylor and G. Veneziano, *Phys. Lett.* **B245** (1990) 409; M. Cvetic, A. Font, L.E. Ibañez, D. Lüst and F. Quevedo, *Nucl. Phys.* **B361** (1991) 194; B. de Carlos, J.A. Casas and C. Muñoz, *Phys. Lett.* **B299** (1993) 234, *Nucl. Phys.* **B399** (1993) 623; A. de la Macorra and G.G. Ross, *Phys. Lett.* **B325** (1994) 85.

36. H.B. Kim and C. Muñoz, *Z. Phys.* **C75** (1997) 367.

37. H.B. Kim and C. Muñoz, *Mod. Phys. Lett.* **A12** (1997) 315.

38. L.E. Ibáñez and D. Lüst, *Nucl. Phys.* **B382** (1992) 305.

39. C.-H. Chen, M. Drees and J.F. Gunion, *Phys. Rev.* **D55** (1997) 330; Y. Kawamura, S. Khalil and T. Kobayashi, *hep-ph/9703239*; A. Love and P. Stadler, *hep-ph/9709234*.

40. Y. Kawamura, T. Kobayashi and M. Watanabe, *DPSU-97-5, hep-ph/9609462*.

41. A. Font, L.E. Ibáñez, D. Lüst and F. Quevedo, *Phys. Lett.* **B249** (1990) 35; A. Sen, *Int.J. Mod.Phys.* **A9** (1994) 3707.

42. See e.g. J.H. Schwarz, *hep-th/9607201*, P.K. Townsend, *hep-th/9612121*, and references therein.

43. E. Witten, *Nucl. Phys.* **B471** (1996) 135.

44. T. Banks and M. Dine, *hep-th/9605136, hep-th/9608197, hep-th/9609046*; E. Caceres, V.S. Kaplunovsky and I.M. Mandelberg, *hep-th/9606036*; P. Horava, *hep-th/9608019*; T. Li, J. Lopez and D.V. Nanopoulos, *hep-ph/9702237, hep-ph/9704247*; E. Dudas and C. Grojean, *hep-th/9704177*; I. Antoniadis and M. Quirós, *hep-th/9705037, hep-th/9707208*; K. Choi, *hep-th/9706171*; H.P. Nilles, M. Olechowski and M. Yamaguchi, *hep-th/9707143*; Z. Lalak and S. Thomas, *hep-th/9707223*; V. Kaplunovsky and J. Louis, *hep-th/9708049*; E. Dudas, *hep-th/9709043*.

REGULARISATION OF SUPERSYMMETRIC THEORIES

I. JACK, D.R.T. JONES

Theoretical Physics Division, Chadwick Building, Peach Street, Liverpool,
L69 3BX, UK

We discuss issues that arise in the regularisation of supersymmetric theories.

1　Beyond the tree approximation

In this chapter we consider issues of both practice and principle that arise
when we take supersymmetric theories and calculate radiative corrections. It
is usually the case that a symmetry of the Lagrangian is still a symmetry of the
full quantum effective action; in which case we say that radiative corrections
preserve the symmetry. There are important exceptions to this rule, however:
for example conformal invariance is in general violated by radiative corrections, and massless quantum electrodynamics has a global U_1 axial symmetry
which is violated at the one-loop level (in accordance with the famous Adler-
Bardeen theorem). It is not a priori obvious, therefore, that supersymmetry
is a symmetry of the full quantum theory in any particular case. Indeed it
has been occasionally claimed that there exists a supersymmetry anomaly. In
some cases these claims have been erroneous, and have occurred because it
is difficult to distinguish between a genuine anomaly and an apparent violation of a supersymmetric Ward identity due to use of a regularisation method
that itself violates supersymmetry. Contrariwise, a detailed formal renormalisation program has been pursued in a series of papers by Piguet and collaborators[1] including one[2] where a proof that supersymmetry is *not* anomalous
was presented. There is no real reason to doubt this conclusion (although
the treatment of infra-red singularities in the program is a possible weakness),
for the class of theories considered; the evidence adduced so far points to supersymmetry being a symmetry of the full quantum theory. (Note, however,
recent suggestions that there may indeed exist a supersymmetry anomaly in
composite operators[3].)

The existence of an anomaly is intimately related to the question of regularisation. Beyond the tree level, certain amplitudes in any given quantum
field theory are not defined, due to divergences caused by the need to integrate
over all momenta for particles in intermediate states. Regularisation is the
process whereby the result of an ill-defined correction to a given amplitude is
separated into a finite part (which is retained) and an "infinite" part (or more

precisely, a part which tends to infinity in the limit that a certain parameter, or parameters, specific to the regularisation method is removed) which is removed from the theory ("subtracted") by introducing a counter-term which precisely cancels it. If the regularised theory fails to respect any given symmetry then the finite amplitudes will fail to satisfy the Ward identities of the symmetry, giving rise to an apparent anomaly and the confusion alluded to above. When the anomaly really is specious, it is possible to restore invariance by modifying the counter-terms by finite amounts. (Obviously the counter-terms are ambiguous, in that their defining role is to cancel something which becomes infinite as the regulator is removed, so that adding a finite quantity to any counter–term leaves its *raison d'être* intact. If no modification of the counter-terms will restore the Ward identity then there is an anomaly.[a]

From a formal point of view, the choice of regularisation scheme made in the implementation of a renormalisation is not of great significance; it is important only that it corresponds to addition of *local* counter-terms. For the extraction of physical predictions, however, the choice becomes a matter of considerable practical significance. It is convenient, for example, to use a regularisation method that preserves symmetries. It should be clear, in fact, from the above discussion that the existence of a regulator consistent with a given symmetry suffices to prove that symmetry to be anomaly–free. In this context the approach of West [4], consisting of higher derivative regularisation supplemented by Pauli-Villars [b] at one loop, is worthy of consideration; but as the author himself remarks, the issue of a possible anomaly is not thereby fully resolved because of the infra-red difficulties already alluded to.

Dimensional regularisation (DREG) is an elegant and convenient way of dealing with the infinities that arise in quantum field theory beyond the tree approximation. [6] It is well adapted to gauge theories because it preserves gauge invariance; it is less well-suited, however, for supersymmetry because invariance of an action with respect to supersymmetric transformations only holds in general for specific values of the space-time dimension d. This is essentially due to the fact that a necessary condition for supersymmetry is equality of Bose and Fermi degrees of freedom. In non-gauge theories it is relatively easy to circumvent this problem, and DREG as usually employed is, in fact, a supersymmetric procedure. Gauge theories are a different matter, however, and the question as to whether there exists a completely satisfactory supersym-

---

[a] Sometimes an anomaly can be apparently removed only to reappear in another guise. This is the case with the Adler–Bardeen anomaly, which is a property of the fermion triangle with two vector and one axial–vector vertices. The anomaly may affect the axial current or the vector current, depending on how the theory is regularised.

[b] Use of Pauli-Villars at one loop in supersymmetry was also advocated by Gaillard [5] in 1995.

metric regulator for gauge theories remains controversial. This fact has been exploited recently to suggest that there may be supersymmetric anomalies.[3]

An elegant attempt to modify DREG so as to render it compatible with supersymmetry was made by Siegel.[7] The essential difference between Siegel's method (DRED) and DREG is that the continuation from 4 to d dimensions is made by *compactification*, or *dimensional reduction*. Thus while the momentum (or space-time) integrals are d-dimensional in the usual way, the number of field components remains unchanged and consequently supersymmetry is undisturbed. (A pedagogical introduction to DRED was given by Capper et al[8].)

As pointed out by Siegel himself,[9] there remain potential ambiguities with DRED associated with treatment of the Levi-Civita symbol, $\epsilon^{\mu\nu\rho\sigma}$. We will address this difficulty and the related one involving γ^5 in Section 3.

We must also address problems which arise only when DRED is applied to non-supersymmetric theories. That DRED is a viable alternative to DREG in the non–supersymmetric case was claimed early on[8]. Subsequently it has been adopted occasionally, motivated usually by the fact that Dirac matrix algebra is easier in four dimensions–and in particular by the desire to use Fierz identities[10,11]. One must, however, be very careful in applying DRED to non-supersymmetric theories because of the existence of *evanescent couplings*. These were first described[12] in 1979, and independently discovered by van Damme and 't Hooft[13]. They argued, in fact, that while DRED is a satisfactory procedure for supersymmetric theories[14,15] (modulo the subtleties alluded to above) it leads to a catastrophic loss of unitarity in the non-supersymmetric case. Evidently there is an important issue to be resolved here–is use of DRED in fact forbidden (except in the supersymmetric case) in spite of its apparent convenience? It has been conclusively demonstrated[16,17] that if DRED is employed in the manner envisaged by Capper et al[8], (which as we shall see differs in an important way from the definition of DRED primarily used by 't Hooft and van Damme) then there is no problem with unitarity. There exist a set of transformations whereby the β–functions of a particular theory (calculated using DRED) may be related to the β–functions of the same theory (calculated using DREG) by means of coupling constant reparametrisation. The key is that a correct description of any non-supersymmetric theory impels us to a recognition of the fundamental fact that in general the evanescent couplings renormalise in a manner different from the "real" couplings with which we may be tempted to associate them. This means that care must be taken as we go beyond one loop; nevertheless it is still possible to exploit the simplifications in the Dirac algebra which have motivated the use of DRED. We will return to this point later.

At this point the reader may wonder why, in a book about supersymmetry, we should worry about renormalising non–supersymmetric theories at all. The main practical reason is that the supersymmetric standard model is an effective theory in which supersymmetry is *explicitly* broken, albeit by terms with non–zero dimension of mass.

The reader may also feel that, given the problems with DRED, we should explore other regulators. For example, there has been some recent work on a new approach known as differential regularisation [18]. The fact is, however, that the convenience of DREG for calculations beyond one loop makes the use of some variant of it very desirable. Use of other proposed regulators is rarely pursued beyond verification of some already known (and usually one-loop) results.

2   Introduction to DRED

As a concrete example, let us consider a non-abelian gauge theory with fermions but no elementary scalars. The theory to be studied consists of a Yang-Mills multiplet $W_\mu^a(x)$ with a multiplet of spin $\frac{1}{2}$ fields[c] $\psi^\alpha(x)$ transforming according to an irreducible representation R of the gauge group G. Of course if ψ *is* Majorana, then R must be a real representation, since the Majorana condition is not preserved by a unitary transformation.

The Lagrangian density (in terms of bare fields) is

$$L_B = -\frac{1}{4}G_{\mu\nu}^2 - \frac{1}{2\alpha}(\partial^\mu W_\mu)^2 + C^{a*}\partial^\mu D_\mu^{ab}C^b + i\overline{\psi}^\alpha \gamma^\mu D_\mu^{\alpha\beta}\psi^\beta \tag{1}$$

where

$$G_{\mu\nu}^a = \partial_\mu W_\nu^a - \partial_\nu W_\mu^a + g f^{abc}W_\mu^b W_\nu^c \tag{2}$$

and

$$D_\mu^{\alpha\beta} = \delta^{\alpha\beta}\partial_\mu - ig(R^a)^{\alpha\beta}W_\mu^a \tag{3}$$

and the usual covariant gauge fixing and ghost terms have been introduced.

The process of dimensional reduction consists of imposing that all field variables depend only on a subset of the total number of space-time dimensions– in this case d out of 4 where $d = 4 - \epsilon$. We will use $\mu, \nu \cdots$ to denote 4-dimensional indices and i, j to denote d-dimensional ones, with corresponding metric tensors $g_{\mu\nu}$ and g_{ij}. It is also convenient to introduce "hatted" quantities (such as $\hat{g}_{\mu\nu}$ and $\hat{\gamma}^\mu$) which are identical to the corresponding d-dimensional quantities $(g_{ij}, \gamma^i \cdots)$ within the d-dimensional subspace, but whose remaining

---

[c] which may be Dirac or Majorana at this stage

components are zero. Momenta p_μ exist only in the d dimensional subspace so we do not bother to "hat" them. Thus we have for example

$$\not{p} = p_\mu \gamma^\mu = p_\mu \hat{\gamma}^\mu \tag{4}$$

and

$$g^{\mu\nu} g_{\mu\nu} = 4, \quad \hat{g}^{\mu\nu} \hat{g}_{\mu\nu} = g^{ij} g_{ij} = d. \tag{5}$$

In particular, we have that

$$\hat{g}^{\mu\nu} g_\nu{}^\lambda = \hat{g}^{\mu\lambda} \qquad \text{and} \qquad \hat{g}^{\mu\nu} \gamma_\nu = \hat{\gamma}^\mu. \tag{6}$$

These apparently innocuous relations will cause us trouble in the next section.

In order to fully appreciate the consequences of DRED for L_B we must make the decomposition

$$W_\mu^a(x^j) = \{ W_i^a(x^j), W_\sigma^a(x^j) \} \tag{7}$$

where

$$\delta_i^i = \delta_j^j = d \qquad \text{and} \qquad \delta_{\sigma\sigma} = \epsilon. \tag{8}$$

It is then easy to show that

$$L_B = L_B^d + L_B^\epsilon \tag{9}$$

where

$$L_B^d = -\frac{1}{4} G_{ij}^2 - \frac{1}{2\alpha}(\partial^i W_i)^2 + C^{a*} \partial^i D_i^{ab} C^b + i\overline{\psi}^\alpha \gamma^i D_i^{\alpha\beta} \psi^\beta \tag{10}$$

and

$$L_B^\epsilon = \frac{1}{2}(D_i^{ab} W_\sigma^b)^2 - g\overline{\psi}\gamma_\sigma R^a \psi W_\sigma^a - \frac{1}{4} g^2 f^{abc} f^{ade} W_\sigma^b W_{\sigma'}^c W_\sigma^d W_{\sigma'}^e. \tag{11}$$

Conventional dimensional regularisation (DREG) amounts to using Eq. 10 and discarding Eq. 11. For DRED , on the other hand we include both. [d] In simple applications it is in general more convenient to eschew the separation performed above and calculate with 4-dimensional and d-dimensional indices rather than d-dimensional and ϵ-dimensional ones. As a simple illustration, consider the following typical calculation:

$$\gamma^\mu \not{p} \gamma_\mu = p_\nu \gamma^\mu \gamma^\nu \gamma_\mu = -2p_\nu \gamma^\nu = -2\not{p} \tag{12}$$

---

[d]The additional contributions from L_B^ϵ are precisely what is required to restore the supersymmetric Ward identities at one loop in supersymmetric theories, as described in section 4.

or equivalently

$$\gamma^\mu \not{p} \gamma_\mu = \gamma^i \not{p} \gamma_i + \gamma^\sigma \not{p} \gamma_\sigma = (2-d)\not{p} + (d-4)\not{p} = -2\not{p}. \qquad (13)$$

From the dimensionally reduced form of the gauge transformations:

$$\begin{aligned} \delta W_i^a &= \partial_i \Lambda^a + g f^{abc} W_i^b \Lambda^c \\ \delta W_\sigma^a &= g f^{abc} W_\sigma^b \Lambda^c \\ \delta \psi^\alpha &= ig(R^a)^{\alpha\beta} \psi^\beta \Lambda^a \end{aligned} \qquad (14)$$

we see that each term in Eq. 11 is separately invariant under gauge transformations. The W_σ-fields behave exactly like scalar fields, and are hence known as ϵ-scalars. The significance of this is that gauge invariance *per se* provides no reason to expect the $\overline{\psi}\psi W_\sigma$ vertex to renormalise in the same way as the $\overline{\psi}\psi W_i$ vertex. In the case of the quartic ϵ-scalar coupling the situation is more complex since in general of course more than one such coupling is permitted by Eq. 14 . In other words, we cannot in general expect the $f-f$ tensor structure present in Eq. 11 to be preserved under renormalisation. This is clear from the abelian case, where there is no such quartic interaction in L_B^ϵ but there is a divergent contribution at one loop from a fermion loop.

In the case of supersymmetric theories, however, these difficulties do not arise. If ψ above is a Majorana fermion in the adjoint representation, then L_B is supersymmetric. This links W_i and W_σ in a way that is not severed by the dimensional reduction. Thus the $\overline{\psi}\psi W_\sigma$ and $\overline{\psi}\psi W_i$ vertices (both equal to g at the tree level) remain equal under renormalisation. We will return in section 6 to the application of DRED to non-supersymmetric theories.

3   DRED ambiguities

With DRED it would seem that necessarily $d < 4$, since the regulated action is, after all, defined by dimensional *reduction*. Then, given $d < 4$, one can define an object $\hat{\epsilon}^{\mu\nu\rho\sigma}$ as follows:

$$\hat{\epsilon}^{\mu\nu\rho\sigma} = \hat{g}^{\mu\alpha} \hat{g}^{\nu\beta} \hat{g}^{\rho\gamma} \hat{g}^{\sigma\delta} \epsilon_{\alpha\beta\gamma\delta} \qquad (15)$$

where $\epsilon_{\alpha\beta\gamma\delta}$ is the usual 4-dimensional tensor. Unfortunately it is now possible to show that algebraic inconsistencies result [9] unless $d = 4$. Let us illustrate these problems in the two dimensional case. The alternating tensor $\epsilon^{\mu\nu}$ satisfies (in two Euclidean dimensions) the relation

$$\epsilon^{\mu\nu} \epsilon^{\rho\sigma} = g^{\mu\rho} g^{\nu\sigma} - g^{\mu\sigma} g^{\nu\rho}. \qquad (16)$$

Using Eq. 6 it is easy to show that

$$\hat{\epsilon}^{\mu\nu}\hat{\epsilon}^{\rho\sigma} = \hat{g}^{\mu\rho}\hat{g}^{\nu\sigma} - \hat{g}^{\mu\sigma}\hat{g}^{\nu\rho} \tag{17}$$

where $\hat{\epsilon}^{\mu\nu}$ is defined similarly to Eq. 15.

However it is trivial to demonstrate that the result of applying Eq. 17 to the tensor

$$A^{\mu\nu} = \hat{\epsilon}^{\mu\nu}\hat{\epsilon}^{\rho\sigma}\hat{\epsilon}_{\rho\sigma} \tag{18}$$

is ambiguous inasmuch that it differs according to which pair of $\hat{\epsilon}$-tensors are selected: the result is the identity

$$(d+1)(d-2)\hat{\epsilon}^{\mu\nu} = 0. \tag{19}$$

A related problem (of course) is the fact that the only mathematically consistent treatment of γ^5 within DREG is predicated [6] on having $d > 4$. Given Eq. 6 and the usual relation

$$\{\gamma_\mu, \gamma^5\} = 0. \tag{20}$$

it follows that

$$\{\hat{\gamma}_\mu, \gamma^5\} = 0. \tag{21}$$

and hence that

$$(d-4)\mathrm{Tr}\left[\gamma^5\hat{\gamma}^\mu\hat{\gamma}^\nu\hat{\gamma}^\rho\hat{\gamma}^\sigma\right] = 0. \tag{22}$$

This is unfortunate since it renders problematic the discussion of the axial anomaly.

For $d > 4$, however, Eq. 6 does not hold and so Eq. 21 no longer follows. Instead we impose

$$[\gamma_\sigma, \gamma^5] = 0, \quad \text{for} \quad 4 < \sigma < d. \tag{23}$$

and this leads to a straightforward and unambiguous derivation of the axial anomaly. It has been verified [19] that this prescription correctly reproduces the Adler-Bardeen theorem at the two-loop level using DREG.

Returning to the DRED prescription, there are a number of possible "fixes" at one loop; [20] at two loops, it was shown [19] that the Adler-Bardeen theorem could indeed still be satisfied if relations like

$$\gamma^i\gamma^5\gamma i = (d-8)\gamma^5 \tag{24}$$

which follow in the $d > 4$ case, are used in conjunction with DRED .

A possible point of view concerning all this [21] is that DRED is terminally inconsistent and should not be used. We believe, however, that the difficulties are essentially technical and can be evaded. For example, one well-defined procedure would be to write

$$\gamma^5 = \frac{1}{4!}\epsilon^{\mu\nu\rho\sigma}\gamma_\mu\gamma_\nu\gamma_\rho\gamma_\sigma \tag{25}$$

and factor all out ϵ-tensors. Renormalised amplitudes may then be calculated, which, being finite as $d \to 4$, are unambiguous when the ϵ-tensors are contracted in. [e] We would claim also that other modes of procedure which would give different answers because of the ambiguities detailed above, correspond nevertheless to *the same physical results*. This assertion has in fact been verified in one particular case [22], where a prescription first suggested by Hull and Townsend [23] was used. This amounted to employing as $\epsilon^{\mu\nu}$ not the usual alternating tensor but instead a structure satisfying

$$\hat\epsilon^\mu{}_\nu\hat\epsilon^{\nu\rho} = (1 + c\epsilon)\hat g^{\mu\rho} \tag{26}$$

(where here $\epsilon = 2 - d$). It turns out [22] that the dependence of the results on the parameter c can be absorbed into redefinitions of the renormalised metric and torsion tensors. In the special case $c = 0$, $\hat\epsilon^{\mu\nu}$ is an almost complex structure [24].

There have been a considerable number of papers discussing the interpretation of γ^5 in both DREG and DRED , and the reader may consult them for further enlightenment [25]. We turn in the next section to another (but again related) problem with DRED , arising from the fact that in spite of the correct counting of degrees of freedom, there are still ambiguities associated with establishing invariance of the action: we will look at this in more detail below in the context of the supersymmetry Ward identity.

4 The supersymmetry Ward identity

The first concrete illustration of the different results provided by DRED and DREG for a supersymmetric theory was as follows. Consider the basic supersymmetric gauge theory in the Wess-Zumino gauge as defined by the Lagrangian L_S where

$$L_S = -\frac{1}{4}G^{\mu\nu}G_{\mu\nu} + i\frac{1}{2}\overline\lambda^\alpha\gamma^\mu D_\mu^{\alpha\beta}\lambda^\beta + \frac{1}{2}D^2. \tag{27}$$

---

[e]This somewhat cumbersome procedure can usually be finessed. For instance, if only even-parity fermion loops are present, then there is no problem with a fully anti-commuting γ^5.

In $d = 4$, L_S is invariant (up to a total derivative) under the transformations

$$\delta W_\mu^a = i\bar{\epsilon}\gamma_\mu\lambda^a, \qquad \delta\lambda^a = \frac{1}{2}G_{\mu\nu}^a\gamma^\mu\gamma^\nu\epsilon - iD^a\gamma^5\epsilon$$

$$\delta D^a = -\bar{\epsilon}\gamma_\mu\gamma^5(D^\mu\lambda)^a. \tag{28}$$

It is an excellent exercise in spinor algebra to verify this invariance. Note the presence in Eq. 28 of γ^5-terms; to obtain invariance one must assume that γ^5 is totally anti-commuting. Of course in this particular case we could set $D^a = 0$, and still have an invariance (not involving γ^5). However this does not escape the Siegel ambiguity, as we shall now show. With due care, one obtains (up to total derivatives)

$$\delta L_S = g\frac{1}{2}f^{abc}\bar{\epsilon}\gamma^\mu\lambda^a\lambda^b\gamma_\mu\lambda^c. \tag{29}$$

This is identically zero for $d = 4$, though this is not obvious even if we rewrite in two–component formalism; a Fierz re-ordering is required. For $d \neq 4$, δL_S is not zero; and the key to the distinction between DRED and DREG lies in the $\gamma^\mu \otimes \gamma_\mu$ contraction, which is d-dimensional for DREG and four-dimensional for DRED . There are important consequences for the regularisation of super-symmetric theories, as we shall now see. Let us add to L_S gauge fixing and ghost terms:

$$L_S \to L_T = L_S + L_G \tag{30}$$

where

$$L_G = -\frac{1}{2\alpha}(\partial^\mu W_\mu)^2 + C^{a*}\partial^\mu D_\mu^{ab}C^b. \tag{31}$$

Then we introduce the functional $Z(J, j, j_D)$ where

$$Z = \int d\{W_\mu\}d\{\lambda\}d\{D\}e^{i\int d^dx[L_T + J^\mu W_\mu + \bar{j}\lambda + j_D D]} \tag{32}$$

and use of Eq. 28 leads to the following Ward identity:

$$0 = <\int d^dx\left[J^\mu\delta W_\mu + \bar{j}\delta\lambda + j_D\delta D + \delta L_S + \delta L_G\right]> \tag{33}$$

where

$$\delta L_G = -\frac{1}{\alpha}\partial^\mu W_\mu\partial^\mu\delta W_\mu + gf^{abc}C^{a*}\partial^\mu\delta W_\mu^c C^b \tag{34}$$

and

$$<X> = \int d\{W_\mu\}d\{\lambda\}d\{D\}Xe^{i\int d^dx[L_T + J^\mu W_\mu + \bar{j}\lambda + j_D D]}. \tag{35}$$

Notice we have included the term δL_S from Eq. 29 to allow for the fact that this may not be zero away from $d = 4$. When this Ward identity was investigated at one loop[8], it was found to be true with DRED and false with DREG; which conclusion was arrived at because the contribution from δL_S was ignored. It is easy to show that this contribution is zero for DRED , and in the case of DREG serves precisely to restore Eq. 34. The distinction between DRED and DREG is manifest in the fact that the contribution of δL_S is zero in the former case. It is in this sense that DRED is more consistent with supersymmetry. In terms of this Ward identity the difference may not seem crucial, but the fact that δL_S is effectively non-zero (with DREG) means that if we employ DREG in a supersymmetric theory then great care must be taken with the formulation of physical predictions. This becomes particularly clear when we generalise to include matter fields; in supersymmetric QCD, for example, the fact that the gluino-quark-squark coupling is equal to the gauge coupling g is a consequence of supersymmetry and will not be preserved under renormalisation if DREG is employed. In fact, this is very similar to the problems that occur when we want to apply DRED to non–supersymmetric theories; once again there are coupling constant relations that are not preserved by radiative corrections.

While DRED was successful in the above application, it does not follow that an insertion of δL_S in a diagram of arbitrary complexity gives zero. It can be shown[26] that such an insertion depends on the quantity Δ where

$$\Delta = \mathrm{Tr}(A\gamma^\mu B\gamma_\mu) + \mathrm{Tr}(A\gamma^\mu)\mathrm{Tr}(B\gamma_\mu) - (-1)^k \mathrm{Tr}(A\gamma^\mu B^R\gamma_\mu). \qquad (36)$$

Here A and B are products of Dirac γ-matrices; k is the number of such matrices in B, and B^R consists of the the same set of matrices as B but written down in reverse order. In strict $d = 4$, Δ is zero; but because A, B may contain d-dimensional γ-matrices (due to contraction with momenta) Δ is non-zero in general. If we set

$$A = \gamma^{\mu_1}\gamma^{\mu_2}\cdots\gamma^{\mu_5} \quad \text{and} \quad B = \gamma_{\nu_1}\gamma_{\nu_2}\cdots\gamma_{\nu_5} \qquad (37)$$

then

$$\Delta = 48\delta^{[\mu_1}_{\nu_1}\delta^{\mu_2}_{\nu_2}\cdots\delta^{\mu_5]}_{\nu_5} \qquad (38)$$

where the brackets $[\cdots]$ denote antisymmetrisation. This is clearly zero for integer $d \leq 4$ but if the various indices are d-dimensional then it is not. For instance if we calculate the trace by contracting Δ with $\delta^{\nu_1}_{\mu_1}\cdots\delta^{\nu_5}_{\mu_5}$ then we obtain $\mathrm{Tr}\Delta = 48d(d-1)(d-2)(d-3)(d-4)$. A diagram with at least four loops is required[26] so that we get enough γ-matrices to activate this problem in the propagator Ward identity[8]. This is clearly the same ambiguity at bottom addressed by Siegel[9].

One might hope that this problem is somehow resolved by use of super-fields. Using superfield perturbation theory Feynman rules maintains super-symmetry in a manifest way; but a crucial part of the calculational procedure relies on the reduction of products of supercovariant D_α and $\overline{D}_{\dot\alpha}$ derivatives to products of four or less, and this is possible only when the fact that the $\alpha, \dot\alpha$ indices are two-valued is used. Thus the same ambiguity must be present, albeit in a somewhat different form. However as we argued in the previous section, the ambiguity will not affect physical results since it is equivalent to the freedom available in choice of regularisation scheme, as long as a systematic procedure is adopted. Despite all difficulties, DRED remains the regulator of choice for supersymmetric theories, and has survived many practical tests.

5 $N=2$ and $N=4$ supersymmetry

$N=2$ supersymmetry corresponds, in the language of $N=1$ superfields, to the special case of a superpotential taking the form:

$$W = \sqrt{2}g\phi^a \xi^T S_a \chi, \tag{39}$$

where ξ, χ, ϕ are multiplets transforming under the S^*, S and adjoint repre-sentations of the gauge group $\mathcal{G}$ respectively. In the special case that S is the adjoint representation we have $N=4$ supersymmetry. $N=2$ theories are extraordinary in that they have only one-loop divergences[27]. This means that for $N=2$, β_g vanishes beyond one-loop if computed using DRED; crucial here is the fact that DRED incorporates *minimal subtraction*. In the $N=4$ case the one-loop contribution also vanishes, so $N=4$ theories are ultra-violet finite to all orders of perturbation theory.

$N=2$ and $N=4$ theories, although obviously of great interest, possess a property that is unfortunate if we want to try and incorporate them into a realistic theory. This property is that since the chiral superfields are ei-ther adjoint or in S, S^* pairs, gauge invariant mass terms are possible for the fermionic components of the multiplets. It is difficult, therefore, to arrange for fermion masses (such as the electron mass) to be much less than the scale of supersymmetry–breaking, at least. Nevertheless there have been occasional attempts to construct phenomenologically viable models[28], and explore their consequences[29].

6   Non–supersymmetric theories

We saw in section 2 that under renormalisation the ϵ-scalars behave differently from the gauge fields, except in supersymmetric theories. On might be tempted to assert that it doesn't matter if Green's functions with external ϵ-scalars are divergent (since they are anyway unphysical) and introduce a common wave function subtraction for W_i and W_σ, a wave function subtraction for ψ and a coupling constant subtraction for g, these being determined (as usual) by the requirement that Green's functions with real particles be rendered finite. This was the procedure adopted in the main by van Damme and 't Hooft [13]. On the other hand we could insist on all Green's functions (including those with external ϵ-scalars) being finite, leading to the introduction of a plethora of new subtractions or equivalently coupling constants. We have shown [16,17] that it is only the latter procedure which leads to a consistent theory; the former manifestly breaks unitarity.

Now in a supersymmetric theory the complications described above can be safely ignored. The wave function renormalisations of W_σ and W_i are equal because of supersymmetry, and the evanescent couplings remain equal to their "natural" values after renormalisation. At first sight, this conclusion also appears to obtain when supersymmetry is softly broken, since the dimensionless couplings renormalise exactly as in the fully supersymmetric theory. This is not quite true, however, since there is nothing to protect the ϵ–scalars acquiring a mass through interacting with the genuine fields; and, indeed, precisely this happens [30]. In other words, the β–function for the ϵ-scalar mass $\tilde{m}$ is inhomogeneous with respect to $\tilde{m}$:

$$\beta_{\tilde{m}^2} = A(g, Y)\tilde{m}^2 + \sum_i B_i(g, Y)m_i^2 + \cdots, \tag{40}$$

where the m_i^2 are the genuine scalar masses, Y represents the Yukawa couplings and the $+\cdots$ denotes terms involving the gaugino mass(es) and the A-parameter(s). Moreover, the two–loop β–functions for the genuine scalar masses depend explicitly on the ϵ–scalar masses, when calculated using DRED. This fact would annoyingly complicate an extension to two loops of the standard running analysis relating the low energy values of the various soft parameters to the corresponding values at gauge unification. Fortunately, however, there exists a hybrid scheme [31] which decouples this ϵ-scalar dependence both from the β–functions and from the threshold corrections to the physical masses. At leading order, this scheme is arrived at from DRED by redefining the masses m_i^2 as follows:

$$m_i^2|_{\text{DRED}'} = m_i^2|_{\text{DRED}} - C_i(g)\tilde{m}^2 \tag{41}$$

where $C_i(g)$ is easily calculated [31]. The resulting β_{m^2} is independent of $\tilde{m}^2$ through two loops; and conveniently the same transformation removes the $\tilde{m}^2$ term from the one-loop relationship between the renormalised and physical scalar masses m^2. So in the DRED$'$ scheme, although $\tilde{m}$ evolves under the renormalisation group, it is decoupled from physical quantities and so can be safely ignored. To sum up, the DRED$'$ procedure for the standard running analysis (at two loops) and extraction of predictions for physical masses is:

(1) Use the DRED$'$ β-functions [30–35] to do the running analysis.

(2) Calculate the one loop corrections to convert the renormalised masses to the physical (pole) masses using DRED but with $\tilde{m}^2 = 0$. The top quark mass would not in any case have $\tilde{m}$ dependence, but note that the result for the one–loop gluon contribution to it is

$$m_t^{\text{pole}} = m_t(\mu)\left[1 + \frac{\alpha_3(\mu)}{3\pi}\left(5 - 3\ln\frac{m_t^2}{\mu^2}\right)\right] \tag{42}$$

not

$$m_t^{\text{pole}} = m_t(\mu)\left[1 + \frac{\alpha_3(\mu)}{3\pi}\left(4 - 3\ln\frac{m_t^2}{\mu^2}\right)\right] \tag{43}$$

as in DREG.

7 The NSVZ β-function

In this section we examine the β-functions for an $N = 1$ supersymmetric theory defined by the superpotential

$$W = \frac{1}{6}Y^{ijk}\Phi_i\Phi_j\Phi_k + \frac{1}{2}\mu^{ij}\Phi_i\Phi_j. \tag{44}$$

The multiplet of chiral superfields Φ_i transforms as a representation R of the gauge group $\mathcal{G}$, which has structure constants f_{abc}. In accordance with the non-renormalisation theorem [36], the β-functions for the Yukawa couplings β_Y^{ijk} are given by

$$\beta_Y^{ijk} = Y^{p(ij}\gamma^{k)}{}_p = Y^{ijp}\gamma^k{}_p + (k \leftrightarrow i) + (k \leftrightarrow j), \tag{45}$$

where $\gamma(g, Y)$ is the anomalous dimension for Φ. There exists an all-orders relation between the gauge β-function $\beta_g(g, Y)$ and γ which was first derived using instanton calculus [37]:

$$\beta_g^{\text{NSVZ}} = \frac{g^3}{16\pi^2}\left[\frac{Q - 2r^{-1}\text{Tr}\left[\gamma^{\text{NSVZ}}C(R)\right]}{1 - 2C(G)g^2(16\pi^2)^{-1}}\right]. \tag{46}$$

Here $Q = T(R) - 3C(G)$, $T(R)\delta_{ab} = \text{Tr}(R_aR_b)$, $C(G)\delta_{ab} = f_{acd}f_{bcd}$, $r = \delta_{aa}$ and $C(R)^i{}_j = (R_aR_a)^i{}_j$.

We have added a "NSVZ" label to both β_g and γ in Eq. 46 because of scheme dependence issues which we will discuss shortly. In the special case $Y = 0$, the fixed point $g^* = 0$, defined by

$$Q = \frac{2}{r}\text{Tr}\left[\gamma^{\text{NSVZ}}C(R)\right] \tag{47}$$

is important for duality, in the context of the conformal window identified by Seiberg [38]. We will return to this fixed point in the context of large-N expansions.

It turns out that if β_g and γ are calculated using DRED, then they begin to deviate from Eq. 46 at three loops [39,40]. The relationship between β_g^{NSVZ} and β_g^{DRED} has been explored recently, with the conclusion that there exists an analytic redefinition of g, $g \to g'(g, Y)$ which connects them. We emphasise that it is quite non-trivial that the redefinition exists at all; in the abelian case for example, the redefinition consists of a single term, but it affects four distinct terms (with different tensor structure) in the β-functions. By exploiting the fact that $N = 2$ theories are finite beyond one loop [27] it was possible to determine β_g^{DRED} at three loops by a comparatively simple calculation, and at four loops in the general case except for one undetermined parameter. What one learns from this is that it is highly non-trivial that the DRED and NSVZ results correspond to schemes which can be related in this manner. Use of what regularisation scheme would lead to the NSVZ result, which is associated with the holomorphic nature of the Wilsonian action? Presumably, for example, a combination of Pauli-Villars and higher derivatives [4]. Notwithstanding the existence of the exact NSVZ result, however, it is still important to have β_g^{DRED} as accurately as possible, because in calculating physical predictions DRED (or more accurately DRED$'$) is the scheme most often used. The three loop results for β_g^{DRED} and γ^{DRED} have found phenomenological applications [41,42]. The results for β_g in supersymmetric QCD (SQCD) with N_f flavours and N_c colours are:

$$16\pi^2\beta_g^{(1)} = (N_f - 3N_c)\,g^3,$$

$$(16\pi^2)^2\beta_g^{(2)} = \left(\left[4N_c - \tfrac{2}{N_c}\right]N_f - 6N_c^2\right)g^5,$$

$$(16\pi^2)^3\beta_g^{(3)} = \left(\left[\tfrac{3}{N_c} - 4N_c\right]N_f^2 + \left[21N_c^2 - \tfrac{2}{N_c^2} - 9\right]N_f - 21N_c^3\right)g^7.$$

$$\tag{48}$$

For $\beta_g^{(4)}$ we have only a partial result[43]:

$$
\begin{aligned}
(16\pi^2)^4 \beta_g^{(4)} \;=\;& \left(-\tfrac{2}{3N_c}N_f^3 + \left[\tfrac{100}{3} + 4\alpha + \tfrac{6\kappa-20}{3N_c^2} - \left(\tfrac{62}{3} + 2\kappa + 8\alpha\right)N_c^2 \right]N_f^2 \right. \\
& + \left[36(1+\alpha)N_c^3 - (34+12\alpha)N_c - \tfrac{8}{N_c} - \tfrac{4}{N_c^3} \right]N_f \\
& \left. - (6+36\alpha)N_c^4 \right) g^9
\end{aligned}
$$

$$(49)$$

where α is an as yet undetermined parameter, and where $\kappa = 6\zeta(3)$. (A recent application of the method of asymptotic Padé approximants[44] suggests that $\alpha \approx 2.4$.)

It is very interesting that the higher order group theory invariants found by van Ritbergen et al[45] in the corresponding calculation for QCD do not appear here. Of course the QCD calculation was done with DREG rather than DRED; but since these group structures first appear at four loops we would expect, for these particular terms, that DRED and DREG should give the same result at this order. It is an excellent check on both calculations, therefore, that when in the QCD case we go to the special case of $N = 1$ supersymmetry (by setting $N_f = \tfrac{1}{2}$ and putting the fermions in the adjoint representation) the new invariants cancel. It is also interesting to note that in the pure gauge theory, these invariants signalled the first contribution from non-planar structures to β_g in QCD; for SQCD, it remains possible that β_g^{DRED} is free of such structures to all orders (this is manifestly so for β_g^{NSVZ} in the absence of chiral superfields, of course).

Recently, some exact results for soft supersymmetry-breaking masses and couplings have been derived by Hisano and Shifman[46] using the holomorphy of the Wilsonian action. Soft breaking terms may be accommodated within the superfield formalism by the introduction of an external "spurion" field[47], $\eta \equiv \theta^2$. The renormalisation-group functions for soft breaking parameters may all be derived from the anomalous dimension γ^η of the chiral fields in the presence of the spurion[35]. γ^η may be expanded as

$$
\gamma^\eta = \gamma + \gamma^{[1]}\eta + \gamma^{[1]\dagger}\bar{\eta} + \gamma^{[2]}\bar{\eta}\eta,
\tag{50}
$$

where γ is the conventional anomalous dimension for Φ, in the absence of the spurion. Furthermore, simple rules[35] may be derived for obtaining γ^η directly from γ. It is possible to derive from the exact results of Hisano and Shifman[46] an elegant formula for the β-function for the gaugino mass M, namely

$$\beta_M^{\text{NSVZ}} = \frac{2}{g}\left[\frac{M\beta_g^{\text{NSVZ}} - 2g^3(16\pi^2 r)^{-1}\text{Tr}\left[\gamma^{[1]\text{NSVZ}}C(R)\right]}{1 - 2C(G)g^2(16\pi^2)^{-1}}\right], \tag{51}$$

where $\gamma^{[1]}$ is as defined in Eq. 50. This result is strikingly similar in form to the primordial NSVZ result for β_g in Eq. 46. The other soft breaking β-functions are also simply related to $\gamma^{[1]}$, as follows (from now on we suppress the "NSVZ" label):

$$\beta_h^{ijk} = \gamma^{(i}{}_l h^{jk)l} - 2\gamma^{[1](i}{}_l Y^{jk)l} \tag{52}$$

where $h^{ijk}\phi_i\phi_j\phi_k$ is the soft ϕ^3 interaction, and

$$\beta_b^{ij} = \gamma^{(i}{}_l b^{j)l} - 2\gamma^{[1](i}{}_l \mu^{j)l} \tag{53}$$

where $b^{ij}\phi_i\phi_j$ is the soft ϕ^2 interaction.

8 Large-N_f supersymmetric gauge theories

The large-N expansion is an alternative to conventional perturbation theory. In both QCD and SQCD, the large N_c expansion is of particular interest [48]; more tractable, however, is the large N_f expansion. Recently the leading and $O(1/N_f)$ terms in β_g and γ have been calculated (using DRED) for a number of supersymmetric theories [49]. We give below the results for SQCD (noting that we have rescaled the gauge coupling, $g \to g/\sqrt{N_f}$):

$$\gamma = -\frac{(N_c^2 - 1)}{N_f N_c}\hat{K}G(\hat{K}), \tag{54}$$

and

$$\begin{aligned}
\beta_g &= g\hat{K} - \tfrac{3N_c}{N_f}g\hat{K} + 4g\hat{K}\tfrac{N_c}{N_f}\int_0^{\hat{K}}(1-x)G(x)\,dx \\[2mm]
&\quad - \tfrac{2g\hat{K}}{N_c N_f}\int_0^{\hat{K}}(1-2x)G(x)\,dx.
\end{aligned} \tag{55}$$

where $\hat{K} = g^2/(16\pi^2)$, and

$$G(x) = \frac{\Gamma(2-2x)}{\Gamma(2-x)\Gamma(1-x)^2\Gamma(1+x)}. \tag{56}$$

These results do not satisfy Eq. 46, because they were calculated using DRED. It is quite remarkable that the $O(1/N_f)$ corrections to the SQCD β-function depend only on simple integrals involving $G(x)$. G has a simple pole at $x = 3/2$ and consequently β_g has a logarithmic singularity at $g^2 = 24\pi^2$ and a finite radius of convergence in g. Using Eq. 55 for $N_f = 6$ and $N_c = 3$, values which lie in the conformal window $3N_c/2 < N_f < 3N_c$, we indeed find an infra-red fixed point in the gauge coupling evolution, corresponding to $g^* \approx 8$. It is interesting that the range of N_f such that $g^* < 24\pi^2$ is given by

$$aN_c < N_f < 3N_c,$$

where $a \approx 1.7$ depends weakly on N_c. This is remarkably close to the exact conformal window.

Of course the result for g^* is scheme dependent. In the NSVZ scheme it is possible to show that the $O(1/N_f)$ contribution to γ is in fact the same as in DRED, with the corresponding result for β_g being easily calculated from Eq. 46. One then finds that the fixed point (for $N_c = 3$) corresponds to $g^* \approx 7$.

Of course it is not clear that this regime is within the region of validity of our approximation: do we believe that the appropriate expansion parameter is N_c/N_f or $3N_c/N_f$? It would obviously be interesting if we could calculate more terms in the $1/N_f$ expansion. Even the $O(1/N_f^2)$ contribution presents considerable technical problems, however.

References

1. O. Piguet, Lectures at the 2nd International Winter School on High-Energy Physics (WHEP 96), Brazil, 1996, hep-th/9611003
2. O. Piguet, M. Schweda and K. Sibold, *Nucl. Phys.* B **174**, 183 (1980)
3. J.A. Dixon, *Comm. Math. Phys.* **140**, 169 (1991); F. Brandt, *Nucl. Phys.* B **392**, 428 (1993)
4. P. West, *Nucl. Phys.* B **268**, 113 (1986)
5. M.K. Gaillard, *Phys. Lett.* B **342**, 125 (1995); *Phys. Lett.* B **347**, 284 (1995)
6. G. 't Hooft and M. Veltman, *Nucl. Phys.* B **44**, 189 (1972)
7. W. Siegel, *Phys. Lett.* B **84**, 193 (1979)
8. D.M. Capper, D.R.T. Jones and P. van Nieuwenhuizen, *Nucl. Phys.* B **167**, 479 (1980)
9. W. Siegel, *Phys. Lett.* B **94**, 37 (1980)
10. J.G. Körner and M.M. Tung, *Z. Phys.* C **64**, 255 (1994)
11. M. Misiak, *Phys. Lett.* B **321**, 113 (1994)
12. D.R.T. Jones (unpublished) 1979; see also W. Siegel, P.K. Townsend and P van Nieuwenhuizen, Proc. 1980 Cambridge meeting on supergravity, ITP-SB-80-65

13. R. van Damme and G. 't Hooft, *Phys. Lett.* B **150**, 133 (1985)

14. I. Jack, *Phys. Lett.* B **147**, 405 (1984);
    G. Curci and G. Paffuti, *Phys. Lett.* B **148**, 78 (1984);
    D. Maison, *Phys. Lett.* B **150**, 139 (1985)

15. I. Jack and H. Osborn, *Nucl. Phys.* B **249**, 472 (1985)

16. I. Jack, D.R.T. Jones and K.L. Roberts, *Z. Phys.* C **62**, 161 (1994)

17. I. Jack, D.R.T. Jones and K.L. Roberts, *Z. Phys.* C **63**, 151 (1994)

18. D.Z. Freedman, K. Johnson and J.I. Latorre, *Nucl. Phys.* B **371**, 353 (1992)

19. D.R.T. Jones and J.P. Leveille, *Nucl. Phys.* B **206**, 473 (1982)

20. H. Nicolai and P.K. Townsend, *Phys. Lett.* B **93**, 111 (1980)

21. G. Bonneau, *Int. J. Mod. Phys.* A **5**, 3831 (1990)

22. R.W. Allen and D.R.T. Jones, *Nucl. Phys.* B **303**, 271 (1988)

23. C.M. Hull and P.K. Townsend, *Phys. Lett.* B **191**, 115 (1987)

24. H. Osborn, *Ann. Phys.* **200**, 1 (1990)

25. S.A. Larin, *Phys. Lett.* B **303**, 113 (1993);
    C. Schubert, *Nucl. Phys.* B **323**, 478 (1989);
    J.G. Körner, D. Kreimer and K. Schilcher, *Z. Phys.* C **54**, 503 (1992)

26. L.V. Avdeev, G.A. Chochia and A.A. Vladimirov, *Phys. Lett.* B **105**, 272 (1981)

27. P.S. Howe, K.S. Stelle and P. West, *Phys. Lett.* B **124**, 55 (1983);
    P.S. Howe, K.S. Stelle and P.K. Townsend, *Nucl. Phys.* B **236**, 125 (1984)

28. F. del Aguila et al, *Nucl. Phys.* B **250**, 225 (1985)

29. I. Antoniadis, J. Ellis and G.K. Leontaris, hep-ph/9701292

30. I. Jack and D.R.T. Jones, *Phys. Lett.* B **333**, 372 (1994)

31. I. Jack, D.R.T. Jones, S.P. Martin, M.T. Vaughn and Y. Yamada, *Phys. Rev.* D **50**, R5481 (1994)

32. S.P. Martin and M.T. Vaughn, *Phys. Lett.* B **318**, 331 (1993)

33. S.P. Martin and M.T. Vaughn *Phys. Rev.* D **50**, 2282 (1994)

34. Y. Yamada, *Phys. Rev. Lett.* **72**, 25 (1994)

35. Y. Yamada, *Phys. Rev.* D **50**, 3537 (1994)

36. M.T. Grisaru, W. Siegel and M. Roček, *Nucl. Phys.* B **159**, 429 (1979)

37. V. Novikov et al, *Nucl. Phys.* B **229**, 381 (1983);
    V. Novikov et al, *Phys. Lett.* B **166**, 329 (1986);
    M. Shifman and A. Vainstein, *Nucl. Phys.* B **277**, 456 (1986)

38. N. Seiberg, *Nucl. Phys.* B **435**, 129 (1995)

39. I. Jack, D.R.T. Jones and C.G. North, *Nucl. Phys.* B **473**, 308 (1996)

40. I. Jack, D.R.T. Jones and C.G. North, *Phys. Lett.* B **386**, 138 (1996)

41. P.M. Ferreira, I. Jack and D.R.T. Jones, *Phys. Lett.* B **387**, 80 (1996)

42. C. Kolda and J. March-Russell, hep-ph/9609480
43. I. Jack, D.R.T. Jones and C.G. North, *Nucl. Phys.* B **486**, 479 (1997)
44. I. Jack, D.R.T. Jones and M.A. Samuel, hep–ph/9706249, *Phys. Lett.* B (to be published)
45. T. van Ritbergen, J.A.M. Vermaseren and S.A. Larin, hep–ph/9701390
46. J. Hisano and M. Shifman, hep–ph/9705417
47. L. Girardello and M.T. Grisaru, *Nucl. Phys.* B **194**, 65 (1982)
48. E. Witten, *Ann. Phys.* **128**, 363 (1980)
49. P.M. Ferreira, I. Jack and D.R.T. Jones, *Phys. Lett.* B **399**, 258 (1997); P.M. Ferreira et al, hep-ph/9705328, *Nucl. Phys.* B (to be published)

SUPERSYMMETRY and STRING THEORY

MICHAEL DINE
Santa Cruz Institute for Particle Physics
University of California
Santa Cruz, CA 95064

Supersymmetry was first discovered in string theory. This essay describes how supersymmetry is crucial to our present understanding of string theory, and why if string theory is correct, low energy supersymmetry is a likely consequence.

1 INTRODUCTION

Many particle theorists are convinced that nature is supersymmetric. This despite the fact that, as of this writing, there is at best only one compelling piece of experimental evidence (the unification of couplings) to support this view, and that evidence is very indirect. We have gotten to this situation because there are at least three reasons to hypothesize a role for supersymmetry in low energy physics:

- The hierarchy problem

- The unification of couplings

- Low energy supersymmetry emerges naturally from string theory, the only known theory which unifies gravity and other interactions in a quantum-mechanically consistent way.

The hierarchy problem and the unification of couplings are discussed thoroughly in many reviews (including those in this volume), and I will not review them here. My focus will be principally on the role of supersymmetry in string theory, and on the role that string theory suggests for supersymmetry.

A lawyer might say that each of these three points above helps build a circumstantial case for supersymmetry. However, each can also be called into question. The hierarchy problem is often described as a problem of fine tuning. At a more primitive level, it is simply a problem of dimensional analysis: why isn't the Higgs mass as large as the largest mass scales in nature? When simple-minded dimensional analysis fails, it is often because there is some underlying symmetry principle. Supersymmetry is the only symmetry we know which could explain the lightness of the Higgs field. While this argument seems compelling, it is perhaps troubling that there is a problem which on its face looks rather similar: the cosmological constant problem. This is another striking failure of dimensional analysis. For this failure of dimensional analysis, we know of no symmetry which can provide an explanation. We can hope that we will yet find such a symmetry, or that alternatively the explanation may be dynamical, either involving a light field, or caught up in the subtleties of quantum gravity (perhaps in string theory). Still, we are left with the uneasy feeling that some fine tuning problems may be solved by mysterious cancellations between high and low energy phenomena. The unification of couplings, while reasonably impressive, involves only one number, and might turn out to be a coincidence. And while many feel that string theory is so extraordinary it could not fail to be correct, we seem to be a long way from understanding how the theory might make contact with nature, and what role supersymmetry might play.

This essay will not provide the reader with any pat answers to these questions. It will be largely descriptive, explaining the role of supersymmetry in our current understanding of string theory. It turns out that even at weak coupling it is difficult to describe string configurations which are not at least approximately supersymmetric, and supersymmetry has been crucial to all of the recent progress in understanding strongly coupled strings. Supersymmetry also figures crucially in all present ideas about string phenomenology. At this point it is clear that the phrase "superstring theory" is really just a code for some larger structure. This structure looks in some regimes like one or another type of string theory, in others like a field theory. It has regimes in which we have no idea how to describe at all. But, within our current understanding, supersymmetry is crucial to understanding the behavior in any of these regimes. The role of supersymmetry in string theory will be the subject of the second section. In the third, we will turn to the issue of relating string theory to nature. Given our limited understanding of the theory, this is clearly a risky endeavor. Still, there seem to be a set of fundamental questions which must be faced in understanding how string theory might describe the universe we observe. We will end with some timid speculations about how supersymmetry and string theory might resolve some of the deep questions of particle physics.

2 Supersymmetry in String Theory

Among the glaring limitations of the standard model is that it does not incorporate general relativity. Einstein's theory of gravitation does not seem to make sense when viewed as a quantum theory. It is non-renormalizable, which means that it almost certainly has to be incorporated in some larger structure. It suffers, as well, from Hawking's black hole information paradox, which seems to have no resolution within a field theory picture[3].

170

Until 1984, it was perhaps appropriate to write off these problems as issues of ultra high energy physics, which would not be accessible to experimental study on any forseeable time scale. But in superstring theory[1], we seem to have stumbled on a framework which does incorporate both general relativity and the gauge interactions of the standard model, and obey the rules of quantum mechanics. String theory has none of the ultraviolet divergences of conventional quantum field theories; it is completely finite. For example, in string theory, it is possible to calculate graviton scattering amplitudes without encountering divergences. Further theoretical support for the theory comes from the realization in recent years that there is only one such theory; all of the known consistent string theories are particular limiting cases of this larger, "M" theory[2] Moreover, in the last few years, a great deal of evidence has accumulated that string theory resolves one of the great puzzles of general relativity, the black hole information pardox[3].

Remarkably, string theory is a structure which we have stumbled on by accident. String theory was discovered as an outgrowth of attempts to understand certain properties of strong interaction amplitudes. These could be explained if one supposed that hadrons were well described as excitations of strings. It is long since clear that the basic degrees of freedom strong interactions are quarks and gluons. The question of why QCD produces stringlike excitations remains an active area of research.But suppose one forgets this motivation, and simply hypothesizes that the fundamental objects in nature are strings, rather than point particles. Then one has a textbook problem, of quantizing a string no different than that described in standard undergraduate classical mechanics texts[4]. Such strings are characterized by a tension, and by a dimensionless number which describes the coupling of three strings. There are an infinite number of vibrational modes of the string, as well as translatonal modes (momenta). In a quantum theory, these states each have a definite energy and momentum, and thus a definite mass. The only subtlety is that one must make sure that the string interacts in a Lorentz invariant fashion (e.g. the states must fall in suitable representations of the Lorentz group), while at the same time the quantum mechanics of the string is unitary. When one imposes these requirements, one finds that the resulting theories possess:

- General coordinate invariance, i.e. general relativity.

- Yang-Mills Gauge Fields

- Supersymmetry – Supersymmetry appears in these theories as a local symmetry, analogous to general coordinate invariance or Yang-Mills gauge invariance, and presumably equally fundamental.

Again, it must be stressed that these features are not imposed on the theories, but emerge automatically, simply from the hypothesis that the fundamental entities are strings.

The most simple-minded treatment yields theories which are ten-dimensional. Five consistent theories emerge in this way: the Type I theory, with open and closed strings, the Type IIA and IIB theories, and two "heterotic" string theories, all theories of closed strings only. These theories are known to have classical solutions corresponding to compactification of some of the "extra" dimensions, and a huge number of four dimensional solutions are known. These include states with:

- General relativity

- Yang-Mills gauge groups, including the standard model gauge group.

- Three generations of quarks and leptons.

- Light Higgs particles

- Gauge coupling unification

- Calculable Yukawa couplings

- $N=1$ Supersymmetry.

It is exciting that supersymmetry appears in this list. When supersymmetry is introduced simply to solve the hierarchy problem, one sometimes has the uneasy feeling that it is simply a trick, concocted to arrange cancellation between various Feynman diagrams. When the symmetry is made local (supergravity), one obtains non-renormalizable theories, and it is not clear whether they make sense. But in superstring theory, supersymmetry is essential to the entire structure. Supersymmetry was actually *discovered* in the framework of string theory. String theory, then, provides a certain rationale for the existence of supersymmetry.

3 String Phenomenology

Given all of these remarkable features, what stands in the way of developing a string phenomenology, determining, for example, the soft breaking parameters, and subjecting the theory to experimental tests? In order to calculate physical quantities, we need a small parameter, and yet string theory is asserted to have no small parameters. What the theories do have is a huge ground state degeneracy. In ten dimensions, for example, in all of these theories, there are several massless states. In addition to the graviton, the gravitinos, gauge bosons and others, there is a scalar field known as the dilaton. There is no potential for this field, so the field can have *any* vacuum expectation value. This is something unfamiliar in conventional, non-supersymmetric field theories, but which makes perfectly good sense in the framework of supersymmetry, where non-renormalization theorems can forbid or suppress a potential. Such fields are called moduli. Now it turns out that for some values of the vev, it is possible to make a perturbation expansion, i.e. the theory is weakly coupled. For others, the theory is strongly coupled. So in some sense there is a free parameter, but, since it is the expectation value of a dynamical field, one can hope that it will be determined by the dynamics of the theory. For example, quantum effects might give rise to a potential for ϕ.

When one considers the compactification of the theory to fewer dimensions (say $3 + 1$), there are typically many more moduli. One can think of these as describing the size and shape of the compact space, as well as the expectation values of gauge fields with indices on the compact space, and others. For some values of these moduli (some "regions of the moduli space") the theory is weakly coupled and one can perform a perturbative expansion. For others it is strongly coupled. What is often called the "second superstring revolution" is really an exploration of this moduli space. Exploiting supersymmetry, one can show that the strong coupling limit of various string theories are the weak coupling limits of others. For example, the strong coupling limit of the $O(32)$ heterotic string in ten dimensions is the weakly coupled Type I theory; the strong coupling limit of the IIB theory is the IIB theory; the strong coupling limit of the IIA theory is a theory, not yet fully understood, whose low energy limit is *eleven dimensionl supergravity*.

It should be stressed that apart from these continuous choices, there are many discrete choices. For example, while there are ground states with three generations of quarks and leptons, there are also ground states with four, five, 100, and many with no chiral generations at all. There are also potential ground states in more, and less, than four dimensions.

3.1 Vacuum Selection and Supersymmetry Breaking

The great question in string theory, then, is what selects among these many possible vacua. In a non-supersymmetric theory, the existence of moduli at the classical level would be merely a curiousity; quantum effects would eliminate any such vacuum degeneracy in low orders of perturbation theory. In supersymmetric theories, these degeneracies often persist even *beyond* perturbation theory. For example, in ten dimensions the dilaton is part of the supergravity multiplet, and it turns out that all of the terms in the effective action with up to two derivatives are completely fixed by supersymmetry[5]. It is simply not possible to write a potential for ϕ consistent with the local supersymmetry. So any value of ϕ seems to be as good as any other. Similar arguments apply to compactifications of the theory with more than $N = 1$ supersymmetry. There are, for example, a large set of compactifications to four dimensions with $N = 4$ supersymmetry, and supersymmetry implies that these are all exact, non-perturbative ground states of the theory.

In four dimensional compactifications with only $N = 1$ supersymmetry, one can develop a potential for the moduli. This is the case in models of "gluino condensation," for example[6]. However, such a potential will always tend to zero as the coupling tends to zero. So it is difficult to see how there can be a minimum of the potential in any region of the moduli space where the coupling is small[7].

The main hope that this problem has a solution is the fact that the couplings we observe in nature are all weak. The question, then, is how can the moduli be stabilized at such a point. The strong coupling problem described above means that, whatever the numerical values of the coupling, weak coupling string theory *cannot* be valid in any ground state which is to describe the real world. We have no tools at the present time with which we could explore such a state.[a] Here, again, supersymmetry can play a significant role, however. In [9], it was argued that if the moduli are stabilized for small numerical values of the coupling, it may be possible to understand why there is a large hierarchy, why the low energy (i.e. multi-TeV) theory is approximately supersymmetric, why the couplings are unified. Even better, one can make precise, numerical predictions, in principle, of some quantities. The point is that in four dimensional string theories, the gauge couplings and superpotential are holomorphic functions of the various moduli. Holomorphy and discrete symmetries greatly constrain the form of these functions. If $e^{-8\pi^2/g^2}$ is numerically small (as it seems to be in nature), then these quantities are hardly corrected from their weak coupling values. Stabilization of the moduli occurs because the Kahler potential of these fields is not restricted by these symmetries. With the limitations of of our current understanding, one can only speculate that such phenomena occur. However, given that the couplings *are* weak, and that string theory has so many impressive features, it seems a good working hypothesis. From this point, one can try and build a phenomenology, by studying particular string compactifications, and calculating at low orders those quan-

---

[a]One might hope that using dualities, the relevant strong coupling region could be related to the weak coupling region of some other string theory. But in this other region, because the coupling is weak, it will not be possible to stabilize the moduli[8]

tities which receive only small corrections as one passes to the weak coupling limit.

This problem of the vacuum degeneracy is perhaps the most serious difficulty in relating string theory to nature. Related to this problem is the problem of the cosmological constant: in string theories in which the vacuum energy can be reliably calculated, it is generally of order the scale of supersymmetry breaking. If string theory does solve this problem, some entirely new ideas are probably required. There are at least two serious (promising is perhaps two strong a word at this point) suggestions for how the cosmological constant problem might be solved. The first is the observation of Susskind that string theory is "holographic," and that, in particular, there are not nearly as many degrees of freedom as one might naively expect[10]. The second, is due to Witten[11], who notes that in three dimensions, it may make sense to have broken supersymmetry with vanishing cosmological constant. No one has yet developed either of these suggestions into a precise explanation of why the cosmological constant would vanish in superstring theory with broken supersymmetry.

4 Recent Progress and Insights

String theory is a mysterious structure. As we have described it, there are five string theories, described by considering the actual motion of strings. There are also numerous solutions, characterized by discrete choices as well as by continuous parameters called moduli. We don't possess (at least in any truly useful sense) a field theory of strings, one whose symmetries would be manifest, and which might permit some direct, non-perturbative attack on questions such as the moduli potential. As a result, we have to use more indirect methods to probe the theory.

One important tool in any such analysis is the low energy effective action for the light fields. This is a particularly powerful tool when the low energy theory possesses a certain amount of supersymmetry, as well perhaps as other symmetries. The light fields might include the graviton, the gravitino and other states related to them by supersymmetry, gauge bosons, fermions, moduli and others. We have already given an example of this sort of reasoning: in ten dimensions, with $N = 1$ supersymmetry, and in four dimensions with $N = 4$ supersymmetry, supersymmetry is enough to prevent the appearance of any potential for the moduli; vacuua which are classically allowed at tree level are therefore exact quantum mechanical ground states of the theory. For $N = 1$ (fortunately) this result does not hold, but it is still possible to derive general results by rather simple reasoning. For example, one can prove that no potential for the moduli is generated in perturbation theory. To do this, one notes, first, that the dilaton, which is the coupling constant of the theory, is in a supermultiplet with an axion. A simple argument[1] shows that string theory always has a "Peccei-Quinn" symmetry under which this axion field shifts by a constant. Because the superpotential must be a holomorphic function of the coupling and must respect this symmetry, it must be independent of the dilaton (coupling), at least in perturbation theory. $N = 1$ symmetry allows one to prove numerous other results. One can show that even if supersymmetry is broken non-perturbatively at weak coupling, the cosmological constant term in the lagrangian cannot vanish[12]. Numerous other results have been proven in the past by these methods.

4.1 Supersymmetry as a Tool for Understanding String Theory

The past three years have seen a great deal of progress in string theory. In particular, under the heading of "duality," many aspects of non-perturbative string theory have been understood. A review of these developments would fill volumes, but supersymmetry has been crucial to all of them. Among the most striking discoveries is that there is really only one string theory. Starting with, say, the weakly coupled heterotic string theory, by varying the moduli one comes to the Type I theory. Similarly, if the heterotic theory is compactified, by varying the moduli one can find the various type II theories. Perhaps most suprising of all, if one takes the strongly coupled limit of the Type IIA theory, one finds a theory whose low energy limit is eleven-dimensional supergravity.

All of these strange connections have been found by exploiting supersymmetry. In particular, in most of the well-understood examples, one has more than $N = 1$ supersymmetry, in terms of four dimensional counting. In such cases, the supersymmetry algebra admits central charges. This, in turn, often means that one can prove exact mass formulas for certain states, known as "BPS" states. Because they are exact, these formulas are true for all values of the moduli. This means that one can determine the spectrum of such states, and their masses, at weak coupling, and then extrapolate reliably to strong coupling. One can then compare, for example, the weak coupling spectrum of one string theory with the strong coupling spectrum of another. Similarly, in many cases, the supersymmetry completely determines the form of the terms with two derivatives in the effective action, so one knows the action even at strong coupling. This, again, has been used to provide evidence for duality.

4.2 Supersymmetry and the Structure of Space-Time

Perhaps the most remarkable recent development is a proposal for a non-perturbative formulation of string theory[13]. The conjecture, in its most basic form, is very simple. Supersymmetry is a crucial ingredient – supersymmetry is essential for the very existance of space-time!

It is easiest to describe how this proposal looks in the eleven dimensional limit. There, the theory is supposed to be described an $SU(N)$ matrix quantum mechanics. This quantum mechanics is supersymmetric; it is the reduction to $0 + 1$ dimensions of 10 dimensional supersymmetric Yang Mills theory. As such, it has 16 supersymmetries in ten dimensions. Without going into great detail, it is not difficult to explain some of the remarkable features of this theory (which has been dubbed "M(atrix) Theory"). The matrices are just the reduction of the ten dimensional gauge fields, i.e.

$$X_i(t) = A_i(t). \tag{1}$$

The potential for the X's is just

$$V(X) = -\frac{-1}{4}\text{Tr}([X_i, X_j])^2. \tag{2}$$

This potential has the feature that it vanishes if the X_i's are diagonal, i.e. if

$$\vec{X} = \text{diag}(\vec{x}_1 \ldots \vec{x}_N). \tag{3}$$

The $\vec{x}_i$'s are naturally interpreted as the location of particles. When the model is examined in more detail, it turns out that these particles have the correct properties to be identified as the graviton, gravitino, and anytisymmetric tensor field of eleven dimensional supergravity. At low energies, they scatter just as do gravitons in eleven dimensions.

There is significant evidence that this conjecture for a non-perturbative formulation of the theory is correct. Many of the intricate properties of string theory are reproduced, and there is hope of computing things which are new. One of the striking features of M(atrix) theory is the crucial role which supersymmetry seems to play. In a conventional quantum mechanical system, the vanishing of the potential described above would be a tree level accident; loop corrections would generate a potential, and the space-time interpretation would collapse. In this example, if there were only bosons, then this would already occur for the leading quantum correction. This is because, for large x, the off-diagonal modes of the matrices $\vec{X}$ have frequencies proportional to $|\vec{x}|$. Their zero point energies, then, grown with x, which is to say that there is a potential for $\vec{x}$. In the supersymmetric theory, however, for each bosonic oscillator of a given frequency, there is a fermionic mode of the opposite frequency. The ground state for this system has a filled, negative energy state which cancels the bosonic term. This can be shown to persist as an exact property of the quantum theory. The supersymmetry of the matrix model is also the space-time supersymmetry. So in these theories, the very existence of space-time (the ability to have large $\vec{x}$'s, i.e. to separate gravitons and other particles) is a consequence of supersymmetry.

4.3   Outlook

Supersymmetry plays a crucial role in all present thinking about superstring theory, to the point that it is hard to imagine that if superstring theory is correct, we will not see evidence for supersymmetry at weak-scale energies.[b] Using supersymmetry as a tool, many remarkable properties of string theory have been elucidated. Perhaps the most remarkable is that all of the known consistent theories of gravity are limits of just one theory. Many find these facts persuasive – they are certainly extremely impressive. This provides strong support for the view that nature is supersymmetric, and that supersymmetry should appear "soon." A skeptic might remind us that ten years ago two-dimensional conformal invariance was thought to be the deep, underlying principal of string theory, one of the interesting outcomes of recent developments has been the overthrow of this view. Supersymmetry, similarly, may turn out to be only a convenient crutch in the study of the theory. Unlike two dimensional conformal invariance, though, supersymmetry is a fundamental new symmetry of space and time, and I think it reasonable to bet that it has a much deeper significance.

Acknowledgments

This work was supported in part by the U.S. Department of Energy.

1. M.B. Green, J.H. Schwarz and E. Witten, *Superstring Theory*, Cambridge University Press, Cambridge (1987).
2. For a recent review, with extensive references, see J. Schwarz, "The Status of String Theory," hep-th/9711029.
3. For an up to date review, see J. Maldacena, "Black Holes in String Theory," hep-th/9705078; "Black Hoes and D-Branes," hep-th/9705078; D. Youm, "Black Holes and Solitons in String Theory," hep-tg.8619946.

---

[b]It can hardly be said that we have an understanding of how supersymmetry breaking in string theory induces the breaking of electroweak symmetry; in making this identification I am relying on the standard ideas about supersymmetry breaking and the hierarchy problem.

4. J.B. Marion and S.T. Thornton, *Classical Dynamics of Particles and Systems*, Saunders, Fort Worth (1995).

5. E. Bergshoeff, M. De Roo, B. de Wit and P. Van Nieuwenhuizen, Nucl. Phys. **B195** (1982) 97.

6. J.P. Derendinger, L.E. Ibanez and H.P. Nilles, Phys. Lett. **155B** (1985) 65; M. Dine, R. Rohm, N. Seiberg and E. Witten, Phys. Lett. **156B** (1985) 55.

7. M. Dine and N. Seiberg, Phys. Lett. **162B** (299) 1985 and in *Unified String Theories*, M. Green and D. Gross, Eds. World Scientific, Singapore (1986).

8. M. Dine and Y. Shirman, Phys. Lett. **B377** (1996) 36, hep-th/9601175.

9. T. Banks and M. Dine, Phys. Rev. **D50** (1994) 7454, hep-th/9406132.

10. L. Susskind, J. Math Phys. **36** (1995) 6377, hep-th/940989.

11. E. Witten, Mod. Phys. Lett. **A10** (1995) 2153, hep-th/9506101.

12. M. Dine and N. Seiberg, Nucl. Phys. **B301** (1988) 357.

13. T. Banks, W. Fischler, S.H. Shenker, and L. Susskind, Phys. Rev. **D55** (1997) 5112, hep-th/9610043.

SUPERSYMMETRIC PARTICLE SEARCHES AT LEP

J.-F. GRIVAZ

Laboratoire de l'Accélérateur Linéaire,
IN2P3-CNRS et Université de Paris-Sud,
F-91405 Orsay, France

Searches for supersymmetric particles performed at LEP 1 and LEP 2 are reviewed. Using the MSSM with R-parity conservation as a reference model, the various analyses are briefly described, and the results are presented in terms of mass and coupling limits. Further implications of these results are discussed, including lower limits on the mass of a neutralino LSP, assuming the MSSM with GUT relations. Less conventional scenarii, among which those involving R-parity violation, are also investigated.

1 Introduction

1.1 The LEP history

From 1989 to 1995, LEP, CERN's large e^+e^- collider, operated at centre-of-mass energies close to the Z mass. Each of the four experiments, ALEPH, DELPHI, L3 and OPAL, collected an integrated luminosity in excess of 150 pb^{-1}, corresponding to more than 4 million hadronic Z decays. Starting in the autumn of 1995, the beam energy was raised in steps with the adjunction of superconducting RF cavities: the centre-of-mass energy reached 136 GeV at the end of 1995, 161 GeV (just above the threshold for W pair production) in the summer of 1996 and 172 GeV in the autumn of 1996. Integrated luminosities around 6, 11 and 11 pb^{-1} were accumulated by each experiment at $130 - 136$, 161 and $170 - 172$ GeV, respectively. It is foreseen that the operation in 1997 will take place at an energy of 183 GeV, with further increases to 192 GeV in 1998 and possibly close to 200 GeV in 1999. In the following, the operation at and near the Z peak will be referred to as LEP 1, the operation at $130 - 136$ GeV as LEP 1.5, and the operation at higher energies as LEP 2.

1.2 General features of supersymmetric particle searches at e^+e^- colliders

Most of the searches for supersymmetric particles at e^+e^- colliders are inspired by the phenomenology of the Minimal Supersymmetric extension of the Standard Model (MSSM), although their results often apply in a broader framework. In particular, it is generally assumed that R-parity is conserved and that the Lightest Supersymmetric Particle (LSP) is the lightest neutralino χ. This leads to the celebrated signature of supersymmetry: missing energy.

Since all charged supersymmetric particles are fairly democratically produced via s channel γ/Z exchange in e^+e^- annihilation, the searches are naturally directed toward the Lightest Charged Supersymmetric Particle (LCSP), typically a scalar lepton or a chargino. This is in contrast to the situation at hadron colliders where only strongly interacting particles such as squarks or gluinos are abundantly produced. Moreover, in actual model calculations, it usually turns out that cascades do not play an important role in the decay of the LCSP. The phenomenology is therefore rather straightforward, again in contrast to the situation at hadron colliders. As an example, if the LCSP is the right smuon $\tilde{\mu}_R$, the only production mechanism is s channel γ/Z exchange, with well defined couplings. The only decay mode is $\tilde{\mu}_R \to \mu\chi$ (at least if $m_{\chi'} > m_{\tilde{\mu}_R}$), which leads to a final state consisting of a $\mu^+\mu^-$ pair with missing energy. The absence of any signal in a given data sample is easily translated into a mass lower limit for the smuon.

This simple approach needs to be refined however, in particular when the production of neutral supersymmetric particles is taken into account. The lightest neutralinos or the sneutrinos can be pair produced via s channel Z exchange, in which case they contribute to the invisible Z width, while the production of heavier neutralinos may lead to observable final states, possibly with an energy threshold lower than the one of LCSP pair production. For instance, it is a rather common feature in supersymmetric models that $m_\chi + m_{\chi'} < 2m_{\chi^+}$, in which case $e^+e^- \to \chi\chi'$ could be kinematically allowed while $e^+e^- \to \chi^+\chi^-$ is not. The value of the $Z\chi\chi'$ coupling is however highly model dependent, and the search result usually cannot be translated into mass limits in a simple way. This unpleasant feature is more than compensated by the fact that, at the expense of fairly simple and general hypotheses, the results from searches in different channels can be combined in a consistent way, sometimes leading to more powerful constraints than could have naively been expected.

1.3 Specific features of the searches at LEP 1 and LEP 2

With the steady increase of the LEP beam energy since 1995, it could be expected that only the most recent results, namely those obtained at the highest energies, are relevant for supersymmetric particle searches. While this *a priori* belief is indeed found to be valid when comparing the results from LEP 2 and LEP 1.5, it does not fully hold when considering the LEP 1 results for two reasons. Firstly, the precision measurement of the Z boson properties, in particular of its total width, allow mass limits on supersymmetric particles to be obtained independently of their decay patterns. Secondly, the large statistics collected at LEP 1 allow Z decay branching ratios at the 10^{-6} level to be probed, which provides irreplaceable constraints in the neutralino sector.

As already mentioned, the main signature of supersymmetry is missing energy, at least when R-parity is conserved. From this point of view, there are big differences between LEP 1 and LEP 2. For processes such as chargino or slepton pair production, the kinematic limit of $m_Z/2$ had been reached at LEP 1 within months, with only a few thousand Z decays collected. The reason is that there is no irreducible background from standard model processes: energy can be lost along the beam axis in $\gamma\gamma$ interactions ($e^+e^- \rightarrow e^+e^-f\bar{f}$) where the final state electrons escape undetected in the beam pipe, or inside jets in the form of neutrinos from heavy flavour semileptonic decays in hadronic final states; both of these backgrounds can be eliminated by simple cuts on the direction and the isolation of the missing momentum. For processes such as Z decays into neutralinos, much smaller signal to background ratios deserve attention so that rare effects have to be considered such as fake missing energy due to instrumental effects or wrong missing momentum direction due to a conspiration of missing energy sources; this renders the analysis much more involved, but the techniques used are identical to those developed for the standard model Higgs boson search [1] in the $e^+e^- \rightarrow H\nu\bar{\nu}$ channel and they will not be detailed here.

At LEP 2, on the contrary, new standard model processes take place which lead to large missing energy in configurations much more difficult to disentangle from those expected from signals of supersymmetry. Examples of such processes are W pair production, with at least one $W \rightarrow \ell\nu$ decay, ZZ or $Z\gamma^*$ pair production, with $Z \rightarrow \nu\bar{\nu}$, or single W production in the process $e^+e^- \rightarrow We\nu$. It has nevertheless been possible to reduce these backgrounds to a level which remains negligible, at least with the modest integrated luminosities collected until now, without giving up too much signal efficiency, as will be explained in some detail further down. As the statistics accumulated increases, it may however become necessary to accept some level of background; but the situation is better from this point of view than at LEP 1 in the sense that this background is due to well calculable processes rather than to less controllable instrumental effects.

1.4 Synopsis and warnings

The rest of this chapter is organized as follows. The conventional scenario, with R-parity conservation and with a neutralino LSP, will be discussed first, starting with the basic facts, namely the mass and coupling limits obtained at LEP 1 and LEP 2, and then proceeding toward the interpretation in the MSSM, with in particular the derivation of mass limits for the LSP. Less conventional scenarii will be considered next, involving for instance a light gravitino or R-parity violation. Finally, a brief outlook toward the future will be given.

Unless otherwise stated, all limits are given at 95% confidence level (CL). The results quoted are, as much as possible, extracted from published papers. However, when no publication was available on a given topic at the time of writing, preliminary results submitted to the Summer '97 conferences, and to which the author had access, have been used. There has been no attempt toward a complete list of references. Only those from which the quoted results have been extracted are explicitly given. Moreover, it has not been judged useful to give theoretical references since the necessary theoretical background can be found in this book. Finally, although a major prediction of supersymmetry is the existence of a light Higgs boson, which could well lie within the reach of LEP 2, no discussion of this issue is presented here since the searches for supersymmetric Higgs bosons at LEP are described in detail in Ref. [1].

2   The conventional scenario: basic facts

In the conventional scenario, R-parity is conserved and the lightest supersymmetric particle is colourless and electrically neutral. This leaves the gravitino, a sneutrino or the lightest neutralino as possible candidates, which implies in turn that the LSP is weakly interacting with ordinary matter, hence the missing energy signature of supersymmetry. The conventional choice for the LSP is the lightest neutralino χ.

Although most of the searches for supersymmetric particles at LEP have been conducted using the MSSM as a reference model, the results obtained are often fairly general. It is the purpose of this section to review these basic facts.

2.1   Constraints from the Z width

One of the principal goals of the LEP 1 run, and in particular of the scans performed in the vicinity of the Z peak, was the precise determination of the parameters of the Z resonance: mass, total and partial decay widths, cross section at the peak. In the standard model, these parameters can be accurately computed as a function of the mass of the top quark and, to a lesser extent, of the Higgs boson, these heavy particles contributing via virtual effects, in particular in the Z boson propagator. This allowed the top quark mass to be predicted where it was finally measured, and now gives indications in favour of a light Higgs boson, a feature in agreement with the expectation from supersymmetry. Contributions from heavy supersymmetric particles to the Z parameters are however too small to allow interesting constraints to be obtained in a similar way.

On the other hand, if supersymmetric particles (or any kind of new particles) are light enough to be produced in Z decays, they will increase, often in a significant fashion, the total Z width with respect to the standard model expectation. Therefore, the agreement between the predicted and measured Z widths allows constraints on supersymmetric particles kinematically accessible in Z decays to be inferred. The Z width is now measured to be [2] 2494.7 ± 2.6 MeV. The prediction is 2502.0 MeV, for a top mass of 175 GeV/c^2, a Higgs mass of 150 GeV/c^2, and a value of 0.118 for the strong coupling constant α_S. This prediction is decreased to 2498.4 MeV, allowing for values as low as 0.115 for α_S and 169 GeV/c^2 for the top mass, corresponding to a one standard deviation change for each.[3] The value of 150 GeV/c^2 for the Higgs mass is an upper limit in supersymmetry, so no further decrease of the predicted Z width is possible from this origin. This leaves a maximum of 3.4 MeV for any contribution to the Z width due to supersymmetric particle production (at 95% CL, using Bayesian statistics). This corresponds to 2% of the partial width Γ_ν of the Z into a single flavour of $\nu\bar\nu$ pair.

Very light charginos would contribute $4.5\Gamma_\nu$ if pure gauginos, or $\sim 0.5\Gamma_\nu$ if pure higgsinos. This is large enough to exclude charginos practically up to $m_Z/2$, irrespective of their field content. This limit is much lower than the typical ones achieved nowadays at LEP 2, but it applies independently of the chargino decay pattern and remains the only one to be valid in some extreme configurations, as will be discussed further down.

Light sneutrinos would contribute $0.5\Gamma_\nu$ for each flavour. Because the phase space factor is less favourable than for charginos, this results into a limit of only 43 GeV/c^2 for a single flavour. For three mass degenerate flavours, the limit is again half the Z mass. Constraints on an invisible sneutrino, applicable if it is the LSP or even if the decay mode $\tilde\nu \to \nu\chi$ is dominant, can be inferred in a similar way from the measurement of the Z partial width into invisible final states (or equivalently from the effective number of neutrinos). In contrast to the situation of a few years ago, they turn out to be hardly stronger than those inferred from the total width measurement.

Somewhat weaker limits can also be derived from the Z width for sleptons or squarks. In the case of neutralinos, the couplings to the Z are highly model dependent (and even parameter dependent within a given model). The coupling is largest for higgsino-like neutralinos, and vanishes for pure gauginos. The best that can be done is therefore to set coupling (or branching ratio) upper limits as a function of the involved neutralino masses. However, these are superseded by those obtained from direct searches, except in the $Z \to \chi\chi$ case where the final state is invisible.

2.2 Constraints from direct searches at LEP 1

As discussed just above, it is only in the case of neutralinos that direct searches at LEP 1 play a major role. The most relevant channels leading to visible final states are $Z \to \chi\chi'$, kinematically the most favourable, and to a lesser extent $Z \to \chi'\chi'$. The main χ' decay mode is $\chi' \to \chi f\bar{f}$ which is accessed through virtual Z or sfermion exchange ($\chi' \to \chi Z^*$; $\chi' \to \tilde{f}^* \bar{f}$). In some particular instances the $\chi' \to \chi\gamma$ mode, which proceeds via loops, can become significant or even dominant. (This happens for small χ'–χ mass differences or if one of the neutralinos is almost purely higgsino and the other one almost purely photino.) The possible final states resulting from $\chi\chi'$ production are therefore: purely invisible ($\chi\chi\nu\bar{\nu}$), lepton pairs ($\chi\chi\ell^+\ell^-$), jets ($\chi\chi q\bar{q}$) or single photons ($\chi\chi\gamma$) with missing energy.

At LEP 1, the only significant standard sources of lepton or jet pairs are Z decays and $\gamma\gamma$ interactions. In the first case there is no missing energy, ignoring for the moment neutrinos involved in τ or heavy flavoured hadron decays. In $\gamma\gamma$ interactions, on the contrary, the spectator electrons (*i.e.* the electrons which radiated the photons participating in the collision) tend to remain undetected in the beam pipe, giving rise to a large amount of missing energy. The direction of the missing momentum is however close to the beam axis and there is only little missing transverse momentum p_T. In both Z decays and $\gamma\gamma$ interactions, the leptons or the jets therefore appear back-to-back in the plane transverse to the beam axis. Even for τ pairs, this (almost) coplanar topology is preserved in the visible decay products, and the same holds in the case of semileptonic decays of heavy flavoured hadrons. This feature is the basis of all searches for "acoplanar" leptons or jets. Of course, the detectors should be as hermetic as possible to ensure that no additional particles, for instance photons radiated at large angle, escape detection, thus inducing fake missing p_T. In the case of single photons, radiative Bhabha events ($e^+e^- \to e^+e^-\gamma$) with both electrons remaining undetected in the beam pipe can be eliminated by a cut on the transverse momentum of the photon corresponding to the maximum p_T that these two electrons can carry without entering the detector acceptance. The same cut also rejects events from $e^+e^- \to \gamma\gamma\gamma$, with two photons close to the beam axis. The actual analyses performed by the LEP experiments follow those general principles, but they ended up being appreciably more involved both because the detectors are not ideal and because standard physics is more complicated. (For instance, while a two-jet event with missing energy in one of the jets remains coplanar, this is no longer the case for a three-jet event.)

Events from irreducible standard model backgrounds are expected to be selected, but at a very low level. Indeed, a few spectacular "monojet" events were found,[4] but they can be explained with not unreasonably small probabilities as originating from four fermion final states such as $\ell^+\ell^-\nu\bar{\nu}$ or $q\bar{q}\nu\bar{\nu}$, reached through the process $e^+e^- \to Z^*\gamma^*$, with $Z^* \to \nu\bar{\nu}$ and $\gamma^* \to \ell^+\ell^-$ or $q\bar{q}$. Such an event is shown in Fig. 1(top). Similarly, no single photon signal was observed beyond the background expected from $e^+e^- \to Z^*\gamma$. With the full LEP 1 statistic, upper limits at the level of a few 10^{-6} have thus been set [5] for the product branching ratios $BR(Z \to \chi'\chi)BR(\chi' \to \chi Z^*)$ and $BR(Z \to \chi'\chi)BR(\chi' \to \chi\gamma)$, as shown in Fig. 2.

2.3 Results from the searches at LEP 2

The searches for supersymmetric particles performed at LEP 2 address sleptons, stops, charginos and neutralinos. Sneutrinos are expected to decay invisibly ($\tilde{\nu} \to \nu\chi$) and cannot be searched for efficiently. Gluinos are not produced directly in e^+e^- collisions and have therefore not been considered. In contrast to the case of generic squarks, for which the limits obtained at the Tevatron cannot be rivaled at LEP, large mixing can be expected in the stop sector, as will be discussed further down. This may lead to a top squark significantly lighter than all other squarks, hence the relevance of stop searches at LEP.

Sleptons and stops

Sleptons are pair produced and have been searched in the decay mode $\tilde{\ell} \to \ell\chi$, *i.e.* in final states consisting of acoplanar lepton pairs of the same flavour. Similarly, pair produced stops have been searched in the acoplanar jet topology expected to arise from the $\tilde{t} \to c\chi$ decay mode. (The normal decay mode, $\tilde{t} \to t\chi$, is kinematically forbidden.) This effectively flavour changing neutral current process occurs at the one loop level, and the corresponding decay width is small enough for the top squark to hadronize into a stop hadron before decaying, a feature which has been implemented by the LEP collaborations in their Monte Carlo generators. In the particular case where the sneutrino is lighter than the stop, the $\tilde{t} \to b\ell\tilde{\nu}$ decay mode is expected to become dominant. This rather peculiar mass hierarchy has also been addressed at LEP, but it will not be considered further here.

A number of backgrounds to these acoplanar lepton and acoplanar jet searches have already been discussed in the context of neutralino searches at LEP 1. Furthermore, abundant fermion pair production occurs through radiative return to the Z ($e^+e^- \to \gamma Z \to \gamma f\bar{f}$), leading mostly to final states

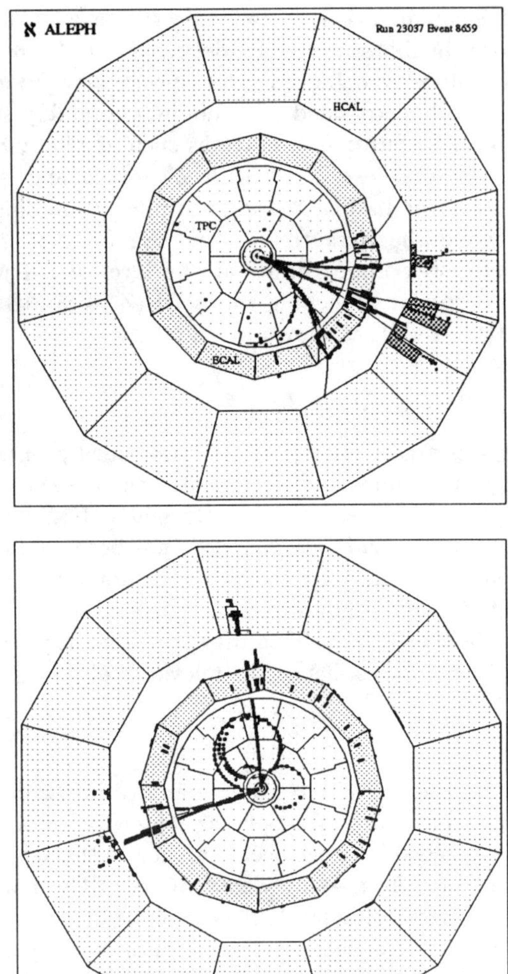

Figure 1: (top): A monojet event observed at LEP 1. This event is probably due to the process $e^+e^- \to Z\gamma^* \to \nu\bar{\nu}q\bar{q}$. (bottom): An acoplanar τ pair observed at LEP 2. The visible τ decay products are a ρ and an a_1. This event cannot be interpreted as resulting from $e^+e^- \to W^+W^-$ with two $W \to \tau\nu$ decays. It is rather due to $e^+e^- \to ZZ^*$, with $Z \to \nu\bar{\nu}$ and $Z^* \to \tau^+\tau^-$.

with large missing energy along the beam axis. All these essentially coplanar backgrounds are reduced at LEP 2 in a way similar to that at LEP 1. In addition, new backgrounds arise from processes such as $e^+e^- \to W^+W^-$, $We\nu$, Zee or ZZ^* which lead to four-fermion final states with missing energy originating from leptonic decays such as $W \to \ell\nu$ or $Z \to \nu\bar{\nu}$ or from electrons escaping undetected in the beam pipe in the case of $We\nu$ or Zee. Although these backgrounds are irreducible at some point, their kinematic properties are nevertheless sufficiently different from those expected from slepton or stop pair production to allow almost background free samples to be selected with the presently accumulated integrated luminosities. In the case of smuon searches for instance, requiring two leptons identified as muons selects only 1% of the W pairs. Moreover, the typical momentum of muons from smuon decay is smaller than that of muons from W decay, a feature which becomes more and more pronounced as the neutralino mass increases. In the case of acoplanar jets, the background from W^+W^- with $W \to \ell\nu$ is reduced by a veto against energetic leptons, and even against isolated particles to cope with $W \to \tau\nu$ decays. An additional requirement that the mass of the visible system should not exceed some value slightly lower than the W mass also helps in reducing the $We\nu$ background. In the end, a few events are selected, compatible with the background from standard model processes, typically W^+W^- with $W \to \tau\nu$, or ZZ^* with $Z \to \nu\bar{\nu}$. An example of an event probably due to this last process, with $Z^* \to \tau^+\tau^-$, is shown in Fig. 1(bottom).

Almost model independent mass limits can be derived for smuons and staus which are produced only via s channel γ/Z exchange, with well defined couplings, under the assumption that $\tilde{\mu} \to \mu\chi$ and $\tilde{\tau} \to \tau\chi$ are the only decay modes. (In the MSSM, this assumption is usually valid for right sleptons within the mass range of interest here, given the other constraints existing in the chargino/neutralino sector.) At the time of writing, a $\tilde{\mu}_R$ mass limit of 59 GeV/c^2 is obtained [6] for $\tilde{\mu}_R$–χ mass differences exceeding 10 GeV/c^2. In the case of selectrons, t channel neutralino exchange also contributes, usually increasing the production cross section. But this contribution is model dependent and no fully general result for selectrons can be extracted. In Fig. 3(left), an example of $\tilde{e}_R$ mass limit is presented,[6] valid in the MSSM for the set of parameters indicated. (Here, account has been taken of decay modes other than $\tilde{e}_R \to e\chi$.)

It is worth mentioning at this point that results from lower energy machines in this sector can still be relevant. The single photon final state has been studied at PEP, PETRA and TRISTAN. In the standard model, it originates through initial state radiation from the reaction $e^+e^- \to \gamma\nu\bar{\nu}$. A supersymmetric contribution could arise in a similar way from the process $e^+e^- \to \gamma\chi\chi$,

188

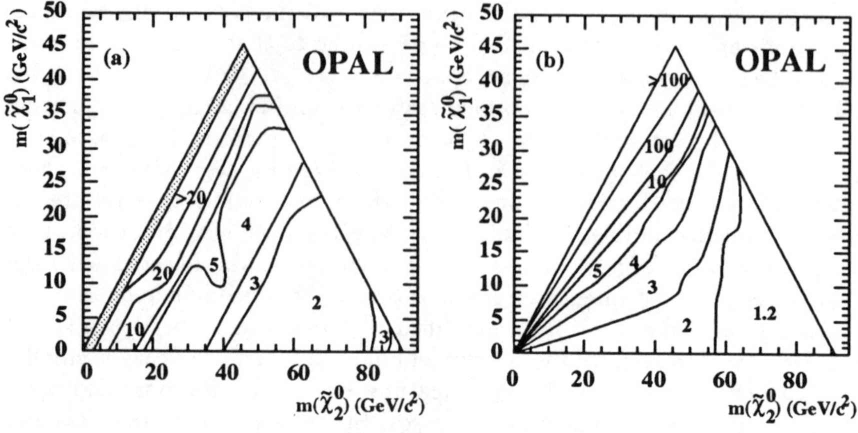

Figure 2: Upper limits on product branching ratios times 10^6 for neutralino production: (left) $\mathrm{BR}(Z \to \chi'\chi)\mathrm{BR}(\chi' \to \chi Z^*)$; (right) $\mathrm{BR}(Z \to \chi'\chi)\mathrm{BR}(\chi' \to \chi\gamma)$.

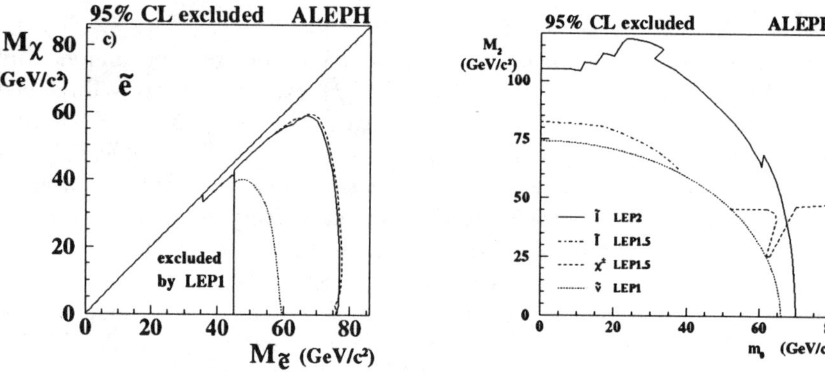

Figure 3: (left): Excluded regions in the plane of the $\tilde{e}_R$ and χ masses, in the MSSM with $\tan\beta = 2$ and for $\mu = -200$ GeV/c^2 (solid) and $\mu = 1000$ GeV/c^2 (dashed). The dotted curve is the LEP1.5 limit. (right): Excluded regions in the (m_0, M_2) plane, for $\tan\beta = 2$ and $\mu = -200$ GeV/c^2 (solid curve).

mediated by t channel selectron exchange. The absence of any excess above the standard model expectation allows selectrons with masses up to 79 GeV/c^2 to be excluded (90% CL) in the case of an almost massless and photino-like LSP.[7]

In the case of stops, the mass limit depends on the value of the mixing angle in the stop sector which controls the coupling of the stop to the Z. This coupling is largest for a pure left stop (null mixing angle) and vanishes for a value of the mixing angle close to 0.98 radians. The results [8] for these two extreme cases are shown in Fig. 4(left). The exclusion domain obtained at the Tevatron is extended toward smaller $\tilde{t}$–χ mass differences.

Charginos and neutralinos

Pair production of charginos is mediated by s channel γ/Z exchange, and by t channel sneutrino exchange, with destructive interference between the two processes. The influence of sneutrino exchange is largest for light sneutrinos and for gaugino-like charginos. The main chargino decay mode is $\chi^+ \to \chi f \bar{f}'$ which is accessed through virtual W or sfermion exchange ($\chi^+ \to \chi W^*$; $\chi^+ \to \tilde{f}^* \bar{f}'$). For gaugino-like charginos and sleptons significantly lighter than squarks, the latter contribution enhances leptonic decays. If sneutrinos are sufficiently light, two-body decays open up ($\chi^+ \to \ell^+ \tilde{\nu}$). The possible final states resulting from chargino pair production are therefore: acoplanar leptons ($\chi\nu\ell^+\chi\bar{\nu}\ell^-$ or $\tilde{\nu}\ell^+\tilde{\nu}\ell^-$), purely hadronic with missing energy ($\chi q\bar{q}'\chi q'\bar{q}$) or mixed ($\chi\nu\ell\chi q\bar{q}'$).

The searches for acoplanar leptons from charginos are similar to those designed for slepton pairs, with minor differences: the two leptons need not be of the same flavour, but their momenta are softer than in the case of sleptons. With four primary quarks, the topology of the purely hadronic final states is not identical to the acoplanar jet topology encountered in the case of stop pair production, which again induces modifications to the analyses: the visible mass cut is somewhat relaxed, but the events are required to exhibit a more spherical pattern. Undoubtedly, the gold plated topology is the mixed one: with an isolated lepton and missing energy in a hadronic environment, the backgrounds from fermion pair production and from $\gamma\gamma$ interactions are easy to reduce to a negligible level; the *a priori* harmful background from $e^+e^- \to W^+W^- \to \ell\nu q\bar{q}'$ can be distinguished from the signal by the mass of the hadronic system (close to the W mass for the W^+W^- background, smaller for a signal in the mass reach of LEP 2) and by the missing mass (close to zero for the W^+W^- background, significantly larger for the signal, a feature which is enhanced as the χ mass increases). In practice, the signal topology and the background conditions are also affected by the $\chi^+-\chi$ mass difference which controls the amount of visible energy: the $\gamma\gamma$ background and the trigger conditions are of greater concern for small mass differences, while the W^+W^- background is most dangerous for large mass differences. Therefore, the LEP collaborations were led to design matrices of analyses according to the type of final state and to the mass difference.

Small numbers of events were selected by the various experiments, but at a level compatible with the expectation from standard model processes. In the absence of signal, results on chargino production have been expressed in an almost model independent way in terms of cross section upper limit as a function of the χ^+ and χ masses, assuming dominance of the decay through virtual W exchange (*i.e.* heavy sleptons and squarks). (The "almost" restriction comes from the fact that, for a given χ^+ mass, the decay kinematics depend not only on the $\chi^+-\chi$ mass difference but also, however to a much lesser extent, on the detailed χ^+ and χ field content.) An example of such limits[9] is given in Fig. 4(right). Further specification of the model is needed to translate these results into chargino mass limits. The statement can however be made in a fairly general way that the kinematic limit of 86 GeV/c^2 is practically reached for gaugino-like charginos when all sfermion masses are very large.

Neutralino production, $e^+e^- \to \chi_i\chi_j$ where χ_i and χ_j stand for any of the neutralinos χ, χ', χ'',...proceeds via s channel Z exchange and via t channel selectron exchange. In contrast to chargino production, the interference is normally constructive. The process with the lowest threshold leading to visible final states is $e^+e^- \to \chi\chi'$. As already discussed in the context of LEP 1, acoplanar lepton pairs and acoplanar jets are the relevant topologies. The analyses are thus similar to those set up to search for sleptons or stops. Further selections were developed for the other processes, involving multileptons or isolated photons. The range of application of these selections depends strongly on the field contents of the various neutralinos involved, as they control both the production cross sections and the decay branching ratios, including cascade decays. It is therefore difficult to express the results in a way at the same time general and meaningful. In the absence of any signal up to a centre of mass energy of 172 GeV, the discussion of the implications of the searches for neutralinos is deferred to the next section.

3   The conventional scenario: interpretation

As indicated a number of times in the previous section, the results obtained by the various LEP collaborations usually cannot be translated directly into general supersymmetric particle mass limits; further specification of the model is needed for that purpose. After specifying the theoretical framework commonly referred to under the logo MSSM by the LEP collaborations, the basic facts presented in the previous section will be turned into constraints on the parameters of this model, and it will be shown how the combination of various inputs allows mass lower limits on the lightest neutralino to be inferred.

3.1 Theoretical framework

In the Minimal Supersymmetric Extension of the Standard Model (MSSM), the gauge group of the minimal standard model is assumed, and only the minimal field content is introduced. In particular, there are just two Higgs doublets, and hence four neutralinos. The parameters needed to specify the model are a set of soft supersymmetry breaking masses and trilinear couplings, m_i and A_i, for all scalar doublets and singlets, and of soft supersymmetry breaking masses, M_i, for the three gauginos. In the Higgs sector, a supersymmetric Higgs mass term, μ, is introduced, and the ratio v_2/v_1 of the vacuum expectation values developed by the Higgs fields coupling to the up-type and down-type quarks is denoted $\tan\beta$.

Additional simplifying assumptions are commonly made, inspired by Supergravity (SUGRA) models: for all matter sfermions, a universal scalar mass m_0 and a universal trilinear coupling A at the Grand Unification (GUT) scale; and for the three gauginos, a universal mass $m_{1/2}$ at the GUT scale. The low energy parameters are then determined using the renormalization group equations. The gaugino masses evolve in the same way as the gauge coupling constants. As a result, the gluino is expected to be much heavier, typically 3.5 times, than the lighter chargino in large regions of the parameter space. It is also expected that sleptons are lighter than squarks, and right sleptons or squarks lighter than their left counterparts; the sneutrino mass should be similar to the left slepton mass, unless $\tan\beta$ is large in which case sneutrinos become lighter. Mixing among the left and right scalars is expected to be small, except in the stop sector because of the large mass of the top quark, and possibly for staus if $\tan\beta$ is large.

In "Minimal SUGRA", the hypothesis of universal scalar masses is extended to the Higgs sector, and the spontaneous breaking of the electroweak symmetry is triggered at low energy by radiative corrections to the Higgs masses. Imposing the proper mass for the Z boson reduces the set of parameters further: the μ parameter is determined from the others, up to its sign.

3.2 Experimental results

An example of results obtained within the framework of the MSSM has already been given for right selectrons in Fig. 3(left). Here, The GUT relation among gaugino masses has been assumed to calculate the contributions of the various neutralino exchanges in the production process and the branching ratio for the decay $\tilde{e}_R \to e\chi$. The results are presented for a specific value of $\tan\beta$, and for two representative choices of μ. The searches for all slepton flavours can be further combined assuming scalar mass universality, and turned into an exclusion region [6] in the (m_0, M_2) plane, as shown in Fig. 3(right). (M_2 and $m_{1/2}$ are related by $M_2 = 0.82m_{1/2}$.) Again, this exclusion is valid for the specified $\tan\beta$ and μ values.

Results on chargino production such as those shown in Fig. 4(right) are commonly translated into exclusion domains in the (M_2, μ) plane, for chosen values of $\tan\beta$. Since the cross section limits depend not only on the chargino mass but also on the mass of the lightest neutralino, the GUT relation between M_1 and M_2 has to be assumed. This allows in turn account to be taken of possible reductions in the detection efficiency due to cascade decays such as $\chi^+ \to \chi'Z^*$. The analysis is simplest with the assumption that all sfermions are sufficiently heavy not to affect the production cross section nor the decay branching ratios. Exclusion domains [10] determined in this way are presented in Fig. 5 for two values of $\tan\beta$, 1.41 and 35. (Such values are typical of the low and large $\tan\beta$ solutions in the so-called infra-red quasi fixed point scenario.) With the same hypotheses, all neutralino production cross sections and decay branching ratios can be calculated from the same set of parameters (M_2, μ and $\tan\beta$, still with the assumption of heavy sfermions). The additional exclusion domains resulting from the neutralino searches are also displayed in Fig. 5.

These exclusion contours can in turn be translated into chargino mass limits, as shown in Fig. 6 where the limit is displayed as a function of μ for gaugino-like charginos and as a function of M_2 for higgsino-like charginos. The decrease in the limit for very large values of M_2 is due to the fact that the χ^+–χ mass difference, and hence the selection efficiency, diminishes as M_2 increases. The impact of the neutralino searches is clearly visible for moderate values of M_2, allowing chargino masses to be indirectly excluded beyond the kinematic limit.

These results are modified for lower sfermion masses. Light sneutrinos reduce the production cross section for gaugino-like charginos, while light selectrons increase the production cross section for gaugino-like neutralinos. Since also the decay branching ratios are affected by the sfermion masses, the assumption of a universal scalar mass m_0 is made so that the various production

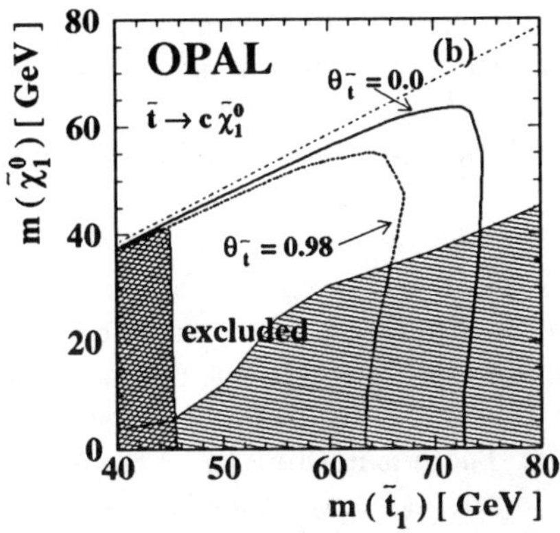

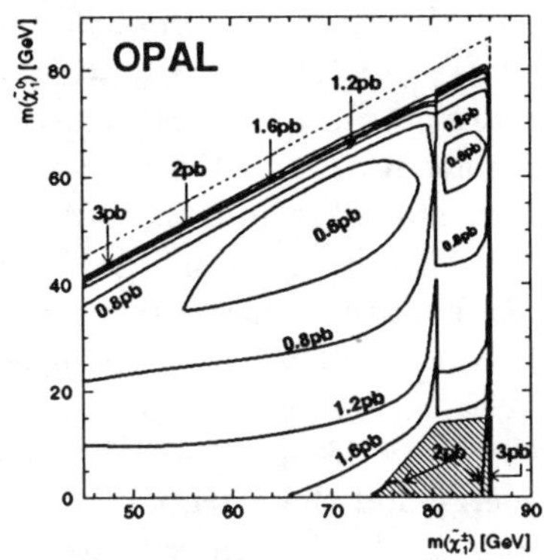

Figure 4: (left): Excluded regions in the plane of the $\tilde{t}$ and χ masses, for two values of the mixing angle in the stop sector, 0.0 and 0.98 rad corresponding to the most and least favourable cases, respectively. The cross hatched region has been excluded at LEP 1. The singly hatched region has been excluded by D0 at the Tevatron. (right): Cross section upper limit for chargino pair production at 172 GeV in the plane of the χ^+ and χ masses.

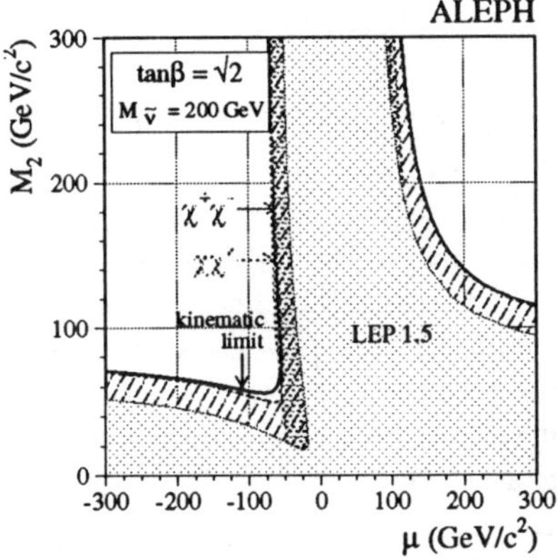

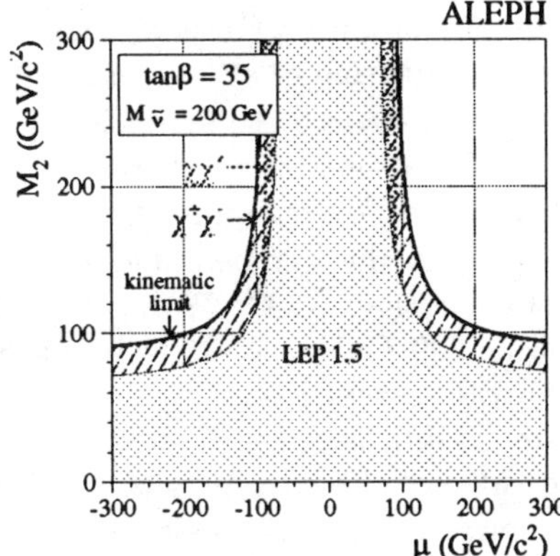

Figure 5: Excluded regions in the (M_2, μ) plane for $\tan\beta = 1.41$ (left) and $\tan\beta = 35$ (right), assuming heavy sfermions ($m_0 = 200\ \mathrm{GeV}/c^2$).

194

and decay processes are correlated. The influence of light sfermions is shown in Fig. 6(left). For $m_0 = 75$ GeV/c^2, for instance, it can be seen that the chargino mass limit is reduced by ~ 10 GeV/c^2, and it would degrade even further for $\mu > -100$ GeV/c^2 if the constraints from the neutralino searches were not taken into account.

Altogether, even if no hard number can be given as a lower mass limit for charginos, masses smaller than 75 to 85 GeV/c^2 are excluded over most of the parameter space. There remains however a loophole when the sneutrino mass is very close to but smaller than the chargino mass. In such a case, the two-body decay $\chi^+ \to \tilde{\nu}\ell^+$ dominates, but the final state lepton has too little energy and the detection efficiency vanishes. There is nothing to be recovered from the neutralino searches because the dominant decay mode, $\chi' \to \tilde{\nu}\nu$, is invisible, and the only limit remaining is the one deduced from the Z width measurement which is insensitive to the chargino decay pattern.

3.3 Neutralino LSP mass limits

Direct searches for the lightest neutralino χ cannot be performed efficiently at LEP 2. The relevant process is $e^+e^- \to \chi\chi$, leading to an invisible final state. At energies well below the Z peak, initial state radiation tagging ($e^+e^- \to \gamma\chi\chi$) had been used, leading to correlated constraints on the neutralino and selectron masses, as already mentioned in the context of selectron mass limits. This technique is however useless at LEP 2 energies because of the overwhelming background from $e^+e^- \to \gamma\nu\bar{\nu}$.

However, once the model is sufficiently specified, indirect limits can be obtained from the constraints on charginos and on the more massive neutralinos. For instance, by scanning the domain remaining unexcluded in Fig. 5, no set of (M_2, μ) value can be found for which the mass m_χ of the lightest neutralino is smaller than 31 GeV/c^2, for $\tan\beta = 1.41$ and assuming heavy sfermions ($m_0 = 200$ GeV/c^2). It is worth noticing that, in this particular example, the LEP 1 exclusion still plays some role, due to the fact that the neutralino search is limited, in the region near the point where the χ mass limit is set, by the value of the couplings to the Z rather than by kinematics. (In a general fashion, the LEP 1 constraints remain useful for low values of $\tan\beta$.) The lower limit obtained [11] for m_χ as a function of $\tan\beta$ and for heavy sfermions ($m_0 = 200$ GeV/c^2) is displayed in Fig. 7(left).

For lower sfermion masses, as already discussed, the assumption of a universal scalar mass m_0 is needed to render the analysis manageable. Since, as has been seen above, the constraints in the chargino/neutralino sector are weaker in that case, it is not surprising that the limits on m_χ also degrade.

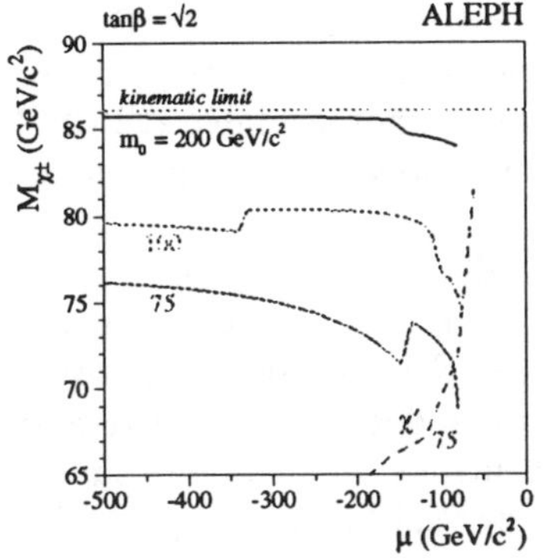

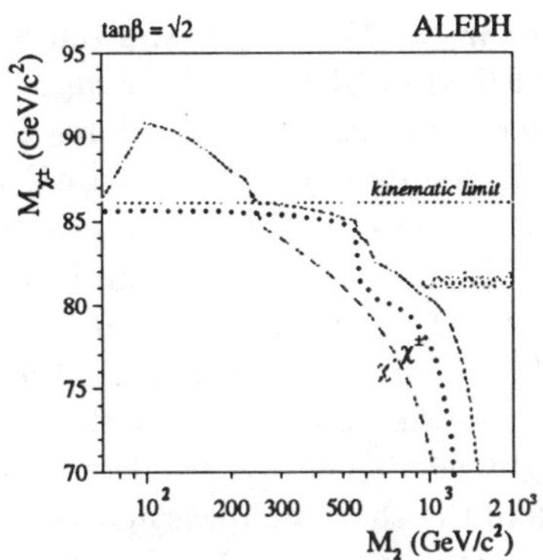

Figure 6: Chargino mass limit as a function of μ (left) and of M_2 (right), for $\tan\beta = 1.41$. On the left, the limits from the chargino searches are given for various values of m_0, and the dashed curve is the limit from the neutralino searches for $m_0 = 75$ GeV/c^2. On the right, the limits from the chargino, neutralino and combined searches are given for $m_0 = 200$ GeV/c^2.

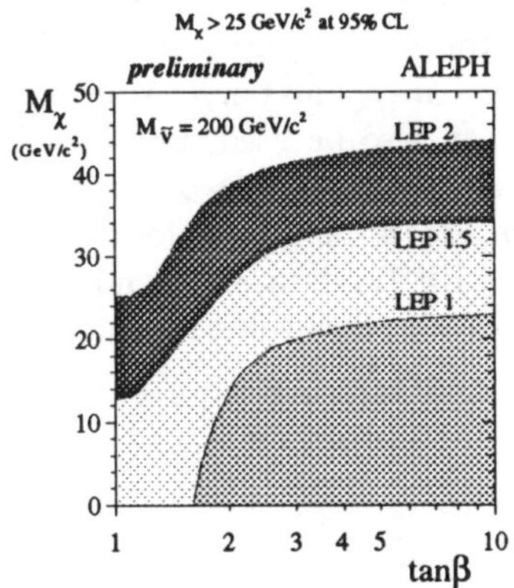

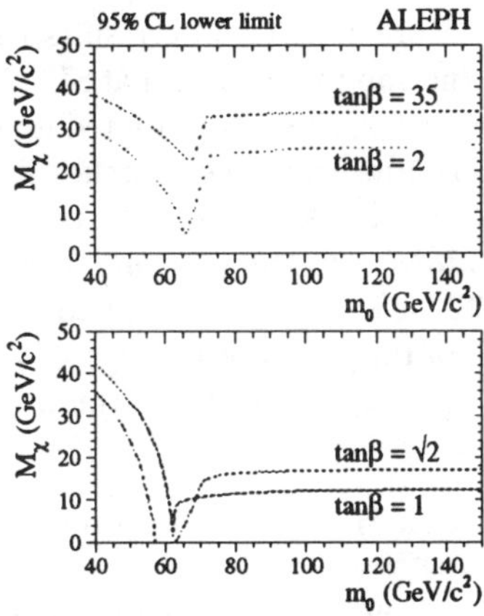

Figure 7: (left) Mass limit for the lightest neutralino as a function of $\tan\beta$ and for heavy sfermions ($m_0 = 200$ GeV/c^2). (right): Mass limit obtained at LEP 1.5 for the lightest neutralino as a function of m_0 and for various values of $\tan\beta$ ($\mu < 0$).

196

For $m_0 = 75$ GeV/c^2 and $\tan\beta = 1.41$, the χ mass lower limit is 21 GeV/c^2 (instead of 31 GeV/c^2 for $m_0 = 200$ GeV/c^2). At the time of writing, a complete analysis, letting both m_0 and $\tan\beta$ vary freely, is not available at LEP 2. The results obtained [12] at LEP 1.5 will therefore be used in the following for the purpose of illustration of the methods.

In Fig. 7(right), the limit obtained for m_χ is displayed, for various values of $\tan\beta$, as a function of m_0. For large values of m_0, they reflect the equivalent of Fig. 7(left) at LEP 1.5. As m_0 decreases, the limit degrades slowly, due to the destructive interference in the chargino production cross section for light sneutrinos. As m_0 decreases further, the sneutrino mass becomes smaller than the chargino mass and the detection efficiency vanishes abruptly, as mentioned previously. However, for very low values of m_0, the results from slepton searches, such as the LEP 1.5 counterpart of Fig. 3(right), can be used (including the sneutrino mass limit from LEP 1 which is most relevant for large values of $\tan\beta$). In general, the exclusions from sleptons and from charginos overlap, and a massless neutralino is excluded irrespective of m_0. This is however not the case for $\tan\beta = 1.41$ and $m_0 \sim 60$ GeV/c^2, as can be seen in Fig. 7(right). Preliminary results [11] from LEP 2 indicate that this small region is now excluded, with a neutralino LSP mass limit of 14 GeV/c^2, independent of m_0 and of $\tan\beta$.

More constraining results can be derived by imposing further constraints on the model. If minimal SUGRA is assumed, the μ value, up to now a free parameter, can be calculated up to its sign from m_0, $m_{1/2}$ and $\tan\beta$. Moreover, the results from the searches for Higgs bosons can also be used, particularly relevant for low values of $\tan\beta$ and for negative μ. Examples of exclusion domains [12] in the $(m_0, m_{1/2})$ plane are shown in Fig. 8. A χ mass lower limit of 22 GeV/c^2 can be inferred this way from the LEP 1.5 data, independent of m_0, of $\tan\beta$ and of the sign of μ. (Here too, this rather low value for the limit on m_χ is entirely due to the loophole discussed earlier.)

4 Unconventional scenarii

4.1 Constraints from LEP 1 on light gluinos

In most of the analyses presented up to now, it has been assumed that gluinos are too heavy to affect the decay pattern of the supersymmetric particles considered. Indeed, stringent mass limits for gluinos have been obtained at hadron colliders, as reported elsewhere in this book. Although some simplifying assumptions had to be made when deriving these limits, for instance regarding the details of cascade decays, it is hard to see how relaxing these assumptions

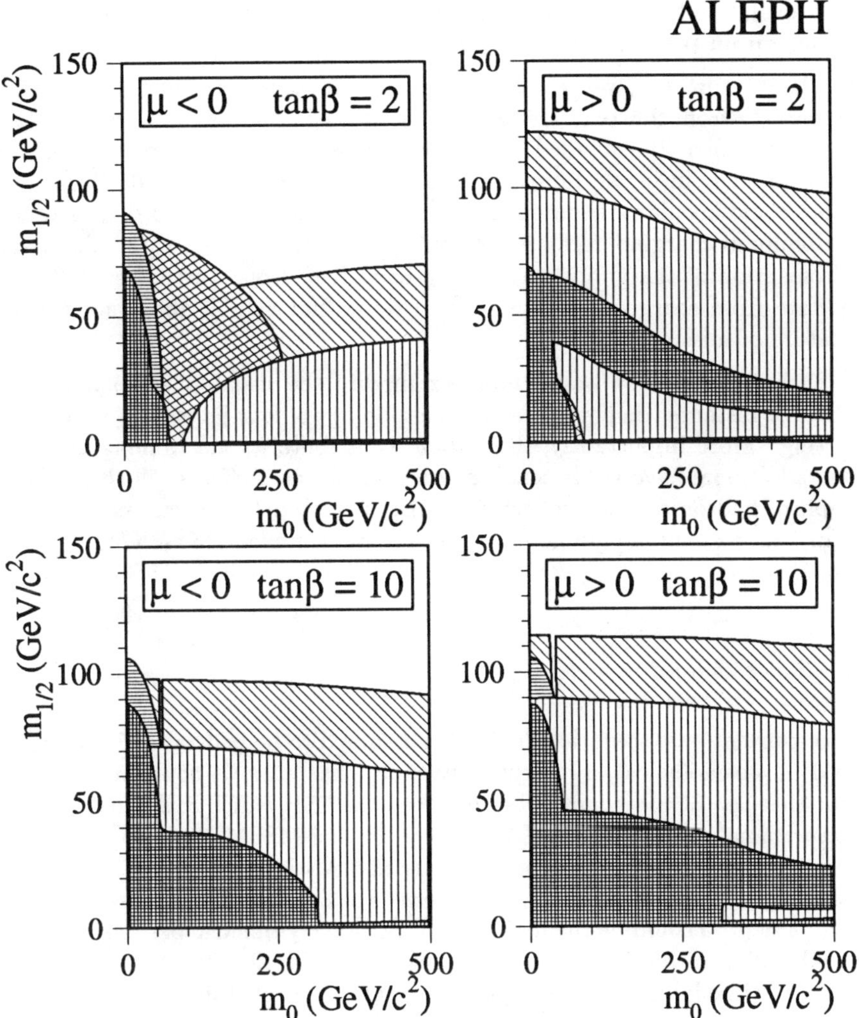

Figure 8: Excluded domains in the $(m_0, m_{1/2})$ plane for $\tan\beta = 2$ (top) and for $\tan\beta = 10$ (bottom), and for $\mu < 0$ (left) and $\mu > 0$ (right). The dark shaded regions are theoretically excluded. The vertical, horizontal, crossed and slanted hatched regions are excluded by charginos at LEP 1, by sneutrinos at LEP 1, by Higgs bosons at LEP 1 and by charginos at LEP 1.5, respectively.

could invalidate the exclusion of gluinos in the mass range of interest at LEP, at least within the MSSM with GUT relations.

There remains however a small mass window for very light gluinos (approximately from 2.5 to 4 GeV/c^2, for squarks in the few hundred GeV/c^2 range) not officially excluded by any of the searches at hadron colliders or elsewhere (*e.g.* in beam dump experiments or in upsilon decays). Such light gluinos could invalidate some of the searches performed at LEP 2. For instance, the dominant chargino decay mode could well be $\chi^+ \to q\bar{q}\tilde{g}$, with subsequent hadronization of the gluino into a long-lived R-hadron. In that case, the final state from chargino pair production would not exhibit the characteristic signature of missing energy.

Such light gluinos would however modify the usual phenomenology of QCD. For instance, they would affect the topology of four-jet events, via $g \to \tilde{g}\tilde{g}$ splitting. More importantly, they would contribute to the running of α_S as three additional flavours, in leading order, up to mass effects. With the large sample of hadronic events collected at LEP 1, such detailed studies have been performed,[13] excluding gluinos with masses smaller than 6.3 GeV/c^2.

4.2 Stable charged particles

In the conventional scenario, the LSP is assumed to be neutral and colourless, based on cosmological arguments. Even in that case, it could nevertheless occur that the LCSP is hardly heavier than the LSP, which could make it long lived without conflicting with cosmology. In the MSSM, this may occur for instance if the GUT relation between M_1 and M_2 is not satisfied. In scenarii with a light gravitino LSP, of which some implications are discussed further down, a slepton could be the next to lightest supersymmetric particle; for large enough values of the supersymmetry breaking scale, this slepton would decay with a lifetime long enough to appear as stable in a LEP detector.

Searches for long-lived weakly interacting heavy charged particles have been performed up to 172 GeV centre-of-mass energy.[14] Pair production of such particles would resemble muon pair production, with a back to back topology but with smaller particle velocities and hence larger specific ionization loss (dE/dx). The absence of any signal excludes charginos almost up to the kinematic limit of 86 GeV/c^2, for heavy sneutrinos, and right smuons or staus up to 67 GeV/c^2. (The selectron production cross section is affected by the details of the neutralino sector.)

4.3 The light gravitino LSP scenario

A class of supersymmetric models in which the gravitino is the lightest supersymmetric particle has recently received renewed attention, as discussed elsewhere in this book. For LEP, the main feature of such models is that the lightest neutralino is expected to decay into a photon and a gravitino, $\chi \to \gamma \tilde{G}$. For practical purposes, the neutralino lifetime is negligibly small as soon as the gravitino mass is smaller than a few eV/c^2.

Pair production of the lightest neutralinos, $e^+e^- \to \chi\chi$, then leads to a final state consisting of two acoplanar photons with missing energy. The background from the process $e^+e^- \to \nu\bar{\nu}\gamma\gamma$ is reduced by the requirement that the mass recoiling to the two photons should not be close to the Z mass; in addition, a minimum energy for both photons is required, the value of which depends on the χ mass considered. In the absence of signal, upper limits[15] are set on the production cross section of χ pairs at the level of 0.25–0.35 pb. To turn these constraints into a χ mass limit, further specification of the model is required (right and left selectron masses, χ field content). Typically, values around 70 GeV/c^2 are obtained.

4.4 R-parity violation

R-parity conservation was introduced to forbid lepton number or baryon number violating terms in the superpotential, otherwise allowed by supersymmetry and by gauge invariance. These terms are of the form $\lambda_{ijk} L_i L_j \overline{E}_k$, $\lambda'_{ijk} L_i Q_j \overline{D}_k$ or $\lambda''_{ijk} \overline{U}_i \overline{D}_j \overline{D}_k$, where i, j, k are generation indices, L and Q are doublet superfields, and U, D and E are singlet superfields. The simultaneous presence of the last two types of term would induce unacceptably fast proton decay. There is however no compelling theoretical justification for this prescription, and there exist alternatives, in which only some subsets of these terms are present, and which are equally viable. In particular, it is sufficient to assume that only one (or more generally only one type) of the R-parity violating terms is present to obtain an acceptable phenomenology.

The main consequence of R-parity violation is that the LSP is no longer stable. For instance, a $\lambda_{ijk} L_i L_j \overline{E}_k$ term may induce the decay of a neutralino LSP to final states such as $\bar{\nu}_i \ell_j^+ \ell_k^-$ (via virtual slepton or sneutrino exchange) or of a sneutrino $\tilde{\nu}_i$ to a lepton pair $\ell_j^- \ell_k^+$. The characteristic signature of supersymmetry therefore no longer consists in missing energy, but rather in multileptonic final states. Searches for supersymmetric particles have been performed at LEP 1[16] and LEP 2[17] under the assumption that R-parity is violated by such a $\lambda_{ijk} L_i L_j \overline{E}_k$ term. A large variety of final state topologies arise, depending on the type of particles produced in the e^+e^- annihilation,

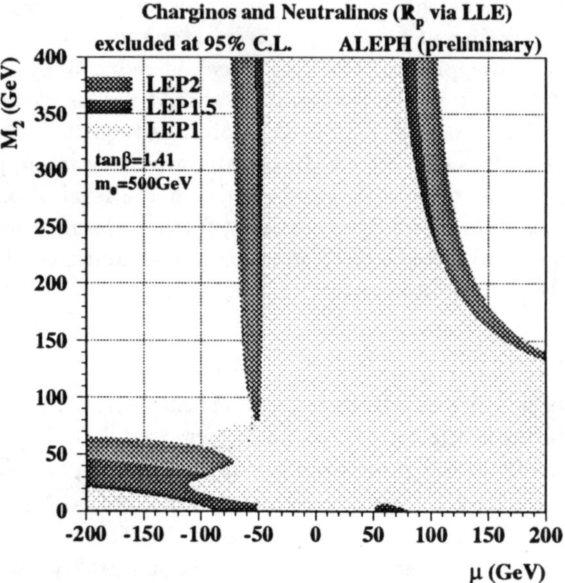

Figure 9: Excluded regions in the (M_2,μ) plane for $\tan\beta = 1.41$ and for heavy sfermions, assuming that R-parity is violated by a $\lambda_{ijk} L_i L_j \overline{E}_k$ term. The dashed line is the kinematic limit for $m_{\chi^+} = 86$ GeV/c^2.

and on whether they decay directly via an R-parity violating interaction or first toward the LSP via a gauge interaction (with the LSP subsequently decaying as indicated above in the case of a neutralino LSP). The generation indices in the λ_{ijk} Yukawa coupling control the lepton flavours appearing in the final state and therefore affect the selection efficiencies; the most favourable case is obtained for $\{ijk\} = \{122\}$, with only electrons and muons as final state leptons, and the worst case for $\{ijk\} = \{133\}$, with taus also appearing in the final state.

No signal above the expected standard model backgrounds was detected in any of the topologies investigated, resulting into mass limits or constraints on the parameters of the MSSM at least as strong as in the case of R-parity conservation, even assuming a dominant λ_{133} coupling. This can be seen, for instance, in Fig. 9 for chargino and neutralino searches. Interestingly, a substantial region is excluded at LEP 1 beyond the kinematic limit for chargino searches at LEP 2. This is due to the large statistics accumulated at the Z peak and to the fact that the $Z \to \chi\chi$ decay leads to visible final states, in contrast to the case of R-parity conservation.

The first results of searches for signals of supersymmetry at LEP under the assumption of a dominant $\lambda'_{ijk} L_i Q_j \overline{D}_k$ coupling were reported as this chapter was being completed. The final states typically involve multijets with leptons or with some missing energy. Here too, the kinematic limit for charginos is almost reached. [17]

5   Outlook

As indicated in the introduction, it is foreseen that the centre of mass energy at LEP 2 will be increased beyond 172 GeV, the energy at which most of the results reported above were obtained. Based on the experience already gathered, it appears that ~ 10 pb^{-1} are enough to reach the kinematic limit for charginos over large regions of the parameter space. Increasing the energy is therefore always beneficial compared to collecting more luminosity at a given energy. Once the ultimate LEP 2 energy is reached, it will nevertheless remain worthwhile accumulating sufficient integrated luminosity to improve the sensitivity to sleptons, stops and neutralinos. More optimistically, large statistics will be welcome to allow the LEP experiments to determine accurately the masses and couplings of the various supersymmetric particles which they will have revealed.

References

1. See for instance: P. Janot, *"Searching for Higgs bosons at LEP 1 and LEP 2"*, to appear in Perspectives on Higgs Physics II, World Scientific, ed. G.L. Kane.

2. The LEP Collaborations, *"A combination of preliminary electroweak measurements and constraints on the standard model"*, CERN-PPE/96-183.

3. M. Schmelling and P. Tipton, review talks at ICHEP96, Warsaw, July 1996, to be published.

4. The ALEPH Collaboration, *"Observation of monojet events and tentative interpretation"*, Phys. Lett. **B 334** (1994) 244.

5. The OPAL Collaboration, *"Topological search for the production of neutralinos and scalar particles"*, Phys. Lett. **B 377** (1996) 273.

6. The ALEPH Collaboration, *"Search for sleptons in e^+e^- collisions at centre-of-mass energies of 161 GeV and 172 GeV"*, CERN-PPE/97-056.

7. The AMY Collaboration, *"New limits on the masses of the selectron and photino"*, Phys. Lett. **B 369** (1996) 86.

8. The OPAL Collaboration, *"Search for scalar top and scalar bottom quarks at $\sqrt{s} = 170$–172 GeV in e^+e^- collisions"*, CERN-PPE/97-046.

9. The OPAL Collaboration¡ *"Search for chargino and neutralino production at $\sqrt{s} = 170$ and 172 GeV at LEP"*, CERN-PPE/97-083.

10. The ALEPH Collaboration, *"Searches for charginos and neutralinos in e^+e^- collisions at $\sqrt{s} = 161$ and 172 GeV"*, contribution 614 to EPS-HEP97, Jerusalem, August 1997.

11. The ALEPH Collaboration, *"Update of the mass limit for the lightest neutralino"*, contribution 594 to EPS-HEP97, Jerusalem, August 1997.

12. The ALEPH Collaboration, *"Mass limit for the lightest neutralino"*, Z. Phys. **C 72** (1996) 549.

13. The ALEPH Collaboration, *"A measurement of the QCD colour factors and a limit on the light gluino"*, CERN-PPE/97-002.

14. The ALEPH Collaboration, *"Search for pair-production of long-lived heavy charged particles in e^+e^- annihilation"*, CERN-PPE/97-041; The DELPHI Collaboration, *"Search for stable heavy charged particles in e^+e^- collisions at $\sqrt{s} = 130$–136, 161 and 172 GeV"*, Phys. Lett. **B 396** (1997) 315.

15. The L3 Collaboration, *"Single and multi=photon events with missing energy in e^+e^- collisions at 161 GeV $< \sqrt{s} < 172$ GeV"*, CERN-PPE/97-76.

16. The ALEPH Collaboration, *"Search for supersymmetric particles with R-parity violation in Z decays"*, Phys. Lett. **B 349** (1995) 238.

17. The ALEPH Collaboration, *"Searches for R-parity violating supersymmetry at LEP II"*,
contribution 621 to EPS-HEP97, Jerusalem, August 1997.

THE SEARCH FOR SUPERSYMMETRY AT THE TEVATRON COLLIDER

M. CARENA,[1] R.L. CULBERTSON,[2] S. ENO,[3] H.J. FRISCH,[2] and S. MRENNA[4]

[1] *Fermi National Accelerator Laboratory*
[2] *University of Chicago*
[3] *University of Maryland*
[4] *Argonne National Laboratory*

Abstract

We review the status of searches for Supersymmetry at the Tevatron Collider. After discussing the theoretical aspects relevant to the production and decay of supersymmetric particles at the Tevatron, we present the current results for Runs Ia and Ib as of the summer of 1997.

1 Introduction

The Tevatron is a $p\bar{p}$ collider located at the Fermi National Accelerator Laboratory (Fermilab) in Batavia, Illinois, USA. Two experimental collaborations, CDF and DØ, have collected data at a $p\bar{p}$ center–of–mass energy of $\sqrt{s} = 1.8$ TeV. The CDF detector[1] features a magnetic solenoidal spectrometer inside the calorimeters and a silicon vertex detector; the DØ detector's[2] strengths are the finely–segmented hermetic calorimeter and the good muon coverage. There have been two major runs of the Tevatron, accumulating approximately 20 pb^{-1} of data in '92–'93 and 100 pb^{-1} in '94–'95. The next run, starting around 2000, will provide 1000 pb^{-1} per year. Table 1 summarizes those CDF and DØ analyses that have been published or presented at conferences. This review addresses all analyses available as of summer 1997. Due to limited space here many details must be left out; a more thorough review is available including lists of cuts for each analysis and the basic theoretical tools necessary to understand the current and potential Tevatron analyses.[3]

The Tevatron Collider has been the world's accelerator–based high energy frontier since it first began taking data in 1987, and has thus been a prime location to search for the final pieces of the Standard Model (SM) and new phenomena beyond. Supersymmetry (SUSY)[4] is a new symmetry which provides a well–motivated extension of the SM. If SUSY is a consistent description of Nature, then the lower range of SUSY particle masses can be within the reach of the Tevatron, motivating a wide range of searches in a large number of channels.[5] The mass of the lightest neutral Higgs boson is strongly constrained within SUSY,[6] and could be within the reach of the upgraded Tevatron.[7,8]

2 The MSSM

In the past two decades, a detailed picture of the Minimal Supersymmetric extension of the Standard Model (MSSM), has emerged.[9] The details are discussed in other chapters of this book. In short, the MSSM contains the known SM particles plus SUSY partners: neutralinos $\widetilde{\chi}^0_{1-4}$, charginos $\widetilde{\chi}^\pm_{1-2}$, gluinos $\widetilde{g}$, squarks $\widetilde{Q}_L$ and $\widetilde{Q}_R$, sleptons $\widetilde{\ell}_L$ and $\widetilde{\ell}_R$, sneutrinos $\widetilde{\nu}$, and 3 neutral Higgses h, H and A, plus the charged Higgs $H^\pm$. In addition to the usual SM parameters, the masses and interactions of the sparticles depend on $\tan\beta,$[a] the Higgsino mass parameter $\mu,$[b] and a number of soft SUSY-breaking (S̶U̶S̶Y̶) mass parameters which are added explicitly to the Lagrangian.

Sparticle Spectrum: The $\widetilde{\chi}^\pm_{1-2}$ and $\widetilde{\chi}^0_{1-4}$ masses and their gaugino and Higgsino composition are determined by M_W, M_Z, $\tan\beta$, μ, and two gaugino soft S̶U̶S̶Y̶ parameters M_1 and M_2, all evaluated at the electroweak scale M_{EW}.[c] The $\widetilde{g}$ mass is determined by the $SU(3)_C$ gaugino mass parameter M_3. Depending on the gaugino/Higgsinos composition, the $\widetilde{\chi}$ couplings to gauge bosons, and to L and R sfermions can differ substantially, strongly affecting production and decay processes.

The mass eigenstates of sfermions are, in principle, mixtures of their L and R components, and the magnitude of mixing is proportional to the mass of the corresponding fermion. The $L-R$ mixing of $\widetilde{u}$, $\widetilde{d}$, $\widetilde{c}$, $\widetilde{s}$, $\widetilde{e}$ and $\widetilde{\mu}$ is thus negligible, and the L and R components are the real mass eigenstates, with masses $m_{\widetilde{Q}_{L,R}}$ and $m_{\widetilde{\ell}_{L,R}}, m_{\widetilde{\nu}_\ell}$ fixed by soft S̶U̶S̶Y̶ parameters. For $\widetilde{t}$, $\widetilde{b}$ and $\widetilde{\tau}$, the $L-R$ mixing can be nontrivial. For example, the $\widetilde{t}$ mixing term is $m_t(A_t - \mu/\tan\beta)$, where A_t is a soft S̶U̶S̶Y̶ parameter.[b] Unless there is a cancellation between A_t and $\mu/\tan\beta$, $\widetilde{t}$ mixing occurs because m_t is large, and it is possible that the lightest stop $\widetilde{t}_1$ is one of the lightest sparticles. For the $\widetilde{b}$ and $\widetilde{\tau}$, $L-R$ mixing becomes important when $m_b\tan\beta$ or $m_\tau\tan\beta$ is $\mathcal{O}(m_t)$, since the $\widetilde{b}$ mixing term, for example, is $m_b(A_b - \mu\tan\beta)$.

At tree level, the lightest CP–even Higgs boson satisfies the relation $M_h \lesssim M_Z$, but this result is strongly modified by radiative corrections that depend on other MSSM parameters.[12] The dominant radiative corrections to M_h grow as m_t^4 and are logarithmically dependent on the $\widetilde{t}$ and $\widetilde{b}$ masses. Within the MSSM, a *general* **upper** bound on M_h can be determined by a careful

---

[a]One Higgs doublet, H_2, couples to u, c, and t, while the other, H_1, couples to d, s, b, e, μ, and τ. The parameter $\tan\beta$ is the ratio of vacuum expectation values $\langle H_2\rangle/\langle H_1\rangle \equiv v_2/v_1$, and $v^2 = v_1^2 + v_2^2$, where v is the order parameter of EWSB $\simeq 246$ GeV.

[b]Beware of different sign conventions for μ and A_f in the literature. Both PYTHIA[10] and ISAJET[11] use the convention stated in S.P. Martin's review in this book.

[c]The electroweak scale M_{EW} is roughly the scale of the sparticle masses themselves. Usually, in the literature, for simplicity, $M_{EW} \simeq M_Z$.

evaluation of the one–loop and dominant two–loop radiative corrections.[6] After setting the masses of all SUSY particles and M_A to values around 1 TeV, setting $\tan\beta > 20$, and varying the $\tilde{t}$ mixing parameters to give the largest possible effect, the upper bound on M_h is maximized, yielding $M_h \lesssim 130$ GeV for $m_t = 175$ GeV. For more moderate values of the MSSM parameters, the upper bound on M_h becomes smaller. Given the general upper bound on M_h of about 130 GeV, the upgraded Tevatron has the potential to provide a crucial test of the MSSM.[7,8]

The SUSY spectrum can be sensitive to the exact range of $\tan\beta$.[13] The measured value $m_t \simeq 175$ GeV defines a lower bound on $\tan\beta$ of about 1.2, provided that the top t Yukawa coupling remains finite up to a scale of the order of 10^{16} GeV. If, instead, the t Yukawa coupling should remain finite only up to scales of order of a TeV, values of $\tan\beta$ as low as .5 would still be possible.[d] Similarly, if $\tan\beta$ becomes too large, large values of the b Yukawa coupling are necessary to obtain values of m_b compatible with experiment. Generically, it can be shown that values of $\tan\beta \geq 60$ are difficult to obtain if the MSSM is expected to remain a valid theory up to scales of order 10^{16} GeV.

Supergravity (SUGRA): At present, the exact mechanism of ~~SUSY~~ is unknown. SUGRA models assume the existence of extra superfields (the so–called "hidden sector") which couple to the MSSM particles only through gravitational–like interactions. These interactions are responsible for inducing the soft ~~SUSY~~ parameters. The number of possible parameters is over 100, but in the minimal SUGRA (mSUGRA) scenario, all scalars (Higgs bosons, sleptons, and squarks) are assumed to have a common squared–mass m_0^2, all gauginos (Bino, Wino, and gluino) have a common mass $m_{1/2}$, and all triple–scalar couplings have the value A_0, all at a scale of order M_{Planck} (or, approximately, M_{GUT}, the scale where the gauge couplings unify). After specifying $\tan\beta$, all that remains is to relate the values of the soft ~~SUSY~~ parameters specified at M_{GUT} to their values at M_{EW}. This is accomplished using renormalization group equations (RGE's). Moreover, μ is determined up to a sign by demanding the correct electroweak symmetry breaking (EWSB). Finally, the physical sparticle masses are determined as a function of the low energy values of the soft ~~SUSY~~ parameters, which can be written in terms of m_0, $m_{1/2}$, A_0, $\tan\beta$, and μ.[14]

The masses of the $\tilde{g}$'s, $\tilde{\chi}^{\pm}$'s and $\tilde{\chi}^0$'s are strongly correlated. Once the RGE evolution is included, μ tends to be larger than M_1 and M_2, becoming the largest as $\tan\beta \to 1$. As a result, $\tilde{\chi}_1^0$, $\tilde{\chi}_2^0$ and $\tilde{\chi}_1^{\pm}$ tend to be gaugino–like,

---

[d]This implies that a perturbative description of the MSSM would only be valid up to the weak scale, which is, of course, not a very interesting possibility.

and $\widetilde{\chi}_1^0$ can be the LSP. The approximate mass hierarchy is $M_{\widetilde{\chi}_2^0} \simeq 2M_{\widetilde{\chi}_1^0} \simeq M_{\widetilde{\chi}_1^\pm} \simeq 1/3M_{\widetilde{g}} \simeq 0.8m_{1/2}.$[e]

Because $\widetilde{\ell}$'s have only EW quantum numbers and the lepton Yukawa couplings are small, the $\widetilde{e}$ and $\widetilde{\mu}$ SUSY mass parameters do not evolve much from M_{GUT} to M_{EW} and are roughly given by $m_{\widetilde{\ell}_L}^2 \simeq m_0^2 + 0.5m_{1/2}^2$ and $m_{\widetilde{\ell}_R}^2 \simeq m_0^2 + 0.15m_{1/2}^2$. The $\widetilde{\nu}$ mass is fixed by a sum rule $m_{\widetilde{\nu}_\ell}^2 = m_{\widetilde{\ell}_L}^2 + M_W^2 \cos 2\beta$, and, when m_0 is small, the $\widetilde{\nu}$ can be the LSP instead of the $\widetilde{\chi}_1^0$. For $\tan\beta \geq 40$, the $\widetilde{\tau}$ SUSY mass parameters also receive non–negligible contributions from the τ Yukawa coupling.

The $\widetilde{Q}$ soft SUSY mass parameters evolve mainly through the strong coupling to the $\widetilde{g}$, so their dependence on the common gaugino mass is stronger than for $\widetilde{\ell}$'s. For $\widetilde{u}$, $\widetilde{d}$, $\widetilde{c}$ and $\widetilde{s}$, the mass eigenstates are roughly given by $m_{\widetilde{Q}_L}^2 \simeq m_0^2 + 6.3m_{1/2}^2$ and $m_{\widetilde{Q}_R}^2 \simeq m_0^2 + 5.8m_{1/2}^2$. The above relations between the mass parameters lead to the general SUGRA prediction, $m_{\widetilde{Q}} \geq 0.85M_{\widetilde{g}}$. In general, the $\widetilde{Q}$'s are heavier than the $\widetilde{\ell}$'s or the lightest $\widetilde{\chi}^0$ and $\widetilde{\chi}^\pm$. For the L and R components and the $L - R$ mixing of $\widetilde{t}$ and $\widetilde{b}$ and for the Higgs soft SUSY parameters, the large t Yukawa coupling (and possibly the b Yukawa coupling for large $\tan\beta$) plays a crucial role in the RGE evolution. As a result, the $\widetilde{t}$ and $\widetilde{b}$ can be light. In this case, the coefficients of m_0 and $m_{1/2}$ have an important dependence on the low–energy values of the b and t Yukawa couplings which, in turn, depend on $\tan\beta$.[3] The coefficients of $m_{1/2}$ depend on the exact values of α_s and the scale of the sparticle masses.

Although mSUGRA is defined in terms of only 5 parameters at a high scale (m_0, $m_{1/2}$, A_0, $\tan\beta$, and the sign of μ), it is natural to question exact universality of the soft SUSY parameters.[15] For example, in a $SU(5)$ SUSY GUT model, the $\widetilde{e}_L$ and $\widetilde{d}_R$ reside in the same 5–multiplet of $SU(5)$, and may naturally have the common mass parameter $m_0^{(5)}$ at the GUT scale. Similarly, $\widetilde{u}_L, \widetilde{d}_L, \widetilde{u}_R$, and $\widetilde{e}_R$, which reside in the same 10–multiplet, may have a common mass $m_0^{(10)}$. The two Higgs bosons doublets reside in different 5– and $\bar{5}$–multiplets, with masses $m_0^{(5')}$ and $m_0^{(\bar{5}')}$. There is no known symmetry principle that demands that all these mass parameters should be the same.

Gauge–Mediated Supersymmetry Breaking: In contrast to SUGRA, the soft SUSY terms in the MSSM Lagrangian, can be generated through gauge interactions at a scale much lower than M_{Planck}, introducing many in-

---

[e]In general, it may occur that $m_{1/2} \simeq 0$. Low–energy gaugino masses are then dominated by contributions of stop–top and Higgs–Higgsino loops. In this case the $\widetilde{g}$ could be the LSP with a mass of a few GeV, and the $\widetilde{\chi}_1^0$ may be somewhat heavier due to contributions from electroweak loops.[16]

208

teresting features. In most models of gauge–mediated, low–energy SUSY, the gaugino and scalar masses are roughly of the same order of magnitude. Even after RGE evolution, sfermions with the same quantum numbers acquire the same masses (ignoring the effects of Yukawa couplings), yielding a natural mass hierarchy between weakly and strongly interacting sfermions; the mass hierarchy of the gauginos is fixed by the gauge couplings (as in SUGRA models). One distinctive feature of these models is that the spin–3/2 superpartner of the graviton, the gravitino $\widetilde{G}$, can be very light and become the LSP,[f] unlike in SUGRA models, where the gravitino has a mass on the order of the electroweak scale and is very weakly interacting.

R–Parity Violation ($\not{R}$): A multiplicative R–parity symmetry is often assumed, but one simple extension of the MSSM is to break it.[19] Presently, neither experiment nor any theoretical argument demands R–parity conservation. There are several effects on the SUSY phenomenology associated with $\not{R}$ couplings: (1) lepton or baryon number violating processes are allowed, including the production of single sparticles (instead of pair production), (2) the LSP is no longer stable, but might decay to SM particles within a collider detector, and (3) because it is unstable, the LSP need not be the $\widetilde{\chi}^0$ or $\widetilde{\nu}$, but can be charged or colored. Although very strong bounds on $\not{R}$ operators can be derived from the present data,[20] but there is still room for study.

Run Ia Parameter Sets (RIPS): Some CDF and DØ SUSY searches are analyzed in the framework of so–called "SUGRA–inspired models." These RIPS are specified by $M_{\widetilde{g}}, m_{\widetilde{Q}}, M_A, \tan\beta$ and the magnitude and sign of μ. $M_{\widetilde{g}}$ defines M_1 and M_2 using SUGRA unification relations. The $\widetilde{\chi}^\pm$ and $\widetilde{\chi}^0$ properties are then fixed by $\tan\beta$ and μ. In practice, the value of μ is set much larger than M_1 and M_2, so the properties of the $\widetilde{\chi}^0$'s, $\widetilde{\chi}^\pm$'s, and $\widetilde{g}$ are similar to those in a pure SUGRA model. The first 5 flavors of $\widetilde{Q}$'s are degenerate in mass, with the value $m_{\widetilde{Q}}$, while the stop masses are $m_{\widetilde{t}_1} = m_{\widetilde{t}_2} = \sqrt{m_{\widetilde{Q}}^2 + m_t^2}$, assuming also the absence of $L-R$ mixing. This may be quite unrealistic, since the $\widetilde{b}$ and $\widetilde{t}$ mass can be naturally lighter. When $m_{\widetilde{Q}} > M_{\widetilde{g}}$, the $\widetilde{\ell}$ masses can be fixed using approximate SUGRA relations, and the RIPS has many features of a SUGRA model. The region $m_{\widetilde{Q}} < M_{\widetilde{g}}$ is very hard to realize in SUGRA models, but is also worth investigating. In this case, for some analyses, a constant value of 350 GeV is used for $m_{\widetilde{\ell}_L}$, $m_{\widetilde{\ell}_R}$, and $m_{\widetilde{\nu}}$. Finally, the Higgs mass M_A is used to determine the Higgs boson sector. In practice, M_A is set to a large value, so that the lightest neutral Higgs boson h has SM–like couplings

---

[f] It is also possible to construct a model where the gluino is a stable LSP with a mass of a few tens of GeV.[18] In this case, the missing energy signal for SUSY disappears, since a stable LSP gluino will form stable hadrons.

to gauge bosons and fermions, and all other Higgs bosons are heavy.

3   The Present Status of Sparticle Searches

3.1   Charginos and Neutralinos

$\widetilde{\chi}^{\pm}$ and $\widetilde{\chi}^0$ pairs can be produced directly at hadron colliders through electroweak processes. The cross sections are functions of the sparticle masses and mixings, with different contributions from t–channel $\widetilde{Q}$ exchange, s–channel vector boson production, and t– and s–channel interference. Figure 1 shows the production cross sections of various $\widetilde{\chi}^{\pm}$ and $\widetilde{\chi}^0$ pairs at the Tevatron in the limits of large and small $|\mu|$ relative to the soft SUSY masses M_1 and M_2. The decay patterns also depend on the masses and mixings. Two–body decays dominate if allowed: $\widetilde{\chi}_i^{\pm} \to W^{\pm}\widetilde{\chi}_j^0$, $H^{\pm}\widetilde{\chi}_j^0$, $\widetilde{\chi}_j^{\pm}h$, $\widetilde{\chi}_j^{\pm}Z$, $\tilde{\ell}_{L,R}\nu$, $\tilde{\nu}\ell$ or $\widetilde{Q}q'$ and $\widetilde{\chi}_i^0 \to Z\widetilde{\chi}_j^0$, $h\widetilde{\chi}_j^0$, $W^{\mp}\widetilde{\chi}_j^{\pm}$, $H^{\mp}\widetilde{\chi}_j^{\pm}$, $\tilde{\ell}_{L,R}\ell$, $\tilde{\nu}\nu$, or $\widetilde{Q}q$. When no 2–body final states are kinematically allowed, 3–body decays like $\widetilde{\chi}_i^{\pm} \to \widetilde{\chi}_j^0 f\bar{f}'$, $\widetilde{\chi}_i^{\pm} \to \widetilde{\chi}_j^{\pm} f\bar{f}$, $\widetilde{\chi}_i^0 \to \widetilde{\chi}_j^{\pm} f\bar{f}'$, and $\widetilde{\chi}_i^0 \to \widetilde{\chi}_j^0 f\bar{f}$ occur through virtual sfermions and gauge bosons. If the sparticles are light enough, a 100% branching ratio to $\ell\nu$, $\ell^+\ell^-$ or jj final states is possible. The one–loop decay $\widetilde{\chi}_2^0 \to \widetilde{\chi}_1^0\gamma$ can be important if the $\widetilde{\chi}_1^0$ is Higgsino–like and $\widetilde{\chi}_2^0$ is gaugino–like, or *vice versa*. For a light enough $\tilde{g}$, the decays $\widetilde{\chi}_i^{\pm} \to \tilde{g}f\bar{f}'$ and $\widetilde{\chi}_i^0 \to \tilde{g}f\bar{f}$ can be important.

The production of $\widetilde{\chi}_1^{\pm}\widetilde{\chi}_2^0$, followed by the decays $\widetilde{\chi}_1^{\pm} \to \widetilde{\chi}_1^0\ell\nu$ and $\widetilde{\chi}_2^0 \to \ell^+\ell^-\widetilde{\chi}_1^0$, is a source of three charged leptons (e or μ) and $\not{E}_T$. This trilepton signal has small SM backgrounds, and is consequently one of the "golden" SUSY signatures.[21]

The results of the CDF[22,23] and DØ searches[24,25] are shown in Fig. 2 analyzed using RIPS. The searches include 4 channels: $e^+e^-e^{\pm}$, $e^+e^-\mu^{\pm}$, $e^{\pm}\mu^+\mu^-$ and $\mu^{\pm}\mu^+\mu^-$. The CDF analysis requires one lepton with $E_T >11$ GeV, passing tight identification cuts, and two other leptons with $E_T >5$ GeV (e) or $p_T >4$ GeV (μ), passing loose identification cuts. All leptons must be isolated, meaning there is little excess E_T in a cone of size $R = 0.4$ in $\eta - \phi$ space centered on the lepton. The event must have two leptons with the same flavor and opposite sign. If two leptons of the same flavor and opposite charge have a mass consistent with the J/Ψ, Υ or Z boson, the event is rejected. After this selection, 6 events remain in the data set, while the expected background, dominated by Drell–Yan pair production plus a fake lepton, is 8 events. After demanding $\not{E}_T > 15$ GeV, no events remain, while 1.2 are expected from SM model sources.

The DØ analysis requires 3 leptons with $E_T > 5$ GeV. However, several different triggers are used, and some lepton categories are required to have a

Table 1: A compilation of results from Run I Tevatron SUSY searches as of the summer of 1997. The symbol b denotes an additional b–tagged jet. Also listed are the references and the section of this chapter where each analysis is discussed. More information is available for DØ at $http://www-d0.fnal.gov/public/new/new\_public.html$, and for CDF at $http://www-cdf.fnal.gov/$

Sparticle	Signature	Expt.	Run	$\int \mathcal{L}dt(\mathrm{pb}^{-1})$	Ref.	Sec.
Charginos	$\not{E}_T$+trilepton	CDF	Ia	19	[22]	3.1
and	$\not{E}_T$+trilepton	CDF	Iab	107	[23]	"
Neutralinos	$\not{E}_T$+trilepton	DØ	Ia	12.5	[24]	"
	$\not{E}_T$+trilepton	DØ	Ib	95	[25]	"
	$\gamma\gamma+\not{E}_T$or jets	CDF	Ib	85	[74]	4.7
	$\gamma\gamma+\not{E}_T$	DØ	Iab	106	[75]	"
Squarks	$\not{E}_T+ \geq 3{,}4$ jets	CDF	Ia	19	[35]	3.2
and	$\not{E}_T+ \geq 3{,}4$ jets	DØ	Ia	13.5	[31]	"
Gluinos	$\not{E}_T+ \geq 3$ jets	DØ	Ib	79.2	[32]	"
	dilepton+ ≥ 2 jets	CDF	Ia	19	[37]	"
	$\not{E}_T$+dilepton+ ≥ 2 jets	CDF	Ib	81	[38]	"
	$\not{E}_T$+dilepton	DØ	Ib	92.9	[36]	"
Stop	$\not{E}_T+\ell+ \geq 2$ jets+b	CDF	Ib	90	[42]	"
	$\not{E}_T+\ell+ \geq 3$ jets+b	CDF	Iab	110	[44]	"
	dilepton+jets	DØ	Ib	74.5	[43]	"
	$\not{E}_T$+2 jets	DØ	Ia	7.4	[41]	"
	$\not{E}_T+\gamma+b$	CDF	Ib	85	[74]	4.11
Sleptons	$\gamma\gamma\ \not{E}_T$	DØ	Iab	106	[75]	3.7
Charged	dilepton + $\not{E}_T$	CDF	Ia	19	[57]	3.4
Higgs	τ+2 jets+$\not{E}_T$	CDF	Ia	19	[56]	"
	$\tau+b+\not{E}_T+(\ell,\tau,\text{jet})$	CDF	Iab	91	[48]	"
	$\tau+b+\not{E}_T+(\ell,\tau,\text{jet})$	DØ	Iab	125	[49]	"
Neutral	$WH \to \ell + \not{E}_T+b+$jet	CDF	Iab	109	[53]	3.5
Higgs	$WH \to \ell + \not{E}_T+b+$jet	DØ	Ib	100	[52]	"
	$WH, ZH \to \gamma\gamma+2$ jets	DØ	Ib	101.2	[55]	"
	$ZH \to b+$jet$+\not{E}_T$	DØ	Ib	20	[58]	"
	$WH, ZH \to 2$ jets+2 b's	CDF	Ib	91	[54]	"
R violating	dilepton+≥ 2 jets	CDF	Iab	105	[61]	3.6
Charged LSP	slow, long–lived particle	CDF	Ib	90	[59]	3.6

larger E_T to pass the various trigger thresholds. All leptons are required to be isolated. To reduce events with mismeasured $\not{E}_T$, the $\not{E}_T$ direction must not be along or opposite to a muon. Additional cuts are tuned for each topology. For

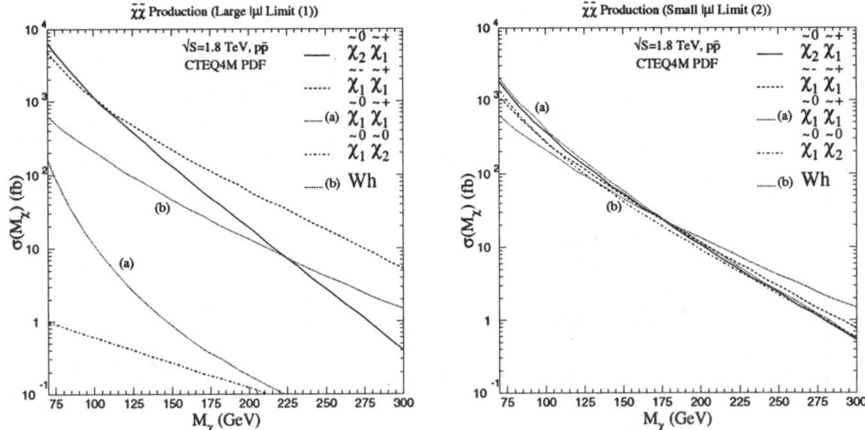

Figure 1: Production cross sections at the Tevatron for $\widetilde{\chi}^{\pm}$ and $\widetilde{\chi}^0$ pair production, assuming $\tan\beta = 2$, $m_{\widetilde{Q}} = 500$ GeV, and $M_1 = \frac{1}{2}M_2$, versus the $\widetilde{\chi}_1^{\pm}$ mass. The left figure is generated by fixing $\mu = -1$ TeV and varying $m_{1/2}$, the right figure by fixing $m_{1/2}=1$ TeV and varying μ. The Wh cross section (curve b) is shown for reference as a function of M_h.

example, the background from Drell–Yan pair production plus a fake lepton is highest in the eee channel, so these events are rejected if an electron pair is back–to–back. The $\not{E}_T$ threshold is 15 GeV for eee, and 10 GeV for the other three topologies. No events are observed in any channel with a total of 1.26 events expected from (i) Drell–Yan production plus a fake lepton and (ii) heavy–flavor production.

The DØ limit is on the "average" of the 4 modes, while the CDF limit is on the sum. After accounting for this difference, the CDF limit is twice as sensitive at a given $\widetilde{\chi}_1^{\pm}$ mass. The CDF limit shown is compared to three RIPS, which have different ratios of $m_{\widetilde{Q}}$ to $M_{\widetilde{g}}$. The DØ limit is compared to a wide variation of possible branching ratios. The DØ theory curve assumes heavy $\widetilde{Q}$'s, which reduces the cross section, but the CDF curves do not. The wide differences in the theory curves in Fig. 2 show the dangers of quoting a mass limit rather than a cross section × branching ratio limit.[27]

212

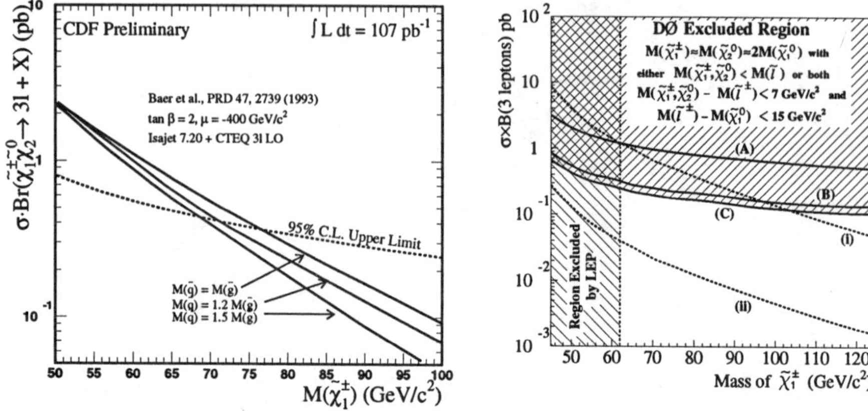

Figure 2: (Left) The CDF 95% C.L. limits on cross section $\times$ branching ratio for $\widetilde{\chi}_1^\pm \widetilde{\chi}_2^0$ production in 107 pb^{-1} of data. The limit is on the *sum* of the final states $eee, ee\mu, \mu\mu\mu$ and $\mu\mu e$ when $\widetilde{\chi}_1^\pm \to \ell\nu\widetilde{\chi}_1^0$ and $\widetilde{\chi}_2^0 \to \ell\ell\widetilde{\chi}_1^0$. The signals expected for three different RIPS scenarios are shown for comparison.[26] (Right) Similar limits from DØ, but for the *average* of all four channels. The curves (A), (B), and (C) show the Run Ia, Run Ib, and combined limits. Curve (i) shows the predicted cross section $\times$ branching ratio assuming BR($\widetilde{\chi}_1^\pm \to \ell\nu\widetilde{\chi}_1^0$)= BR($\widetilde{\chi}_2^0 \to \ell^+\ell^-\widetilde{\chi}_1^0$)=1/3 ($\ell = e, \mu, \tau$). Curve (ii) assumes BR($\widetilde{\chi}_1^\pm \to \ell\nu\widetilde{\chi}_1^0$)=0.1 and BR($\widetilde{\chi}_2^0 \to \ell^+\ell^-\widetilde{\chi}_1^0$)=0.033. For both CDF and DØ, kinematic efficiencies are calculated using the production cross section from ISAJET.

3.2 Squarks and Gluinos

The Tevatron is a hadron collider, so it can produce $\tilde{g}$'s and $\widetilde{Q}$'s through their $SU(3)_C$ couplings to quarks and gluons. The dominant production mechanisms are $gg, q\bar{q} \to \tilde{g}\tilde{g}$ or $\widetilde{Q}\widetilde{Q}^*$, $qq \to \widetilde{Q}\widetilde{Q}$ and $qg \to \widetilde{Q}\tilde{g}, \bar{q}g \to \widetilde{Q}^*\tilde{g}$, and the cross sections can be calculated as a function of only the Q and $\tilde{g}$ masses (ignoring EW radiative corrections). Figure 3 shows the cross sections for $\widetilde{Q}$'s and $\tilde{g}$'s as a function of the sparticle masses at $\sqrt{s} = 1.8$ TeV (left) and 2 TeV (right), where NLO SUSY QCD corrections have been included.[28] The NLO corrections are in general significant, positive (evaluated at a scale equal to the average mass of the two produced particles), and much less sensitive to the choice of scale than a LO calculation.

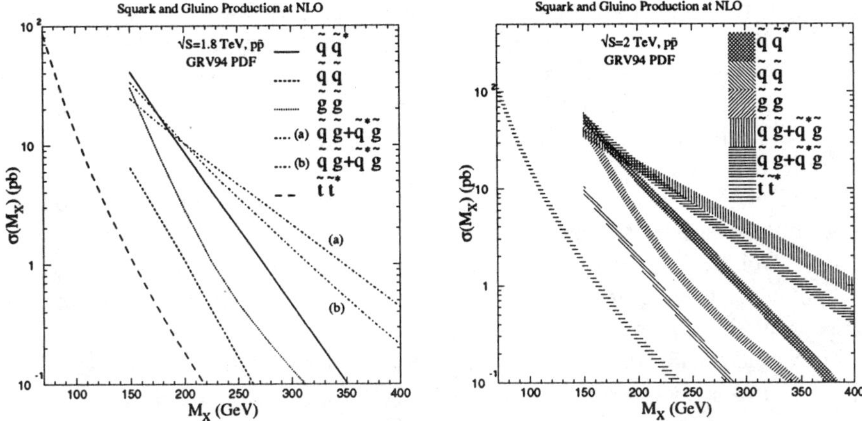

Figure 3: (Left) Production cross sections for $\tilde{g}$'s and $\tilde{Q}$'s versus sparticle mass M_X at the Tevatron, $\sqrt{s} = 1.8$ TeV, assuming degenerate masses for 5 flavors of $\tilde{Q}$'s. For $\tilde{Q}\tilde{Q}^*$ and $\tilde{Q}\tilde{Q}$ production, M_X is the squark mass and $M_{\tilde{g}} = 200$ GeV. For $\tilde{g}\tilde{g}$ production, M_X is the $\tilde{g}$ mass and $M_{\tilde{Q}} = 200$ GeV. For $\tilde{Q}\tilde{g}$ production (a), M_X is the $\tilde{Q}$ mass and $M_{\tilde{g}} = 200$ GeV; for (b), M_X is the $\tilde{g}$ mass and $M_{\tilde{Q}} = 200$ GeV. For $\tilde{t}\tilde{t}^*$ production, M_X is the stop mass. All cross sections are evaluated at a scale equal to the average mass of the two produced sparticles. (Right) The same curves with $\sqrt{s} = 2$ TeV. The bands show the change in rate from varying the scale from 1/2 to 2 times the average mass of the produced particles.

Since $\tilde{Q}$'s and $\tilde{g}$'s decay into $\tilde{\chi}^{\pm}$'s and $\tilde{\chi}^0$'s, their signatures can be similar to $\tilde{\chi}^{\pm}$ and $\tilde{\chi}^0$ pair production, but with accompanying jets. If $m_{\tilde{Q}} > M_{\tilde{g}}$, then the $\tilde{Q}$ has the 2–body decay $\tilde{Q} \to \tilde{g}q$. The $\tilde{g}$ has then the possible decays $\tilde{g} \to q\bar{q}\tilde{\chi}^0_i$ or $\tilde{g} \to q\bar{q}'\tilde{\chi}^{\pm}_i$, where q can stand for t or b as well, or even $\tilde{g} \to t\tilde{t}^*$ or $\tilde{t}\tilde{t}$ if kinematically allowed. The $\tilde{g}$ can also decay at one–loop like $\tilde{g} \to g\tilde{\chi}^0_i$. If, instead, $m_{\tilde{Q}} < M_{\tilde{g}}$, then the $\tilde{g}$ has the 2–body decay $\tilde{g} \to \tilde{Q}q$. The $\tilde{Q}$'s can then decay as $\tilde{Q}_{L,R} \to q\tilde{\chi}^0_i$, $\tilde{u}_L \to d\tilde{\chi}^+_i$, and $\tilde{d}_L \to u\tilde{\chi}^-_i$. The $\tilde{g}$'s and $\tilde{Q}$'s may also be produced in association with $\tilde{\chi}^{\pm}$'s or $\tilde{\chi}^0$'s (analogous to $W + j$ and $Z + j$ production). Event signatures are similar to $\tilde{Q}$ and $\tilde{g}$ production, but possibly with fewer jets. Promising signatures for $\tilde{Q}$ and $\tilde{g}$ production are (i) multiple jets and $\not{E}_T$[29] and (ii) isolated leptons and jets and $\not{E}_T$.[30]

Jets $+$ $\not{E}_T$: Both CDF and DØ have performed searches for events with jets and $\not{E}_T$. This signature has significant physics and instrumental backgrounds. The three dominant physics backgrounds are (i) $Z \to \nu\bar{\nu}$ + jets, (ii) $W \to \tau\nu$ + jets, where the τ decays hadronically, and (iii) $t\bar{t} \to \tau$ +

jets, where the τ decays hadronically. The $\not{E}_T$ in leptonic W decays peaks at $M_W/2 \simeq 40$ GeV, with a long tail at high $\not{E}_T$ due to off–shell or high–p_T W's and energy mismeasurements, so a large $\not{E}_T$ cut is needed to remove these events. Instrumental backgrounds come from mismeasured vector boson, t, and QCD multijet events. Backgrounds from vector boson and t production occur for $W \to e\nu, \mu\nu$ + jets events when the lepton is lost in a crack or is misidentified as a jet. QCD multijet production is a background when jet energy mismeasurements cause false $\not{E}_T$.

The DØ Run Ia analysis[31] searches for events with 3 or more jets and $\not{E}_T$ and with 4 or more jets and $\not{E}_T$. Jets have E_T >20 GeV and cannot point along the $\not{E}_T$ to avoid backgrounds from energy mismeasurement. The $\not{E}_T$ threshold is 65 GeV and leptons are vetoed to remove W backgrounds. The resulting mass limits on $\widetilde{Q}$'s and $\tilde{g}$'s are shown in Fig. 4 ((right), the plot containing the CDF results also shows the DØ Ia results); these limits were set using a RIPS model with the parameters $M_{H\pm} = 500$ GeV, $\tan\beta = 2$, $\mu = -250$ GeV, and $M_{\tilde{l}} = m_{\widetilde{Q}}$.[g]

DØ also has a 3–jet analysis [32] based on 79.2 pb^{-1} of Run Ib data. Jets must have E_T >25 GeV; the leading jet must have E_T >115 GeV because the only available unbiased sample to study the QCD multijet background had this requirement. The $\not{E}_T$ cut ranges from $75 - 100$ GeV and the H_T (scalar sum of the non–leading jet E_T's) cut ranges from $100 - 160$ GeV, optimized for each point in parameter space. Vector boson backgrounds are estimated using VECBOS,[33] while the $t\bar{t}$ background uses HERWIG[34] normalized to the DØ measured $t\bar{t}$ cross section. Two techniques were used to calculate the QCD multijet background. One compares the opening angle between the two leading jets and the $\not{E}_T$ in the signal sample to the analogous distribution in a generic multi–jet sample. The other selects events from a single jet trigger which pass all the selection criteria except for the $\not{E}_T$ requirement. The $\not{E}_T$ distribution is fit in the low $\not{E}_T$ region, and extrapolated into the signal region. The background estimates can be found in Table 2.

The DØ data have been analyzed in the context of a mSUGRA model. For fixed $\tan\beta$, A_0, and sign of μ, exclusion curves are plotted in the $m_0 - m_{1/2}$ plane (Fig. 4 (left)). The limits are from the 3–jet, 79.2pb^{-1}, analysis only.[h] For each point in the limit plane, the $\not{E}_T$ and H_T cuts are reoptimized based on the predicted background and SUSY signal. These results are robust within the mSUGRA framework.[32]

The CDF analysis of the Run Ib data set is not yet finished, but the Run Ia

---

[g]The efficiency and theoretical cross sections were calculated using ISAJET assuming 5 flavors of mass degenerate $\widetilde{Q}$'s without $\tilde{t}$ production.

result based on 19 pb^{-1} has been published.[35] The basic requirements are 3 or 4 jets with $E_T > 15$ GeV and $\not{E}_T > 60$ GeV. The vector boson backgrounds are estimated using VECBOS normalized to the CDF Wjj data. The $t\bar{t}$ backgrounds are determined using ISAJET normalized to the CDF measured cross section. The QCD background is estimated using an independent data sample based on a trigger that required one jet with $E_T > 50$ GeV. First all analysis cuts are applied to this sample except for the S cut, the $\not{E}_T$ cut, and the 3 or 4 jets cut. Next the $\not{E}_T$ distribution is fit and the number of events expected to pass the $\not{E}_T$ cut is derived. Finally the efficiency of the last three cuts is applied to arrive at the final background estimate, shown in Table 2.

The limits derived from the CDF analysis are shown in Fig. 4 (right) within the RIPS framework. In RIPS, a heavy $\tilde{g}$ implies a heavy $\tilde{\chi}_1^0$, so a light $\tilde{Q}$ ($m_{\tilde{Q}} \approx M_{\tilde{\chi}_1^0}$) decay will not produce much $\not{E}_T$. The consequence is an apparent hole in the CDF limit for small $m_{\tilde{Q}}$ and large $M_{\tilde{g}}$. However, lighter $\tilde{g}$'s produce a large $\not{E}_T$ because of the enforced mass splitting between the $\tilde{g}$ and $\tilde{\chi}_1^0$. The results of this analysis do not change substantially as parameters are varied within the RIPS framework.[35]

Table 2: The number of expected and observed events for Tevatron $\tilde{Q}$ and $\tilde{g}$ searches in the jets+$\not{E}_T$ channels.

	DØ		CDF	
Analysis	3 jets	4 jets	3 or 4 jets	4 jets
$\int \mathcal{L}dt(\mathrm{pb}^{-1})$	79.2	13.5	19	19
$W^{\pm}$	$1.56 \pm .67 \pm .42$	4.2 ± 1.2	$13.9 \pm 2.1 \pm 6.0$	$2.6 \pm 0.9 \pm 1.7$
$Z \to \ell\bar{\ell}, \nu\bar{\nu}$	$1.11 \pm .83 \pm .36$	1.0 ± 0.4	$5.0 \pm 0.9 \pm 2.7$	$0.4 \pm 0.2 \pm 0.4$
$t\bar{t}$	$3.11 \pm .17 \pm 1.35$	$-$	$4.2 \pm 0.3 \pm 0.5$	$2.2 \pm 0.2 \pm 0.4$
QCD multijets	3.54 ± 2.64	1.6 ± 0.9	$10.2 \pm 10.7 \pm 4.2$	$3.2 \pm 3.8 \pm 1.3$
Total Background	$9.3 \pm 0.8 \pm 3.3$	6.8 ± 2.4	$33.5 \pm 11 \pm 16$	$8 \pm 4 \pm 4$
Events Observed	**15**	**5**	**24**	**6**

Dileptons+$\not{E}_T$: If, in the cascade decay chain of the $\tilde{Q}$'s and $\tilde{g}$'s, there are two decays $\tilde{\chi}_1^{\pm} \to \ell\nu\tilde{\chi}_1^0$, or one decay $\tilde{\chi}_2^0 \to \ell^+\ell^-\tilde{\chi}_1^0$, the final state can contain 2 leptons, jets, and $\not{E}_T$, which is a relatively clean experimental signature. The requirement of two leptons significantly reduces jet backgrounds and removes most of the W backgrounds. Cutting on lepton pairs with the Z mass removes most of the Z backgrounds. If the leptons are required to have $p_T > 20$ GeV, the major background from physics processes is $t\bar{t} \to bW^+\bar{b}W^- \to b\bar{b}\ell^+\ell^-\not{E}_T$. As the cut on lepton p_T is lowered, $Z \to \tau^+\tau^-$, where the τ's decay semileptonically, also becomes an important background. The instrumental backgrounds

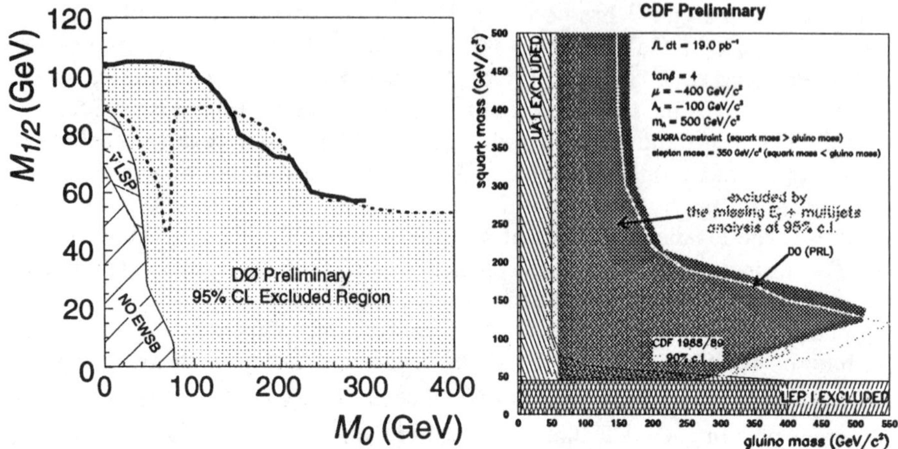

Figure 4: (Left) The DØ Excluded region in the $m_0 - m_{1/2}$ plane with fixed parameters $\tan\beta = 2$, $A_0 = 0$, and $\mu < 0$. The heavy solid line is the limit contour of the DØ jets and $\not{E}_T$ analysis. The dashed line is the limit contour of the DØ dielectron analysis. The lower hashed area is a region where mSUGRA does not predict EWSB correctly. The hashed region above is where the $\tilde{\nu}$ is the LSP. (Right) The CDF mass limits on $\tilde{Q}$'s and $\tilde{g}$'s from the search in jets and $\not{E}_T$[35] using 19 pb^{-1} of data and the ISAJET 7.06 Run I Parameter Set (RIPS) with the indicated values (solid area). For $m_{\tilde{Q}} < M_{\tilde{g}}$, the cross section used is LO, and 3 or more jets are required. For $M_{\tilde{g}} < m_{\tilde{Q}}$, the cross section is NLO,[28] and four jets are required. The line labelled "DØ PRL" is the DØ result from Run Ia using 13.5 pb^{-1} of data.[31]

are small. The spectacular signature of *like–sign*, isolated dileptons, which is difficult to produce in the SM, can occur whenever a $\tilde{g}$ is produced directly or in a cascade decay, since the $\tilde{g}$ is a Majorana particle. This property is exploited in the CDF dilepton searches.

The DØ analysis (e only) requires two leptons, $E_T >15$ GeV, of any sign, while the CDF analysis (e and μ) requires two leptons $E_T >11,5$ GeV, with the same sign. Both analyses require $\not{E}_T > 25$ GeV. Figure 5 shows the the DØ[36] results from Run Ib, plotted for mSUGRA models in the m_0-$m_{1/2}$ plane. Figure 6 ((left), dark shading) shows the CDF[38,37] result plotted in the $M_{\tilde{g}} - m_{\tilde{Q}}$ plane for a RIPS model, and (right) a mapping of the DØ mSUGRA results from Fig. 5 (left) into the same coordinates. The CDF limit is based on NLO cross sections,[28] and the DØ limit on LO cross sections. The DØ limits on m_0 and $m_{1/2}$ are calculated including contributions from the production of all sparticles (for instance, associated production of $\tilde{\chi}^0$'s or $\tilde{\chi}^\pm$'s with $\tilde{Q}$'s or $\tilde{g}$'s), while the CDF result considers only $\widetilde{Q}$ and $\tilde{g}$ production.

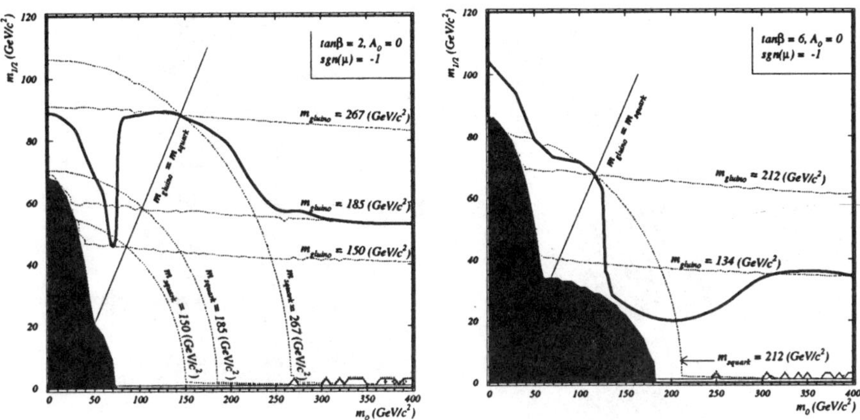

Figure 5: (Left) The DØ limits on the SUGRA parameters m_0 and $m_{1/2}$ from the 2 leptons, 2 jets, and $\not{E}_T$ search[36] for $\tan\beta=2$, $A_0=0$, and $\mu < 0$. (Right) The same plot for $\tan\beta=6$, $A_0=0$, and $\mu < 0$. In both plots, the dark shaded area is the region in which electroweak symmetry breaking is not realized. Selected contours of $\widetilde{Q}$ and $\tilde{g}$ mass are also shown.

For $m_{\widetilde{Q}} \gg M_{\tilde{g}}$ or, equivalently, for $m_0 \gg m_{1/2}$, $\tilde{g}\tilde{g}$ pair production is the dominant SUSY process. As $m_0(m_{\widetilde{Q}})$ is varied with the other parameters fixed, the branching ratios for the 3–body $\tilde{g}$ decays to $\widetilde{\chi}^{\pm}$'s or $\widetilde{\chi}^0$'s and jets become fairly constant, so the production rate of leptonic final states becomes constant. For large enough values of the $\tilde{g}$ mass, the leptons easily pass the experimental cuts, and the experimental limit approaches a constant value asymptotically, as can be seen in both the DØ and CDF plots shown in Fig. 6.

The relation $m_{\widetilde{Q}} \ll M_{\tilde{g}}$ is not possible in SUGRA, and is treated in an *ad hoc* manner in RIPS. There is no limit in this region for either opposite– or like–sign dilepton pairs because the large, fixed $\tilde{\ell}$ masses (as assumed in this analysis) limit the branching ratios to leptonic final states. The possibility of like–sign dilepton pairs is further reduced because both the $\tilde{g}\tilde{g}$ and $\tilde{g}\widetilde{Q}$ cross sections (which produce like–sign leptons because the $\tilde{g}$ is a Majorana particle) and the $\widetilde{Q}\widetilde{Q}$ cross section (which produces like–sign leptons because the $\widetilde{Q}$'s have the same charge) are small in this region.

When $m_{\widetilde{Q}} \simeq M_{\tilde{g}}$, the $\tilde{g}\tilde{g}$ cross section is supplemented by the $\tilde{g}\widetilde{Q}$ cross section. Just above the diagonal line at $M_{\tilde{g}} = m_{\widetilde{Q}}$ (*i.e.* $m_{\widetilde{Q}}$ just larger than $M_{\tilde{g}}$) in Fig. 6 there are "noses" in the limit plots, with the limit becoming

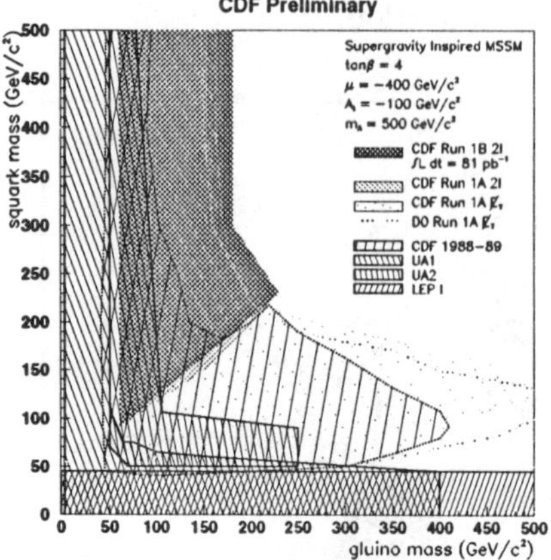
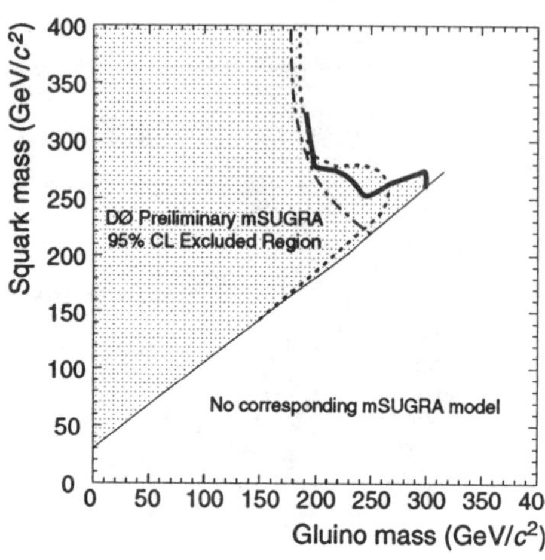

Figure 6: (Left) CDF limits on the $\widetilde{Q}$ and $\tilde{g}$ masses from the 2 like–sign leptons, 2 jets, and $\not{E}_T$ search in 81 pb^{-1} (dark shading). The limits were set using the ISAJET 7.06 Run I Parameter Set (RIPS) with the indicated values. (Right) Excluded region from various DØ analyses in the $m_{\widetilde{Q}} - M_{\tilde{g}}$ plane with fixed mSUGRA parameters $\tan\beta = 2$, $A_0=0$, and $\mu < 0$. There are no mSUGRA models in the region to the right of the diagonal thin line. The heavy solid line is the limit contour of the DØ Run Ib 3 jets and $\not{E}_T$ analysis. The dashed line is the limit contour of the DØ Run Ib dielectron analysis. The dot–dashed line is the limit contour of the DØ Run Ia 3 and 4 jets and missing transverse energy analysis shown only in the region with valid mSUGRA models.

stronger close to the diagonal.

The limits in Fig. 6 are for a specific choice of parameters within RIPS or mSUGRA. If μ, A_t and $\tan\beta$ are varied, the branching ratios into $\widetilde{\chi}^{\pm}$'s or $\widetilde{\chi}^0$'s can vary strongly. The sensitive dependence on the parameters can be seen within mSUGRA from the DØ limits in Fig. 5. The dip in the $\tan\beta=2$ limit (left), around $m_0= 70$ GeV, is a point where $m_{\tilde{l}} > M_{\widetilde{\chi}^0_2} > m_{\tilde{\nu}}$ and $\mathrm{BR}(\widetilde{\chi}^0_2 \to \nu\bar{\nu}\widetilde{\chi}^0_1) \simeq 1$, so the detection efficiency is very sensitive to the choice of high energy parameters m_0 and $m_{1/2}$. In Fig. 5 (right), with $\tan\beta=6$, the limits are severely reduced compared to Fig. 5 (left), with $\tan\beta=2$, in the region where $m_{\widetilde{Q}} \gg M_{\tilde{g}}$. For large $\tan\beta$, the mass splittings are reduced, and the leptons from the $\widetilde{\chi}^{\pm}_1$ and $\widetilde{\chi}^0_2$ decays are softer. The non–trivial shape of the limit curves results from an interplay between the cross section being larger when

m_0 and $m_{1/2}$ are smaller (sparticle masses are smaller) and the mass splittings being smaller. Consequently, although the dileptons+jets+$\not{E}_T$ signature is an excellent discovery channel with little SM background, it is hard to set significant parameter limits even using mSUGRA models.

From the present analyses in the $\not{E}_T$+jets and dileptons+$\not{E}_T$ channels, some preliminary conclusions can be drawn on the $\tilde{Q}$ and $\tilde{g}$ masses. These depend, however, on the assumed SUSY parameters. The DØ limit on the $\tilde{g}$ mass effectively develops a plateau at 185 GeV for large m_0 and $\tan\beta= 2$, and at 134 GeV for $\tan\beta= 6$. The CDF limit on the $\tilde{g}$ mass is 180 GeV for $\tan\beta= 4$ and large $m_{\tilde{Q}}$. For equal $\tilde{Q}$ and $\tilde{g}$ masses, the DØ mass limit for $\tan\beta=2$ is 267 GeV. From the CDF RIPS analyses and $m_{\tilde{Q}} \simeq M_{\tilde{g}}$, the limit is about 220 GeV for $\tan\beta = 4$. A direct comparison of all the above results is rather difficult since DØ and CDF have done analyses assuming different sets of MSSM parameters.

Stop Squarks: The top squark (stop) is a special case.[39,40] The mass degeneracy in the $\tilde{t}$ sector is expected to be strongly broken, and, for sufficiently large mixing, the lightest stop $\tilde{t}_1$ can be rather light, possibly lighter than the $\tilde{\chi}_1^{\pm}$. The $\tilde{t}_1$ has about a tenth the production cross section[28] of a t quark of the same mass, because the cross section behaves as β^3 at threshold (compared to β for fermion pairs), where β is the squark velocity in the rest frame of the pair, and only half the scalar partners are being considered.

The stop can be produced directly as $\tilde{t}\tilde{t}^*$ pairs or, depending on the $\tilde{t}$ mass, indirectly in decays $t \to \tilde{t}\tilde{\chi}_1^0$, or $\tilde{\chi}_i^{\pm} \to b\tilde{t}$. Also depending on the $\tilde{t}$ mass, one of three decay modes is expected to dominate. If (a) $m_{\tilde{t}_1} > m_{\tilde{\chi}_1^{\pm}} + m_b$, then $\tilde{t}_1$ can decay into $b\tilde{\chi}_1^+$, followed by the decay of the $\tilde{\chi}_1^{\pm}$. This can look similar to the decay $t \to bW$, but with different kinematics and branching ratios for the final state. Instead, if $\tilde{t}_1$ is the lightest charged SUSY particle, it is expected to decay exclusively through a $\tilde{\chi}^{\pm} - \tilde{b}$ loop as (b) $\tilde{t}_1 \to c\tilde{\chi}_1^0$, which looks quite different from SM top decays. Finally, the $\tilde{t}$ can decay (c) $\tilde{t} \to bW\tilde{\chi}_1^0$ or, if it is quite heavy, $\tilde{t} \to t\tilde{\chi}_1^0$

DØ has searched for $\tilde{t}\tilde{t}^*$ production with $\tilde{t} \to c\tilde{\chi}_1^0$ using 7.4 pb^{-1} of Run Ia data.[41] The signature used is two acollinear jets, $E_T >30$ GeV and $\not{E}_T>40$ GeV. The dijet cross section is large, and thus this signature has large instrumental backgrounds. It also has backgrounds from vector boson production. The multijet backgrounds can be controlled by requiring $\Delta\phi > 45°$ between the $\not{E}_T$ and each jet, and that the jets not be back–to–back. The vector boson backgrounds are controlled by requiring that the two leading jets are separated by at least $\Delta\phi > 90°$. After these cuts, the dominant backgrounds are from W and Z boson production and decay, with the largest being $W \to \tau\nu$.

The efficiency is largest when the stop is heavy compared to the $\widetilde{\chi}_1^0$ (near the kinematic boundary for the decay $\tilde{t}_1 \to bW^+\widetilde{\chi}_1^0$), reaching a maximum value of only 4%. The mass difference $m_{\tilde{t}} - M_{\widetilde{\chi}_1^0}$ determines the E_T of the charm jet and rapidly limits this search mode as the c–jets become too soft (see Fig. 7). With the assumption that $\mathrm{BR}(\tilde{t}_1 \to c\widetilde{\chi}_1^0)=1$, the predicted SUSY final state depends only on $M_{\widetilde{\chi}_1^0}$ and $m_{\tilde{t}_1}$. The result of this search is a 95% C.L. exclusion limit on a region in the $M_{\widetilde{\chi}_1^0} - m_{\tilde{t}_1}$ plane, shown in Fig. 7.[h]

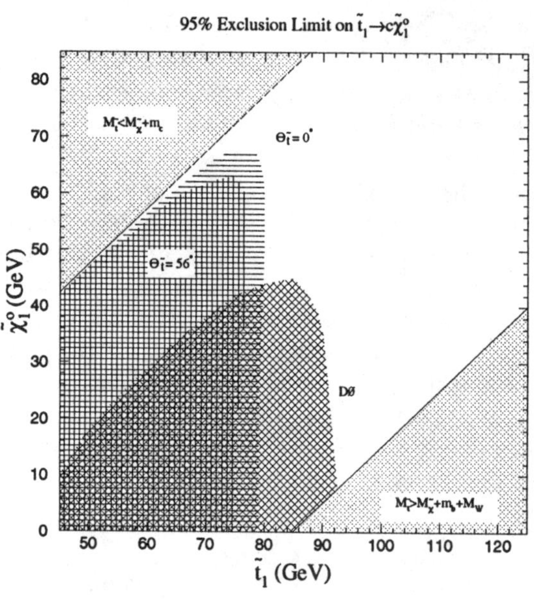

Figure 7: Mass limits from the DØ search for $\tilde{t}\tilde{t}^*$ production with the decay $\tilde{t} \to c\widetilde{\chi}_1^0$ at the Tevatron.[41] The decay is kinematically forbidden in the two solid grey regions. The hashed regions marked $\Theta_{\tilde{t}}$ show the LEP excluded regions as a function of the $\tilde{t}$ mixing angle, which determines the strength of the $\tilde{t}$ coupling to the Z. The mixing does not affect the tree level process at hadron colliders.

CDF and DØ have also presented results from a search for $\tilde{t}_1\tilde{t}_1^*$ production, with $\tilde{t}_1 \to b\widetilde{\chi}_1^\pm$. The CDF search is in the lepton+jets channel, and uses a shape analysis of the transverse mass of the lepton ($E_T > 20$ GeV) and $\not{E}_T(>20$ GeV).[42] The results of the CDF search are shown in Fig. 8 (left).

---

[h]The production rate has been calculated using only LO production cross sections from ISAJET.

The decay $\widetilde{\chi}_1^{\pm} \to W^* \widetilde{\chi}_1^0$ is assumed using the masses (i) $M_{\widetilde{\chi}_1^{\pm}} = 80$ GeV and $M_{\widetilde{\chi}_1^0} = 30$ GeV and (ii) $M_{\widetilde{\chi}_1^{\pm}} = 70$ GeV and $M_{\widetilde{\chi}_1^0} = 30$ GeV.[i] Given these mass choices, there is little other parameter dependence. Presently, the cross section limits are above the predicted cross sections due to the high E_T cuts.

DØ searches in the dilepton channel[43] ($E_T >16,8$ GeV) and $\not{E}_T >22$ GeV. The results are shown in Fig. 8 (right), assuming $M_{\widetilde{\chi}_1^{\pm}} = 47$ GeV and $M_{\widetilde{\chi}_1^0} = 28.5$ GeV. A substantial background comes from $Z \to \tau^+\tau^-$, again requiring a high threshold for the E_T cuts, and no limit can be set.[j]

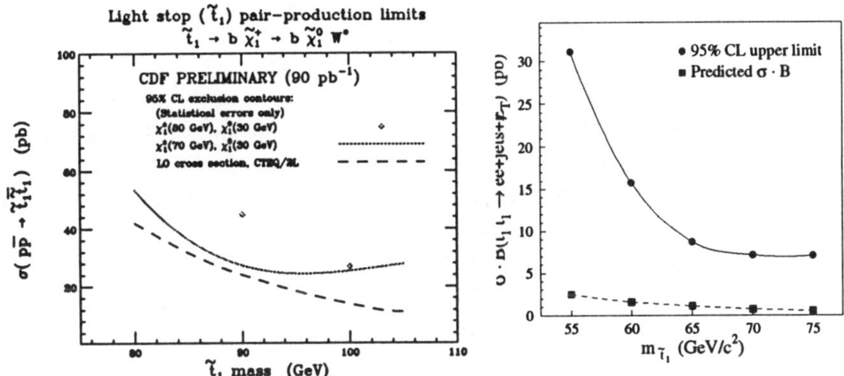

Figure 8: (Left) The CDF cross section limit on direct production of the top squark using 90 pb^{-1} of data. The decay mode is $\tilde{t} \to b\tilde{\chi}_1^+(\to W^*\tilde{\chi}_1^0)$. One W must decay semi–leptonically giving a signature of a lepton, $\not{E}_T$, and jets. The theoretical cross section is from ISAJET 7.06. (Right) The DØ 95% confidence level cross section limit on the cross section for stop production times the branching ratio to a final state containing 2 electrons as a function of the mass of the $\tilde{t}$ is shown as a solid line.[43] The mass of the lightest chargino is assumed to be 47 GeV. The predicted cross section times branching ratio from ISAJET is also shown as a dashed line.

CDF has presented another analysis using the SVX–tagged lepton+jets sample to search for the decay $t \to \tilde{t}_1\tilde{\chi}_1^0$, with $\tilde{t}_1 \to b\tilde{\chi}_1^{\pm}$.[44] If one t in a $t\bar{t}$ event decays $t \to bW(\to \ell\nu)$ and the other $t \to \tilde{t}_1\tilde{\chi}_1^0$ followed by $\tilde{t}_1 \to b\tilde{\chi}_1^{\pm}(\to jj\tilde{\chi}_1^0)$ or $t \to bW(\to jj)$ and $t \to \tilde{t}_1\tilde{\chi}_1^0$ followed by $\tilde{t}_1 \to b\tilde{\chi}_1^{\pm}(\to \ell\nu\tilde{\chi}_1^0)$, the signature is $b\bar{b}\ell\nu jj+\not{E}_T$, the same as in the SM, but where the $\not{E}_T$ includes the momentum of the $\tilde{\chi}_1^0$. The cuts, lepton $E_T >20$ GeV, $\not{E}_T >45$ GeV, and one SVX b–tag, are optimized for acceptance of the SUSY decay and rejection

---

[i]This analysis was done in regions of MSSM parameter space later excluded by LEP. It demonstrates, however, the procedures to be followed in performing these studies in other regions.

of W+jets background. A likelihood function is computed for each event reflecting the probability that the jets with the 2nd and 3rd highest E_T in the event are consistent with the stiffer SM distribution (as compared to the SUSY distribution). The distribution of this likelihood function shows a significant separation of these two hypotheses. After applying the cuts, 9 events remain, all of which fall outside of the SUSY signal region. For $\tilde{t}$ masses between 80 and 150 GeV and $\tilde{\chi}^{\pm}$ masses between 50 and 135 GeV, a BR($t \to \tilde{t}_1 \tilde{\chi}_1^0$)=50% is excluded at the 95% C.L., provided that $M_{\tilde{\chi}_1^0} = 20$ GeV.[j] Because $M_{\tilde{\chi}_1^0}$ is fixed in this manner, it is not related to $M_{\tilde{\chi}_1^{\pm}}$ as in SUGRA.

3.3 Sleptons

At hadron colliders, $\tilde{\ell}$'s and $\tilde{\nu}$'s can only be directly produced through their electroweak couplings to the γ, Z and W bosons. The production cross sections are at most a few tens or hundreds of fb at the Tevatron,[45] and the physics and instrumental backgrounds are numerous. So far neither collaboration has presented results on searches for sleptons in the SUGRA or RIPS framework (we describe limits in gauge–mediated SUSY models later).

A (stable) charged slepton is not a viable LSP candidate, so the decays $\tilde{\ell}^{\pm}_{L,R} \to \ell^{\pm} \tilde{\chi}_i^0$ or $\tilde{\ell}^{\pm}_L \to \nu \tilde{\chi}_i^{\pm}$ are expected. A $\tilde{\nu}$, instead, can be the LSP, or it can decay invisibly $\tilde{\nu} \to \nu \tilde{\chi}_1^0$, or visibly $\tilde{\nu} \to \tilde{\chi}_i^{\pm} \ell^{\mp}$. If $m_{\tilde{\nu}} < m_{\tilde{\ell}} < M_{\tilde{\chi}_1^0}$, then the decay $\tilde{\ell} \to \ell' \nu' \tilde{\nu}$ (or $\tilde{\ell} \to q\bar{q}\tilde{\nu}$) is possible. Promising signatures are (i) $e^+ e^-, \mu^+ \mu^-, \tau^+ \tau^- + \not{E}_T$, (ii) $e\mu, e\tau, \mu\tau + \not{E}_T$, and (iii) e, μ or τ+jets + $\not{E}_T$ (or jets + $\not{E}_T$). Although charged slepton production can lead to charged leptons in the final state, there is no guarantee.

3.4 Charged Higgs Bosons

Even though Higgs bosons are not sparticles, the discovery of one or more could be considered *indirect* evidence for SUSY. If it is light enough, the $H^{\pm}$ can be produced in the decay $t \to bH^+$.[46] The branching fraction for this decay depends on the $H^{\pm}$ mass and $\tan\beta$, and is larger than 50% for $\tan\beta$ less than approximately 0.7 or greater than approximately 50, but very small or large values of $\tan\beta$ are theoretically disfavored. In general, at reasonably small values of $\tan\beta$, $H^+ \to c\bar{s}$; at large $\tan\beta$, $H^+ \to \tau^+ \nu_\tau$.

CDF has searched for the decay $t \to bH^+$ using both direct[47,48] and indirect[23] methods. Direct searches look for an excess over SM expectations of events with τ leptons from the decay $H^+ \to \tau^+ \nu_\tau$ (dominant for large $\tan\beta$). The signature for hadronically decaying τ's is a narrow jet associated with one or three tracks with no other tracks nearby. Indirect searches are "disappearance" experiments, relying on the fact that decays $t \to bH^+$ will deplete the SM decays $t \to bW$, decreasing the number of events in the dilepton and lepton+jets channels.

The selection criteria for the CDF direct search are either a single τ, $E_T > 20\,\text{GeV}$, $\not{E}_T > 30\,\text{GeV}$ and a SVX b-tagged jet, or two τ's, with $E_T > 30\,\text{GeV}$. Values of $m_{H\pm}$ and $\tan\beta$ can be excluded based on two methods: either the model is inconsistent with the observation of of τ's, (Fig. 9 (left)), or the model is inconsistent with the combination of the number of τ events and the number of lepton+jets events. The second method has the advantage that a $t\bar{t}$ cross section $\sigma_{t\bar{t}}$ does not need to be assumed; details are presented elsewhere.[48] The reader should be aware that there are subtleties in analyses that assume cross sections.[3]

A direct search at small $\tan\beta$ is difficult since the $H^\pm$ would decay into two jets. However, the indirect method can be applied to both small and large $\tan\beta$ searches. If a choice of $\sigma_{t\bar{t}}$, $m_{H\pm}$, and $\tan\beta$ predicts a number of dilepton and lepton+jets events that is inconsistent with the observations, then that set of values is excluded (Fig. 9 (left)). The area in Fig. 9 (left) labeled "ratio method" is the exclusion region for an indirect search that does not make an assumption for $\sigma_{t\bar{t}}$. The limit is set by comparing the ratio of dilepton to lepton+jets events, since $t \to bH^+$ decays always decreases this ratio, regardless of $\sigma_{t\bar{t}}$.

Recent studies have shown that quantum SUSY effects (SUSY QCD and electroweak radiative corrections) to the decay mode $t \to bH^+$ (with subsequent decays into τ's) may be important and should be considered in future analyses.[50]

3.5 Neutral Higgs Bosons

The lightest CP–even Higgs boson h can be produced at the Tevatron in the channels Wh or Zh.[7] These channels are relevant for large values of M_A (the SM limit) or for small M_A and small $\tan\beta$. The heavier Higgs H could become relevant for searches at an upgraded Tevatron through ZH, WH production, in some restricted region of parameter space, complementary to the one relevant for the h searches. In addition, the enhancement of the b Yukawa coupling for large $\tan\beta$ can enhance $hb\bar{b}$, $Ab\bar{b}$, and $Hb\bar{b}$ production.[51]

Both collaborations have searched for $q\bar{q}' \to W^* \to W(\to e\nu, \mu\nu)h(\to b\bar{b})$. DØ has searched in 100 pb^{-1} of data using a data sample containing a lepton ($E_T > 20$ GeV), $\not{E}_T$ (>20 GeV) and two jets.[52] One of the jets must have a muon associated with it for b-tagging. Twenty–seven events pass the selection

224

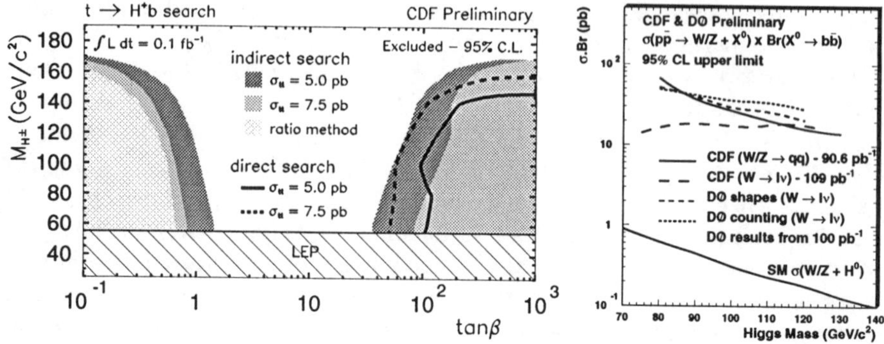

Figure 9: (Left) Exclusion space for the CDF searches for decays $t \to bH^+$ in $t\bar{t}$ events. The region excluded with solid lines at high $\tan\beta$ is from a direct search for events where one or both t's in a $t\bar{t}$ event decay to $bH^+(\to \tau^+\nu)$ and information from the SM channels is ignored. The shaded regions are from the indirect searches. For the regions labeled $\sigma_{t\bar{t}} = 5.0$ and 7.5 pb, $\sigma_{t\bar{t}}$ is assumed and points are excluded if the predicted SUSY decays have depleted the SM channels to an extent that they are inconsistent with the data. The "ratio method" is an indirect method comparing the ratio of lepton+jets to dilepton events and no $\sigma_{t\bar{t}}$ is assumed. (Right) Limits from CDF and DØ for the associated production of Wh or Zh. The CDF limits are shown for the final states of $\ell\nu b\bar{b}$ and $jjb\bar{b}$, and the DØ limit is for the final state $\ell\nu b\bar{b}$. The limit is set using a simple counting method and by fitting the $b\bar{b}$ spectrum ("shapes").

criteria; 25.5 ± 3 events are expected from Wjj and $t\bar{t}$. The limits shown in Fig. 9 (right) are set by a simple event–counting method and by fitting the $b\bar{b}$ dijet mass spectrum.

CDF has recently completed a similar search for the same decay mode using $109\,\mathrm{pb}^{-1}$ of data.[53] Leptons must have $E_T > 25$ (e) or 20 GeV (μ) and the event must have $\not{E}_T > 25$ (e) or 20 GeV (μ) and one SVX b–tag. These events are split into single–tagged (one SVX tag) and double–tagged samples (two SVX tags or one SVX and one lepton (e or μ) tag). The 36 (6) single–tagged (double–tagged) events are consistent with the 30 ± 5 (3.0 ± 0.6) expected from SM W+jets and $t\bar{t}$. Both the single– and double–tagged dijet mass distributions are fit simultaneously to set the limits shown in Fig. 9 (right).

The process $q\bar{q} \to Z^* \to Zh$ occurs at a comparable rate to the W^* process. CDF has searched for both processes assuming $W/Z \to jj$.[54] All events must have 4 jets with $E_T > 15$ GeV and two SVX b–tags. In $91\,\mathrm{pb}^{-1}$ of data,

589 events remain, consistent with the expectation from QCD heavy–flavor production and fake tags. To set limits, the $b\bar{b}$ dijet mass spectrum is fit. Also shown in Fig. 9 (right) is the SM production cross section for Wh and Zh as a function of the Higgs boson mass. The present experimental limits are roughly two orders–of–magnitude away from the predicted cross section.

DØ has also searched for an h with suppressed couplings to fermions,[55] so that $h \to \gamma\gamma$ can be dominant (see the first reference of [7]). Events are selected containing two photons with $E_T >$ 20 and 15 GeV, and two jets with $E_T >$ 20 and 15 GeV. No evidence of a resonance is seen in the mass distribution of the 2 photons, and DØ excludes such a Higgs with a mass less than 81 GeV at a 95% C.L.

3.6 R–Parity Violation

Allowing for $\not{R}$ in the MSSM opens a host of possibilities at the Tevatron.[60] The possible excess of HERA events at large Q^2 has triggered interest in studying the consequences of the interaction of a light $\widetilde{Q}$ (preferably a $\tilde{t}$ or $\tilde{c}$) with an electron and a d quark.[62] If the $\tilde{g}$ were heavier than this $\widetilde{Q}$, then $\tilde{g}$ pair production at the Tevatron and the decay $\tilde{g} \to \bar{c}\tilde{c}_L$ through R–conserving couplings, followed by the $\not{R}$ decay $\tilde{c}_L \to e^+ d$, would yield the signature of two electrons and 4 jets.[j]

CDF has performed a search[61] considering the $\not{R}\ \widetilde{Q}$ decays with the signature of two like–sign electrons ($E_T >$15 GeV) and two jets ($E_T >$15 GeV). In 105 pb^{-1} of Run Ia and Ib data, no events remain after all cuts are applied. Varying the masses of the SUSY particles does not alter the acceptance significantly since they are heavy enough for the decay products to easily pass the E_T thresholds. Because of this, the limit on the cross section times branching ratio is approximately constant at 0.19 pb. For $m_{\tilde{c}_L} = 200$ GeV, this excludes $M_{\tilde{g}} < 230$ GeV, assuming BR$(\tilde{g}\tilde{g} \to e^{\pm}e^{\pm}X) = 1/8$.

Allowing for R–parity conserving $\widetilde{Q}$ decays, the decay $\widetilde{Q} \to q\tilde{\chi}_1^0$ is possible, where $\tilde{\chi}_1^0$ is the LSP. Since the LSP has no R–parity conserving decays kinematically accessible, the $\not{R}$ decay $\tilde{\chi}_1^0 \to c\bar{d}e^-$ or $\bar{c}de^+$ occurs through a virtual $\tilde{c}$ or $\tilde{d}$, while $\tilde{\chi}_1^0 \to d\bar{s}\nu$ or $\bar{d}s\bar{\nu}$ occurs through a virtual $\tilde{s}$ or $\tilde{d}$. For the analysis, 5 $\widetilde{Q}$ masses are assumed to be degenerate, and $\widetilde{Q}$ masses less than 210 GeV are excluded if the mass of the $\tilde{\chi}_1^0$ is more than half of the $\widetilde{Q}$ mass and the $\tilde{g}$ is heavy.

If R–parity is violated, and the LSP is charged (*e.g.* a $\tilde{\tau}$), it can be long–lived and appear as a heavy stable particle. The particle can be identified by

---

[j] If the above $\not{R}$ decay is allowed, then the same $\not{R}$ coupling will induce the decays $\tilde{s}_L \to \nu_e d$, $\tilde{d}_R \to e^- c$ and $\tilde{d}_R \to \nu_e s$.

measuring the dE/dx energy loss as it passes through the CDF SVX and CTC detectors. For a given momentum, a heavy particle has a slower velocity and hence a greater energy loss than a relativistic particle ($\beta \simeq 1$). If the particle is weakly interacting or massive enough to kinematically suppress showering, it will penetrate the detectors and be triggered on and reconstructed as a muon with too much energy loss. A result using part of the Run I data has been presented by CDF [59] and is now updated with the full data set. In 90 pb^{-1} of inclusive muon triggers ($p_T >30$ GeV), CDF searches for particles with ionization consistent with $\beta\gamma < 0.6$ and finds 12 events depositing more than twice the energy expected from a minimum ionizing muon. This is consistent with the number of events expected from muons which overlap with other tracks to fake a large dE/dx signal.

3.7 Photon and $\not{E}_T$ Signatures

SUSY has so many parameters that the full range of its allowed signatures may be hard to predict. In April 1995, the CDF experiment recorded an event with a very unusual topology[63] which may have SUSY interpretations. It has four electromagnetic clusters, which pass the typical cuts for two electrons and two photons, and $\not{E}_T$.

There have been two main proposals for a possible SUSY explanation of the event: the Gravitino LSP and the Higgsino LSP model (there are also non–SUSY explanations[64]). In gauge–mediated SUSY models, the $\tilde{G}$ is the LSP and the next–to–lightest superpartner (NLSP) decays into its SM partner plus the Goldstino component of the $\tilde{G}$.[69] If $\tilde{\chi}_1^0$ is the NLSP, the only modification to SUGRA phenomenology, where all sparticles decay down to $\tilde{\chi}_1^0$, is that $\tilde{\chi}_1^0$ then decays to a photon and $\not{E}_T$. In particular, the CDF event can be interpreted as either $\tilde{e}\tilde{e}^*$ production[65,68] or $\tilde{\chi}_1^+ \tilde{\chi}_1^-$ production.[68]

The Higgsino LSP model[66] involves a region of MSSM parameter space in which the $\tilde{\chi}_2^0$ is photino–like and the $\tilde{\chi}_1^0$ is Higgsino–like, so the radiative decay $\tilde{\chi}_2^0 \to \gamma\tilde{\chi}_1^0$ dominates over other $\tilde{\chi}_2^0$ decay modes.[72] The event can be again interpreted as $\tilde{e}\tilde{e}^*$ production or $\tilde{\chi}_1^+ \tilde{\chi}_1^-$ production.

Both proposals also suggest other signatures that should be expected within these models.[67] The $\tilde{G}$ LSP model predicts a large number of events with many jets, leptons, and γ's, and the fact than none of these other signatures has been detected makes the above LSP $\tilde{G}$ explanation of the CDF $ee\gamma\gamma\not{E}_T$ event unlikely.[70,71] In Higgsino LSP models, γ's only arise from the decay of $\tilde{\chi}_2^0$, and there is no guarantee that there will be other substantial signals. A logical starting place for searches is in the inclusive two photon and $\not{E}_T$ channel.[70] The generic $\gamma\gamma\not{E}_T +X$ signature has no significant background from real γ's. The main backgrounds are caused by jets and electrons faking γ's. The SM production of $W(\to e\nu)\gamma$ + jets can fake some of the signatures if the electron is misidentified as a γ. These events have a $\not{E}_T$ spectrum typical of W events, peaked at about $M_W/2 \simeq 40$ GeV, with a long tail to high $\not{E}_T$. The dominant instrumental background, however, is from di–jet and γ+jet production, where the large production cross section overcomes the small probability $(\simeq 10^{-4} - 10^{-3})$ that a jet fakes a γ.

Figure 10 shows the $\not{E}_T$ distributions from DØ (left) and CDF (right) diphoton events [74,75] after imposing the selection criteria. The DØ analysis requires two γ's with $E_T >$20,12 GeV and $\not{E}_T>$ 25 GeV, while CDF requires two γ's with $E_T >$25 GeV and $\not{E}_T>$ 35 GeV. For the DØ analysis, the shape of the $\not{E}_T$ spectra agrees well with backgrounds containing two electromagnetic-like clusters, where at least one of the two clusters fails the γ selection criteria. Two events satisfy all selection criteria, with a predicted background, dominated by jets faking γ's, of 2.3 $\pm$0.9 events. For the CDF analysis, the shape of the $\not{E}_T$ distribution is in good agreement with the resolution of the $Z \to e^+e^-$ control sample. The event on the tail in $\not{E}_T$ is the "$ee\gamma\gamma\not{E}_T$" event. If the source of this event is an anomalously large $WW\gamma\gamma$ production cross section that yields one event in $\ell\ell\gamma\gamma\not{E}_T$, CDF would expect dozens of events with two γ's and several observed jets. However, the jet multiplicity spectrum in diphoton events is well–modeled by an exponential, and there are no diphoton events with 3 or 4 jets.

DØ presents limits[75] in the framework of the $\tilde{G}$ LSP scenario by considering $\tilde{\chi}^0$ and $\tilde{\chi}^\pm$ pair production. Assuming $M_2 \simeq 2M_1$ and large values of $m_{\tilde{Q}}$, the signatures are a function of only M_2, μ, and $\tan\beta$. Figure 11 shows the limit on the cross section for $\tilde{\chi}_1^+\tilde{\chi}_1^-$ and $\tilde{\chi}_1^\pm\tilde{\chi}_2^0$ production as a function of the $\tilde{\chi}_1^\pm$ mass when $|\mu|$ is large and thus the $\tilde{\chi}_1^\pm$ mass is approximately twice the $\tilde{\chi}_1^0$ mass. The figure also shows, more generally, the excluded region in the M_2–μ plane, along with a prediction for the region that might explain the CDF $ee\gamma\gamma\not{E}_T$ event as $\tilde{\chi}_1^+\tilde{\chi}_1^-$ production. The latter explanation requires 100 GeV $< M_{\tilde{\chi}_1^\pm} <$ 150 GeV with $M_{\tilde{\chi}_1^0} < 0.6M_{\tilde{\chi}_1^\pm}$ to produce one event with a reasonable probability. As can be seen from Fig. 11, the cross section limit is typically 0.24 pb for either $\tilde{\chi}_1^+\tilde{\chi}_1^-$ or $\tilde{\chi}_1^\pm\tilde{\chi}_2^0$ production. By combining all $\tilde{\chi}^\pm$ and $\tilde{\chi}^0$ pair production processes, a $\tilde{\chi}_1^\pm$ with mass below 150 GeV is excluded. Hence, to keep the $\tilde{\chi}_1^+\tilde{\chi}_1^-$ interpretation of the $ee\gamma\gamma\not{E}_T$ event, it would be necessary to expand the analysis of Ref. [68].

DØ also has a limit on the cross section for $\tilde{e}\tilde{e}^* \to e^-e^+\tilde{\chi}_2^0\tilde{\chi}_2^0$, $\tilde{\nu}\tilde{\nu}^* \to \nu\bar{\nu}\tilde{\chi}_2^0\tilde{\chi}_2^0$, and $\tilde{\chi}_2^0\tilde{\chi}_2^0 \to \gamma\gamma\tilde{\chi}_1^0\tilde{\chi}_1^0$ using the same analysis as for the $\tilde{G}$ LSP search. Such signatures might also be expected in Higgsino LSP models. The limit on

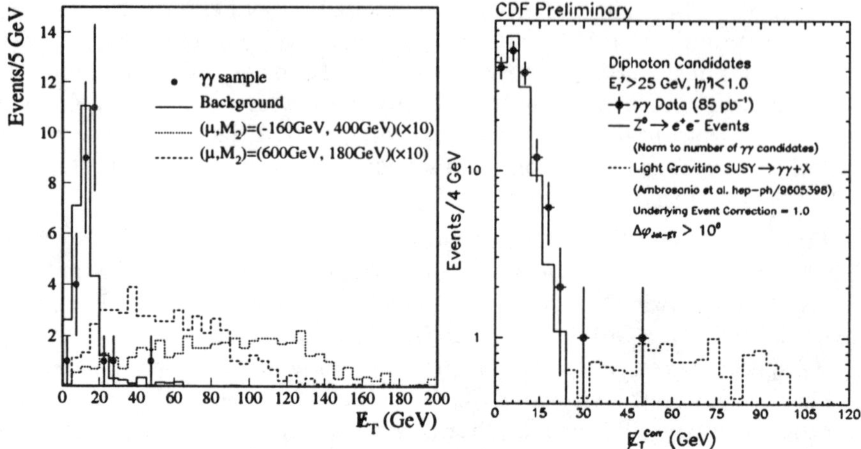

Figure 10: (Left) The $\not{E}_T$ spectra in the DØ search for events with 2 γ's, one with $E_T >$ 20 GeV, the other with $E_T >$ 12 GeV.[75] The points are the data, the solid line is the estimated background from di–jet events and direct γ events. The dotted lines are for gaugino production within gauge–mediated models using the parameters listed and $M_1 \simeq 2M_2$. (Right) The CDF $\not{E}_T$ spectrum for events with two central γ's with $E_T > 25$ GeV. Events which have any jet with $E_T > 10$ GeV pointing within 10 degrees in azimuth of the $\not{E}_T$ are removed. The solid histogram shows the resolution from the $Z \to e^+e^-$ control sample. The dashed line shows the expected distribution from all SUSY production in a model [70] with
$$M_2 = 225 \text{ GeV}, \ \mu = 300 \text{ GeV}, \ \tan\beta = 1.5, \text{ and } M_{\tilde{Q}} = 300 \text{ GeV}.$$

the cross section for such processes is about 0.35 pb for $M_{\tilde{\chi}_2^0} - M_{\tilde{\chi}_1^0} > 30$ GeV, which is close to the maximum cross section predicted in these models.

CDF has searched for the signature $\gamma bc\not{E}_T$, as predicted in Higgsino LSP models with a light $\tilde{t}$.[73] The data sample of 85 pb⁻¹ contains events with an isolated γ with $E_T^\gamma > 25$ GeV and a jet with an SVX b–tag. After requiring $\not{E}_T > 20$ GeV, 98 events remain.[74] The estimated background to the 98 events is $77 \pm 23 \pm 20$ events. The shape is consistent with background. About 60% of the background is due to jets faking γ's, 13% to real γ's and fake b–tags, and the remainder to SM $\gamma b\bar{b}$ and $\gamma c\bar{c}$ production; all of these sources require fake $\not{E}_T$. When the $\not{E}_T$ cut is increased to 40 GeV, 2 events remain. More than 6.4 events of anomalous production in this topology is excluded. The efficiency used in the limits is derived from a "baseline" model with $M_{\tilde{\chi}_1^0} = 40$ GeV, $M_{\tilde{\chi}_2^0} = 70$ GeV, $m_{\tilde{t}_1} = 60$ GeV, $m_{\tilde{Q}} = 250$ GeV, and $M_{\tilde{g}} = 225$ GeV. The baseline model predicts 6.65 events, so this model is excluded (at the 95% C.L.). This result does not rule out the Higgsino LSP model with a light $\tilde{t}$ in

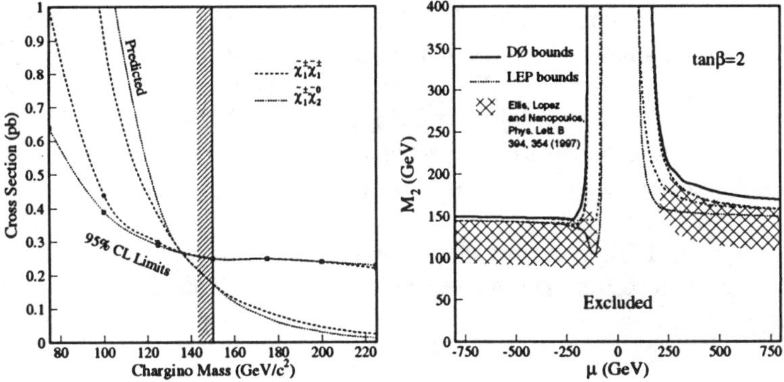

Figure 11: (Left) The DØ cross section limit on $\tilde{\chi}_1^{\pm}\tilde{\chi}_1^{\pm}$ and $\tilde{\chi}_1^{\pm}\tilde{\chi}_2^0$ production, assuming $M_{\tilde{\chi}_1^{\pm}} \approx 2M_{\tilde{\chi}_1^0}$ and $\mathrm{BR}(\tilde{\chi}_1^0 \to \gamma\tilde{G}) = 100\%$. The top dotted (dashed) curve is the cross section from PYTHIA for $\tilde{\chi}_1^{\pm}\tilde{\chi}_2^0$ ($\tilde{\chi}_1^+\tilde{\chi}_1^-$) production. The bottom dotted (dashed) curve is the cross section limit from the DØ collaboration[75] on $\tilde{\chi}_1^{\pm}\tilde{\chi}_2^0$ ($\tilde{\chi}_1^+\tilde{\chi}_1^-$) production. The vertical, hatched line marks the 95% C.L. lower limit on the lightest chargino mass from considering all chargino and neutralino pair production processes and all values of μ. (Right) The limits on the parameters M_2 and μ in gauge–mediated models based on PYTHIA for $\tan\beta=2$ and $M_{\tilde{Q}}=800$ GeV.[75] The hatched area is the region proposed[68] to explain the CDF $ee\gamma\gamma\not{E}_T$ event. The solid line shows the DØ bounds. The long–dashed line shows a contour with $M_{\tilde{\chi}_1^{\pm}} = 150$ GeV and the dash–dotted line shows a contour with $M_{\tilde{\chi}_1^0} = 75$ GeV. The dotted lines show an interpretation of preliminary LEP results at an energy of 161 GeV.

general, only one version with a fairly light mass spectrum. A more general limit can be set by holding the lighter sparticle masses constant and varying the $\tilde{Q}$ and $\tilde{g}$ masses. In this case $\tilde{Q}$'s and $\tilde{g}$'s less than 200 GeV and 225 GeV, respectively, are excluded.

4 Conclusions

As can be seen from Table 1, there has been much effort directed into SUSY searches at the Tevatron. However, given the wide range of possible experimental signatures in the MSSM, there is still work in progress and much to be done. Many Run I analyses are under way.

In Run II, two upgraded detectors at the Tevatron will collect more data at a higher energy of 2 TeV. The nominal integrated luminosity is 2 fb^{-1}, with a possible extension to 10 or even 30 fb^{-1}. The production cross sections for heavy sparticles will increase significantly with the higher energy, and the

$\widetilde{\chi}^{\pm}$ and $\widetilde{\chi}^0$ searches, as well as $\widetilde{Q}$ and $\widetilde{g}$ searches, will cover a wide range of SUSY parameter space. The experience gained from Run I analyses will greatly increase the quality of the Run II searches.[76,77] New triggering capabilities will open up previously inaccessible channels, particularly those involving τ's and heavy flavor. Increased b–tagging efficiency and $\not{E}_T$ resolution will enhance many analyses. By extending Run II up to an integrated luminosity of about 20 fb^{-1} and combining search channels, the Tevatron can perform a crucial test of the MSSM Higgs boson sector. A factor of 20 or more data combined with improved detector capabilities makes the next Run at the Tevatron an exciting prospect.

Acknowledgments

The authors thank G.L. Kane for suggesting this review, and the following people for useful discussions and comments: H. Baer, A. Beretvas, J. Berryhill, B. Bevensee, S. Blessing, A. Boehnlein, D. Chakraborty, P. Chankowski, M. Chertok, D. Claes, R. Demina, J. Done, E. Flattum, C. Grosso–Pilcher, J. Hobbs, M. Hohlmann, T. Kamon, S. Lammel, A. Lyon, D. Norman, M. Paterno, S. Pokorski, J. Qian, A. Savoy–Navarro, H.C. Shankar, M. Spira, D. Stuart, B. Tannenbaum, X. Tata, D. Toback, C. Wagner, N. Whiteman, P. Wilson and P. Zerwas.

References

1. F. Abe, *et al.*, (CDF), Nucl. Instrum. Meth. **A271**, 387 (1988).
2. S. Abachi, *et al.*, (DØ), Nucl. Instrum. Meth. **A338**, 185 (1994), and references therein.
3. M. Carena, R.L. Culbertson, S. Eno, H.J. Frisch, and S. Mrenna, submitted to Rev. Mod. Phys., hep-ex/9712022.
4. Refs. [4–9] in Ref. [3], this chapter.
5. S. Dawson, E. Eichten and C. Quigg, Phys. Rev **D31**, 1581 (1985), and references on particular signatures given afterwards in this chapter.
6. M. Carena, J.R. Espinosa, M. Quirós and C.E.M. Wagner, Phys. Lett. **B335**, 209 (1995); M. Carena, M. Quirós and C.E.M. Wagner, Nucl. Phys. **B461**, 407 (1996); H. Haber, R. Hempfling and H. Hoang, Z. Phys. **75**, 539 (1997); M. Carena, P. Zerwas and the Higgs Physics Working Group, *Physics at LEP2*, Vol. 1, eds. G. Altarelli, T. Sjöstrand and F. Zwirner, CERN Report No. 96–01.
7. A. Stange, W. Marciano and S. Willenbrook, Phys. Rev. **D49** 1354 (1994) and Phys. Rev. **D50**, 4491 (1994); J. Dai J. Gunion and R,

Vega, Phys. Rev. Lett **71**, 2699 (1993); Report of the TeV 2000 Study Group: Light Higgs Physics, FERMILAB–PUB–96/046; W.M. Yao, FERMILAB–CONF–96–383–E, 1996 Snowmass proceedings; S. Kim, S. Kuhlmann and W.M. Yao, 1996 Snowmass proceedings.

8. S. Mrenna and G.L. Kane, hep–ph/9406337; S. Mrenna, in *Perspectives on Higgs Physics*, edited by G.L. Kane (World Scientific, Singapore, 1997).

9. S.P. Martin, this book; Refs. [15–19] in Ref. [3], this chapter.

10. T. Sjöstrand, Comput. Phys. Commun. **82**, 74 (1994). S. Mrenna, Comput. Phys. Commun. **101**, 232 (1997).

11. F. Paige and S.D. Protopopescu, *Supercollider Physics*, ed. D. Soper, 41 (World Scientific, 1986); H. Baer, F. Paige, S. Protopopescu and X. Tata, in *Proceedings of the Workshop on Physics at Current Accelerators* and Brookhaven National Laboratory report No. 38304, 1986 (unpublished).

12. Ref. [20] in Ref. [3], this chapter.

13. Refs. [17,22–26] in Ref. [3], this chapter.

14. Refs. [26,28–30] in Ref. [3], this chapter.

15. Ref. [31] in Ref. [3], this chapter.

16. R. Barbieri and L. Maiani, Nucl. Phys. **B243**, 129 (1989). R. Barbieri, N. Cabbibo, L. Maiani and S. Petrarca, Phys. Lett. **B127**, 458 (1983). G.R. Farrar and A. Masiero, hep–ph/9410401. G.R. Farrar, hep–ph/9508291.

17. Refs. [36,108–114] in Ref. [3], this chapter.

18. Ref. [37] in Ref. [3], this chapter.

19. Ref. [21] in Ref. [3], this chapter.

20. H. Dreiner, An introduction to Explicit R–parity Violation, to be published in *Perspectives on Supersymmetry*, ed. G.L. Kane, (World Scientific, Singapore, 1997), hep–ph 9707435 and references therein; G. Bhattacharyya, A Brief Review on R–Parity–Violating couplings, invited talk at Beyond the Desert, Castle Ringberg, Tengernsee, Germany, June 1997, hep–ph/9709395.

21. Refs. [54,56–58] in Ref. [3], this chapter.

22. F. Abe, *et al.*, (CDF), Phys. Rev. Lett. **76**, 4307 (1996).

23. B. Bevensee, (CDF), *Proceedings of The International Workshop on Quantum Effects in the MSSM*, (Barcelona, Spain, 1997), FERMILAB–CONF–97/405–E.

24. S. Abachi, *et al.*, (DØ), Phys. Rev. Lett. **76**, 2228 (1996).

25. B. Abbott, *et al.*, (DØ), submitted to Phys. Rev. Lett., FERMILAB PUB–97/201–E, hep–ex/9705015.

26. H. Baer and X. Tata, Phys. Rev. **D47**, 2739 (1993).

27. S. Mrenna, G.L. Kane, G.D. Kribs and J.D. Wells, Phys. Rev. **D53**, 1168 (1996).

28. W. Beenakker, R. Hopker, M. Spira and P.M. Zerwas, Z. Phys. **C69**, 163 (1995), Phys. Rev. Lett. **74** 2905 (1995), and Nucl. Phys. **B492**, 51 (1997); W. Beenakker, R. Hopker, and M. Spira, hep–ph/9611232; W. Beenakker, M. Kramer, T. Plehn, M. Spira and P.M. Zerwas, hep–ph/9710451.

29. H. Baer, X. Tata and J. Woodside, Phys. Rev. Lett. **63**, 352 (1989). Phys. Rev. **D41**, 906 (1990), and Phys. Rev. **D44**, 207 (1991).

30. V. Barger, Y. Keung and R.J. Philips, Phys. Rev. Lett. **55**, 166 (1985). R.M. Barnett, J. Gunion and H. Haber, Phys. Lett. **B315**, 349 (1993). H. Baer, C. Kao and X. Tata, Phys. Rev. **D48**, 2978 (1993).

31. S. Abachi *et al.*, (DØ), Phys. Rev. Lett. **75**, 618 (1995).

32. S. Abachi *et al.*, (DØ), FERMILAB–CONF–97/357–E.

33. F.A. Berends, H. Kuijf, B. Tausk and W.T. Giele, Nucl. Phys. **B357**, 32 (1991).

34. G. Marchesini and B. R. Webber, Nucl. Phys. **B238**, 1 (1984) and Nucl. Phys. **B310**, 461 (1988).

35. F. Abe, *et al.*, (CDF), Phys. Rev. **D56**, 1357 (1997).

36. S. Abachi, *et al.*, (DØ), *Proceedings of the 28th International Conference on High Energy Physics*, eds. Z. Ajduk and A. K. Wroblewski (World Scientific, Singapore, 1997), FERMILAB–CONF–96/427–E.

37. F. Abe, *et al.*, (CDF), Phys. Rev. Lett. **76**, 2006 (1996).

38. J. Done, (CDF), *Proceedings of 1996 Divisional Meeting of the Division of Particles and Fields*, (DPF '96 Minneapolis, MN, 1996), FERMILAB–CONF–96/371–E.

39. H. Baer, M. Drees, R. Godbole, J. Gunion and X. Tata, Phys. Rev. **D44**, 725 (1991).

40. H. Baer, J. Sender and X. Tata, Phys. Rev. **D50**, 4517 (1994).

41. S. Abachi, *et al.*, (DØ), Phys. Rev. Lett. **76**, 2222 (1996).

42. P. Azzi, (CDF), *Proceedings of XXXII Rencontres de Moriond, QCD and High Energy Hadronic Interactions*, Les Arcs, France, 1997, FERMILAB–CONF–97/148–E (to be published).

43. S. Abachi, *et al.*, (DØ), FERMILAB–PUB–96/449–E, hep–ex/9612009.

44. P.J. Wilson, (CDF), *Proceedings of Les Rencontres de Physique de La Vallee D'Aosta*, La Thuile, Italy, 1997, FERMILAB–CONF–97–241–E (to be published).

45. H. Baer, C. Chen, F. Paige and X. Tata, Phys. Rev. **D49**, 3283 (1994).

46. R.M. Godbole and D.P. Roy, Phys. Rev. **D43**, 3640 (1991).

M. Guchait and D.P. Roy, Phys. Rev. **D55**, 7263 (1997).

47. F. Abe, *et al.*, (CDF), Phys. Rev. Lett. **72**, 1977 (1994); see also C. Jessop, Ph.D thesis, Harvard University, 1994. This analysis uses 4.2 pb^{-1} from the 1989 run and a signature of a hadronically–decaying τ + $\geq$ 1 jet +$\not{E}_T$.

48. F. Abe, *et al.*, (CDF), Phys. Rev. **D54**, 357 (1997).

49. S. Abachi, *et al.*, (DØ), FERMILAB–CONF–97/386–E.

50. J. Gausch, R. Jimenez and J. Sola, Phys. Lett. **B360**, 47 (1995); J. Coarasa, D. Garcia, J. Guasch, R. Jimenez, J. Sola, to appear in Z. Phys. C; J. Guasch and J. Sola, to appear in Phys. Lett. **B**, hep–ph/9707535; J. Sola, hep–ph/9708494.

51. J. Dai, J. F. Gunion and R. Vega, Phys. Lett. **B387** 801 (1996).

52. S. Abachi, *et al.*, (DØ), *Proceedings of the 28th International Conference on High Energy Physics*, eds. Z. Ajduk and A. K. Wroblewski (World Scientific, Singapore, 1997), FERMILAB–CONF–96/427–E.

53. F. Abe, *et al.*, (CDF), submitted to Phys. Rev. Lett., FERMILAB–PUB–97–247–E.

54. J. A. Valls, (CDF), *Proceedings 32nd Rencontres de Moriond*, Les Arcs, France, March, 1997, FERMILAB–CONF–97/135–E.

55. S. Abachi, *et al.*, (DØ), FERMILAB–CONF–97/325–E.

56. F. Abe, *et al.*, (CDF), Phys. Rev. **D54**, 735 (1996). This analysis uses 18.7 pb^{-1} from Run Ia and a signature of a hadronically–decaying τ + $\geq$ 1 jet +$\not{E}_T$.

57. F. Abe, *et al.*, (CDF), Phys. Rev. Lett. **73**, 2667 (1994); J. Wang, Ph.D thesis, University of Chicago, 1994. This analysis uses 19.3 pb^{-1}from Run Ia and a signature of two leptons + $\not{E}_T$.

58. S. Abachi, *et al.*, (DØ), FERMILAB–CONF–97/325–E.

59. K. Maeshima, (CDF), *Proceedings of the 28th International Conference on High Energy Physics*, eds. Z. Ajduk and A. K. Wroblewski (World Scientific, Singapore, 1997), FERMILAB–CONF–96–412–E.

60. Ref. [103] in Ref. [3], this chapter.

61. X. Wu, (CDF), talk at International Europhysics Conference on High Energy Physics, Jerusalem, Israel, 1997.

62. D. Choudhury and S. Raychaudhuri, Phys. Rev. **D56**, 1778 (1997).

63. S. Park, (CDF), *Proceedings of the 10th Topical Workshop on Proton–Antiproton Collider Physics*, eds. R. Raja and J. Yoh, (AIP Press, Woodbury, NY, 1996).

64. G. Bhattacharyya and R.N. Mohapatra, Phys. Rev. **D54**, 4204 (1996). J. L. Rosner, Phys. Rev. **D55**, 3143 (1997).

65. S. Dimopoulos, M. Dine, S. Raby and S. Thomas, Phys. Rev. Lett. **76**,

3494 (1996) and Phys. Rev. **D55**, 1372 (1997).

66. S. Ambrosanio, G.L. Kane, G.D. Kribs, S.P. Martin and S. Mrenna, Phys. Rev. Lett. **76**, 3498 (1996).

67. S. Dimopoulos, S. Thomas and J. Wells, Phys. Rev. **D54**, 3283 (1996) and Nucl. Phys. **B488**, 39 (1997).
S. Dimopoulos, M. Dine, S. Raby, S. Thomas and J. Wells, Nucl. Phys. Proc. Suppl. **A52**, 38 (1997).
D. Dicus, B. Dutta and S. Nandi, Phys. Rev. Lett. **78**, 3055 (1997) and Phys. Rev. **D56**, 5748 (1997).

68. J. Ellis, J.L. Lopez and D.V. Nanopoulos, Phys. Lett. **B394**, 354 (1997).

69. P. Fayet, Phys. Lett. **B70**, 461 (1977); P. Fayet, Phys. Lett. **B84**, 416 (1979); R. Casalbuoni, S. de Curtis, D. Dominici, F. Feruglio and R. Gatto, Phys. Lett. **B215**, 313 (1988).

70. S. Ambrosanio, G.L. Kane, G.D. Kribs, S.P. Martin and S. Mrenna, Phys. Rev. **D54**, 5395 (1996).

71. H. Baer, M. Brhlik, C–H. Chen and X. Tata, Phys. Rev. **D55**, 4463 (1997).

72. H.E. Haber, G.L. Kane and M. Quiros, Phys. Lett. **B160**, 297 (1985) and Nucl. Phys. **B273**, 333 (1986).

73. S. Mrenna and G.L. Kane, Phys. Rev. Lett. **77**, 3502 (1996).

74. R.L. Culbertson, (CDF and DØ), to be published in Nucl. Phys. **B**, *Proceedings of the 5th International Conference on Supersymmetries in Physics* (SUSY '97), Philadelphia, PA, 1997, FERMILAB-CONF-97-277-E.

75. S. Abachi, *et al.*, (DØ), submitted to Phys. Rev. Lett., FERMILAB-PUB-97/273-E.

76. D. Amidei, *et al.*, *Future Electroweak Physics at the Fermilab Tevatron: Report of the TeV-2000 Study Group*, FERMILAB-PUB-96-082.

77. J. Amundson, *et al.*, *Report of the Supersymmetry Theory Subgroup* in *New Directions for High-Energy Physics*, eds. D. Cassel, L. Gennari and R. Siemann, 644 (SLAC, Stanford, 1997).

LOW-ENERGY SUPERSYMMETRY AT FUTURE COLLIDERS

JOHN F. GUNION

Davis Institute for High Energy Physics,
Department of Physics,
University of California, Davis, CA 95616

HOWARD E. HABER

Santa Cruz Institute for Particle Physics,
University of California, Santa Cruz, CA 95064

We classify the variety of low-energy supersymmetric signatures that can be probed at future colliders. We focus on phenomena associated with the minimal supersymmetric extension of the Standard Model. The structure of the supersymmetry-breaking introduces additional model assumptions. The approaches considered here are supergravity-mediated and gauge-mediated supersymmetry-breaking. Alternative phenomenologies arising in non-minimal and/or R-parity-violating approaches are also briefly examined.

1 Introduction

In this chapter, we focus on the signatures for low-energy supersymmetry at future colliders in the context of the minimal supersymmetric extension of the Standard Model (MSSM) [1,2]. In its most general form (with the assumption of R-parity conservation), the MSSM is a 124-parameter theory [3,4]; most of the parameter freedom is associated with the supersymmetry-breaking sector of the model [a]. This huge parameter space can be reduced by: (i) imposing phenomenological constraints, and (ii) imposing theoretical assumptions on the structure of supersymmetry-breaking. In addition, the scale of supersymmetry-breaking, $\sqrt{F}$, must be specified. It determines the properties of the gravitino, $\tilde{g}_{3/2}$. In previous chapters, two broad model categories for supersymmetry-breaking were discussed, gravity-mediated supersymmetry breaking (SUGRA) and gauge-mediated supersymmetry breaking (GMSB).

In most SUGRA models [5], $\sqrt{F}$ is so large that the $\tilde{g}_{3/2}$ interactions are too weak for it to play any role in collider phenomenology. In the *minimal* supergravity (mSUGRA) framework, the soft-supersymmetry-breaking parameters at the Planck scale take a particularly simple form and depend on essentially five new parameters. These include m_0 (a flavor universal soft

---

[a]The notation for the supersymmetric parameters used in this paper for the most part follows that of Ref. [2]. The notation for supersymmetric particle names follows that of Ref. [1].

236

supersymmetry-breaking scalar mass), $m_{1/2}$ (a universal gaugino mass), and A_0 (a flavor universal tri-linear scalar interaction). In particular, gaugino mass unification implies that at the unification scale (M_X), the U(1), SU(2) and SU(3) gaugino Majorana mass parameters are equal, *i.e.*, $M_1(M_X) = M_2(M_X) = M_3(M_X) = m_{1/2}$. This implies that the *low-energy* gaugino mass parameters satisfy:

$$M_3 = \frac{g_3^2}{g_2^2} M_2 \simeq 3.5 M_2, \qquad M_1 = \tfrac{5}{3} \tan^2 \theta_W M_2 \simeq 0.5 M_2. \tag{1}$$

The other two mSUGRA parameters are the supersymmetric Higgs mass parameter μ and an off-diagonal soft Higgs squared-mass. After the imposition of electroweak symmetry breaking, these two parameters can be traded in for the two Higgs vacuum expectation values (modulo a sign ambiguity in μ). By fixing the Z mass, the remaining mSUGRA parameters are determined by the ratio of Higgs vacuum expectation values ($\tan \beta$) and the sign of μ. The lightest supersymmetric particle (LSP) is nearly always the lightest neutralino (denoted in this chapter by $\widetilde{\chi}_1^0$). Non-minimal extensions of the mSUGRA model have also been considered in which some of the parameter universality assumptions have been relaxed.

In GMSB models [6], $\sqrt{F}$ is suffiently small that the $\widetilde{g}_{3/2}$ is almost always the lightest supersymmetric particle (LSP) and plays a prominent phenomenological role. Then, different choices for the next-to-lightest supersymmetric particle (NLSP) lead to different phenomenologies. In the simplest GMSB models, the gaugino and scalar soft-supersymmetry-breaking masses are given by SU(3), SU(2) and U(1) gauge group factors times an overall scale Λ, while the A parameters are expected to be negligible. [The low-energy values of the gaugino mass parameters also satisfy Eq. (1).] The parameter set is then completed by $\tan \beta$ and $\text{sign}(\mu)$.

Finally, one can also consider alternative low-energy supersymmetric approaches. For example, if R-parity violation (RPV) is present [7], additional supersymmetric parameters are introduced. These include parameters λ_L, λ'_L and λ_B which govern new lepton and baryon number violating scalar-fermion Yukawa couplings derived from the following supersymmetic interactions:

$$(\lambda_L)_{pmn} \widehat{L}_p \widehat{L}_m \widehat{E}_n^c + (\lambda'_L)_{pmn} \widehat{L}_p \widehat{Q}_m \widehat{D}_n^c + (\lambda_B)_{pmn} \widehat{U}_p^c \widehat{D}_m^c \widehat{D}_n^c, \tag{2}$$

where p, m, and n are generation indices, and gauge group indices are suppressed. In the notation above, the "superfields" $\widehat{Q}$, $\widehat{U}^c$, $\widehat{D}^c$, $\widehat{L}$, and $\widehat{E}^c$ respectively represent $(u,d)_L$, u_L^c, d_L^c, $(\nu, e^-)_L$, and e_L^c and the corresponding superpartners. The Yukawa interactions are obtained from Eq. (2) by taking all possible combinations involving two fermions and one scalar superpartner.

2 Classes of Supersymmetric Signals

The lack of knowledge of the origin and structure of the supersymmetry-breaking parameters implies that the predictions for low-energy supersymmetry and the consequent phenomenology depend on a plethora of unknown parameters. Nevertheless, we can broadly classify supersymmetric signals at future colliders by considering a variety of theoretical approaches. In this section, we delineate the possible supersymmetric signatures, and in the next section we explore their consequences for experimentation at future colliders.

2.1 Missing energy signatures

In R-parity-conserving low-energy supersymmetry, supersymmetric particles are produced in pairs. The subsequent decay of a heavy supersymmetric particle generally proceeds via a multistep decay chain [8–10], ending in the production of at least one supersymmetric particle that (in conventional models) is weakly interacting and escapes the collider detector. Thus, supersymmetric particle production yields events that contain at least two escaping non-interacting particles, leading to a missing energy signature. At hadron colliders, it is only possible to detect missing transverse energy (E_T^{miss}), since the center-of-mass energy of the hard collision is not known on an event-by-event basis.

In conventional SUGRA-based models, the weakly-interacting LSP's that escape the collider detector (which yields large missing transverse energy) are accompanied by energetic jets and/or leptons. This is the "smoking-gun" signature of low-energy supersymmetry. However, there are two unconventional approaches in which the smoking-gun signature is absent. First, consider a model in which the $\widetilde{\chi}_1^0$ is the LSP but the lightest neutralino and chargino are nearly degenerate in mass. If the mass difference is $\lesssim 100$ MeV, then $\widetilde{\chi}_1^+$ is long-lived and decays outside the detector [11,12]. In this case, some supersymmetric events would yield *no* missing energy and two semi-stable charged particles that pass through the detector. Second, there are models in which a gluino (more precisely, the $R^0 = \widetilde{g}g$ bound state) is the LSP [b]. A massive R^0 is likely to simply pass through the detector without depositing significant energy. Even when it is light enough to be stopped, the hadronic calorimeter will measure only the kinetic energy of the R^0. In either case, there would be

---

[b]Farrar has advocated the existence of a very light gluino with a mass less than a few GeV [13]. Recent experimental data [14] show no evidence for a such a light gluino, although the assertion that light gluinos are definitively ruled out is still in dispute. The possibility of a more massive LSP gluino in SUGRA-based models has been considered in Ref. [12].

substantial missing energy [15]. However, there would be no jets arising from $\tilde{g}$ decays in such models.

In conventional GMSB models with a gravitino-LSP[c], all supersymmetric events contain at least two NLSP's, and the resulting signature depends on the NLSP properties. Four physically distinct possible scenarios emerge:

- The NLSP is electrically and color neutral and long-lived, and decays outside of the detector to its associated Standard Model partner and the gravitino.

- The NLSP is the sneutrino and decays invisibly into $\nu \tilde{g}_{3/2}$ either inside or outside the detector.

In either of these two cases, the resulting missing-energy signal is similar to that of the SUGRA-based models where $\tilde{\chi}_1^0$ or $\tilde{\nu}$ is the LSP.

- The NLSP is the $\tilde{\chi}_1^0$ and decays inside the detector to $N\tilde{g}_{3/2}$, where $N = \gamma$, Z or a neutral Higgs boson.

In this case, the gravitino-LSP behaves like the neutralino or sneutrino LSP of the SUGRA-based models. However, in contrast to SUGRA-based models, the missing energy events of the GMSB-based model are characterized by the associated production of (at least) two N's, one for each NLSP[d]. Note that if $\tilde{\chi}_1^0$ is lighter than the Z and h^0 then BR$(\tilde{\chi}_1^0 \to \gamma\tilde{g}_{3/2}) = 100\%$, and all supersymmetric production will result in missing energy events with at least two associated photons.

- The NLSP is a charged slepton (typically $\tilde{\tau}_R$ in GMSB models if $m_{\tilde{\tau}_R} < m_{\tilde{\chi}_1^0}$), which decays to the corresponding lepton partner and gravitino.

If the decay is prompt, then one finds missing energy events with associated leptons (taus). If the decay is not prompt, one observes a long-lived heavy semi-stable charged particle with *no* associated missing energy (prior to the decay of the NLSP).

There are also GMSB scenarios in which there are several nearly degenerate so-called co-NLSP's [17], any one of which can be produced at the penultimate step of the supersymmetric decay chain[e]. The resulting supersymmetric signals

---

[c]It is also possible to construct a GMSB scenario in which the $\tilde{g}$ is the LSP [16]. The resulting phenomenology corresponds to that of the massive gluino LSP discussed above.

[d]If the decay of the NLSP is not prompt, it is possible to produce events in which one NLSP decays inside the detector and one NLSP decays outside of the detector.

[e]For example, if $\tilde{\tau}_R^\pm$ and $\tilde{\chi}_1^0$ are nearly degenerate in mass, then neither $\tilde{\tau}_R^\pm \to \tau^\pm\tilde{\chi}_1^0$ nor $\tilde{\chi}_1^0 \to \tilde{\tau}_R^\pm \tau^\mp$ are kinematically allowed decays. In this case, $\tilde{\tau}_R^\pm$ and $\tilde{\chi}_1^0$ are co-NLSP's, and each decays dominantly into its Standard Model superpartner plus a gravitino.

would consist of events with two (or more) co-NLSP's, each one of which would decay according to one of the four scenarios delineated above. For additional details on the phenomenology of the co-NLSP's, see Ref. [17].

In R-parity violating SUGRA-based models the LSP is unstable. If the RPV-couplings are sufficiently weak, then the LSP will decay outside the detector, and the standard missing energy signal applies. If the LSP decays inside the detector, the phenomenology of RPV models depends on the identity of the LSP and the branching ratio of possible final state decay products. If the latter includes a neutrino, then the corresponding RPV supersymmetric events would result in missing energy (through neutrino emission) in association with hadron jets and/or leptons. Other possibilities include decays into charged leptons in association with jets (with no neutrinos), and decays into purely hadronic final states. Clearly, these latter events would contain little missing energy. If R-parity violation is present in GMSB models, the RPV decays of the NLSP can easily dominate over the NLSP decay to the gravitino. In this case, the phenomenology of the NLSP resembles that of the LSP of SUGRA-based RPV models.

2.2 Lepton (e, μ and τ) signatures

Once supersymmetric particles are produced at colliders, they do not necessarily decay to the LSP (or NLSP) in one step. The resulting decay chains can be complex, with a number of steps from the initial decay to the final state [9]. Along the way, decays can produce real or virtual W's, Z's, charginos, neutralinos and sleptons, which then can produce leptons in their subsequent decays. Thus, many models yield large numbers of supersymmetric events characterized by one or more leptons in association with missing energy, with or without hadronic jets.

One signature of particular note is events containing like-sign di-leptons [18]. The origin of such events is associated with the Majorana nature of the gaugino. For example, $\tilde{g}\tilde{g}$ production followed by $\tilde{g} \to q\bar{q}\tilde{\chi}_1^\pm \to q\bar{q}\ell^\pm\nu\tilde{\chi}_1^0$ can result in like-sign leptons since the $\tilde{g}$ decay leads with equal probability to either ℓ^+ or ℓ^-. If the masses and mass differences are both substantial (which is typical in mSUGRA models, for example), like-sign di-lepton events will be characterized by fairly energetic jets and isolated leptons and by large E_T^{miss} from the LSP's. Other like-sign di-lepton signatures can arise in a similar way from the decay chains initiated by the heavier neutralinos.

Distinctive tri-lepton signals [19] can result from $\tilde{\chi}_1^\pm \tilde{\chi}_2^0 \to (\ell^\pm\nu\tilde{\chi}_1^0)(\ell^+\ell^-\tilde{\chi}_1^0)$. Such events have little hadronic activity (apart from initial state radiation of jets off the annihilating quarks at hadron colliders). These events can have a variety of interesting characteristics depending on the fate of the final state neutralinos.

If the soft-supersymmetry breaking slepton masses are flavor universal at the high energy scale M_X (as in mSUGRA models) and $\tan\beta \gg 1$, then the $\tilde{\tau}_R$ will be significantly lighter than the other slepton states. As a result, supersymmetric decay chains involving (s)leptons will favor $\tilde{\tau}_R$ production, leading to a predominance of events with multiple τ-leptons in the final state.

In GMSB models with a charged slepton NLSP, the decay $\tilde{\ell} \to \ell \tilde{g}_{3/2}$ (if prompt) yields at least two leptons for every supersymmetric event in association with missing energy. In particular, in models with a $\tilde{\tau}_R$ NLSP, supersymmetric events will characteristically contain at least two τ's.

In RPV models, decays of the LSP (in SUGRA models) or NLSP (in GMSB models) mediated by RPV-interactions proportional to λ_L and λ'_L will also yield supersymmetric events containing charged leptons. However, if the only significant RPV-interaction is the one proportional to λ'_L, then such events would *not* contain missing energy (in contrast to the GMSB signature described above).

2.3 b-quark signatures

The phenomenology of gluinos and squarks depends critically on their relative masses. If the gluino is heavier, it will decay dominantly into $q\tilde{q}$ [f], while the squark can decay into quark plus chargino or neutralino. If the squark is heavier, it will decay dominantly into a quark plus gluino, while the gluino will decay into the three-body modes $q\bar{q}\tilde{\chi}$ (where $\tilde{\chi}$ can be either a neutralino or chargino, depending on the charge of the final state quarks). A number of special cases can arise when the possible mass splitting among squarks of different flavors is taken into account. For example, models of supersymmetric mass spectra have been considered where the third generation squarks are lighter than the squarks of the first two generations. If the gluino is lighter than the latter but heavier than the former, then the only open gluino two-body decay mode could be $b\tilde{b}$ [g]. In such a case, all $\tilde{g}\tilde{g}$ events will result in at least four b-quarks in the final state (in associated with the usual missing energy signal, if appropriate). More generally, due to the flavor independence of the strong interactions, one expects three-body gluino decays into b-quarks in at least

---

[f] In this section, we employ the notation $q\tilde{q}$ to mean either $q\tilde{\bar{q}}$ or $\bar{q}\tilde{q}$.

[g] Although one top-squark mass-eigenstate $(\tilde{t}_1)$ is typically lighter than $\tilde{b}$ in models, the heavy top-quark mass may result in a kinematically forbidden gluino decay mode into $t\tilde{t}_1$.

20% of all gluino decays[h]. Additional b-quarks can arise from both top-quark and top-squark decays, and from neutral Higgs bosons produced somewhere in the chain decays [20]. Finally, at large $\tan\beta$, the enhanced Yukawa coupling to b-quarks can increase the rate of b-quark production in neutralino and chargino decays occurring at some step in the gluino chain decay.

These observations suggest that many supersymmetric events at hadron colliders will be characterized by b-jets in association with missing energy [10,21,22].

2.4 Signatures involving photons

In mSUGRA models, most supersymmetric events do not contain isolated energetic photons. However, some areas of low-energy supersymmetric parameter space do exist in which final state photons can arise in the decay chains of supersymmetric particles. If one relaxes the condition of gaugino mass unification, then the *low-energy* gaugino mass parameters no longer must satisfy Eq. (1). As a result, interesting alternative supersymmetric phenomenologies can arise. For example, if the low-energy mass parameters satisfy $M_1 \simeq M_2$, then the branching ratio for $\tilde{\chi}_2^0 \to \tilde{\chi}_1^0 \gamma$ can be significant [23]. In the model of Ref. [24], the $\tilde{\chi}_1^0$-LSP is dominantly higgsino, while $\tilde{\chi}_2^0$ is dominantly gaugino. Thus, many supersymmetric decay chains end in the production of $\tilde{\chi}_2^0$, which then decays to $\tilde{\chi}_1^0 \gamma$. In this picture, the pair production of supersymmetric particles often yields two photons plus associated missing energy. At LEP-2, one can also produce $\tilde{\chi}_1^0 \tilde{\chi}_2^0$ which would then yield single photon events in association with large missing energy.

In GMSB models with a $\tilde{\chi}_1^0$-NLSP, all supersymmetric decay chains would end up with the production of $\tilde{\chi}_1^0$. Assuming that $\tilde{\chi}_1^0$ decays inside the collider detector, one possible decay mode is $\tilde{\chi}_1^0 \to \gamma \tilde{g}_{3/2}$. In many models, the branching ratio for this radiative decay is significant (and could be as high as 100% if other possible two-body decay modes are not kinematically allowed). In the latter case, supersymmetric pair production would also yield events with two photons in associated with large missing energy. The characteristics of these events differ in detail from those of the corresponding events expected in the model of Ref. [24].

2.5 Kinks and long-lived heavy particles

In most SUGRA-based models, all supersymmetric particles in the decay chain decay promptly until the LSP is reached. The LSP is exactly stable and escapes

---

[h]Here we assume the approximate degeneracy of the first two generations of squarks, as suggested from the absence of flavor-changing neutral-current decays. In many models, the b-squarks tend to be of similar mass or lighter than the squarks of the first two generations.

the collider detector. However, exceptions are possible. In particular, if there is a supersymmetric particle that is just barely heavier than the LSP, then its (three-body) decay rate to the LSP will be significantly suppressed and it could be long lived. For example, in the models with $|\mu| \gg M_1 > M_2$ [11,12] implying $m_{\widetilde{\chi}_1^\pm} \simeq m_{\widetilde{\chi}_1^0}$, the $\widetilde{\chi}_1^\pm$ can be sufficiently long lived to yield a detectable vertex, or perhaps even exit the detector.

In GMSB models, the NLSP may be long-lived, depending on its mass and the scale of supersymmetry breaking, $\sqrt{F}$. The NLSP is unstable and eventually decays to the gravitino. For example, in the case of the $\widetilde{\chi}_1^0$-NLSP (which is dominated by its U(1)-gaugino component), one finds $\Gamma(\widetilde{\chi}_1^0 \to \gamma\widetilde{g}_{3/2}) = m_{\widetilde{\chi}_1^0}^5 \cos^2\theta_W / 16\pi F^2$. It then follows that

$$(c\tau)_{\widetilde{\chi}_1^0 \to \gamma\widetilde{g}_{3/2}} \simeq 130 \left(\frac{100 \text{ GeV}}{m_{\widetilde{\chi}_1^0}} \right)^5 \left(\frac{\sqrt{F}}{100 \text{ TeV}} \right)^4 \mu\text{m} . \qquad (3)$$

For simplicity, assume that $\widetilde{\chi}_1^0 \to \gamma\widetilde{g}_{3/2}$ is the dominant NLSP decay mode. If $\sqrt{F} \sim 10^4$ TeV, then the decay length for the NLSP is $c\tau \sim 10$ km for $m_{\widetilde{\chi}_1^0} = 100$ GeV; while $\sqrt{F} \sim 100$ TeV implies a short but vertexable decay length. A similar result is obtained in the case of a charged NLSP. Thus, if $\sqrt{F}$ is sufficiently large, the charged NLSP will be semi-stable and may decay outside of the collider detector.

Finally, if R-parity violation is present, the decay rate of the LSP in SUGRA-based models (or the NLSP in R-parity-violating GMSB models) could be in the relevant range to yield visible secondary vertices.

3 Supersymmetry searches at future colliders

In this section, we consider the potential for discovering low-energy supersymmetry at future colliders. A variety of supersymmetric signatures have been reviewed in Section 2, and we now apply these to supersymmetry searches at future colliders. Ideally, experimental studies of supersymmetry should be as model-independent as possible. Ultimately, the goal of experimental studies of supersymmetry is to measure as many of the 124 MSSM parameters (and any additional parameters that can arise in non-minimal extensions) as possible. In practice, a fully general analysis will be difficult, particularly during the initial supersymmetry discovery phase. Thus, we focus the discussion in this section on the expected phenomenology of supersymmetry at the various future facilities under a number of different model assumptions. Eventually, if candidates for supersymmetric phenomena are discovered, one would utilize precision experimental measurements to map out the supersymmetric parameter space and uncover the structure of the underlying supersymmetry-breaking.

3.1 SUGRA-based models

We begin with the phenomenology of mSUGRA. Of particular importance are the relative sizes of the different supersymmetric particle masses. Generic properties of the resulting superpartner mass spectrum are discussed in Ref. [2]. An important consequence of the mSUGRA mass spectrum is that substantial phase space is available for most decays occurring at each step in a given chain decay of a heavy supersymmetric particle.

Extensive Monte Carlo studies have examined the region of mSUGRA parameter space for which direct discovery of supersymmetric particles at the Tevatron and the LHC will be possible [25]. At the hadron colliders, the ultimate supersymmetric mass reach is determined by the searches for both the strongly-interacting superpartners (squarks and gluinos) and the charginos/neutralinos. Cascade decays of the produced squarks and gluinos lead to events with jets, missing energy, and various numbers of leptons. Pair production of charginos and/or neutralinos can produce distinctive multilepton signatures. The chargino/neutralino searches primarily constrain the mSUGRA parameter $m_{1/2}$, which can be translated into an equivalent bound on the gluino mass. As a result, gluino and squark masses up to about 400 GeV can be probed at the upcoming Tevatron Run-II; further improvements are projected at the proposed TeV-33 upgrade [26], where supersymmetric masses up to about 600 GeV can be reached. The maximum reach at the LHC is generally attained by searching for the $1\ell + \text{jets} + E_T^{\text{miss}}$ channel; one will be able to discover squarks and gluinos with masses up to several TeV [20]. Some particularly important classes of events include:

- $pp \to \tilde{g}\tilde{g} \to \text{jets} + E_T^{\text{miss}}$ and $pp \to \tilde{g}\tilde{g} \to \ell^\pm \ell^\pm + \text{jets} + E_T^{\text{miss}}$ (the like-sign dilepton signal [18]). The mass difference $m_{\tilde{g}} - m_{\tilde{\chi}_1^\pm}$ can be determined from jet spectra end points, while $m_{\tilde{\chi}_1^\pm} - m_{\tilde{\chi}_1^0}$ can be roughly determined by analyzing various distributions of kinematic observables in the like-sign channel [18,21,27]. An absolute scale for $m_{\tilde{g}}$ can be estimated (within an accuracy of roughly $\pm 15\%$) by separating the like-sign events into two hemispheres corresponding to the two $\tilde{g}$'s [18], by a similar separation in the $\text{jets} + E_T^{\text{miss}}$ channel [20], or variations thereof [21,27].

- $pp \to \tilde{\chi}_1^\pm \tilde{\chi}_2^0 \to (\ell^\pm \nu \tilde{\chi}_1^0)(\ell^+ \ell^- \tilde{\chi}_1^0)$, which yields a tri-lepton + E_T^{miss} final state. The mass difference $m_{\widetilde{\chi}_2^0} - m_{\widetilde{\chi}_1^0}$ is easily determined [i] if enough events are available [19].

- $pp \to \widetilde{\ell}\widetilde{\ell} \to 2\ell + E_T^{\text{miss}}$, detectable at the LHC for $m_{\widetilde{\ell}} \lesssim 300$ GeV [20].

- Squarks will be pair produced and, for $m_0 \gg m_{1/2}$, would lead to $\widetilde{g}\widetilde{g}$ events with two extra jets emerging from the primary $\widetilde{q} \to q\widetilde{g}$ decays.

The LHC provides significant opportunities for precision measurements of the mSUGRA parameters [21]. In general, one expects large samples of supersymmetric events with distinguishing features that allow an efficient separation from Standard Model backgrounds. The biggest challenge in analyzing these events may be in distinguishing one set of supersymmetric signals from another. Within the mSUGRA framework, the parameter space is small enough to permit the untangling of the various signals and allows one to extract the mSUGRA parameters with some precision.

Important discovery modes at the NLC include the following [28]:

- $e^+e^- \to \widetilde{\chi}_1^+ \widetilde{\chi}_1^- \to (q\bar{q}\widetilde{\chi}_1^0 \text{ or } \ell\nu\widetilde{\chi}_1^0) + (q\bar{q}\widetilde{\chi}_1^0 \text{ or } \ell\nu\widetilde{\chi}_1^0)$;

- $e^+e^- \to \widetilde{\ell}^+\widetilde{\ell}^- \to (\ell^+\widetilde{\chi}_1^0 \text{ or } \bar{\nu}\widetilde{\chi}_1^+) + (\ell^-\widetilde{\chi}_1^0 \text{ or } \nu\widetilde{\chi}_1^-)$.

In both cases, the masses of the initially produced supersymmetric particles as well as the final state neutralinos and charginos will be well-measured. Here, one is able to make use of the energy spectra end points and beam energy constraints to make precision measurements of masses and determine the underlying supersymmetric parameters. Polarization of the beams is an essential tool that can be used to enhance signals while suppressing Standard Model backgrounds. Moreover, polarization can be employed to separate out various supersymmetric contributions in order to explore the inherent chiral structure of the interactions. The supersymmetric mass reach is limited by the center-of-mass energy of the NLC. For example, if the scalar mass parameter m_0 is too large, squark and slepton pair production will be kinematically forbidden. To probe values of $m_0 \sim 1$—1.5 TeV requires a collider energy in the range of $\sqrt{s} \gtrsim 2$—3 TeV. It could be that such energies will be more easily achieved at a future $\mu^+\mu^-$ collider.

The strength of the lepton colliders lies in the ability to analyze supersymmetric signals and make precision measurements of observables. Ideally, one would like to measure the underlying supersymmetric parameters without

---

[i]In some cases, $m_{\widetilde{\chi}_2^0} - m_{\widetilde{\chi}_1^0}$ can still be determined if $\widetilde{\chi}_2^0$ is produced at some step in a supersymmetric decay chain.

prejudice. One could then test the model assumptions, and study possible deviations. The most efficient way to carry out such a program is to set the lepton collider center-of-mass energy to the appropriate value of $\sqrt{s}$ in order to first study the light supersymmetric spectrum (lightest charginos and neutralinos and sleptons). In this way, one limits the interference among competing supersymmetric signals. Experimentation at the lepton colliders then can provide model-independent measurements of the associated underlying supersymmetric parameters. Once these parameters are ascertained, one can analyze with more confidence events with heavy supersymmetric particles decaying via complex decay chains. Thus, the NLC and LHC supersymmetric searches are complementary.

Beyond mSUGRA, the MSSM parameter space becomes more complex. It is possible to perturb the mSUGRA model by adding some non-universality among the scalar mass parameters without generating phenomenologically unacceptable flavor changing neutral currents. There has been no systematic analysis of the resulting phenomenology at future colliders. (The implications of non-universal scalar masses for LHC phenomenology were briefly addressed in Ref. [22].) Nevertheless, the possible non-degeneracy of squarks could have a significant impact on the search for squarks at hadron colliders. In particular, in mSUGRA models one typically finds that four flavors of squarks (with two squark eigenstates per flavor) and $\tilde{b}_R$ are nearly mass-degenerate, while the masses of $\tilde{b}_L$ and the top-squark mass eigenstates could be significantly different ^{j}. This means that the *observed* cross-section for the production of squark pairs at hadron colliders would be enhanced by a multiplicity factor of eight or larger (depending on the number of approximately mass-degenerate squark species). Clearly, if some of the first and second generation squarks are split in mass, the relevant effective cross-sections are smaller. This could lead to more background contamination of squark signals at hadron colliders. The impact of squark non-degeneracy on the discovery mass reach for squarks at the Tevatron and LHC has not yet been analyzed.

It is also possible to introduce arbitrary non-universal gaugino mass parameters (at the high-energy scale). For example, suppose that the non-universal gaugino masses at the high-energy scale imply that the gaugino mass parameters at the low-energy scale satisfy $M_2 < M_1$, *i.e.*, the SU(2)-gaugino component is dominant in the lightest chargino and neutralino [11,12]. In this case, the $\tilde{\chi}_1^0$ and $\tilde{\chi}_1^\pm$ can be closely degenerate, in which case the visible decay products in $\tilde{\chi}_1^+ \to \tilde{\chi}_1^0 + X$ decays will be very soft and difficult to detect. Con-

---

^{j}If $\tan\beta \gg 1$, then $\tilde{b}_L$–$\tilde{b}_R$ mixing can be significant, in which case the two bottom-squark mass eigenstates could also be significantly split in mass from the first two generations of squarks.

sequences for chargino and neutralino detection in e^+e^- and $\mu^+\mu^-$ collisions, including the importance of the $e^+e^- \to \gamma\tilde{\chi}_1^+\tilde{\chi}_1^-$ production channel, are discussed in Refs. [11,12]. There is also the possibility that $m_{\tilde{g}} \sim m_{\tilde{\chi}_1^\pm} \simeq m_{\tilde{\chi}_1^0}$. The decay products in the $\tilde{g}$ decay chain would then be very soft, and isolation of $\tilde{g}\tilde{g}$ events would be much more difficult at hadron colliders than in the usual mSUGRA case. In particular, hard jets in association with missing energy would be much rarer, since they would only arise from initial state radiation. The corresponding reduction in supersymmetric parameter space coverage at the Tevatron Main Injector is explored in Ref. [12].

As a second example, consider the case where the low-energy gaugino mass parameters satisfy $M_2 \sim M_1$ [k]. If we also assume that $\tan\beta \sim 1$ and $|\mu| < M_1$, M_2 [l], then the lightest two neutralinos are nearly a pure photino and higgsino respectively, *i.e.*, $\tilde{\chi}_2^0 \simeq \tilde{\gamma}$ and $\tilde{\chi}_1^0 \simeq \tilde{H}$. For this choice of MSSM parameters, one finds that the rate for the one-loop decay $\tilde{\chi}_2^0 \to \gamma\tilde{\chi}_1^0$ dominates over all tree level decays of $\tilde{\chi}_2^0$ and $\mathrm{BR}(\tilde{e} \to e\tilde{\chi}_2^0) \gg \mathrm{BR}(\tilde{e} \to e\tilde{\chi}_1^0)$. Clearly, the resulting phenomenology [24] differs substantially from mSUGRA expectations. This scenario was inspired by the CDF $ee\gamma\gamma$ event [29]. Suppose that the $ee\gamma\gamma$ event resulted from $\tilde{e}\tilde{e}$ production, where $\tilde{e} \to e\tilde{\chi}_2^0 \to e\gamma\tilde{\chi}_1^0$. Then in the model of Ref. [24], one would expect a number of other distinctive supersymmetric signals to be observable at LEP-2 (running at its maximal energy) and at Run-II of the Tevatron. In particular, LEP-2 would expect events of the type: $\ell\ell + X + E_T^{\mathrm{miss}}$ and $\gamma\gamma + X + E_T^{\mathrm{miss}}$, while Tevatron would expect events of the type: $\ell\ell + X + E_T^{\mathrm{miss}}$, $\gamma\gamma + X + E_T^{\mathrm{miss}}$, $\ell\gamma + X + E_T^{\mathrm{miss}}$, $\ell\ell\gamma + X + E_T^{\mathrm{miss}}$, $\ell\gamma\gamma + X + E_T^{\mathrm{miss}}$, and $\ell\ell\ell + X + E_T^{\mathrm{miss}}$. In the above signatures, X stands for additional leptons, photons, and/or jets. These signatures can also arise in GMSB models, although the kinematics of the various events can often be distinguished.

3.2 GMSB-based models

The collider signals for GMSB models depend critically on the NLSP identity and its lifetime (or equivalently, its decay length). Thus, we examine the phenomenology of both promptly-decaying and longer-lived NLSP's. In the latter case, the number of decays where one or both NLSP's decay within a radial distance R is proportional to $[1 - \exp(-2R/c\tau)] \simeq 2R/(c\tau)$. For large $c\tau$, most decays would be non-prompt, with many occurring in the outer parts

---

[k] We remind the reader that gaugino mass unification at the high-energy scale would predict $M_2 \simeq 2M_1$.

[l] To achieve such a small μ-parameter requires, *e.g.*, some non-universality among scalar masses of the form $m_{H_1}^2 \neq m_{\tilde{q}}^2, m_{\tilde{\ell}}^2$.

of the detector or completely outside the detector. To maximize sensitivity to GMSB models and fully cover the $(\sqrt{F}, \Lambda)$ parameter space, we must develop strategies to detect decays that are delayed, but not necessarily so delayed as to be beyond current detector coverage and/or specialized extensions of current detectors.

In the discussion below, we focus on various cases, where the NLSP is a neutralino dominated by its U(1)-gaugino ($\widetilde{B}$) or Higgsino ($\widetilde{H}$) components, and where the NLSP is the lightest charged slepton (usually the $\widetilde{\tau}_R$). We first address the case of prompt decays, and then indicate the appropriate strategies for the case of the longer-lived NLSP.

• Promptly-decaying NLSP: $\widetilde{\chi}_1^0 \simeq \widetilde{B}$

We focus on the production of the neutralinos, charginos, and sleptons since these are the lightest of the supersymmetric particles in the GMSB models. The possible decays of the NLSP in this case are: $\widetilde{B} \to \gamma \widetilde{g}_{3/2}$ or $\widetilde{B} \to Z \widetilde{g}_{3/2}$. The latter is only relevant for the case of a heavier NLSP (and moreover is suppressed by $\tan^2 \theta_W$). It will be ignored in the following discussion.

At hadronic colliders, the $\widetilde{\chi}_1^0 \widetilde{\chi}_1^0$ production rate is small, but rates for $\widetilde{\chi}_1^+ \widetilde{\chi}_1^- \to W^{(\star)} W^{(\star)} \widetilde{\chi}_1^0 \widetilde{\chi}_1^0 \to W^{(\star)} W^{(\star)} \gamma\gamma + E_T^{\text{miss}}$, $\widetilde{\ell}_R \widetilde{\ell}_R \to \ell^+ \ell^- \widetilde{\chi}_1^0 \widetilde{\chi}_1^0 \to \ell^+ \ell^- \gamma\gamma + E_T^{\text{miss}}$, $\widetilde{\ell}_L \widetilde{\ell}_L \to \ell^+ \ell^- \widetilde{\chi}_1^0 \widetilde{\chi}_1^0 \to \ell^+ \ell^- \gamma\gamma + E_T^{\text{miss}}$, etc. will all be substantial. Implications for GMSB phenomenology at the Tevatron can be found in Refs. [30–32]. It is possible to envision GMSB parameters such that the $ee\gamma\gamma + E_T^{\text{miss}}$ CDF event [29] corresponds to selectron pair production followed by $\widetilde{e} \to e \widetilde{\chi}_1^0$ with $\widetilde{\chi}_1^0 \to \gamma \widetilde{g}_{3/2}$ [30,24,32,33]. However, in this region of GMSB parameter space, other supersymmetric signals should be prevalent, such as $\widetilde{\nu}_L \widetilde{\ell}_L \to \ell\gamma\gamma + E_T^{\text{miss}}$ and $\widetilde{\nu}_L \widetilde{\nu}_L \to \gamma\gamma + E_T^{\text{miss}}$. The $\widetilde{\chi}_2^0 \widetilde{\chi}_1^\pm$ and $\widetilde{\chi}_1^+ \widetilde{\chi}_1^-$ rates would also be significant and lead to $X\gamma\gamma + E_T^{\text{miss}}$ with $X = \ell^\pm, \ell^+ \ell'^-, \ell^+ \ell^- \ell'^\pm$. Limits on these event rates from current CDF and D0 data already eliminate much, if not all, of the parameter space that could lead to the CDF $ee\gamma\gamma$ event [34].

At LEP-2/NLC [35], the rate for the simplest signal, $e^+ e^- \to \widetilde{\chi}_1^0 \widetilde{g}_{3/2} \to \gamma + E_T^{\text{miss}}$, is expected to be very small. A more robust channel is $e^+ e^- \to \widetilde{\chi}_1^0 \widetilde{\chi}_1^0 \to \gamma\gamma + E_T^{\text{miss}}$ with a (flat) spectrum of photon energies in the range $\frac{1}{4}\sqrt{s}(1 - \beta) \leq E_\gamma \leq \frac{1}{4}\sqrt{s}(1 + \beta)$.

• Promptly decaying NLSP: $\widetilde{\chi}_1^0 \simeq \widetilde{H}$

The possible decays of the NLSP in this case are: $\widetilde{H} \to \widetilde{g}_{3/2} + h^0, H^0, A^0$, depending on the Higgs masses. If the corresponding two-body decays are not kinematically possible, then three-body decays (where the corresponding Higgs state is virtual) may become relevant. However, in realistic cases, one expects $\widetilde{\chi}_1^0$ to contain small but non-negligible gaugino components, in which case the rate for $\widetilde{\chi}_1^0 \to \widetilde{g}_{3/2}\gamma$ would dominate all three-body decays. In what follows, we assume that the two-body decay $\widetilde{H} \to \widetilde{g}_{3/2}h^0$ is kinematically allowed and dominant. The supersymmetric signals that would emerge at both Tevatron/LHC and LEP-2/NLC would then be $4b + X + E_T^{\mathrm{miss}}$ final states, where X represents the decay products emerging from the cascade chain decays of the more massive supersymmetric particles. Of course, at LEP-2/NLC direct production of higgsino pairs, $e^+e^- \to \widetilde{H}\widetilde{H}$ (via virtual s-channel Z-exchange) would be possible in general, leading to pure $4b + E_T^{\mathrm{miss}}$ final states.

• Promptly decaying NLSP: $\widetilde{\ell}_R$

The dominant slepton decay modes are: $\widetilde{\ell}_R^\pm \to \ell^\pm \widetilde{g}_{3/2}$ and $\widetilde{\ell}_L^\pm \to \ell^\pm \widetilde{\chi}_1^{0\star} \to \ell^\pm(\widetilde{\ell}_R^\pm \ell^\mp)' \to \ell^\pm(\ell^\pm \ell^\mp)'\widetilde{g}_{3/2}$. The $\widetilde{\chi}_1^0$ will first decay to $\ell\widetilde{\ell}_L$ and $\ell\widetilde{\ell}_R$, followed by the above decays.

At both the Tevatron/LHC and LEP-2/NLC, typical pair production events will end with $\widetilde{\ell}_R\widetilde{\ell}_R \to \ell^+\ell^- + E_T^{\mathrm{miss}}$, generally in association with a variety of cascade chain decay products. The lepton energy spectrum will be flat in the $\widetilde{\ell}_R\widetilde{\ell}_R$ center of mass. Of course, pure $\widetilde{\ell}_R\widetilde{\ell}_R$ production is possible at LEP/NLC and the $\widetilde{\ell}_R\widetilde{\ell}_R$ center of mass would be the same as the e^+e^- center of mass. Other simple signals at LEP/NLC, would include $\widetilde{\ell}_L\widetilde{\ell}_L \to 6\ell + E_T^{\mathrm{miss}}$.

If a slepton is the NLSP, it is most likely to be the $\widetilde{\tau}_R$. If this state is sufficiently lighter than the $\widetilde{e}_R$ and $\widetilde{\mu}_R$, then $\widetilde{e}_R \to e\widetilde{\tau}_R\tau$ and $\widetilde{\mu}_R \to \mu\widetilde{\tau}_R\tau$ decays (via the $\widetilde{B}$ component of the mediating virtual neutralino) might dominate over the direct $\widetilde{e}_R \to e\widetilde{g}_{3/2}$ and $\widetilde{\mu}_R \to \mu\widetilde{g}_{3/2}$ decays, and all final states would cascade to τ's. The relative importance of these different possible decays has been examined in Ref. [17]. A study of this scenario at LEP-2 has been performed in Ref. [36].

• Longer-lived NLSP: $\widetilde{\ell}_R$

If the $\widetilde{\ell}_R$ mainly decays before reaching the electromagnetic calorimeter, then one should look for a charged lepton that suddenly appears a finite distance from the interaction region, with non-zero impact parameter as measured by either the vertex detector or the electromagnetic calorimeter. Leading up to this decay would be a heavily ionizing track with $\beta < 1$ (as could be measured if a magnetic field is present).

If the $\widetilde{\ell}_R$ reaches the electromagnetic and hadronic calorimeters, then it behaves much like a heavy muon, presumably interacting in the muon chambers or exiting the detector if it does not decay first. Limits on such objects should be pursued. There will be many sources of $\tilde{\ell}_R$ production, including direct slepton pair production, and cascade decays resulting from the production of gluinos, squarks, and charginos [37]. Based on current Tevatron data, a charged pseudo-stable $\tilde{\ell}_R$ can be ruled out with a mass up to about 80–100 GeV. Similar limits can probably be extracted from LEP-2 data.

• Longer-lived NLSP: $\tilde{\chi}_1^0$

This is a much more difficult case. As before, we assume that the dominant decay of the NLSP in this case is $\tilde{\chi}_1^0 \to \gamma \tilde{g}_{3/2}$. Clearly, the sensitivity of detectors to delayed γ appearance signals will be of great importance. If the $\tilde{\chi}_1^0$ escapes the detector before decaying, then the corresponding missing energy signatures are the same as those occurring in SUGRA-based models.

At the Tevatron, standard supersymmetry signals (*e.g.*, jets or tri-leptons plus $E_T^{\rm miss}$) are viable if $\Lambda \lesssim 30$–70 TeV (given an integrated luminosity of $L = 0.1$–30 fb^{-1}) independent of the magnitude of $\sqrt{F}$ [38,39]. Meanwhile, the prompt $\tilde{\chi}_1^0 \to \gamma \tilde{g}_{3/2}$ decay signals discussed earlier are viable only in a region defined by $\sqrt{F} \lesssim 500$ TeV at low Λ, rising to $\sqrt{F} \lesssim 1000$ TeV at $\Lambda \sim 120$ TeV [38,39]. This leaves a significant region of $(\sqrt{F}, \Lambda)$ parameter space that can only be probed by the delayed $\tilde{\chi}_1^0 \to \gamma \tilde{g}_{3/2}$ decays [38,39].

The ability to search for delayed-decay signals is rather critically dependent upon the detector design. The possible signals include the following [38,39]: (i) looking for isolated energy deposits (due to the γ from the $\tilde{\chi}_1^0$ decay) in the outer hadronic calorimeter cells of the D0 detector; (ii) searching for events where the delayed-decay photon is identified by a large (transverse) impact parameter as it passes into the electromagnetic calorimeter; and (iii) looking for delayed decays where the photon first emerges outside the main detector and is instead observed in a scintillator array (or similar device) placed at a substantial distance from the detector. The observed signal will always contain missing energy from one or more emitted gravitinos and/or from $\tilde{\chi}_1^0$'s that do not decay inside the detector. Thus, by requiring large missing energy, the backgrounds can be greatly reduced while maintaining good efficiency for the GMSB signal. In combination, the above techniques may[m] allow the detection of supersymmetric particle production at the Tevatron in the GMSB parameter region $\sqrt{F} \lesssim 3000$ TeV and $\Lambda \lesssim 150$ TeV.

---

[m] Event rates are significant even after very strong cuts on jets, photon energy and missing energy, but detailed background calculations remain to be done.

3.3 R-parity violating (RPV) models

In R-parity violating models, the LSP is no longer stable[n]. The relevant signals depend upon the nature of the LSP decay. The phenomenology depends on which R-parity violating couplings [Eq. (2)] are present. Only a brief discussion will be given here; for further details, see Ref. [7].

At the Tevatron and LHC [40], consider $\tilde{g}\tilde{g}$ production followed by gluino decay via the usual set of possible decay chains ending up with the LSP plus Standard Model particles. Until this point, all decays have involved only R-parity conserving interactions[o]. The RPV-interactions now enter in the decay of the LSP. We shall assume in the following discussion that the $\tilde{\chi}_1^0$ is the LSP, although other possible choices can also be considered.

If $\lambda_B \neq 0$, then the dominant decay of $\tilde{\chi}_1^0$ would result in the production of a three-jet final state ($\tilde{\chi}_1^0 \to jjj$). The large jet backgrounds imply that we would need to rely on the like-sign dilepton signal (which would still be viable despite the absence of missing energy in the events). In general, this signal turns out to be sufficient for supersymmetry discovery out to gluino masses somewhat above 1 TeV. However, if the leptons of the like-sign dilepton signal are very soft, then the discovery reach would be much reduced[p]. This is one of the few cases where one could miss discovering low-energy supersymmetry at the LHC. If λ_L dominates $\tilde{\chi}_1^0$ decays, $\tilde{\chi}_1^0 \to \mu^\pm e^\mp \nu, e^\pm e^\mp \nu$, and there would be many very distinctive multi-lepton signals. If λ_L' is dominant, then $\tilde{\chi}_1^0 \to \ell jj$ and again there would be distinctive multi-lepton signals.

More generally, many normally invisible events become visible. An important example is sneutrino pair production. Even if the dominant decay of the sneutrino is $\tilde{\nu} \to \nu \tilde{\chi}_1^0$ (which is likely if $m_{\tilde{\nu}} > m_{\tilde{\chi}_1^0}$), a visible signal emerges from the $\tilde{\chi}_1^0$ decay as sketched above. Of course, for large enough λ_L or λ_L' the $\tilde{\nu}$'s would have significant branching ratio for decay to charged lepton pairs or jet pairs, respectively. Indeed, such decays might dominate if $m_{\tilde{\nu}} < m_{\tilde{\chi}_1^0}$.

At LEP-2, NLC or the muon collider [41–43], the simplest process is

$$e^+ e^- \to \tilde{\chi}_1^0 \tilde{\chi}_1^0 \to \underbrace{(jjj)(jjj)}_{\lambda_B}, \quad \underbrace{(\ell\ell\nu)(\ell\ell\nu)}_{\lambda_L}, \quad \underbrace{(\ell jj)(\ell jj)}_{\lambda_L'} \tag{4}$$

(or the $\mu^+\mu^-$ collision analogue), where the relevant RPV-coupling is indicated

---

[n]We assume that the gravitino is not relevant for RPV phenomenology, as in SUGRA-based models.

[o]By assumption, the strengths of the R-parity conserving interactions are significantly larger than the corresponding RPV-interaction strengths.

[p]Soft leptons would occur in models where $m_{\tilde{\chi}_1^\pm} \sim m_{\tilde{\chi}_1^0}$, which requires non-universal gaugino masses [11,22].

below the corresponding signal. Substantial rates for equally distinctive signals from production of more massive supersymmetric particles (including sneutrino pair production) would also be present. All these processes (if kinematically allowed) should yield observable supersymmetric signals. Some limits from LEP data already exist [44]. Of particular potential importance for non-zero λ_L is s-channel resonant production of a sneutrino in e^+e^- [42] and $\mu^+\mu^-$ [43] collisions. In particular, at $\mu^+\mu^-$ colliders this process is detectable down to quite small values of the appropriate λ_L, and could be of great importance as a means of actually determining the R-parity-violating couplings. Indeed, for small R-parity-violating couplings, absolute measurements of the couplings through other processes are extremely difficult. This is because such a measurement would typically require the R-parity-violating effects to be competitive with an R-parity-conserving process of known interaction strength. (For example, R-parity-violating neutralino branching ratios constrain only ratios of the R-parity-violating couplings.) Since sneutrino pair production would have been observed at the LHC, NLC and/or the muon collider, it would be easy to center on the sneutrino resonance in order to perform the crucial sneutrino factory measurements.

We end this section with two additional remarks. First, if the RPV coupling strengths are very small, then the RPV-violating decay of the LSP (*e.g.*, $\widetilde{\chi}_1^0$) could occur a substantial distance from the primary interaction point, but still within the detector (or at least not far outside the detector). The general techniques for detecting such delayed decays outlined at the end of Section 3.2 would again be relevant. It is particularly important to note that observation of the delayed decays would allow a determination of the absolute strengths of the RPV couplings. Second, one should not neglect the possibility that RPV couplings could be present in GMSB models. If the RPV couplings are substantial, then the RPV decays of the NLSP will dominate its R-parity-conserving decays into $\widetilde{g}_{3/2} + X$ [45], and all the RPV phenomenology described in this section will apply. For smaller RPV couplings, there could be competition between the RPV decays and the $\widetilde{g}_{3/2} + X$ decays of the NLSP.

4 Summary and Conclusions

Much effort has been directed at trying to develop strategies for precision measurements to establish the underlying supersymmetric structure of the interactions and to distinguish among models. However, we are far from understanding all possible facets of the most general MSSM parameter space (even restricted to those regions that are phenomenologically viable). Moreover, the phenomenology of non-minimal and alternative low-energy supersymmetric models (such as models with R-parity violation) and the consequences for collider physics have only recently begun to attract significant attention. The variety of possible non-minimal models of low-energy supersymmetry presents an additional challenge to experimenters who plan on searching for supersymmetry at future colliders.

If supersymmetry is discovered, it will provide a plethora of experimental signals and theoretical analyses. The many phenomenological manifestations and parameters of supersymmetry suggest that many years of experimental work will be required before it will be possible to determine the precise nature of supersymmetry-breaking and its implications for a more fundamental theory of particle interactions.

Acknowledgements

This work was supported in part by the Department of Energy. JFG is supported in part by the Davis Institute for High Energy Physics.

References

1. H.E. Haber and G.L. Kane, *Phys. Rep.* **117** (1985) 75.
2. S.P. Martin, hep-ph/9709356, chapter in this volume.
3. S. Dimopoulos and D. Sutter, *Nucl. Phys.* **B452** (1995) 496; D.W. Sutter, Stanford Ph. D. thesis, hep-ph/9704390.
4. H.E. Haber, *Nucl. Phys. B (Proc. Suppl.)* **62A-C** (1998) 469.
5. H.P. Nilles, *Phys. Rep.* **110** (1984) 1; P. Nath, R. Arnowitt, and A.H. Chamseddine, *Applied $N = 1$ Supergravity* (World Scientific, Singapore, 1984).
6. For a review of gauge-mediated supersymmetry breaking, see G.F. Giudice and R. Rattazzi, CERN-TH/97-380 [hep-ph/9801271], submitted to *Physics Reports*; and chapter in this volume
7. For a recent review and guide to the literature, see H. Dreiner, hep-ph/9707435, chapter in this volume.
8. G.L. Kane and J.P. Leveille, *Phys. Lett.* **112B** (1982) 227.
9. H. Baer, J. Ellis, G. Gelmini, D. Nanopoulos and X. Tata, *Phys. Lett.* **161B** (1985) 175; G. Gamberini, *Z. Phys.* **C30** (1986) 605; H. Baer, V. Barger, D. Karatas and X. Tata, *Phys. Rev.* **D36** (1987) 96; R.M. Barnett, J.F. Gunion and H.E. Haber, *Phys. Rev. Lett.* **60** (1988) 401; *Phys. Rev.* **D37** (1988) 1892;
10. H. Baer, X. Tata and J. Woodside, *Phys. Rev.* **D41** (1990) 906; *Phys. Rev.* **D45** (1992) 142.

11. C.H. Chen, M. Drees, and J.F. Gunion, *Phys. Rev. Lett.* **76** (1996) 2002.

12. C.H. Chen, M. Drees, and J.F. Gunion, *Phys. Rev.* **D55** (1997) 330.

13. G.R. Farrar, *Phys. Rev. Lett.* **76** (1996) 4111, 4115; *Phys. Rev.* **D51** (1995) 3904. For a recent review, see G.R. Farrar, *Nucl. Phys. B (Proc. Suppl.)* **62A-C** (1998) 485.

14. F. Csikor and Z. Fodor, *Phys. Rev. Lett.* **78** (1997) 4335; preprint ITP-BUDAPEST-538 [hep-ph/9712269]; Z. Nagy and Z. Trocsanyi, hep-ph/9708343; hep-ph/9712385; J. Adams *et al.* [KTeV Collaboration], *Phys. Rev. Lett.* **79** (1997) 4083; P. Abreu *et al.* [DELPHI Collaboration], *Phys. Lett.* **B414** (1997) 401; R. Barate *et al.* [ALEPH Collaboration], *Z. Phys.* **C96** (1997) 1.

15. J.F. Gunion, to appear in Proceedings of the *International Workshop on Quantum Effects in the MSSM*, UAB, Barcelona, 9–13 September 1997, edited by J. Sola (World Scientific, Singapore).

16. S. Raby, *Phys. Rev.* **D56** (1997) 2852; preprint OHSTPY-HEP-T-97-024 [hep-ph/9712254].

17. S. Ambrosanio, G.D. Kribs, and S.P. Martin, SLAC-PUB-7668 (1997) [hep-ph/9710217].

18. R.M. Barnett, J.F. Gunion and H.E. Haber, *Phys. Lett.* **B315** (1993) 349.

19. For an LHC study, see H. Baer, C.H. Chen, F. Paige and X. Tata, *Phys. Rev.* **D50** (1994) 4508. Tevatron studies by theorists include: H. Baer and X. Tata, *Phys. Rev.* **D47** (1993) 2739; H. Baer, C. Kao and X. Tata, *Phys. Rev.* **D48** (1993) 5175; H. Baer, C.-H. Chen, C. Kao and X. Tata, *Phys. Rev.* **D52** (1995) 1565; S. Ambrosanio, G.L. Kane, G.D. Kribs, S.P. Martin and S. Mrenna, *Phys. Rev.* **D54** (1996) 5395. The most recent supersymmetric limits based on Tevatron tri-lepton searches are given in B. Abbott *et al.* [D0 Collaboration], FERMILAB-PUB-97-153-E (1997) [hep-ex/9705015]; and, F. Abe *et al.* [CDF Collaboration], *Phys. Rev. Lett.* **76** (1996) 4307.

20. H. Baer, C.-H. Chen, F. Paige and X. Tata, *Phys. Rev.* **D52** (1995) 2746; *Phys. Rev.* **D53** (1996) 6241.

21. I. Hinchliffe, F.E. Paige, M.D. Shapiro, J. Soderqvist and W. Yao, *Phys. Rev.* **D55** (1997) 5520.

22. G. Anderson, C.H. Chen, J.F. Gunion, J. Lykken, T. Moroi, Y. Yamada, in *New Directions for High-Energy Physics*, Proceedings of the 1996 DPF/DPB Summer Study on High Energy Physics, Snowmass '96, edited by D.G. Cassel, L.T. Gennari and R.H. Siemann (Stanford Linear Accelerator Center, Stanford, CA, 1997) pp. 669–673.

23. H. Komatsu and J. Kubo, *Phys. Lett.* **157B** (1985) 90; *Nucl. Phys.* **B263** (1986) 265; H.E. Haber, G.L. Kane and M. Quiros, *Phys. Lett.* **160B** (1985) 297; *Nucl. Phys.* **B273** (1986) 333; R. Barbieri, G. Gamberini, G.F. Giudice and G. Ridolfi, *Nucl. Phys.* **B296** (1988) 75; H.E. Haber and D. Wyler, *Nucl. Phys.* **B323** (1989) 267.

24. S. Ambrosanio, G.L. Kane, G.D. Kribs, S.P. Martin and S. Mrenna, *Phys. Rev.* **D55** (1997) 1372.

25. H. Baer *et al.*, in *Electroweak Symmetry Breaking and New Physics at the TeV Scale*, edited by T.L. Barklow, S. Dawson, H.E. Haber and J.L. Siegrist (World Scientific, Singapore, 1996) pp. 216–291.

26. See, *e.g.*, H. Baer. C.-H. Chen, F. Paige and X. Tata, *Phys. Rev.* **D54** (1996) 5866; D. Amidei *et al.* [TeV-2000 Study Group], FERMILAB-PUB-96-082; and references therein.

27. A. Bartl *et al.*, in *New Directions for High-Energy Physics*, Proceedings of the 1996 DPF/DPB Summer Study on High Energy Physics, Snowmass '96, edited by D.G. Cassel, L.T. Gennari and R.H. Siemann (Stanford Linear Accelerator Center, Stanford, CA, 1997) pp. 693–707.

28. T. Tsukamoto, K. Fujii, H. Murayama, M. Yamaguchi and Y. Okada, *Phys. Rev.* **D51** (1995) 3153; J.L. Feng, M.E. Peskin, H. Murayama and X. Tata, *Phys. Rev.* **D52** (1995) 1418; H. Baer, R. Munroe and X. Tata, *Phys. Rev.* **D54** (1996) 6735; J.L. Feng and M.J. Strassler, *Phys. Rev.* **D55** (1997) 1326; H. Murayama and M.E. Peskin, *Ann. Rev. Nucl. Part. Sci.* **46** (1996) 533.

29. F. Abe *et al.* [CDF Collaboration], FERMILAB-PUB-98-024-E [hep-ex/9801019].

30. S. Dimopoulos, S. Thomas and J.D. Wells, *Phys. Rev.* **D54** (1996) 3283; *Nucl. Phys.* **B488** (1997) 39.

31. H. Baer, M. Brhlik, C.-H. Chen and X. Tata, *Phys. Rev.* **D55** (1997) 4463.

32. S. Ambrosanio, G.L. Kane, G.D. Kribs, S.P. Martin, *Phys. Rev.* **D54** (1996) 5395.

33. S. Dimopoulos, M. Dine, S. Raby and S. Thomas, *Phys. Rev. Lett.* **76** (1996) 3494; S. Ambrosanio, G.L. Kane, G.D. Kribs, S.P. Martin and S. Mrenna, *Phys. Rev. Lett.* **76** (1996) 3498.

34. B. Abbott *et al.* [D0 Collaboration], *Phys. Rev. Lett.* **80** (1998) 442; E. Flattum [for the CDF and D0 Collaborations], FERMILAB-Conf-97/404-E.

35. J. Lopez, D. Nanopoulos and A. Zichichi, *Phys. Rev.* **D55** (1997) 5813; *Phys. Rev. Lett.* **77** (1996) 5168; S. Dimopoulos, M. Dine, S. Raby and S. Thomas, *Phys. Rev. Lett.* **76** (1996) 3494; S. Ambrosanio, G.D. Kribs

and S.P. Martin, *Phys. Rev.* **D56** (1997) 1761; D.R. Stump, M. Wiest and C.-P. Yuan, *Phys. Rev.* **D54** (1996) 1936; J. Bagger, K. Matchev, D. Pierce and R.-J. Zhang, *Phys. Rev. Lett.* **78** (1997) 1002.

36. D.A. Dicus, B. Dutta and S. Nandi, *Phys. Rev. Lett.* **78** (1997) 3055.

37. J.L. Feng and T. Moroi, LBNL-41133 [hep-ph/9712499]; C.-H. Chen and J.F. Gunion, work in progress.

38. C.-H. Chen and J.F. Gunion, UCD-97-15 (1997) [hep-ph/9707302].

39. C.-H. Chen and J.F. Gunion, UCD-98-3 [hep-ph/9802252].

40. P. Binetruy and J.F. Gunion, in *Heavy Flavors and High Energy Collisions in the 1—100 TeV Range*, Proceedings of the INFN Eloisatron Project Workshop, Erice, Italy, June 10–27, 1988, edited by A. Ali and L. Cifarelli (Plenum Press, New York, 1989) p. 489; H. Dreiner and G.G. Ross, *Nucl. Phys.* **B365** (1991) 597; H. Dreiner, M. Guchait and D.P. Roy, *Phys. Rev.* **D49** (1994) 3270; V. Barger, M.S. Berger, P. Ohmann, R.J.N. Phillips, *Phys. Rev.* **D50** (1994) 4299; H. Baer, C. Kao and X. Tata, *Phys. Rev.* **D51** (1995) 2180; H. Baer, C.-H. Chen and X. Tata, *Phys. Rev.* **D55** (1997) 1466; A. Bartl *et al.*, *Nucl. Phys.* **B502** (1997) 19.

41. G. Bhattacharyya, D. Choudhury and K. Sridhar, *Phys. Lett.* **B355** (1995) 193; G. Bhattacharyya, J. Ellis and K. Sridhar, *Mod. Phys. Lett.* **A10** (1995) 1583; J.C. Romao, F. de Campos, M.A. Garcia-Jareno, M.B. Magro, and J.W.F. Valle, *Nucl. Phys.* **B482** (1996) 3; K. Huitu, J. Maalampi and K. Puolamaki, preprint HIP-1997-24-TH [hep-ph/9705406]; F. de Campos, O.J.P. Eboli, M.A. Garcia-Jareno, and J.W.F. Valle, preprint IFUSP-1278 [hep-ph/9710545].

42. S. Dimopoulos and L.J. Hall, *Phys. Lett.* **B207** (1988) 210; V. Barger, G.F. Giudice and T. Han, *Phys. Rev.* **D40** (1989) 2987; R.M. Godbole, P. Roy and X. Tata, *Nucl. Phys.* **B401** (1993) 67; J. Erler, J.L. Feng and N. Polonsky, *Phys. Rev. Lett.* **78** (1997) 3063; J. Kalinowski, R. Ruckl, H. Spiesberger and P.M. Zerwas, *Phys. Lett.* **B406** (1997) 314.

43. J.L. Feng, J.F. Gunion and T. Han, UCD-97-25 (1997) [hep-ph/9711414].

44. See, *e.g.*, D. Buskuli *et al.* [ALEPH Collaboration], *Phys. Lett.* **B384** (1996) 461.

45. M. Carena, S. Pokorski and C.E.M. Wagner, CERN-TH-97-373 [hep-ph/9801251].

SUPERSYMMETRIC EVENT GENERATION AND THE REACH OF COLLIDERS FOR SUPERSYMMETRY

HOWARD BAER

*Dept. of Physics, Florida State University, Tallahassee,
FL 32306 USA*

STEVE MRENNA

*Argonne National Laboratory, 9700 South Cass Avenue, Argonne,
IL 60439 USA*

Models of particle physics that include weak scale supersymmetry predict a rich spectra of new particle states lying at or below the TeV energy scale. These new states ought to be accessible to present or future collider experiments. Event generator programs provide the crucial tool for linking models of weak scale supersymmetry to collider search experiments. In this chapter we discuss the current status of event generator ISAJET and PYTHIA programs. In addition, we show how these can be used for collider searches at e^+e^-, $p\bar{p}$ or pp colliders. As examples, we show for several popular models the regions of parameter space where SUSY signals ought to be discoverable above Standard Model backgrounds for the LEP2, Tevatron, LHC and NLC colliders.

1 Introduction

Perhaps the most compelling models of particle physics today are those that incorporate some form of weak scale supersymmetry (SUSY).[1] Virtually all model builders who make use of superstring theory, supergravity theory or grand unification theory these days take for granted the existence of a spectrum of supersymmetric particles with masses $\tilde{m} \sim M_{weak} \sim 250$ GeV. On the other hand, many of their colleagues (both experimental and theoretical) wisely choose to remain skeptical about weak scale supersymmetry until definitive evidence can be found. Conclusive evidence will most likely come from colliding beam experiments, where superpartners, should they exist, can be produced in abundance, and where their properties can be measured.

Theories of weak scale supersymmetry, with a plethora of new particles and interactions, can seem quite separated from the experimental measurement of pions, electrons, muons, *etc.*, at collider detector facilities. The crucial link between theory and experiment is made by computer programs known as event generators. Our task in this Chapter is twofold. First, we will survey the present status of event generator programs for simulating SUSY. Then, we will demonstrate their utility in assessing the ability of various collider facilities to discover supersymmetry by measuring event rates of distinctive supersymmetry signatures at levels beyond those expected from background processes predicted by the Standard Model (SM).

To make predictions for sparticle production and decay rates, a specific particle physics model of supersymmetry is needed. The most conservative approach is to adopt the so-called Minimal Supersymmetric Standard Model, or MSSM.[2] In this model, one assumes a superpartner for each particle of the SM, with the same quantum numbers under the gauge symmetries $SU(3)_C \times SU(2)_L \times U(1)_Y$. By necessity, the minimal Higgs sector contains two Higgs doublets, which leads to two neutral physical scalar bosons (h and H, with $m_h < m_H$), a neutral pseudoscalar A, and a pair of charged Higgs bosons $H^\pm$. Because of the absence of superpartner states degenerate in mass with SM particles, it is clear that SUSY must be broken. The effects of SUSY breaking are encompassed in the theory by adding all allowable soft supersymmetry breaking terms to the MSSM Lagrangian. These have the general form of either mass terms for scalars, mass terms for the sfermion partners to the gauge bosons, or scalar trilinear and bilinear couplings. An important implication of *minimality* is that no unnecessary interactions be added to the model, such as baryon or lepton number violating trilinear terms. This leads to a conserved multiplicative R parity, which implies that sparticles must be pair produced at colliders, and that sparticles must decay to other sparticles. Thus, the lightest SUSY particle (LSP) would be absolutely stable. Arguments from cosmology and searches for exotic nuclei and atoms then imply that the LSP should be charge and color neutral. LSP's produced in sparticle cascade decays will then escape detection at collider experiments, resulting in more missing energy/momentum than that expected from SM neutrino production and detector resolution effects. Based on these considerations, excess missing energy is the classic signature for SUSY particle production at collider experiments. Without further assumptions, the above considerations lead to a model with over 100 free parameters. While there are many parameters, there is no fine tuning needed to stabilize the Higgs boson mass, provided the superpartners are not too heavy. This latter property is by far the most attractive feature of weak scale SUSY. However, given such a model, how should we proceed in exploring its ramifications?

One approach is to treat the MSSM Lagrangian as an effective Lagrangian and let various experimental measurements constrain the parameters. Unless there are some relations between them and some limits on their sizes, many things will go awry. For example, some of the terms in the Lagrangian allow flavor changing neutral currents at a large rate, which must be suppressed. This can be avoided by assuming a mass degeneracy between squarks of the first two generations. Also, some testable features of a particular model depend on only a few parameters. For example, the D0 search for top squark pair production and decay at the Tevatron depends on only the lightest top squark mass, the lightest neutralino mass, and the assumption that all other superpartners are relatively heavy.[3] On the other hand, an effective Lagrangian is not very satisfying, and certainly cries out for a deeper understanding. If some of the parameters are constrained to be small or vanishing, we want to understand the reason.

Another approach is to appeal to higher principles or symmetries to reduce the number of parameters needed to uniquely specify a model. A popular model is based on the ideas of supergravity and grand unification. The minimal supergravity model (mSUGRA)[4] is a possible low energy ($E \lesssim 10^{16}$ GeV) effective theory that would arise in models where gravity acts as the messenger to communicate supersymmetry breaking between the hidden and observable sectors of the model. The gravitino (the superpartner of the graviton) has a weak scale mass, but an extremely weak coupling to matter. The mSUGRA model assumes the particle content of the MSSM, and a desert between the weak and GUT scales. The assumption is then made that all scalar soft breaking masses are equal to m_0 and all gaugino soft masses equal $m_{1/2}$ at M_{GUT}. In addition, trilinear A terms unify to A_0. The weak scale sparticle masses are computed by solving the renormalization group equations linking parameters at M_{GUT} to those at M_{weak}. Electroweak symmetry is broken radiatively by the large top quark Yukawa coupling. The parameter space of the model reduces essentially to four continuous plus one discrete parameter:

$$m_0, \ m_{1/2}, \ A_0, \ \tan\beta \text{ and } sgn(\mu). \tag{1}$$

Here, $\tan\beta$ is the ratio of the vacuum expectation values (vev's) of the two Higgs fields at the weak scale, and μ is the superpotential Higgsino mass term. The magnitude of μ is determined by constraining the minimization of the Higgs sector effective potential to obtain the correct value for M_Z.

A recently popular alternative is given by models with gauge mediated communication of supersymmetry breaking.[5] In these models, supersymmetry is again broken in a hidden sector, but now the breaking is communicated by gauge interactions. To facilitate this mechanism, additional matter multiplets are hypothesized which couple to visible matter through the SM gauge couplings, but which also couple to the hidden sector. The model parameters for the minimal gauge-mediated (MGM) model are taken to be

$$\Lambda, \ M, \ \tan\beta \text{ and } sgn(\mu). \tag{2}$$

Additional sets of messenger fields are also possible. In the above, M is the messenger scale, taken to be $M \gtrsim 100$ TeV, and Λ is a ratio of singlet field vev's which sets the scale for the SUSY particle masses at scale $Q = M$. Electroweak symmetry is again broken radiatively. In this model, the gravitino ($\tilde{G}$) can have a mass $m_{\tilde{G}} \sim 1\text{eV}{-}1$ keV, so that $\tilde{G}$ is the LSP.

Numerous other models are possible. Many of these reduce to the MSSM at the weak scale, but have different boundary conditions, scale choices or intermediate scale matter or gauge fields. Other models can have additional matter or gauge fields appearing at the weak scale, so that additional production and decay mechanisms beyond those of the MSSM are needed. A survey of several alternatives can be found in this volume or in the references.[6,7,8]

Currently, most event generators are set up to accommodate the physics of the MSSM, with extra switches for favorable models, such as mSUGRA and MGM. We turn next to a brief discussion of event generator programs in general, and then to specific details of the event generators ISAJET and PYTHIA.

2   General facets of event generation

There exist two approaches to event generation, which we characterize as exact tree-level calculations (ETC) and approximate factorized calculations (AFC). To calculate sparticle production and decay using ETC methods, one starts with a *specific* hard scattering production process (usually $2 \to 2$), as well as *specific* decay modes for the particles or sparticles. One may then proceed to write down the complete production and decay amplitude, including any spin correlations between initial and final state particles for the complete $2 \to n$ parton level process. The matrix element can in principle be explicitly squared. However, there can exist many Feynman diagrams for production as well as decay, and in general the decays proceed through a cascade, for which there may again exist spin correlations. For such complicated configurations, it is usually more straightforward to proceed with direct numerical evaluation of helicity amplitudes. This approach has been adopted for chargino/neutralino production at the Tevatron,[9] and more recently for chargino production for e^+e^- colliders.[10,11] The advantage is that one has a complete tree level expression for production and decay. Such calculations will be necessary for precision measurements of, for instance, final state particle angular distributions once SUSY is discovered. The disadvantage is that a new calculation is needed for each specific correlated production and decay mechanism. Hence, this has been the technique of choice for studies of precision SUSY particle measurements at e^+e^- machines, where one is focussed on a single production mechanism and a simple decay mode. Such ETC calculations can be interfaced with parton shower and hadronization routines for use in detector simulations to make them more realistic.

The AFC based programs, such as ISAJET,[12] PYTHIA[13] and HERWIG[14] (to list only the most popular) are based upon building up successive independent production and decay mechanisms. Thus, in AFC based programs, numerous different production mechanisms can be included. Particle decays are assumed to factorize into strings of 2 or 3 body decays. In this way, many strings can be patched together so that a multitude of different decay chains can be generated probabilistically. Of course, since spin correlations are neglected, the sparticle decay distributions won't be quite right. However, in many cases the sparticle decays are dominated by kinematics instead of dynamics, or in the case of scalar sparticles, there is no spin correlation. Exact decay matrix elements can be included, but are often neglected in favor of pure phase space weighting. For AFC based programs, initial and final state parton showers are included, as well as particle hadronization routines. AFC based programs are usually the program of choice for hadron colliders, where very many production mechanisms are simultaneously present, and parton distribution functions (PDF's) tend to wash out the effect of spin correlations. Since the initial state partons are colored, there are large QCD corrections to event shapes that are reproduced by parton showering. Also, AFC based programs are useful for e^+e^- colliders when i.) one is producing scalar sparticles, ii.) one has very many competing production mechanisms, iii.) the produced sparticles are so heavy that very many or very long decay cascades are operating, or iv.) the effects of spin correlations do not affect the particular observables one is interested in. The AFC based programs can be broken down into four distinct calculational steps which we now briefly review.

2.1 Hard scattering

The hard scattering is almost always a partonic $2 \to 2$ or $2 \to 3$ production subprocess calculated perturbatively using tree level Feynman diagrams. For a given kinematical configuration of the initial and final state partons, numerous possible cross sections can be computed. A random number can select which particular subprocess is to be generated based on the relative production probabilities. Of course, one must always convolute initial state quarks and gluons with an appropriate set of PDF's. In addition, some generators convolute initial state electrons and photons with distribution functions to describe the most important effects of initial state photon radiation. The latter effect can be significant especially at very high energy e^+e^- colliders. For $e^\pm p$ colliders, the hadronic structure of the photon can also be included.

2.2   Parton showers

To properly describe jet broadening from the produced quarks and gluons, approximate all orders QCD corrections must be built in. In event generators, parton shower (PS) algorithms are used. In this case, one describes radiative quark or gluon emission using matrix elements in the collinear approximation, but with exact kinematics. These approximate matrix elements include the soft and collinear pole terms, which are the most important part of the dynamics. The squared matrix elements for multiple parton emissions actually factorize in the collinear limit, so that parton emission can be programmed probabilistically. For final state showers, ISAJET uses the original algorithm of Fox and Wolfram,[15] while JETSET/PYTHIA and HERWIG use different algorithms incorporating angle ordering, which accounts for some extra subleading dynamics due to interference terms in multiple parton emissions. Initial state showers develop using the backward shower method of Sjöstrand[16], where the Q^2 dependent PDF's are actually used to compute the quark and gluon emission probabilities. The showering schemes can also differ in their choice of evolution variables, which can depend on the virtuality of the showering parton or the energy fraction carried away by an emitted parton. Finally, all of the showering algorithms rely on an energy cutoff that fixes the minimum energy of a showering parton.

So far, no specification has been made as to the type of emitted parton. The emission of photons can be included, as well as that of gluons and quarks. In fact, it is easier to deal with photon emission, since they are produced on shell and the gauge theory is Abelian. In practice, however, initial state showering of photons is only included for lepton colliders.

2.3   Hadronization schemes

The process of hadronization– turning quarks and gluons into observable hadrons– is not perturbatively calculable. Instead, hadronization algorithms can be constructed using various QCD inspired ideas, and by fitting functions to data. The simplest approach fragments a parton into collinear hadrons carrying a fraction of the initial parton energy or momentum. The probability of obtaining a given hadron with a given fraction of a parton's energy is determined by a phenomenological function called a fragmentation function, which must be fit to data. Transverse momentum can also be added based on another function also fit to data. In the Field-Feynman (FF) independent hadronization model used in ISAJET, each parton hadronizes independently of the others. While FF generates the bulk features of hadronization, it fails to maintain color correlations and conserve energy–momentum, and the particle multiplicity is not

Lorentz invariant. These problems are remedied by string fragmentation used in JETSET/PYTHIA, where partons are assumed to be linked to other partons via color strings. The string hadronization model suffers however from being non-local. The cluster hadronization model of HERWIG improves upon this by converting partons into hadrons locally in phase space.

2.4 Beam remnants

For hadron colliders, the beam remnants of partons not engaged in the hard scattering can have measurable consequences. This facet is again non-perturbative, so that one generally appeals to models which are fit to data. In ISAJET, the remnants scatter via a Pomeron exchange model. After scattering, the remnants hadronize according to a modified independent fragmentation model fit to data. In PYTHIA, a multiple scattering algorithm is used, where the multiple scatterings are adjusted again to fit data. Finally, in HERWIG, a beam remnant model based on data from experiment UA6 is used.

2.5 Other Considerations

PYTHIA and HERWIG keep track of the color flow for parton showering and hadronization purposes. In principle, different production or decay diagrams have different color flow, and can be distinguishable, so that quantum interference terms are not used. To implement this effect, one needs to know the cross section for a specific color configuration. Using a $1/N_c$ approximation, where N_c is the number of colors (3 in this case), the exact matrix element can be projected into the different color combinations. However, there are some terms left over of order $1/N_c^2$ which must be treated. These extra terms are redistributed according to their pole structure. This means that the scattering cross sections in PYTHIA may be different from those in ISAJET.

3 Overview of ISAJET

The program ISAJET, created by Frank Paige and Serban Protopopescu,[12] was one of the first multi-purpose event generator programs to appear. It was originally constructed to give an idea of what jet events would look like at the ISABELLE pp collider. H. Baer and X. Tata are co-authors of the supersymmetry code in ISAJET, for versions 7.00 and greater.

3.1 Availability

Currently, ISAJET version 7.29 is available. ISAJET is encoded using FORTRAN 77, but is managed using the PATCHY code management system developed at CERN. The file ISAJET.CAR containing source code and PATCHY commands is available by anonymous ftp at ftp://penguin.phy.bnl.gov:/pub/isajet. It can also be obtained via the HEP Decnet from bnlajc::user03:[isajet.isalibrary]. There is also a Unix makefile and a VMS isamake.com in this directory; these require that PATCHY be installed on one's system. Upon creating ISAJET, a number of files are created, including isajet.for (the main ISAJET code), ISASUSY (which generates a decay table with MSSM inputs), ISASUGRA (which solves for the particle spectra in the mSUGRA model, and lists sparticle decay modes) and ISAJET.DOC (where detailed documentation and sample programs are stored).

3.2 The ISAJET SUSY models and sparticle decays

Upon running the program ISASUSY, the user is prompted for an output file name, and then various sets of input parameters from the MSSM:

$$\text{MSSMA}: \quad m_{\tilde{g}}, \ \mu, \ m_A, \ \tan\beta$$

$$\text{MSSMB}: \quad m_{Q_1}, \ m_{d_R}, \ m_{u_R}, \ m_{L_1}, \ m_{e_R}$$

$$\text{MSSMC}: \quad m_{Q_3}, \ m_{b_R}, \ m_{t_R}, \ m_{L_3}, \ m_{\tau_R}, \ A_t, \ A_b, \ A_\tau.$$

In the above, $m_{\tilde{g}}$ is the physical gluino mass. It undergoes a pole mass correction to derive the $SU(3)$ soft breaking gaugino mass M_3, which is then used to calculate the $SU(2)$ and $U(1)$ soft-breaking gaugino masses M_2 and M_1 via the grand unification relation. The terms in MSSMB and MSSMC above are 1st and 3rd generation soft SUSY breaking parameters: the physical masses are derived by including electroweak D-terms, fermion masses and mixings (mixings for just the 3rd generation masses). Second generation soft masses are taken by default to be equal to 1st generation soft masses.

Optionally, one may enter the parameters

$$\text{MSSMD}: m_{Q_2}, \ m_{s_R}, \ m_{c_R}, \ m_{L_2}, \ m_{\mu_R}$$

if the model of interest has different 1st and 2nd generation soft scalar masses. Optionally, one may also enter

$$\text{MSSME}: M_1, \ M_2$$

if one wants independent $U(1)$ and $SU(2)$ gaugino masses as input; here, M_1 and M_2 can be either positive or negative (M_3 is assumed to be positive).

The ISASUSY program then calculates the complete sparticle and Higgs boson mass spectra, mixing angles and couplings, and dumps out the complete decay table for sparticles, Higgs bosons and the top quark, including the partial widths in GeV and branching fractions. The Higgs boson masses and couplings are calculated using the renormalization group improved one loop effective potential, including just 3rd generation contributions. These results agree with Ref. [17] in predicting m_h to within 2-3 GeV.

Alternatively, one can run the ISASUGRA program. In this case, the inputs are

$$\text{SUGRA} : m_0, \ m_{1/2}, \ A_0, \ \tan\beta, \ sgn(\mu) \text{ and } m_t.$$

In this case, ISASUGRA begins with GUT scale boundary conditions of the mSUGRA model and calculates the weak scale soft masses via iterative running of 26 renormalization group equations,[18] using 2-loop RGE's for gauge couplings and 1-loop RGE's for Yukawas and soft terms. SUSY particle threshold effects are included. The scalar potential is minimized at scale $Q = \sqrt{m_{\tilde{t}_L} m_{\tilde{t}_R}}$, which yields a prediction of m_A which is stable under scale variations even at large $\tan\beta$. The program then outputs various GUT and weak scale parameters, SUSY and Higgs boson masses, and the complete decay table.

ISASUSY and ISASUGRA print out warning messages if parameter choices violate various theoretical constraints, or violate constraints from precision LEP data.

ISAJET version 7.29 includes as well all 3rd generation Yukawa and mixing effects for stops, sbottoms and staus, and their associated decays. It also includes a complete set of Feynman diagrams for the 3-body $\tilde{g}$, $\tilde{\chi}_1^{\pm}$ and $\tilde{\chi}_2^0$ decays, including all 3rd generation Yukawa and mixing effects. Finally, helicity information for subsequent tau lepton decays from W's, Higgs bosons and SUSY particles is kept and used for subsequent event generation,[19] since ISAJET includes full matrix elements for τ-lepton decays.

Alternative models of weak-scale supersymmetry are usually easy to simulate in ISAJET provided they reduce to the MSSM at the weak scale. For models with gauge-mediated supersymmetry breaking, the user must provide a program to generate the appropriate sparticle masses, which can then be entered into ISAJET via MSSMi commands. Decays of the NLSP can be entered either via the FORCE command, or through explicit addition to the decay table in ISADECAY.DAT. A gravitino state is defined in ISAJET.

Models with R parity violation can be simulated provided the R violating couplings can be neglected in sparticle production and decay reactions, other than for LSP decay. For the LSP, the R violating decays can be explicitly added to `ISADECAY.DAT`.

3.3 Event generation with ISAJET: pp or $p\bar{p}$ collisions

For hadron colliders, along with SM processes, `ISAJET` contains the following SUSY production reactions:

- $qq,\ gg,\ qg \to \tilde{g}\tilde{g},\ \tilde{g}\tilde{q},\ \tilde{q}\tilde{q}$, (strong production)

- $qq,\ qg \to \tilde{g}\tilde{\chi}_i^0,\ \tilde{g}\tilde{\chi}_i^\pm,\ \tilde{q}\tilde{\chi}_i^0,\ \tilde{q}\tilde{\chi}_i^\pm$ (associated production)

- $qq \to \tilde{\chi}_i^\pm \tilde{\chi}_j^\mp,\ \tilde{\chi}_i^\pm \tilde{\chi}_j^0,\ \tilde{\chi}_i^0 \tilde{\chi}_j^0$ ($\tilde{\chi}$ pair production)

- $qq \to \tilde{\ell}\tilde{\nu},\ \tilde{\ell}\tilde{\ell},\ \tilde{\nu}\tilde{\nu}$ (slepton pair production).

(Bars have been omitted on anti-particle labels.) Specific reactions can be generated via use of the `ISAJET JETTYPEi` command; if `JETTYPEi='ALL'`, then *all* the above subprocesses are generated according to their relative probabilities. Once produced, the sparticles decay to other sparticles via the pre-calculated branching fractions until the LSP state is reached. At this stage in `ISAJET`, sparticle decay distributions are generated according to phase space (although exact matrix elements are included for various SM particles such as heavy quarks and tau leptons).

In addition, the Higgs bosons of the MSSM can be produced via the direct s-channel subprocess,

- $qq,\ gg \to h,\ H,\ A,\ H^\pm$.

The Higgs bosons can then decay to SM or SUSY particle modes.

Finally, $t\bar{t}$ production can also be generated, including SUSY decay modes of the top quark.

3.4 Event generation with ISAJET: e^+e^- collisions

At e^+e^- colliders, the following production mechanisms are included:

- $e^+e^- \to \tilde{q}\bar{\tilde{q}}$ (squark pair production)

- $e^+e^- \to \tilde{\ell}\bar{\tilde{\ell}}$ (slepton pair production)

- $e^+e^- \to \tilde{\chi}_i^\pm \tilde{\chi}_j^\mp,\ \tilde{\chi}_i^0 \tilde{\chi}_j^0$ ($\tilde{\chi}$ pair production)

- $e^+e^- \to hZ,\ HZ, hA,\ HA,\ H^+H^-$ (Higgs production).

For e^+e^- collisions, the beam polarization may be specified via the ISAJET keyword EPOL, which can take values from -1 to +1, where -1 corresponds to a pure left polarized e beam, and +1 corresponds to pure right polarization. Both e^- and e^+ beam polarization may be specified. The ISAJET e^+e^- generator can be used to generate $\mu^+\mu^-$ collider events for the above processes since initial state radiation effects are not included. However, ISAJET does not at this time include s-channel Higgs boson production for $\mu^+\mu^-$ collisions.

4 Overview of PYTHIA

Like ISAJET, PYTHIA and JETSET are programs for the generation of high-energy physics events. The development of JETSET, the first member of the "Lund Monte Carlo" family, was begun by members of the Lund theory group in 1978. PYTHIA was originated by Hans–Uno Bengtsson, and developed in parallel with T. Sjöstrand. Over the years, the two programs have come to be maintained in common. In the most recent version they have been merged into one, under the PYTHIA label, with T. Sjöstrand as the principal author and developer. The supersymmetric extension was developed by S. Mrenna.

4.1 Availability

PYTHIA version 6.1 is the new standard distribution, including the supersymmetric extension, an upgrade to double precision variables, and a change of some subroutine, function, and variable names from the incorporation of JETSET. Past and present versions, documentation, examples of use, and other useful information about PYTHIA are available on the World Wide Web at the address http://thep.lu.se/tf2/staff/torbjorn/Pythia.html. PYTHIA is written in the FORTRAN 77 computing language. The code is available in text form, gzipped text, or uuencoded gzipped text.

4.2 PYTHIA *SUSY Models*

PYTHIA is run in a master–slave format, where the user is the master and PYTHIA is the slave. The user writes a main program that initializes PYTHIA by calling the subroutine PYINIT, then generates individual events by calling the subroutine PYEVNT. Examples are available at the PYTHIA webpage. The PYTHIA parameters are accessed through common blocks, or, alternatively, at initialization time, through a special subroutine PYGIVE which converts characters into commands (e.g., the FORTRAN statement CALL PYGIVE('IMSS(1)=1') will set the variable IMSS(1)=1).

In general, various options for the simulation of supersymmetry are set by the array IMSS, while numerical values are fixed by the array RMSS. The level of supersymmetry simulation is set by the parameter IMSS(1)=0,1,2, which means no SUSY, a general MSSM model where all parameters must be fixed, or a mSUGRA model based on only five parameters, respectively.

For a general MSSM model (IMSS(1)=1), the elements of the array RMSS have the following meaning (the array element number is in parentheses):

$$(1)\ M_1,\ (2)\ M_2,\ (3)\ M_{\tilde{g}},\ (4)\ \mu,\ (5)\ \tan\beta,$$
$$(6)\ M_{\tilde{\ell}_L},\ (7)\ M_{\tilde{\ell}_R},\ (8)\ M_{\tilde{q}_L},\ (9)\ M_{\tilde{q}_R}$$
$$(10)\ M_{\tilde{b}_L},\ (11)\ M_{\tilde{b}_R},\ (12)\ M_{\tilde{t}_R},\ (13)\ M_{\tilde{\tau}_L},\ (14)\ M_{\tilde{\tau}_R},$$
$$(15)\ A_b,\ (16)\ A_t,\ (17)\ A_\tau,\ (19)\ M_A.$$

Note that all of these parameters have units of mass (with the exception of $\tan\beta$, which is the ratio of mass parameters). Units of GeV are used in the program, unless otherwise specified. Different values for the IMSS array allow many other options, such as fixing relations between M_1, M_2 and $M_{\tilde{g}}$ or determining the third generation squark properties from the lightest sbottom and stop masses and their mixing angles. ¿From these input parameters, the complete sparticle spectrum and decay modes are calculated. The Higgs boson properties are determined using effective potential methods including the effects of third generation squarks and the chargino.[17] All third generation Yukawa and mixing effects for stops, sbottoms and staus, and their associated decays, are included. 3-body $\tilde{g}$ decays also include these effects, but 3-body $\tilde{\chi}_i^\pm$ and $\tilde{\chi}_i^0$ decays do not include virtual Higgs bosons.

The phenomenology of mSUGRA models can also be studied using approximate analytic formulae.[21] When IMSS(1)=2, the elements of the array RMSS have the meaning:

$$(1)\ m_{1/2},\ (4)\ sgn(\mu),\ (5)\ \tan\beta,\ (8)\ m_0,\ (16)\ A_0.$$

All the other MSSM parameters are fixed by these few parameters, and the simulation then continues as before. For a more accurate study of mSUGRA models, the output spectra of a program such as ISASUGRA can be input as a general MSSM model (i.e. IMSS(1)=1).

As in ISAJET, alternative models of weak scale supersymmetry are usually easy to simulate provided they reduce to the MSSM at the weak scale. For models with gauge mediated supersymmetry breaking, the user must provide the effective MSSM parameters and the gravitino mass. All additional two body decays to the gravitino LSP are automatically calculated.

Also similarly to ISAJET, models with R parity violation can be simulated provided the R violating couplings can be neglected in sparticle production and decay reactions, other than for the LSP. For the LSP, the R violating decays can be explicitly simulated by updating the PYTHIA decay table with the subroutine PYUPDA.

4.3 Event generation with PYTHIA

PYTHIA simulates the same superpartner pair production processes as listed in the ISAJET section. However, the number of production processes for the different Higgs bosons is more extensive, including the associated production processes $q\bar{q}, gg \rightarrow Q\bar{Q}\phi$, where $Q = b, t$ and $\phi = h, H, A$, $q\bar{q} \rightarrow V\phi$, where $V = W, Z$, and higher order processes.

Also, PYTHIA allows the simulation of a wide range of colliders, including $\mu^+\mu^-$ and ep. For lepton colliders, however, beam polarization is not included as in ISAJET.

5 Other event generators

A widely used event generator by LEP2 groups is called SUSYGEN.[22] SUSYGEN includes the production and decay of superpartner pairs of interest to LEP2 experiments (e^+e^- collisions only). It is set up to interface with the JETSET/PYTHIA program for parton showering and hadronization. One advantage of SUSYGEN is that it includes matrix elements for sparticle decay distributions.[22]

6 Applications

As an application, event generators can be used to estimate the reach in model parameter space for various colliders, assuming some typical integrated luminosity. Below, we present results for the LEP2, Tevatron, LHC and NLC colliders. Most of the work presented is for the mSUGRA model, although some is for the MGM model.

6.1 LEP2

At LEP2, the most promising signals for supersymmetry in the mSUGRA model are i.) light Higgs boson production $e^+e^- \rightarrow Zh$ or $e^+e^- \rightarrow hA$, ii.) chargino pair production $e^+e^- \rightarrow \tilde{\chi}_1^\pm \tilde{\chi}_1^\mp$, iii.) slepton pair production, $e.g.$ $e^+e^- \rightarrow \tilde{e}_R\bar{\tilde{e}}_R$ and neutralino pair production iv.) $e^+e^- \rightarrow \tilde{\chi}_1^0 \tilde{\chi}_2^0$. Backgrounds come from various SM processes, especially WW and ZZ production.

In Fig. 1, we show the regions of the m_0 *vs.* $m_{1/2}$ plane where various SUSY particle reactions are observable at the 5σ level above SM and SUSY backgrounds for an integrated luminosity of 500 pb^{-1}, and for $A_0 = 0$, $\mu < 0$ and $m_t = 170$ GeV. The study has been performed using ISAJET.[23] The energy of LEP2 is assumed to be $\sqrt{s} = 190$ GeV. The regions labelled by TH are excluded by lack of radiative electroweak breaking or a charged or colored LSP. The EX regions are excluded by SUSY searches from LEP1. We see that most of the region below $m_{1/2} = 100$ GeV should be explorable via the single lepton plus jets channel from chargino searches; this corresponds to a reach in $m_{\tilde{\chi}_1^\pm} \simeq 95$ GeV, *i.e.* saturating the kinematic limit. In addition, the lower left region of the plane can be explored via the di-electron plus missing energy channel from selectrons. There exists a small intermediate region where the dilepton signal from $\tilde{\chi}_1^0 \tilde{\chi}_2^0$ production is visible. Not shown is the reach for SUSY in the Zh channel. In fact, the reach of LEP2 in this channel exceeds $m_{1/2} \simeq 350$ GeV, thus filling the whole plane shown. For other $\tan\beta$ and A_0 values, the reach is similar for the chargino and selectron detection, while the h mass is typically heavier, making it more difficult to detect.[23] Other detailed studies for more general models can be found in Ref.[24]

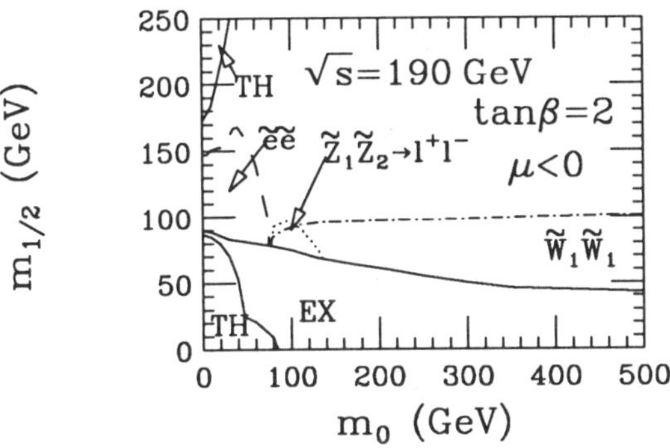

Figure 1: Reach of LEP2 collider for SUSY

6.2 *Tevatron*

The most favorable mSUGRA search modes for the Fermilab Tevatron $p\bar{p}$ collider are *i.*) the missing energy plus jets mode, from squark and gluino production and *ii.*) the clean trilepton search mode $\tilde{\chi}_1^\pm \tilde{\chi}_2^0 \to 3\ell + \not{E}_T$.[9,25] In Fig. 2, we show the parameter space points where SUSY signals ought to be observable above SM backgrounds, using ISAJET.[26] The cross hatched region is excluded by LEP1.5 searches. The black squares denote mSUGRA model points discoverable beyond the 5σ level with only 0.1 fb^{-1} of integrated luminosity (corresponding to the data sample from Run I). In this case, the discovery channel is from the jets $+\not{E}_T$ channel. For 2 fb^{-1}, the discovery region is denoted by grey squares; in this case, most of the reach actually comes from the clean trilepton channel. The uppermost grey squares at $m_{1/2} = 250$ GeV correspond to a gluino mass of 650 GeV and a chargino of mass 190 GeV— well beyond the reach of LEP2. The white squares denote points accessible to the TeV33 upgrade, with projected integrated luminosity of 25 fb^{-1}. The TeV33 has an expanded reach mainly via the clean trilepton channel. Note however that there exist points with large m_0 where the $\tilde{\chi}_2^0$ leptonic branching fraction drops to very low values, so that the reach of TeV33 is worse than that from LEP2 at $\sqrt{s} = 190$ GeV in this region.

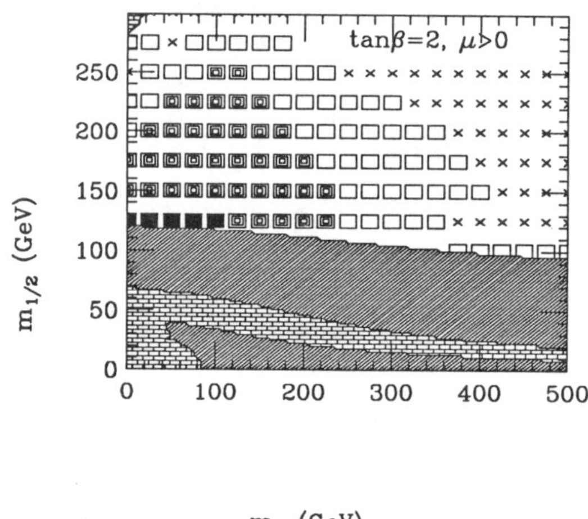

Figure 2: Reach of Tevatron collider for SUSY

For the MGM model, the reach of the Tevatron can be exhibited in terms of the parameter Λ, which sets the scale for all the SUSY particle masses at the effective SUSY breaking scale M. A lightest chargino mass of roughly 100 GeV in these models can be excluded in a fairly model independent manner by studying $\gamma\gamma\not{E}_T$ events.[27,28] In Ref. [29], the reach of the Tevatron has been estimated for the minimal gauge mediated model in terms of isolated photon plus jets plus leptons plus missing energy channels. Their calculated reach corresponds to $\Lambda \sim 100$ (135) TeV for the Tevatron MI (TeV33). This corresponds to a reach in $m_{\tilde{g}}$ of 800 (1000) GeV!

6.3   LHC

The CERN Large Hadron Collider (LHC) is a *pp* collider which is under construction. It is expected to turn on around the year 2005, with projected energy of 14 TeV and initial luminosity of about 10 fb^{-1} per year. If weak-scale supersymmetry exists, a wide variety of signals are expected to be observable. For the mSUGRA model, these consist mainly of various multi-jet plus multi-lepton plus missing transverse energy events. Various signals and SM backgrounds have been calculated in the mSUGRA model using ISAJET, and the reach contours are presented in Fig. 3 in the m_0 *vs.* $m_{1/2}$ plane, for $A_0 = 0$, $\tan\beta = 2$ and $\mu > 0$.[30] The region in the lower-left corner denoted by $\tilde{l}$ is where the dilepton plus missing energy signal from direct production of sleptons should be visible: this essentially covers the region where $m_{\tilde{l}} < 300$ GeV, which is the region favored by neutralino relic density calculations.[31] The region denoted by $\tilde{\chi}_1^{\pm}\tilde{\chi}_2^0$ is where jet-free $\tilde{\chi}_1^{\pm}\tilde{\chi}_2^0 \to 3\ell$ signals ought to be visible above SM and SUSY backgrounds. Note, as for the Tevatron cases, the gap around $m_0 \sim 500$ GeV and $m_{1/2} < 200$ where this signal is not visible.

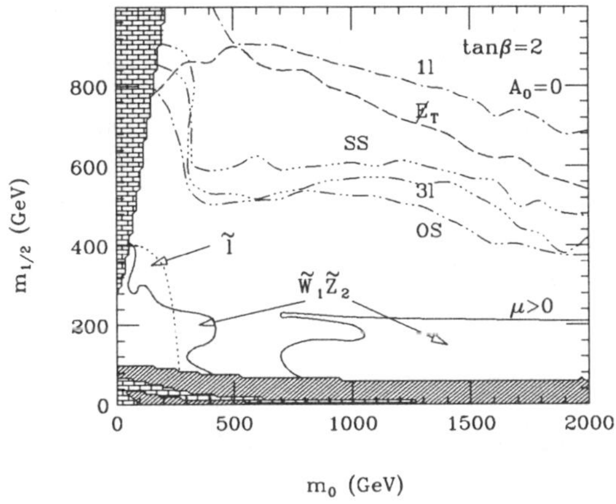

Figure 3: Reach of LHC *pp* collider for SUSY

The highest contour labelled by 1ℓ is where single isolated lepton plus jets plus $\not{E}_T$ is visible above SM background, and gives the maximum reach of the LHC— all the way out to $m_{\tilde{g}} > 2$ TeV for the LHC low luminosity option. Comparing this reach to expectations from fine-tuning estimates shows that the LHC should be able to either discover or rule out weak scale supersymmetry, at least in its mSUGRA manifestation! The contour labelled $\not{E}_T$ is where multi-jet plus $\not{E}_T$ signals (no isolated leptons) should be visible. The SS, OS and 3ℓ

contours all show the reach for mSUGRA in the same-sign dilepton, opposite sign dilepton and trilepton channels, where each of these are accompanied by jets and missing transverse energy. By comparing various combinations of energy and mass distributions with model expectations, various constraints on SUSY particle masses and parameters ought to be obtainable: see Ref. [32] for some detailed case studies.

6.4 NLC

Many physicists are interested in constructing a very high energy linear e^+e^- collider (the NLC, or Next Linear Collider), with possible center of mass energy options of 0.5, 1 and 1.5 TeV. The reach of such a collider for the mSUGRA model has been calculated in Ref.[33] using ISAJET. For instance, assuming an NLC with $\sqrt{s} = 0.5$ TeV and integrated luminosity of 20 fb^{-1}, the lightest Higgs scalar h should be visible over the whole parameter space. Thus, if a light scalar Higgs is not found at the NLC, then the mSUGRA model would be excluded. A reach contour for NLC for SUSY particle discovery is shown in Fig. 4. NLC500 should see chargino pairs at 5σ above SM background for $m_{1/2} \lesssim 300$ GeV and $m_0 \gtrsim 200$ GeV. In addition, selectron pairs ought to be visible in the low m_0 region up to $m_{1/2} \simeq 520$ GeV. The bulge in the NLC500 contour around $m_0 \sim 300$ GeV and $m_{1/2} \sim 350$ GeV is where $\tilde{\chi}_1^0 \tilde{\chi}_2^0 \to b\bar{b} + \not{H}$ is visible. The reach contour is nearly independent of whether the NLC achieves the capability of beam polarization.

Approximate reach contours are plotted in Fig. 4 for NLC at $\sqrt{s} = 1000$ GeV (NLC1000) and $\sqrt{s} = 1500$ GeV (NLC1500). For these cases, detailed signal-to-background studies were not made, but instead the reach was obtained by just plotting where $m_{\tilde{l}}$ and $m_{\tilde{\chi}_1^\pm}$ were equal to the beam energy.

6.5 Comparison of reach in the mSUGRA model

In Fig. 4, we show for comparison purposes the reach of the Tevatron MI, TeV33, the LHC with 10 fb^{-1} of integrated luminosity, and the NLC500, NLC1000 and NLC1500 options. For perspective, we also show contours where $m_{\tilde{g}}$ and $m_{\tilde{q}}$ equal 1000 GeV. On this expanded scale, the Tevatron options appear to only have a small reach for SUSY; however, these regions are exactly those most favored by fine tuning constraints.[34] The LHC collider can explore well beyond the 1 TeV mass scale for supersymmetry (note this statement holds even if R violating LSP decays occur.[35]) If LHC fails to find SUSY, it will be very difficult to understand how weak scale supersymmetry can exist. We also see from Fig. 4 that the reach of LHC is similar to that of a NLC collider at

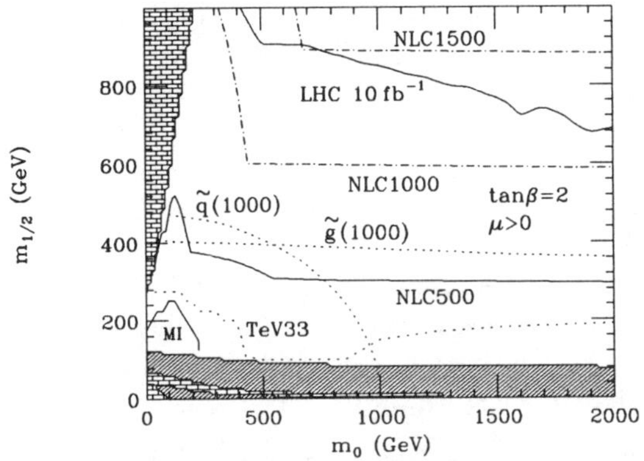

Figure 4: Reach of Tevatron Main Injector, TeV33, LHC and various energy options for a NLC linear e^+e^- collider for SUSY

$\sqrt{s} \simeq 1000 - 1500$ GeV. The NLC collider options have the advantage in that if superparticles are discovered, then relatively model independent precision measurements of sparticle masses and parameters can be made.[10,33,36]

7 Conclusion

It is only within the past several years that realistic event generation of supersymmetric particles has been possible. The capability of supersymmetric event generation in ISAJET and PYTHIA is relatively new, and both programs are undergoing evolution and change to improve on the physics input and the variety of models available for use. However, the significant progress that has occurred has allowed reliable calculations to be made of how supersymmetry would be revealed in colliding beam experiments.

Acknowledgments

We thank X. Tata for comments on the manuscript. This report was supported in part by the U. S. Department of Energy under grant number DE-FG-87ER40319. The work of SM was supported by DOE grant W–31–109–ENG–38.

References

1. See, *e.g.* H. Haber in *Woodlands Superworld*, hep-ph/9308209 (1993).

2. For a review of the minimal model and SUSY phenomenology, see H. Haber and G. Kane, Phys. Rep. **117**, 75 (1985); X. Tata, UH-511-872-97, lectures presented at *The IX Jorge A. Swieca Summer School*, Campos do Jordao, Brazil, February, 1997, hep-ph/9706307; M. Drees, APCTP-05 (lectures given at Inauguration Conference of the Asia Pacific Center for Theoretical Physics (APCTP), Seoul, Korea (1996), hep-ph/9611409.

3. D. Claes (D0 Collaboration), FERMILAB-CONF-95-186-E, (1995), to be published in *The Proceedings of 10th Topical Workshop on Proton-Antiproton Collider Physics*, Batavia, IL, May 1995.

4. A. Chamseddine, R. Arnowitt and P. Nath, Phys. Rev. Lett. **49**, 970 (1982); R. Barbieri, S. Ferrara and C. Savoy, Phys. Lett. **B119**, 343 (1982); L.J. Hall, J. Lykken and S. Weinberg, Phys. Rev. **D27**, 2359 (1983); for a review, see H. P. Nilles, Phys. Rep. **110**, 1 (1984).

5. M. Dine and A. Nelson, Phys. Rev. **D48**, 1277 (1993); M. Dine, A. Nelson, Y. Shirman, Phys. Rev. **D51**, 1362 (1995); M. Dine, A. Nelson, Y. Nir and Y. Shirman, Phys. Rev. **D53**, 2658 (1996).

6. J. Amundson *et al.*, hep-ph/9609374 (1996).

7. M. Peskin, hep-ph/9604339.

8. J. L. Hewett and T. Rizzo, Phys. Rept. **183**, 193 (1989).

9. H. Baer, K. Hagiwara and X. Tata, Phys. Rev. Lett. **57**, 294 (1986) and Phys. Rev. **D35**, 1598 (19987).

10. T. Tsukamoto, K. Fujii, H. Murayama, M. Yamaguchi and Y. Okada, Phys. Rev. **D51**, 3153 (1995); see also JLC-1, KEK Report 92-16 (1992).

11. J. L. Feng and D. E. Finnell, Phys. Rev. **D49**, 2369 (1994).

12. F. Paige and S. Protopopescu, in *Supercollider Physics*, p. 41, ed. D. Soper (World Scientific, 1986); H. Baer, F. Paige, S. Protopopescu and X. Tata, in *Proceedings of the Workshop on Physics at Current Accelerators and Supercolliders*, ed. J. Hewett, A. White and D. Zeppenfeld, (Argonne National Laboratory, 1993), hep-ph/9305342.

13. T. Sjöstrand, Computer Phys. Commun. **82**, 74 (1994).

14. G. Marchesini, B. Webber, G. Abbiendi, I. Knowles, M. Seymour and L. Stanco, Computer Phys. Commun. **67**, 465 (1992).

15. G. Fox and S. Wolfram, Nucl. Phys. **B168**, 285 (1980).

16. T. Sjöstrand, Phys. Lett. **157B**, 321 (1985); T. Gottschalk, Nucl. Phys. **B277**, 700 (1986).

17. M. Carena, J. Espinosa, M. Quiros and C. Wagner, Phys. Lett. **B355**, 209 (1995); M. Carena, M. Quiros and C. Wagner, Nucl. Phys. **B461**, 407 (1996); H. Haber, R. Hempfling and A. Hoang, hep-ph/9609331 (1996).

18. H. Baer, C. H. Chen, R. Munroe, F. Paige and X. Tata, Phys. Rev. **D51**, 1046 (1995).

19. H. Baer, C. H. Chen, M. Drees, F. Paige and X. Tata, hep-ph/9704457 (1997).

20. S. Mrenna, to appear in *Comput. Physics. Commun.* (1997).

21. M. Drees and S. Martin in *Electroweak Symmetry Breaking and New Physics at the TeV Scale*, edited by T. Barklow, S. Dawson, H . Haber and J. Seigrist, (World Scientific) 1995, hep-ph/9504324.

22. S. Katsanevas and S. Melachroinos, SUSYGEN.

23. H. Baer, M. Brhlik, R. Munroe and X. Tata, Phys. Rev. **D52**, 5031 (1995).

24. G. F. Giudice *et al.*, hep-ph/9602207 (1996)

25. See Ref. [9] and also R. Arnowitt and P. Nath, Mod. Phys. Lett. **A2**, 331 (1987); R. Barbieri, F. Caravaglios, M. Frigeni and M. Mangano, Nucl. Phys. **B367**, 28 (1993); H. Baer and X. Tata, Phys. Rev. **D47**, 2739 (1993); J. Lopez, D. Nanopoulos, X. Wang and A. Zichichi, Phys. Rev. **D48**, 2062 (1993) and Phys. Rev. **D52**, 142 (1995); J. Lopez, D. Nanopoulos, G. Park, X. Wang and A. Zichichi, Phys. Rev. **D50**, 2164 (1994); H. Baer, C. Kao and X. Tata, Phys. Rev. **D48**, 5175 (1993); S. Mrenna, G. Kane, G. Kribs and J. Wells, Phys. Rev. **D53**, 1168 (1996).

26. H. Baer, C. H. Chen, C. Kao and X. Tata, Phys. Rev. **D52**, 1565 (1995); H. Baer, C. H. Chen, F. Paige and X. Tata, Phys. Rev. **D54**, 5866 (1996).

27. S. Ambrosanio, G.L. Kane, G.D. Kribs, S.P. Martin, and S. Mrenna, Phys. Rev. **D54**, 5395 (1996).

28. S. Abachi *et. al.*, (D0 Collaboration), Phys. Rev. Lett. **78**, 2070 (1997).

29. H. Baer, M. Brhlik, C. H. Chen and X. Tata, Phys. Rev. **D55**, 4463 (1997).

30. H. Baer, C. H. Chen, F. Paige and X. Tata, Phys. Rev. **D52**, 2746 (1995) and Phys. Rev. **D53**, 6241 (1996).

31. H. Baer and M. Brhlik, Phys. Rev. **D53**, 597 (1996).

32. A. Bartl *et. al.*, LBL-39413 (1996) and I. Hinchliffe, F Paige, M. Shapiro, J. Soderqvist and W. Yao, Phys. Rev. **D55**, 5520 (1997).

33. H. Baer, R. Munroe and X. Tata, Phys. Rev. **D54**, 6735 (1996).

34. For fine-tuning constraints, see *e.g.* G. Anderson and D. Castaño, Phys. Rev. **D52**, 1693 (1995).

35. H. Baer, C. H. Chen, and X. Tata, Phys. Rev. **D55**, 1466 (1997).

36. M. N. Danielson *et al.*, *Supersymmetry at the NLC*, proceedings of 1996 Snowmass workshop.

MASS DENSITY OF NEUTRALINO DARK MATTER

JAMES D. WELLS[a]

Stanford Linear Accelerator Center, Stanford University
Stanford, California 94309, USA

The lightest supersymmetric particle (LSP) is stable in an R-parity conserving theory. In this article the steps needed to calculate the present day mass density of such a particle are detailed. It is shown that there can be a significant amount of LSP dark matter in the universe. Furthermore, relic abundance considerations put an upper bound on how large supersymmetry breaking masses can be without resorting to finetuning arguments.

1 Introduction

The most general gauge invariant superpotential will generally lead to unacceptable fast proton decay. This problem is avoided if a discrete symmetry exists that banishes baryon and/or lepton violating operators contributing to this decay. The simplest such discrete symmetry is R-parity[1]. This symmetry can arise from string theories or from grand unified theories. Exact R-parity conservation makes the lightest supersymmetric partner (LSP) stable, thereby introducing many interesting phenomena. The most studied phenomenon is the large missing energy signature associated with production of superpartners and subsequent decay into the LSP plus jets, photons, leptons, etc. For experimental reasons[2] and theoretical reasons[3] the LSP is now expected to be the lightest neutralino.

A stable LSP has more than collider physics consequences. If they are created in the hot and violent early days of the universe, then there should be some left over today, and perhaps they could have significant cosmological and astrophysical consequences. If it turns out that the LSP constitutes a significant mass fraction of the universe then it should be possible to witness large-scale gravitational effects of these particles on galaxies and clusters. Experiments have been testing large-scale gravity for decades now, and the current consensus states that there are non-luminous sources of gravitational import beyond the ordinary baryonic matter that makes up planets and stars[4]. Part of the evidence of additional mass-energy beyond our luminous matter includes rotation curves of galaxies and infall of clusters. The evidence for non-baryonic dark matter can be attributed to the successful agreement between measurement and theory in big bang nucleosynthesis (BBN). BBN tells

---

[a]Work supported by DOE under Contract DE-AC03-76SF00515. SLAC-PUB-7606.

us that baryonic mass fraction of the universe is probably less than 10%[5]. (I am making the usual assumption that the total energy density of the universe is the critical energy density in accord with a general inflation scenario.)

One is left wondering what the rest of the universe is made of. One suggestion is a weakly interacting massive object, or WIMP. The acronym WIMP is a good general label, but it is a bit misleading. Upon closer inspection a particle of $\mathcal{O}(m_W)$ mass which has full-strength $SU(2)$ (weak) interactions is generally not a good dark matter candidate, and yields a relic abundance less than is required to have astrophysical significance. Some additional suppressions are generally needed in the annihilation cross-section. As we will see in subsequent sections, the LSP generally has the right suppressions to make it a good dark matter candidate while at the same time having mass near the weak-scale. For this reason, I will not use the generic word WIMP, and instead refer to the dark matter candidate as the LSP. LSPs are excellent candidates for the dark matter because *if enough can be around* they allow conformance to experimentally determined properties of galactic rotation curves, structure formation and big bang nucleosynthesis. Furthermore, calculation of their relic abundance indicates that LSPs could contribute most of the energy density of the universe. This is a non-trivial separate test that confirms that *enough can be around* to solve the observational problems.

In the lightest neutralino, supersymmetry provides a natural dark matter candidate. In fits of optimism one could even declare the galactic rotation curve anomalies, etc. as positive experimental evidences for supersymmetry. However, in this chapter the focus will be on two main topics. First, and foremost, I will provide the details on how to calculate the relic abundance of a weakly interacting particle. Since obtaining the correct relic abundance is a somewhat involved calculation, but also an important one, it is useful to have detailed discussion. Second, I will tailor some additional remarks about the calculation to the lightest neutralino, and show how the results constrain *other* supersymmetric particle masses. In general it can be shown that there is an upper limit to superpartner masses due to relic abundance considerations alone. The limit is based on physical necessity and not on arbitrary fine-tuning considerations. This insight is as important as the realization that supersymmetry can cure the "dark matter problem."

The subsequent sections reflect the goals presented in the previous paragraph. Much of the emphasis will be placed in the techniques of calculating the relic abundance. I have attempted to include in this one source all the necessary general relativity, statistical mechanics, and particle physics knowledge needed to follow a precise calculation. I have also included a section which derives an accurate and approximate solution to the Boltzmann equation. These results will then be used to analyze quantitatively how the supersymmetric spectrum is affected by the relic abundance constraint.

2 Solving the Boltzmann equation

The starting point is the Boltzmann equation,

$$\frac{dn}{dt} = -3Hn - \langle \sigma v \rangle (n^2 - n_{eq}^2) \tag{1}$$

where H is the Hubble constant, n is the particle number density in question, n_{eq} is the particle number equilibrium density, and $\langle \sigma v \rangle$ is the thermal averaged cross-section. The Boltzmann equation is simple in form but somewhat subtle to solve. The differentiating parameter is time t, however n and n_{eq} are most easily characterized by temperature. Furthermore, the Hubble constant evolution is best traced by the relative time change in the scale parameter $H = \dot{a}/a$. Only one parameter is independent, and so the first step will be to cast the Boltzmann equation into a purely temperature dependent relation. The two equations that will allow us to do that are the "Friedmann equation" and the "conservation of entropy equation."

The use of Einstein's equation of general relativity is necessary to reveal the explicit time dependence of $\dot{a}/a$. This familiar equation states that

$$R_{\mu\nu} - \frac{1}{2}Rg_{\mu\nu} = 8\pi G_N T_{\mu\nu}. \tag{2}$$

To do explicit calculations the metric and stress-energy must be defined. We use the Robertson-Walker metric

$$ds^2 = dt^2 - a^2(t)\left[dx^2 + dy^2 + dz^2\right] \tag{3}$$

which assumes the universe to be spatially flat, homogeneous, and isotropic. This is equivalent to the metric tensor

$$g_{\mu\nu} = \text{diag}(1, -a^2(t), -a^2(t), -a^2(t)). \tag{4}$$

The general stress-energy consistent with an homogeneous and isotropic universe is

$$T_{\mu\nu} = \text{diag}(\rho, -p, -p, -p). \tag{5}$$

Einstein's equation is of course valid for each $\{\mu\nu\}$, however we only need the 00 component. From the metric tensor it is straightforward[6] to compute the

Ricci tensor component R_{00} and the Ricci scalar R:

$$R_{00} = -3\frac{\ddot{a}}{a} \tag{6}$$

$$R = -6\left(\frac{\ddot{a}}{a} + \frac{\dot{a}^2}{a^2}\right). \tag{7}$$

(Dots indicate time derivative.) The 00 component of the Einstein equation under the flat Robertson-Walker metric is then simply,

$$\frac{\dot{a}}{a} = \frac{\sqrt{\rho}}{\kappa} \quad \text{where} \quad \kappa = \sqrt{\frac{3}{8\pi G_N}}. \tag{8}$$

This equation is often called the Friedmann equation. This will be quite useful to get rid of the scale factor in the Boltzmann equation. The value of ρ on the right hand side of the equation will be calculated later and is conveniently parameterized by its temperature dependence. Thus, the Friedmann equation provides a nice connection between the time dependent relative scale factor (the Hubble constant) and the temperature.

In thermal equilibrium entropy is conserved, and using the first law of thermodynamics we can identify the entropy as $S(T) = (\rho + p)V/T$ up to an irrelevant constant. With $V = a^3$ we define the entropy density $s(T)$ as

$$s(T) = \frac{S(T)}{a^3} = \frac{\rho + p}{T}. \tag{9}$$

The conservation of entropy means that $S(T) = s(T)a^3$ is time independent:

$$\frac{d}{dt}(s(T)a^3) = 0 \implies \dot{T} = -3\frac{\dot{a}}{a}\frac{s(T)}{s'(T)}. \tag{10}$$

(The prime on $s'(T)$ indicates a temperature derivative.) This equation is a direct result of the conservation of entropy in thermal equilibrium and so is called the "conservation of entropy equation". Its utility is relating the time derivative of the temperature ($\dot{T}$) to the scale factor time derivative ($\dot{a}/a$).

We actually have enough information from the above paragraphs to construct the temperature dependent Boltzmann equation. First, we rewrite $dn/dt = (dn/dT)\dot{T}$ and use the "conservation of entropy" equation to replace $\dot{T}$ in favor of $\dot{a}/a$. Then we use the "Friedmann equation" to replace $\dot{a}/a$ in favor of the energy density $\rho(T)$. The result is

$$\frac{dn}{dT} = \frac{s'(T)}{s(T)}\left\{n + \frac{\kappa J(t)}{3\sqrt{\rho(T)}}[n^2 - n_{eq}^2(T)]\right\} \tag{11}$$

280

where $J(T) = \langle \sigma v \rangle(T)$ is the thermal averaged cross-section.

This equation might not look like much progress; however, all non-trivial dependences from $\rho(T)$ and $s(T)$ are easily calculated as functions of temperature. This will be demonstrated below. At sufficiently high temperature (T_H), where all relevant particles are in thermal equilibrium, then we know as a boundary condition that $n(T_H) = n_{eq}(T_H)$. One then need only integrate down to today's temperature $(T \simeq 0)$ to obtain the current number density in the universe $n(0)$. The mass density is then just $\rho_\chi = m_\chi n(0)$ where m_χ is the mass of the relic particle of interest.

Often it is of interest to compare the mass density of our relic particle to the critical density $\rho_c = \kappa^2 H^2$ needed for a flat universe. The relevant formula is

$$\Omega_\chi = \frac{m_\chi n(0)}{\kappa^2 H^2}. \tag{12}$$

For a flat universe the sum of all contributing Ω's (baryons, neutrinos, cosmological constant, neutralinos, etc.) must be equal to 1. If a massive stable particle makes up a significant fraction of the total critical density then it is an interesting cold dark matter candidate. If the calculated critical density is too high, then it is said to "overclose the universe", meaning that the universe became matter dominated too early and it is impossible to reconcile the current mass density and the Hubble constant with the age of the universe.

In order to effectively solve the Boltzmann equation we must have an understanding of the thermodynamic quantities which enter the equation. These are the equilibrium number density $n_{eq}(T)$, the energy density $\rho(T)$, and the entropy density $s(T)$. Since $s(T) = (\rho + p)/T$ we can focus on calculating the pressure $p(T)$ rather than computing $s(T)$ directly. Calculating thermodynamic quantities is standard statistical mechanics and can be found in numerous sources. Here I will merely argue the most salient points that will lead to workable equations.

The density of states in a phase space volume $d^3\vec{x}d^3\vec{k}$ is $1/(2\pi)^3$. Integrating over the volume, the phase space volume density of states is

$$d\xi = \frac{V}{(2\pi)^3}d^3\vec{k} = \frac{V}{(2\pi)^3}(4\pi)k^2 dk, \tag{13}$$

where $k = |\vec{k}|$ here. We are interested to begin with in the number of particles per unit volume (number density). It will be necessary to multiply the density of states times the mean occupation number for a given momentum state $|\vec{k}\rangle$. The mean occupation number is expressed by the Fermi and Bose distribution functions,

$$f_\eta(k,T) = \frac{1}{e^{\sqrt{k^2+m^2}/T} + \eta} \tag{14}$$

where $\eta = -1, 1$ for bosons and fermions respectively. Thus, the total number density integrated over all momentum modes is

$$n_{eq}(T) = \frac{g}{V} \int d\xi f_\eta(k,T) = \frac{g}{2\pi^2} \int_0^\infty dk k^2 f_\eta(k,T) \tag{15}$$

where g is the number of internal spin degrees of freedom.

Similarly, we can calculate the energy density and pressure as

$$\rho(T) = \frac{g}{2\pi^2} \int_0^\infty dk k^2 \sqrt{k^2 + m^2} f_\eta(k,T) \tag{16}$$

$$p(T) = \frac{g}{2\pi^2} \int_0^\infty dk \frac{k^4}{\sqrt{k^2 + m^2}} f_\eta(k,T). \tag{17}$$

Since $s(T) = (\rho(T)+p(T))/T$ we are done. We should keep in mind that $n_{eq}(T)$ in the Boltzmann equation only applies for the one relic particle. However, the $\rho(T)$ and $s(T)$ in the equation are the total energy density and entropy density summed over all particles.

We always want T to be identified with the photon temperature. The thermodynamic quantities technically should be calculated at the temperature of each particle T_i and then the contributions of each particle to the density and pressure should be summed. This creates a subtlety when a stable particle decouples from the photon thermal bath yet still contributes to the mass density, pressure, and entropy of the universe. When a particle decouples its entropy density is separately conserved from that of the photon bath. However, at the decoupling temperature T_d the sum of the two entropies for $T < T_d$ must be equal to the total entropy for $T > T_d$. This is just entropy conservation. Mathematically this is expressed as $s(T_A)T_A^3 = s(T_B)T_B^3$ where $T_B = T_d + \delta$ before decoupling of particle χ and $T = T_\gamma = T_\chi$, and $T_A = T_d - \delta$ is the temperature after decoupling and $T = T_\gamma \neq T_\chi$. However when $\delta \to 0^+$ we can identify $T_A = T$ and $T_B = T_\chi$ to obtain

$$T_i = T \left[\frac{s(T_d - \delta)}{s(T_d + \delta)} \right]^{1/3}. \tag{18}$$

For stable particles which have decoupled then T_i should be substituted into the formulas for $\rho(T)$ and $s(T)$.

When additional particles "distribution decouple" from the thermal bath (W, h, etc. whose density goes to zero as $T \ll m$) the T in Eq. 18 is no longer the current photon temperature but rather the photon temperature before the "distribution decoupling". Using conservation of entropy again at the decoupling boundary, we can find the relation between T_χ and the current photon temperature ($T = T_\gamma$):

$$T_\chi = \lim_{\delta \to 0^+} T \left[\frac{s(T_d - \delta)}{s(T_d + \delta)} \right]^{1/3} \prod_{T_\gamma < m_i < T_d} \left[\frac{s(T_d^i - \delta)}{s(T_d^i + \delta)} \right]^{1/3} \tag{19}$$

$$= T \left[\frac{s(T_\gamma + \delta)}{s(T_d + \delta)} \right]^{1/3} . \tag{20}$$

The thermodynamic quantity integrals are straightforward to calculate numerically for any mass and any temperature. They also can be solved analytically in the limits that $T \gg m$ and $T \ll m$. For example, if $T \gg m$ then the boson energy density is $gT^4 \pi^2/30$, and if $T \ll m$ then energy density decreases rapidly as $\rho \sim T^{3/2} e^{-m/T}$. Doing the same for fermions we can construct the following approximation for $\rho(T)$ and $s(T)$:

$$\rho(T) = \frac{\pi^2}{30} \bar\rho(T) T^4 \tag{21}$$

$$s(T) = \frac{2\pi^2}{45} \bar{s}(T) T^3 \tag{22}$$

where

$$\bar\rho(T) \simeq \sum_{i=bosons} g_i \theta(T_i - m_i) \left(\frac{T_i}{T} \right)^4 + \frac{7}{8} \sum_{i=fermions} g_i \theta(T_i - m_i) \left(\frac{T_i}{T} \right)^4 \tag{23}$$

$$\bar{s}(T) \simeq \sum_{i=bosons} g_i \theta(T_i - m_i) \left(\frac{T_i}{T} \right)^3 + \frac{7}{8} \sum_{i=fermions} g_i \theta(T_i - m_i) \left(\frac{T_i}{T} \right)^3 \tag{24}$$

Again, for most particles $T_i = T$; stable particles decoupled from the photon will have $T_i \neq T$. The equilibrium number density is $n_{eq}(T) \simeq 1.2 g T^3/\pi^2$, $0.9 g T^3/\pi^2$ (bosons, fermions) for $T \gg m$ and decreases rapidly for $T \ll m$.

3 Approximating the relic abundance

Often it is useful to employ approximation techniques to calculate efficiently and reliably the relic abundance[7,8,9,6]. It is based primarily on dividing thermal history into two distinct eras: before freeze-out, and after freeze-out. Freeze-out means that the annihilation rate of the particle, $\Gamma = n_{eq} \langle \sigma v \rangle$, is less than the Hubble expansion rate. Once this occurs the particle can no longer remain in equilibrium.

From previous sections we have the tools to solve for this freeze-out temperature given the condition that $\Gamma = H$. Massive weakly interacting particles are non-relativistic when they freeze-out, and so the equilibrium number density can be well approximated as

$$n_{eq} = g \left(\frac{mT}{2\pi} \right)^{3/2} e^{-m/T}, \tag{25}$$

where g is the number of degrees of freedom of the relic particle (2 for a neutralino). It is convenient to reëxpress all temperatures as the dimensionless variable $x \equiv T/m_\chi$ such that

$$n_{eq} = \frac{gm^3}{(2\pi)^{3/2}} x^{3/2} e^{-1/x}. \tag{26}$$

Furthermore, from our discussion earlier we know that the Hubble constant is

$$H = \frac{\dot{a}}{a} = \sqrt{\frac{8\pi \rho_\chi G}{3}} = \sqrt{N_F} T^2 A \tag{27}$$

$$= \sqrt{N_F} m^2 x^2 A \tag{28}$$

where $A = \sqrt{8\pi^3 G/45}$. The freeze-out condition that $n_{eq}\langle \sigma v \rangle = H$ at $x = x_F$ yields a transcendental equation for x_F,

$$x_F^{-1} = \ln \left[\frac{gm\langle \sigma v \rangle(x_F)}{\sqrt{N_F}\sqrt{x_F} A (2\pi)^{3/2}} \right]. \tag{29}$$

Eq. 29 is only logarithmically dependent on $\langle \sigma v \rangle$ and yields a value of $x_F \simeq 1/20$ quite consistently for massive weakly interacting particles.

After freeze-out the actual number density remains high above the subsequent would-be equilibrium number density. It is therefore appropriate to make the approximation $n^2 - n_{eq}^2 \simeq n^2$. The Boltzmann equation can then be conveniently rewritten as

$$\frac{df}{dx} = \frac{m_\chi}{A\sqrt{N_F}} \langle \sigma v \rangle f^2 \tag{30}$$

where $f = n/T^3$ subject to the boundary condition $f(x_F) = n_{eq}/T_F^3$. To find the current number density we must integrate this equation from x_F (freeze-out) down to $x \simeq 0$ (today):

$$f(0) = \frac{A\sqrt{N_F}}{m_\chi J(x_F) + A\sqrt{N_F} f^{-1}(x_F)} \simeq \frac{A\sqrt{N_F}}{m_\chi J(x_F)} \tag{31}$$

$T_F = m_\chi x_F$	N_F
m_s to m_c	494/8
m_c to m_τ	578/8
m_τ to m_b	606/8
m_b to m_W	690/8
m_W to m_Z	738/8
m_Z to m_t	762/8
above m_t	846/8

Table 1: Degrees of freedom corresponding to freeze-out temperature $T_F = x_F m_\chi \simeq m_\chi/20$.

where

$$J(x_F) = \int_0^{x_F} dx \langle \sigma v \rangle(x). \tag{32}$$

The calculated relic abundance is then simply

$$\rho_\chi = m_\chi T_\chi^3 f(0) = \frac{A\sqrt{N_F}}{J(x_F)} \left(\frac{T_\chi}{T_\gamma}\right)^3 T_\gamma^3. \tag{33}$$

The ratio $T_\chi^3/T_\gamma^3 < 1$ is the photon reheating effect from entropy conservation, which was calculated earlier. Scaling this to the critical density $\rho_c = 3H_0^2/8\pi G$ yields the relic's fraction of critical density. It is customary to define the measured value of the Hubble constant as $H_0 = 100\,h\,\mathrm{km\ s^{-1}\,Mpc^{-1}}$. Then

$$\Omega_\chi h^2 = \frac{A\sqrt{N_F}}{(\rho_c/h^2)J(x_F)} \left(\frac{T_\chi}{T_\gamma}\right)^3 T_\gamma^3. \tag{34}$$

We can further reduce the expression for $\Omega_\chi h^2$ by noting that

$$\left(\frac{T_\chi}{T_\gamma}\right)^3 = \frac{s(T_\gamma)}{s(T_d)} \simeq \frac{N_F(T_\gamma)}{N_F(T_d)} = \frac{2}{N_F}. \tag{35}$$

Using this relation and evaluating all numerical constants we get the convenient form

$$\Omega_\chi h^2 = \frac{1}{\mu^2 \sqrt{N_F} J(x_F)} \tag{36}$$

where $\mu = 1.2 \times 10^5$ GeV. The above formula is an accurate approximation to the Boltzmann equation solution to within about 15%. Table 1 lists the value of N_F for various ranges of the freeze-out temperature $T_F = x_F m_\chi$.

Since a massive weakly interacting particle decouples in the non-relativistic regime, one is able to expand the annihilation cross section in powers of the relative velocity

$$\sigma v = a + \frac{b}{6}v^2 + \cdots \tag{37}$$

The thermal average[10] of this expansion yields

$$\langle \sigma v \rangle = a + \left(b - \frac{3}{2}a\right) x + \cdots. \tag{38}$$

With $x_F \simeq 1/20$ this is a quickly converging expansion. The constants a and b can be evaluated straight-forwardly with knowledge of the squared-matrix element of the annihilation process[11]. Helicity amplitudes also exist for all MSSM processes[12].

The power series expansion of the thermal averaged cross-section breaks down, however, when the relic particle annihilates into a pole (e.g., a $\chi\chi \to Z$ resonance)[13,14]. One then needs to use more careful techniques. Also, the Boltzmann equation becomes more complicated if there are other particles around with mass slightly higher than the stable relic (within a "thermal distance" $|m_\chi - m_i| \lesssim T_F$)[14]. In this case, the close-by particles can help restore thermal equilibrium through coannihilation channels, effectively lowering the relic abundance. These potentially important cases will not be considered further here.

4 Neutralino dark matter

From the previous sections we have found that $\Omega_\chi h^2$ depends on $1/\langle \sigma v \rangle$. By dimensional analysis we can identify $\langle \sigma v \rangle \sim 1/m_{susy}^2$, which implies that $\Omega_\chi h^2 \propto m_{susy}^2$. Since the age of the universe constraints are compatible only with $\Omega_\chi h^2 < 1$ (very conservative requirement) then it should not surprise us that relic abundance considerations put a upper limit to how large m_{susy} can be. We shall see quantitatively the results of these constraints in the following paragraphs. Before going straight to that, a few general comments about the lightest neutralino are useful to review.

The neutralino is a majorana particle, meaning that the particle and conjugate are the same. Therefore when two of them come together to annihilate, Fermi statistics requires each to be in a different helicity state[15]. Thus $\chi\chi \to f\bar{f}$ requires a final state helicity flip for the external fermions, and the total annihilation rate in the s-wave is suppressed by m_f^2/m_W^2 compared to processes that do not require helicity flips. This is often referred to as "p-wave suppression": the p-wave configuration doesn't have this Fermi statistics requirement, although it is suppressed by powers of v^2. However, at higher values of m_χ other channels start opening up, such as W^+W^- and $t\bar{t}$ which do not have this suppression (although they may still have other coupling constant suppressions).

The majorana nature of the neutralino is one important property of supersymmetric dark matter. The other important realization is that numerous other supersymmetric particles play a role in the neutralino annihilation channels. The lightest neutralino, being the LSP, only annihilates into standard model particles when the temperature is near freeze-out. However, many important diagrams of these annihilation processes depend on intermediate supersymmetric states. For example, $\chi\chi \to e^+e^-$ depends not only on an intermediate s-channel Z boson, but also on t-channel slepton exchange. Similarly, $\chi\chi \to q\bar{q}$ depends on t-channel squark exchange, and $\chi\chi \to W^-W^+$ can depend crucially on the chargino spectrum. In general, the neutralino annihilation rate and therefore the LSP relic abundance depends on the entire supersymmetric low energy spectrum – this includes particle content, masses, and mixing angles.

Upon closer inspection most choices of the supersymmetric spectrum ultimately give rise to an LSP relic abundance dependent on only a few parameters in the theory. Relic abundance in the minimal model depends most sensitively on only two parameters, the bino mass and the right-handed slepton mass. As discussed in Martin's introductory article in this volume, this minimal model is described partly by a common scalar mass m_0 at the high scale and a common gaugino mass $m_{1/2}$. As long as m_0 is not much greater than $m_{1/2}$ one typically finds in the low energy spectrum a bino (superpartner of the hypercharge gauge boson) as the lightest supersymmetric particle. One also typically finds the squarks significantly heavier than the sleptons, with the right-handed sleptons being lighter than the left-handed sleptons.

Taking our hints from the minimal model, a useful initial exercise is to declare the LSP as a *pure* bino. Then, the largest contribution to the neutralino annihilation rate will be from t-channel $\tilde{l}_R$ exchange in $\chi\chi \to l^-l^+$. This case has been studied in detail [12], where it was shown that

$$\Omega_\chi h^2 = \frac{\Sigma^2}{M^2 m_\chi^2} \left[\left(1 - \frac{m_\chi^2}{\Sigma}\right)^2 + \frac{m_\chi^4}{\Sigma^2} \right]^{-1} \tag{39}$$

where $M \simeq 1\,\mathrm{TeV}$ and $\Sigma = m_\chi^2 + m_{\tilde{l}_R}^2$. A good approximation to the requirement that $\Omega_\chi h^2 < 1$ yields

$$\frac{m_{\tilde{l}_R}^2}{m_\chi} \lesssim 200\,\text{GeV}. \tag{40}$$

This result has two important consequences. One, it satisfies the general observation that the relic abundance constraint puts an upper limit on at least one superpartner. And second, in this important example of pure bino, both the bino and the right-handed lepton superpartner must have masses below about 200 GeV. *A priori* the upper bound could have been any number. The result is compatible with weak scale supersymmetry $(m_{susy} \sim m_W)$, long thought to be important in electroweak symmetry breaking.

It is not expected that the lightest supersymmetric particle is a pure bino, but rather mostly bino. The coupling of the neutralino to the Z boson depends on the higgsino components. If the LSP is partly Higgsino then efficient annihilations through the Z boson could be possible, thereby reducing the relic abundance for a particular m_χ value. Limits on the superpartner spectrum still exist; they just happen to be at different scales than the 200 GeV we found for the pure bino case.

Our example realistic particle spectrum is a precise formulation of the minimal model with electroweak symmetry breaking enforced. The full 4×4 neutralino mass matrix is diagonalized numerically, and the relic abundance is calculated. The four contour plots [16] of Fig. 1 demonstrate the effects of the relic abundance on the parameter space in the m_0 vs. $m_{1/2}$ plane. The value of the top quark mass for this contour plot is 170 GeV and the sign of the supersymmetric μ parameter is chosen to be negative. In (a) $\tan\beta = 5$, and $A_0/m_0 = 0$, (b) $\tan\beta = 5$ and $A_0/m_0 = -2$, in (c) $\tan\beta = 20$ and $A_0/m_0 = 3$, and in (d) $\tan\beta = 10$, $A_0/m_0 = -2$, and the sign of μ has been changed to positive. The region inside the solid curve is allowed by all constraints. Letter labels are placed outside the solid curve to demonstrate which constraint makes the region phenomenologically unacceptable: A indicates $\Omega_\chi h^2 > 1$; B $B(B \to s\gamma) > 5.4\times10^{-4}$; C $m_{\chi_1^+} > 47\,\text{GeV}$; E electroweak symmetry breaking problems (full one loop effective potential becomes unbounded from below); and, L the LSP becomes charged $(m_{\tilde{l}_R} < m_\chi)$. Current limits on $m_{\chi_1^+}$ and $B(b \to s\gamma)$ are somewhat more restrictive than the limits quoted above, however they will only shrink the allowed parameter space a little bit near B and C. Inside the dotted line corresponds to one particular definition of acceptable finetuning allowed in the electroweak symmetry breaking solutions [16]. It is not important for our further discussion.

The peak or spike in the allowed region for $m_{1/2} \simeq 130\,\text{GeV}$ corresponds to $\chi\chi$ annihilations through a Z pole. Since the LSP is not just a pure bino,

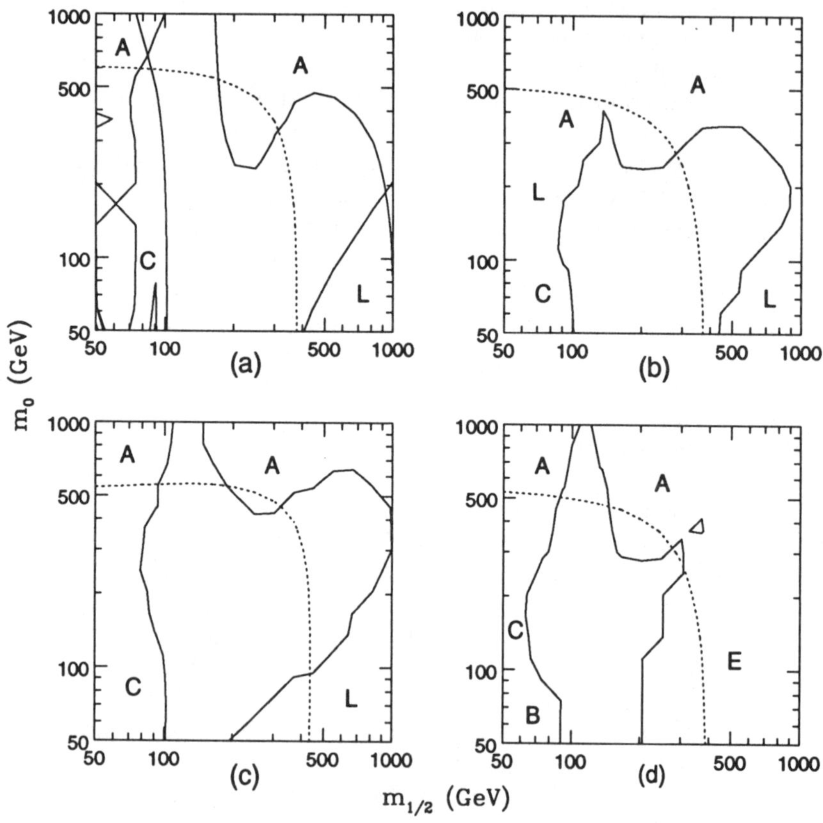

Figure 1: Contours of allowed parameter space. $m_t = 170\,\text{GeV}$ for each frame. In frame (a) $\tan\beta = 5$, $\text{sgn}(\mu) = -$, and $A_0/m_0 = 0$, (b) $\tan\beta = 5$, $\text{sgn}(\mu) = -$ and $A_0/m_0 = -2$, in (c) $\tan\beta = 20$, $\text{sgn}(\mu) = -$ and $A_0/m_0 = 3$, and in (d) $\tan\beta = 10$, $\text{sgn}(\mu) = +$, and $A_0/m_0 = -2$. See the text for explanation of letter labels.

and has some higgsino component to it, the Z pole annihilation rate becomes extremely important in this region. Also, it appears obvious from the figures that the relic abundance constraint has the most effect in the increasing m_0 direction. This is understandable for several reasons. Away from the Z and Higgs poles, the LSP annihilations occur most efficiently through the t-channel scalar diagrams. For a fixed m_χ value, which is roughly equivalent to fixed $m_{1/2}$, the annihilation cross-section decreases rapidly as the scalar mass m_0 increases. Hence, a cutoff in the allowed region must occur as m_0 increases.

Forays into the large $m_{1/2}$ region with m_0 fixed usually end up with other problems, the most common of which is $\tilde{l}_R$ becoming the LSP. This is simply a result of the renormalization group equation dependences on the $U(1)_Y$ couplings. A quick inspection of the renormalization group equations for $m_{\tilde{B}}$ and for $m_{\tilde{l}_R}$ proves that if $m_{1/2} \gg m_0, m_W$, then the right-handed slepton must be lighter than the bino. Ruling out this region of parameter space can be considered a relic abundance constraint, since the dark matter probably cannot be charged[17].

Realization that a weak-scale higgsino could be a legitimate dark matter particle is a rather recent development. One way to obtain an higgsino as the lightest neutralino is to make $|\mu|$ much less than the gaugino parameters in the neutralino mass matrix. A very low value of μ will lead to degenerate multiplets of higgsinos. The charged higgsino and the neutral higgsinos can all coannihilate together with full $SU(2)$ strength, allowing the LSP to stay in thermal contact with the photons more effectively, thereby lowering the relic abundance of the higgsino LSP to an insignificant level. These coannihilation channels are often cited as the reason why higgsinos are not viable dark matter candidates. This claim is true in general, but there are two specific cases that I would like to summarize below that allow the higgsino to be a good dark matter candidate.

Drees *et al.* have pointed out that potentially large one-loop splittings among the higgsinos can render the coannihilations less relevant[18]. Under some conditions with light top squark masses, one-loop corrections to the neutralino mass matrix will split the otherwise degenerate higgsinos. If the mass difference can be more than about 5% of the LSP mass, then the LSP will decouple from the photons alone and not with its other higgsino partners, thereby increasing its relic abundance.

Another possibility[19] relating to a higgsino LSP is to set equal the bino and wino mass to approximately m_Z. Then set the $-\mu$ term to less than m_W. This non-universality among the gauginos and particular choice for the higgsino mass parameter, produces a light higgsino with mass approximately equal to μ, a photino with mass at about m_Z, and the rest of the neutralinos and both charginos with mass above m_W. There are no coannihilation channels to worry about with this higgsino dark matter candidate since no other chargino or neutralino mass is near it. The value of $\tan\beta$ is also required to be near one so that the lightest neutralino is an almost pure symmetric combination of $\tilde{H}_u$ and $\tilde{H}_d$ higgsino states. The exactly symmetric combination does not couple to Z boson (at tree level). The annihilation cross section near $\tan\beta \sim 1$ is proportional to $\cos^2 2\beta$. The relic abundance scales inversely proportional to this, and so the nearly symmetric higgsino in this case is a very good dark matter candidate. Note that there are no t-channel slepton or squark diagrams since higgsinos couple to sfermions proportional to the fermion mass. Because the higgsino mass is below m_W, the top quark final state is kinematically inaccessible, and so the large top Yukawa cannot play a direct role in the higgsino annihilations.

This non-minimal higgsino dark matter candidate described in the previous paragraph was motivated by the $e^+e^-\gamma\gamma$ event reported by the CDF collaboration at Fermilab [20]. The non-minimal parameters [21] which leads to a radiative decay of the second lightest neutralino (photino) into the lightest neutralino (symmetric higgsino) and photon also miraculously yield a model with a good higgsino dark matter candidate.

5 Conclusion

The minimal model bino and the higgsino described above work as dark matter candidates both qualitatively and quantitatively. Nature, of course, might not conform to either of these specific possibilities, but it is straight-forward to catalog the non-minimal possibilities. There are, of course, numerous ways to go beyond what is presented here [22]. Scalar mass universality could be relaxed [23]. One could in fact suppose that R-parity is not exactly conserved and the LSP is not a stable particle. Perhaps one could allow small R-parity violations which create meta-stable LSPs whose lifetimes are greater than the age of the universe to solve the dark matter dilemma. However, it is not enough to just make the lifetime greater than the age of the universe. Remarkably, the measurements of positrons and photons in cosmic rays require that the LSP lifetimes into these particles be many orders of magnitude beyond the age of the universe [24]. In other words, it is not easy to make the LSP a meta-stable dark matter candidate. If R-parity were broken in nature, it appears necessary to look elsewhere for the dark matter candidate.

One common theme exists in non-minimal models which accommodate a supersymmetric solution to the dark matter problem. It is the requirement that $\chi\chi$ not couple to the Z boson. A full strength $SU(2)$ coupling to the Z boson generally allows too efficient LSP annihilation, and therefore too small relic abundance to be relevant to the dark matter problem. This theme can be found in several non-minimal examples such as the symmetric Higgsino[19,18] described above, the $\tilde{Z}$[25], and the sterile neutralino[26] dark matter candidates. Of course, the minimal model also provides a low-strength coupling to the Z naturally with the mostly bino dark matter candidate[27]. The older supersymmetry dark matter literature considered the photino, which also does not couple to the Z, as the primary dark matter candidate[15].

Theoretical ideas about the low-energy superpartner spectrum *always* have dark matter implications. The supersymmetric solution to the dark matter problem is perhaps only superseded by gauge coupling unification in experimentally based indications that nature might be described by softly broken supersymmetry. It is mainly for this reason that much experimental effort is expended on the search for supersymmetric relics[28]. Likewise, theoretical efforts on understanding the origin of LSP stability and the prediction of dark matter properties should continue to enlighten us.

References

1. G. Farrar, P. Fayet, Phys. Lett. B **76**, 575 (1978); S. Dimopoulos, H. Georgi, Nucl. Phys. B **193**, 150 (1981); N. Sakai, T. Yanagida, Nucl. Phys. B **197**, 83 (1982); S. Weinberg, Phys. Rev. D **26**, 287 (1982).
2. T. Falk, K. Olive, M. Srednicki, Phys. Lett. B **339**, 248 (1994).
3. E. Diehl, G.L. Kane, C. Kolda, J.D. Wells, Phys. Rev. D **52**, 4223 (1995).
4. V. Trimble, Ann. Rev. Astron. Astrophys. **25**, 425 (1987); P. Sikivie, Nucl. Phys. Proc. Suppl. **43**, 90 (1995).
5. D. Schramm, M. Turner, astro-ph/9706069.
6. R. Wald, *General Relativity*, Chicago: University of Chicago Press (1984); E. Kolb, M. Turner, *The Early Universe*, Redwood City, USA: Addison-Wesley (1990).
7. B. Lee, S. Weinberg, Phys. Rev. Lett. **39**, 165 (1977); M. Vysotskii, A. Dolgov, Ya. Zeldovich, Pisma Zh.Eksp.Teor.Fiz. **26**, 200 (1977); P. Hut, Phys. Lett. B **69**, 85 (1977).
8. K. Olive, D. Schramm, G. Steigman, Nucl. Phys. B **180**, 497 (1981).
9. J. Ellis, J. Hagelin, D. Nanopoulos, M. Srednicki, Nucl. Phys. B **238**, 453 (1984).
10. M. Srednicki, R. Watkins, K. Olive, Nucl. Phys. B **310**, 693 (1988); P. Gondolo, G. Gelmini, Nucl. Phys. B **360**, 145 (1991).
11. J.D. Wells, hep-ph/9404219; L. Roszkowski, Phys. Rev. D **50**, 4842

(1994).

12. M. Drees, M. Nojiri, Phys. Rev. D **47**, 376 (1993).

13. G. Kane, I. Kani, Nucl. Phys. B **277**, 525 (1986).

14. K. Griest, D. Seckel, Phys. Rev. D **43**, 3191 (1991).

15. H. Goldberg, Phys. Rev. Lett. **50**, 1419 (1983).

16. G.L. Kane, C. Kolda, L. Roszkowski, J.D. Wells, Phys. Rev. D **49**, 6173 (1994).

17. P. Smith, in *Proceedings of the First International Symposium on Sources of Dark Matter in the Universe*, edited by D. Cline, World Scientific, 1995. J. Basdevant, R. Mochkovitch, J. Rich, M. Spiro, A. Vidal-Madjar, Phys. Lett. B **234**, 395 (1990).

18. M. Drees, M. Nojiri, D. Roy, Y. Yamada, Phys. Rev. D **56**, 276 (1997).

19. G.L. Kane, J.D. Wells, Phys. Rev. Lett. **76**, 4458 (1996); K. Freese, M. Kamionkowski, Phys. Rev. D **55**, 1771 (1997).

20. S. Park, "Search for new phenomena at CDF," 10th Topical Workshop on Proton-Antiproton Collider Physics, edited by Rajendran Raja and John Yoh, AIP Press, 1995.

21. S. Ambrosanio, G. Kane, G. Kribs, S. Martin, S. Mrenna, Phys. Rev. Lett. **76**, 3498 (1996); Phys. Rev. D **55**, 1372 (1997).

22. I do not discuss models of low-energy supersymmetry breaking which could have the gravitino as the LSP, or even exotic messenger particles as the cold dark matter. See for example, S. Dimopoulos, G. Giudice, A. Pomarol, Phys. Lett. B **389**, 37 (1996).

23. P. Nath, R. Arnowitt, hep-ph/9701301.

24. J. Ellis, G. Gelmini, J. Lopez, D. Nanopoulos, S. Sakar, Nucl. Phys. B **373**, 399 (1992); G. Kribs, I. Rothstein, Phys. Rev. D **55**, 4435 (1997).

25. A. Gabutti, M. Olechowski, S. Cooper, S. Pokorski, L. Stodolsky, Astropart. Phys. **6**, 1 (1996).

26. B. de Carlos, J.R. Espinosa, hep-ph/9705315.

27. L. Roszkowski, Phys. Lett. B **262**, 59 (1991).

28. G. Jungman, M. Kamionkowski, K. Griest, Phys. Rept. **267**, 195 (1996).

SUPERSYMMETRY AND INFLATION

L. RANDALL

Massachusetts Institute of Technology, Center for Theoretical Physics
77 Massachusetts Avenue, Cambridge, MA 02139

Inflation is a promising solution to many problems of the standard Big-Bang cosmology. Nevertheless, inflationary models have proved less compelling. In this chapter, we discuss why supersymmetry has led to more natural models of inflation. We pay particular attention to multifield models, both with a high and a low Hubble parameter.

1 Introduction

Supersymmetric cosmology is necessarily a speculative subject, since the evolution of the universe is sensitive not only to the observed light degrees of freedom and their superpartners, but also to the as yet undetected heavy particle spectrum. The heavy degrees of freedom only decouple at low temperatures; in the early universe they can be very relevant. In fact, degrees of freedom which are heavy could have been light in the early universe, and vice versa. Furthermore, the vacuum structure of supersymmetric and superstring theories can be very rich and complex; we do not yet know how our vacuum is determined. Nevertheless, despite our ignorance of many aspects of high-energy particle physics, there are certain features of supersymmetric theories which have been shown in recent years to be relevant to cosmology. If the world is supersymmetric, it is clearly important for cosmology, both because of the many new particles which would be present and because of the many flat direction (moduli) fields. These are fields which have no potential in the supersymmetric limit. They can however get a small potential due to supersymmetry breaking, higher dimension operators, or interactions with other fields. These flat directions which only occur naturally in supersymmetric theories can provide large amounts of energy as they will almost certainly not start their evolution from their minimum. Many recent models of inflation are based on this observation, though often in different contexts. In this chapter, we will see how supersymmetric theories might provide more compelling models of inflation. We will consider some examples which demonstrate that supersymmetric theories might provide viable inflaton candidates. Even without knowing the correct particle physics model at high energy, we can identify what might be desirable features of this model if they are to simultaneously account for an earlier epoch of inflation.

In this chapter we will briefly review the motivation for inflation and the requirements for a successful inflationary cosmology[3,4]. We will then discuss

294

in some detail the multifield models of inflation which potentially succeed in meeting the requirements of inflation with little or no fine-tuning. We will discuss several particular models, but cannot attempt a complete enumeration of all models to date-this list is changing very rapidly! We will instead focus on what we think are the additional requirements of the supersymmetric inflation models, their possible predictions, and important questions which remain and attempts to address them.

The standard Big Bang cosmology has many important successes. Most notable are the measured Hubble expansion of the universe, the predictions of the light element abundances from nucleosynthesis in the early universe, and the prediction of the 2.7° microwave background radiation spectrum. More recently, the measurment of the anisotropy in the cosmic microwave background is an indication that theories of structure formation are on the right track. The standard cosmology is simple and successful, but very likely incomplete. As with the standard model of particle physics, the major reason this is believed is that the model as it stands is unnatural, in that it requires very fine-tuned initial conditions.

The shortcomings of the standard cosmology are the problems of the large-scale smoothness of the universe, the spatial flatness problems, the origins of small inhomogeneities, and the potential presence of unwanted relics. Inflationary cosmology successfully resolves the first two problems for a sufficiently long-lived inflationary phase. If inflation involves the correct mass scales and/or parameters, inflation can also lead to the observed density perturbations. For relics which do not get produced late in the universe, inflation can solve the problem of unwanted relics, although it should be noted that the problem of unwanted relics is a serious consideration for most supersymmetric theories, even with an early inflationary epoch.

Most successful inflationary models are based on slow-roll inflation [1,2]. In its earliest implementation it is phenomenologically successful as a model of inflation but requires fine-tuning, either of the potential or of the initial conditions.[a] The requirements on the potential for the inflaton field ϕ for slow roll to be valid are $|V''(\phi)| < 9H^2$ and $|V'M_P/V| < \sqrt{48\pi}$. While these constraints are met, the potential is approximately constant as is the Hubble parameter, $H = \sqrt{8\pi V/3M_P^2}$. During the period of slow-roll, the universe can expand by an exponential factor, the value of which is determined by the time for which the slow-roll conditions are valid. Inflation ends when these conditions cease to apply.

---

[a]There is debate over whether an initial condition should be considered fine-tuned, particularly in an eternal inflationary scenario. We will not discuss this here but refer the reader to Ref. [4].

So far, we see that the important condition for inflation is to have an approximately constant energy density for a finite interval. This in and of itself is not a serious constraint, particularly in a supersymmetric theory for which many light or massless scalars might be present. What makes the construction of inflationary models tricky (or fine-tuned) is that inflation needs to end, so that reheating can produce the known matter content of the universe. With the further requirement that inflation accounts for the density fluctuations in the microwave background (here given in the slow-roll approximation), $V^{3/2}/V'M_P^3 = 5.4 \cdot 10^{-4}$, one is led to the introduction of small parameters. These problems have been reviewed elsewhere [3,4]. In general,

$$\frac{\delta\rho}{\rho} \sim \delta N \sim \frac{H^2}{\dot\phi} \sim \frac{H^3}{V'} \sim \frac{H^3}{m^2\phi} \tag{1}$$

where use has been made of the slow-roll equation of motion $3H\dot\phi = -V'(\phi)$ and N is the number of e-folds. It is important to notice that the Hubble parameter which will give the correct magnitude of density perturbations is determined not only by m, the mass of the inflaton, but also by ϕ, which in this case means the magnitude of the field ϕ during the time density fluctuations are formed. Therefore, although it is conventionally assumed that H during inflation is determined to get the density fluctuations right, this is not necessarily the case. By constructing an inflationary model with a different value of ϕ, one can obtain more than one scale for H which can yield sensible density perturbations.

We note that it is not an essential requirement that density fluctuations formed during inflation account for the observed structure. Other suggestions for producing density perturbations have been given [6]. However, it would certainly be more economical and greatly desirable to have density perturbations taken care of during inflation, since inflation automatically produces fluctuations. This is the assumption which we make here. Moreover, recent papers indicate a substantial vector and tensor contribution to the CMBR implying too low anisotropies on small angular scales in cosmic defect models [7]. Ultimately, measurements should conclusively distinguish inflationary perturbations from others [8]. Current evidence seems to favor inflation [7].

In this chapter, we will give an overview of some recent ideas for implementing inflationary models in the context of supersymmetry. Most of them are based on "hybrid" [9,14] inflationary models, although there have been a few recent suggestions which try to implement slow-roll in the context of a single inflaton field. We will first review the motivation behind multifield inflation models, and discuss some examples. We will then briefly discuss recently suggested single field models.

2 Hybrid Inflation and Supersymmetry

Before introducing supersymmetry into our discussion, let us first consider the potential advantage to multifield inflation models. To do so, we consider a toy model[b] with potential

$$V = (\phi^2 - M^2)^2 + \lambda\phi^2\psi^2 + m^2\psi^2 \tag{2}$$

Notice that the ψ potential is minimized when $\psi = 0$, at which point the ϕ potential is minimized with $\phi = M$, where the potential energy $V = 0$. On the other hand, it is unlikely that ψ starts at its vacuum value. In fact, when the temperature exceeds the ψ mass, the potential is negligible and ψ evolves extremely slowly towards its minimum. Therefore in the early universe, one can reasonably expect large values of ψ. If ψ is greater than $\psi_c = \sqrt{2}M/\sqrt{\lambda}$, and ϕ is sufficiently small, ϕ will rapidly move towards the origin where it will sit leading to a vacuum energy of approximately $V = M^4$ (assuming this dominates the ψ contribution to the energy). This nonzero vacuum energy permits an inflationary stage.

In this model, inflation will end at around the time when the ϕ potential turns over, when $\psi = \psi_c$. In other implementations of "hybrid" inflation[4], it could be that inflation ends when the ψ potential ceases to correspond to a slow-roll situation. We will see an example of this shortly.

The above toy model is fine as a model of inflation. In fact, multifield models seem to resolve very nicely one of the major problems with the standard slow-roll potentials; how can the potential be very flat and then give rise to a rapid end and reheat? Having two fields to control inflation solves this problem beautifully. One field controls the vacuum energy, whereas the other field essentially acts as a "switch" for inflation.

However, there are several obvious questions. First, why should the mass parameters be small? And what sets the mass scales in the first place? Non-supersymmetric field theory cannot in general address this question. Only for a Goldstone boson is there a reason to believe the mass of a scalar is small; in general we would not expect a slow-roll potential for the ψ field.

Why can supersymmetry change this picture? First of all, flat directions are natural in supersymmetric theories. Not only does supersymmetry protect against radiative corrections; supersymmetric theories in general have a large moduli space of flat directions which need not be put in by hand. We should qualify what we mean by flat directions. In general, we are referring to fields

---

[b]I will generally use the GRS[31] conventions for the slow-rolling inflaton field ψ and the field which controls the energy density will be denoted ϕ. The reader should be aware of other conventions existing.

with no potential in the supersymmetric limit, with no other fields away from their vacuum expectation value, and with nonrenormalizable terms neglected. The presence of any of these terms will in fact generate a potential, but one which can in general be consistent with the requirements of inflation if the curvature of the potential is sufficiently small (when compared to the scale set by the vacuum energy).

The other interesting aspect of supersymmetric theories is that in general they require at least one mass scale which is distinct from the Planck scale. This is necessary to account for the supersymmetry breaking scale, which is lower than the Planck scale if the standard low-energy picture of supersymmetry breaking as accounting for stabilization of the electrweak scale is correct. The precise value of this scale is model dependent. In hidden sector models, the new scale will be of order 10^{11}GeV, while in models of supersymmetry breaking based on more direct communication, the supersymmetry breaking scale will be lower. The supersymmetry breaking scale, and in particular, the intermediate scale,seems to be an obvious candidate for application to inflationary models since it is associated with nonvanishing vacuum energy density. In some supersymmetric models, other scales can appear. Notable among these is the Grand Unification (GUT) scale. Many models try to associate directly particle physics models which incorporate a grand unified gauge theory to inflation.

To account for density perturbations, it is clear that there needs to be some small number in the particle physics theory, which could be a ratio of masses. Much of the work on supersymmetric inflation has been focussed on trying to exploit these mass scales to realize the necessary requirements of inflation.

It is useful to divide these efforts into two categories. In one class of theories, the density fluctuations are roughly of order $(M_G/M_p)^2$, whereas the second class of theories only exploits the intermediate scale, and obtains density fluctuations either as (M_I/M_P) or as a result of various parameters which might appear. Here $M_G \approx 10^{16}$GeV is the GUT scale and $M_I \approx \sqrt{M_W M_{Pl}}$ is the intermediate scale of order 10^{11}GeV which determines the soft supersymmetry breaking parameters in a hidden sector scenario for the communication of supersymmetry breaking. Notice that the first class of models involves the scale M_G, which is the VEV of some field and may or may not set the magnitude of the potential energy density. In the second case, M_I, we know it is associated with a vacuum energy density. We will discuss each of these models in turn.

Many of the models we discuss are given in the context of global supersymmetry, although some models are incorporated into a supergravity theory. Before proceeding, we mention two potential problems with supergravity inflaton models; only some of the models presented below address these issues.

298

The first issue is that if W is the superpotential and K is the Kahler potential, the potential takes the form

$$e^{|K|}\left((W_i + K_i W)K_{ij}^{-1}(\bar{W}_{\bar{j}} + K_{\bar{j}}\bar{W}) - 3|W|^2\right) + D - \text{terms} \qquad (3)$$

Since during inflation, the term in parentheses is nonzero if inflation is due to nonvanishing F terms, one will in general find a large potential for any field appearing in the Kahler potential, which will of course include the inflaton. This can destroy the flatness of the inflaton potential and thereby destroy inflation [10,11,12,13,14].

The other potential problem is that in the supergravity theory where the cosmological contant at the desired minimum is cancelled by a constant, one generically finds a deeper minimum out at values of the field larger than M_{Pf}. However, if the superpotential is purely cubic in the fields this problem will not arise[d]. In general though, it is difficult to know how to take this problem, as one is generally treating the potential as a Taylor expansion, and at field values beyond M_p the theory is presumably no longer valid. Furthermore, it is not clear that our world is in the global minimum of the potential.

3 Hybrid Inflation and High Scale Models

The idea of hybrid inflation in supersymmetric theories was studied in a seminal paper by Liddle, Lyth, Stewart, and Wands[14]. Dvali, Shaeffer, and Shafi [DSS][15] pointed out the importance of considering quantum supersymmetry breaking effects during inflation[e], which they then exploited to introduce an interesting model of inflation. However, their model still required arbitrary mass scales. There were subsequent models in which the authors tried to identify an appropriate mass scale. None of these models are perfect, but might nonetheless have the germ of truth.

We first discuss the DSS model. They have the superpotential

$$W = \kappa S\bar{\phi}\phi - \mu^2 S \qquad (4)$$

where S is a singlet and ϕ and $\bar{\phi}$ transform under a GUT group. Notice that when ϕ and $\bar{\phi}$ vanish, the S field is a flat direction. The model also contains an R-symmetry under which S transforms which forbids an S^3 term in the superpotential. This is important as it is essential that S is a flat direction.

---

[c]I thank Paul Langacker for stressing this problem.

[d]I thank Gia Dvali for this comment.

[e]Classical supersymmetry breaking effects during inflation had been pointed out in Ref. [14] and [16].

Let us now consider the potential for this model. We have

$$V(S, \phi, \bar{\phi}) = \kappa^2 |S|^2 (|\bar{\phi}|^2 + |\bar{\phi}|^2) + |\kappa \bar{\phi}\phi - \mu^2|^2 + D - terms \qquad (5)$$

where the D-terms depend on the gauge representation of ϕ and $\bar{\phi}$.

Now at the supersymmetry preserving minimum, the D-term requirement imposes $\phi = \bar{\phi}$, whereas the superpotential imposes $\phi = \bar{\phi} = \mu/\sqrt{\kappa}$ and $S = 0$. However, in the early universe it is very likely that not all fields were at their supersymmetry-preserving minimum. In fact, S might have started off at a value $S > S_c = \mu/\sqrt{\kappa}$, in which case the ϕ potential is minimized at vanishing ϕ and $\bar{\phi}$, where $V = \mu^4$. In other words, this is looking precisely like a hybrid inflation model, where S plays the role of ψ and ϕ and $\bar{\phi}$ play the role of ϕ in our toy model.

Now naively it looks like S is exactly flat, which would be bad, since there would be no potential driving ϕ and inflation would never end. However, this neglects the fact that supersymmetry is broken during inflation! Here the nonzero breaking is due to the nonvanishing F term; however we know this is generally true since inflation relies on nonvanishing vacuum energy. In fact, in general this can be a problem in models with more than one mass scale. Since supersymmetry is broken during inflation, this can be unnatural as in a nonsupersymmetric model. However, in this model, the quantum corrections introduce a potential for the S field which is desired.

The consequence of the nonvanishing F term and the breaking of super-symmetry is that a potential for the S field will be generated through radiative corrections. The one-loop effective potential for S is

$$\Delta V(S) = \Sigma \frac{(-1)^F}{64\pi^2} M_i(S)^4 \log \left(\frac{M_i(S)}{\Lambda} \right)^2 \qquad (6)$$

Here $M_i(S)$ are the S-dependent masses of the fields. This effective inter-action introduces a slope to the S potential. In fact, in this type of model, inflation generally ends when slow-roll ceases to apply, rather than when the "ϕ" potential turns over.

Let us consider this in more detail. Because of the supersymmetry breaking F_S, the ϕ and $\bar{\phi}$ spectrum do not respect supersymmetry. The scalars have mass $\kappa^2 S^2 \pm \kappa \mu^2$, whereas the fermion has mass κS. Substituting these S-dependent masses into the effective potential, one derives the S potential at one-loop to be

$$V_{eff}(S) = \mu^4 + \frac{\kappa^2 \mu^4}{32\pi^2} \left(\log \frac{\kappa^2 S^2}{\Lambda^2} + \frac{3}{2} \right) \qquad (7)$$

This model succeeds as a hybrid inflation model. However, there are some important open questions. First, what is the origin of the scale μ. To get the correct magnitude of density fluctuations, it turns out that μ is of order the GUT scale. This means one might want to tie ϕ to the field with GUT mass and VEV. However, if we take ϕ to be an adjoint, there is too much global symmetry and one obtains too many Goldstone bosons. Interactions which violate this symmetry can also destroy inflation. Alternatively, one can take a GUT group like SU(6) and let ϕ and $\bar{\phi}$ be Higgs fields in the 6 and $\bar{6}$. However, the VEV of this field is likely to be too low for a successful GUT model. So in summary, although the fact that the scale μ is of order the GUT scale is intriguing, in this basic model it is tricky to realize the connection.

Another potential problem when the scale μ is high is that the reheat temperature is likely to be too high, and can cause problems with the gravitino constraint [27]. This is readily seen by a simple estimate assuming instantaneous reheat. Reheat occurs when the Hubble parameter H is of order of the inflaton width Γ. Since $H^2 \sim \rho/M_p^2 \sim T_R^4/M_p^2 \sim \Gamma^2$, we find $T_R \sim \sqrt{\Gamma M_p}$. The bound on the reheat temperature depends on the mass of the gravitino, but is generally of order 10^{10}GeV. If the inflaton decays perturbatively, one expects $\Gamma \sim \frac{\alpha}{4\pi} M_{inf}$. If $M_{inf} \sim M_G$, this is clearly too big. Even if the reheat occurs through higher dimension Planck-suppressed operators, the reheat temperature is probably too high, since it is of order $\sqrt{M_G^3/M_P}$. This is not necessarily an insuperable problem, but it generally requires a more complicated model. In the context of reheat, it should be mentioned that there is still debate over the role of parametric resonance in the decay of the inflaton; however this would generally only increase the reheat temperature. A further point is that the reheat bound assumes only gravitino couplings suppressed by M_{Pl}. If the inflaton decays to particles in the sector in which supersymmetry is broken, the rate for gravitino production can be even larger and the reheat bound even stricter. One can also estimate a reheat bound in gauge-mediated models of supersymmetry breaking [18]. One generally finds even more stringent bounds in this case, since the gravitino is more strongly coupled.

A third problem is that our discussion so far has been in the context of global supersymmetry. Planck-suppressed operators however cannot be neglected, since the Hubble parameter itself is Planck-suppressed. Without some tuning, supergravity corrections can invalidate the conditions for slow-roll.

Subsequent models have tried to address the first type of problem, namely the origin of the scale μ and some have also addressed the third problem. Various authors [17] (see also [19]) suggested D-term inflation. The idea is to generate the scale "μ" though a Fayet-Iliopoulos D-term. They envision a model with an anomalous field content in the low-energy theory, with n_+ fields of charge 1 and n_- fields of charge -1. The model contains a superpotential

$$W = \lambda_A X \phi_+^A \phi_-^A \tag{8}$$

The potential for this model is then

$$V = \lambda^2 |X|^2 \left(|\phi_-|^2 + |\phi_+|^2\right) + \lambda_A^2 |\phi_+ \phi_-|^2 + \frac{g^2}{2} \left(|\phi_+^i|^2 + |\phi_+^A|^2 - |\phi_-^A|^2 + \xi\right)^2 \tag{9}$$

If one looks at the potential along $\phi_+ = 0$, one sees that this potential takes precisely the form required for a successful hybrid inflation model. Here, X plays the role of the ψ field, and ϕ_- (or some linear combination) the role of the ϕ field. One can work out the requirements for sufficiently long slow-roll and for sufficient density fluctuations.

In this model, the vacuum energy density during inflation is given by

$$V = \frac{g^2}{2}\xi^2 \tag{10}$$

This model has the nice feature that because it is D-term inflation, one can control supergravity corrections which would destroy slow-roll. Recall that there is no symmetry to prevent the quadratic terms in the Kahler potential, which, in the presence of a nonzero F-term, will generate an inflaton mass. The situation is better with D-term inflation; however Lyth [20] has pointed out that even in this case, there can be large corrections is there is no symmetry preventing quadratic corrections to the holomorphic function of fields appearing in the gauge kinetic term.

One problem with D-term inflation is that it is difficult to get the scale right. If the parameter ξ arises due to the Green-Schwartz mechanism, it is about $g^2 \mathrm{Tr} Q M_{Pl}^2 / 192\pi^2$, and is probably too big for the scale set by density fluctations. One would need a small parameter to get the correct mass scale.

Some authors have tried to address the question of getting the correct size of the D-term. One possible solution is that the D-term is generated at a scale below the Planck scale. However, it is difficult to see how this can be done without large F-term contributions to the energy as well, destroying the initial motivation for these models.

Matsuda [21] pointed out that the strength of the gauge coupling determines the magnitude of the D-term, and if this coupling is dynamically determined, the D-term at the time of inflation might be of a different size. He shows various ansatzes for the dependence of the coupling on mass scale; unfortunately however these are not motivated by any underlying physics. However, a realistic model of the scale dependence of the coupling would require a solution to the problem of dilaton stabilization.

March-Russell [22] suggests that in a model in which the string scale is reconciled with the GUT scale, one could obtain better numbers for the size of the D-term.

302

Lyth and Riotto[23] also tried to address the discrepancy of scales required for D-term inflation. They point out that the normalization of the magnitude of the D-term which is required to agree with density perturbations depends on the slope of the potential at the end of inflation. In some cases, this slope is given by the one-loop effective potential, so by altering the number of fields coupled to the inflaton, the slope can be increased. However adjusting the slope by this (or any other mechanism) will only buy you at most an order of magnitude (once consistency with the observations on the spectral index n are imposed) if one does not take the coupling g to be small.

Another interesting attempt to tie the μ scale to a physical scale (here a GUT scale) in the problem was made by Dimopoulos, Dvali, and Rattazzi[24]. Their model is based on a quantum corrected moduli space, where the strong interaction scale Λ provides the scale for the overall energy density. The model they give has a gauged SU(2) group with four flavors, so there is a quantum modified moduli space. The superpotential, including the constraint, is

$$W_{eff} = A(DetM - \bar{B}B - \Lambda^4) + S(\mathrm{Tr}M + \frac{g'}{2}\mathrm{Tr}\Sigma^2) + \frac{h}{3}\mathrm{Tr}\Sigma^3 \qquad (11)$$

where the last terms arise due to a tree-level superpotential, and $Q\bar{Q}$ has been replaced by the confined meson field M. If the field Σ is the adjoint field which breaks $SU(5)$ down to the standard model, the scale Λ should be of order the GUT scale, which works well for producing density fluctuations.

Inflation does not involve the Σ field until the end. Initially, the model works as with other hybrid inflation models. S is a flat direction. When S is big, there is a mass, and M sits at zero, so there is nonzero vacuum energy. In this model, inflation ends at a nonzero value of S and Σ and SU(5) is broken. In principle, inflation could also end with Σ zero although the authors of Ref. [24] argue that this is not the case.

There are other models which try to incorporate hybrid inflation into a GUT model. For example, Covi, Mangano, Masiero, and Miele[25] implement hybrid inflation in an SU(5) model with an additional singlet, and an arbitrary parameter μ which they need to take of order the GUT scale. Another model is given by Lazarides, Panagiotakopoulos, and Vlachos[26], who use a nonrenormalizable potential to introduce a slope to the inflaton field (which was given by supersymmetry breaking parameters in other models).

One potential worry with any model based on SU(5) is that most such models do not solve the doublet-triplet splitting problem, and are therefore unrealistic unless a severe fine-tuning is imposed. Since the point of inflationary model building is to eliminate small parameters, this is a less than satisfactory situation.

An even more severe problem in models which really try to tie inflation to an SU(5) GUT of the real world was pointed out by Dvali, Krauss, and Liu [28]. They point out that in SU(5) models with an adjoint which gets a nonzero VEV, there will be two choices of vacua, one in which $SU(4) \times U(1)$ is preserved, and one in which $SU(3) \times SU(2)$ is preserved They parameterize the vacua with three parameters, the overall scale of the symmetry breaking, and two angles (or orbit parameters). When the inflaton field (here we mean the field whose potential generates the vacuum energy density) begins to evolve away from its inflationary value, only one potential minimum is present, namely that corresponding to the bad $SU(4) \times U(1)$ vacuum. They argue further that the field will never make its way to the desired $SU(3) \times SU(2)$ vacuum. Furthermore, whatever vacuum is chosen, the transition happens *after* inflation, so the monopole problem is not solved. So without embellishment, the simplest hybrid inflationary models based on SU(5) GUTS are not successful. These authors suggest possible resolutions which involve somewhat more complicated theories. It is also possible that in a model such as that of Ref. [24] that a noncanonical Kahler potential invalidates the energy argument and that the suitable vacuum is obtained.

Most authors do not address the question of reheat. Lazarides [29] suggests a decay to a second generation neutrino to avoid too big renormalizable couplings. It is hard to think of a natural decay mode without small coupling which can avoid a high reheat temperature and overproduction of gravitinos. An alternative proposal of Dimopoulos and Dvali is that reheat is delayed by rolling along a flat direction [30]; however it is necessary to ensure that no other dangerous perturbations will be produced. It could be that there is some late entropy release which invalidates the gravitino bound; this might be required to solve the Polonyi problem [32] in any case. It is clear that the question of the high reheat temperature should be addressed in these high scale models.

So to summarize, models of hybrid inflation based on a high H scale seem close to working. D-term inflation doesn't quite get the scale right, but is close. Models based on SU(5) generically suffer from the problem outlined in [28] which is unfortunate since it makes the very nice coincidence of scales less useful. However, it is not impossible to make these models work, and further advances might be forthcoming.

4 Low-Scale Models

We now go on to discuss another very promising class of models, which do not introduce a high scale Hubble parameter, but try to implement successful inflation only by assuming the existence of soft-supersymmetry breaking in a hidden sector. These models are intended as illustrations of how the moduli fields can be employed in a hybrid inflation scenario without strong or unnatural assumptions on the particle physics model. With the specific cases that were discussed, one can identify distinguishing characteristics of this class of model, which should be testable.

The goal of Randall, Soljačíc, and Guth (RSG) was to construct models of inflation using *only* the intermediate mass scale M_I and the Planck scale M_p. The aim was to see whether a natural model could be simply constructed which employed moduli fields with soft supersymmetry breaking masses. The essential observation is that the temperature fluctuations observed by COBE do not necessarily require high scale inflation, since the formula for density fluctuations actually depends both on the magnitude of the potential and its slope. Taking the potential quadratic at the time density fluctuations relevant to physical scales are formed, the formula for density fluctuations is

$$\frac{\delta\rho}{\rho} \approx \frac{H^2}{\dot\psi} \approx \frac{H^3}{m^2\psi} \tag{12}$$

where ψ is the slow rolling field. Now if the energy density during inflation is M^4, the Hubble parameter is of order M^2/M_p. On the other hand, if the ψ mass arises from hidden sector supersymmetry breaking, it is $m \approx M_I^2/M_{Pl}$. The assumption in this class of models is that $M \sim M_I$, in which case

$$\frac{\delta\rho}{\rho} \approx \frac{H}{\psi} \tag{13}$$

From this equation, it is clear that the magnitude of density fluctuations depends on the value of ψ when inflation ends, which for hybrid inflation models is essentially ψ_c.

In Ref. [31], two types of potentials were considered. The first class of models assumed the fields were coupled through a higher dimension operator derived from the superpotential

$$W = \frac{\phi^2\psi^2}{2M'} \tag{14}$$

where M' is a relevant physical mass scale. Perhaps the most natural possibility is M_p. However, it is conceivable there are heavy particle exchanges so that M' can be identified with the GUT scale M_G or M_I.

The other class of model assumed there was a potential coupling the two fields involved in hybrid inflation of the form

$$V = \frac{\lambda}{4}\psi^2\phi^2 \tag{15}$$

Such a coupling could arise for example if the superpotential coupled together three fields $W = \chi\phi\psi$, where $\chi = 0$ during inflation. This is in fact something which happens quite naturally, even in the context of the MSSM. For example, if $\psi = \bar{u}d\bar{d}$ and $\phi = H_u H_d$, $W = \lambda_u Q_u H_u \bar{u}$, one realizes this situation. In fact, nonstandard GUT models [31,33] can realize this potential in a way consistent with the inflationary constraints. In this model, the magnitude of density fluctuations will be set by a Yukawa coupling; it is important to recognize that this might well be significantly less than unity.

The complete specification of the model requires the potential for the ϕ and ψ fields (apart from their mutual coupling). Note that these potentials arise due to soft supersymmetry breaking and therefore should be characterized by potentials of the form $M_I^4 g(\phi/M_p)$ where g is a function with a Taylor expansion with coefficients of order unity. To realize the hybrid inflationary scenario, the potential for ϕ is taken as $V = M^4 \cos^2(\phi/\sqrt{2}f)$ and the potential for ψ is taken as $\frac{1}{2}m_\psi^2\psi^2$. To be consistent with the requirement that these are moduli fields with a potential generated by soft supersymmetry breaking, we would want to find $M \sim M_I$ and $m_\psi \sim M_I^2/M_p$. There has been much confusion over the very specific-looking form taken for the ϕ potential. Indeed, all that is relevant to inflation are the first two terms in the Taylor expansion! Only when inflation ends and ϕ moves from zero are the other terms relevant. This is simply a compact way of writing a function which has the correct negative curvature at the origin and zero vacuum energy for the true vacuum. It is interesting however that exactly such a potential could be produced by a nontrivially coupled pseudo-Goldstone mode[f]. However, this precise form of the potential is not at all essential, so the field ϕ can be any moduli field with negative mass squared and a supersymmetry-breaking source for its potential.

This model realizes very nicely the hybrid inflation scenario. Depending on the form of the soft-supersymmetry breaking potential and the couplings between moduli fields, it is very likely one can find suitable candidates for inflation. The major distinguishing characteristic of this type of model is that the ϕ field is light. Therefore, the dynamics controlling the end of inflation is very different. In many other models, the ϕ mass is large, so it very quickly rolls to its true minimum once inflation has stopped. In the RSG models, the field ϕ spends more time moving primarily due to de Sitter fluctuations, subsequent to which it rolls classically. Because of the initial motion when the field is moving relatively slowly, there is a spike in the density fluctuation spectrum. This spike can be interesting (or dangerous) in that it will lead to more structure on small scales. However, it is too large to be present in observed density fluctuations on scales from about 1 Mpc to 10^4 Mpc, which gives a constraint on how

---

[f] I thank Gia Dvali for sharing this observation of Lawrence Krauss and himself.

quickly inflation must end. Including this constraint as well as the constraint from density fluctuations, one finds that the parameters of this model are such that it works rather well, with mild tuning depending on the particular model (numbers as small as 0.01 might be required; see [31] for details). That is, the original goal, to motivate the parameters by supersymmetry breaking scales, can be reasonably well accomodated.

It should be noted that the derivation of the parameters of the spike in the density perturbation spectrum is subtle. In Ref. [31], the calculation was based on using the Fokker-Planck equation to establish the ϕ mean ϕ distribution and the time delay given by a fluctuation in the ϕ field. Garcia-Bellido, Linde, and Wands[34] objected to the calculational method of[31] and instead calculated the fluctuations associated with each element of the ensemble assuming it was classical. However, it can be shown[35] that the method used is not valid, though the original[31] calculation needed to be improved to account for deviation from slow-roll and for a more exact calculation of the fluctuations for a massive field.

Stewart [36,37] has also constructed low-scale models for which the small tuning required to get a sufficiently flat potential is not required and for which the spike will have different properties. He points out that once quantum corrections are incorporated, there can be a special point (or more than one) where the potential is particularly flat. Initial conditions are probably different for this class of model; one relies on entering a phase of eternal inflation from which one will enter the desired hybrid inflationary phase.

In summary, the low-scale inflation models can have quite distinctive features, and do not have the problems associated with introducing a high scale in a particle physics context. However, they might involve a small amount of fine-tuning; on the other hand they might also involve a small parameter which is present. The spike is generically a test of the models; however if the field is rolling quickly at the phase transition, as is true for Ref. [37], this might be lessened; futher work is needed for these models to establish the detailed form of the spike. In general, the low-scale models are well motivated and worthy of further investigation.

Before discussing further models, it is worth noting a distinguishing feature of many hybrid inflation models, those with a mass for the inflaton (like the GRS models), namely the fact that the index n is generally bigger than 1. Furthermore, by measuring the ratio of tensor to scalar perturbations, one can in principle (because it is a difficult measurement) distinguish high and low scale models. The deviation of the index n from 1 measures the scale dependence of density fluctuatons. It can be determined from the potential at the time the relevant perturbation leaves the horizon from the formula

$$n = 1 - 3 \left(\frac{V'}{V}\right)^2 + 2\frac{V''}{V} \tag{16}$$

whereas R, the ratio of tensor to scalar perturbations, is given by

$$R \approx 6 \left(\frac{V'}{V}\right)^2 \tag{17}$$

where in these equations we have set M_p to unity. One can obtain interesting qualitative information from these formulae. First consider the quantity R. If we can approximate V' by a mass term near the end of inflaton, we have $V'/V \sim m^2\psi M_p/H^2 M_p^2$. So if $H \sim m$, which is often the case, we find that R is negligible unless $\psi \sim M_p$. As we have argued, models with low H can achieve adequate density perturbations if ψ is small (much less than M_p) at the end of inflation. We conclude that these type of models will always have neglibile R. For other models, with ψ closer to M_p, it is a detailed question whether R can be measurable.

Notice also that when V'/V is neglibible, the sign of the mass squared term at the end of inflation determines whether n is bigger, or less than unity. So models for which the inflaton field rolling towards the origin will have n bigger than 1. The RSG models are of this type, as are hybrid inflation models with a mass term determining the evolution of the inflaton.

It should be noted that there are many models where the potential for the inflaton is not determined by a mass term. An interesting example of hybrid inflation which he dubbed "Mutated Hybrid Inflation" for which the index n is less than 1 was given by Stewart [38]. He considers a toy model where inflation occurs along a nontrivial trajectory in field space. The net result is that along this trajectory, the potential can be written as a polynomial function of the inverse field, and the index n can be shown to be less than 1. Generalizations of this idea and other mechanisms for producing an index less than 1, again in toy models, was given in Ref. [39].

5 Single Field Models

Aside from the many models of hybrid inflation based on supersymmetry, there are a couple of single field inflationary models worthy of note. By single field, we do not mean there is only one field in the potential, but that the inflationary dynamics can be viewed in terms of a single field (as opposed to hybrid inflation models). Garcia-Bellido [40] observed that the potential given by an $N = 2$ SU(2) gauge theory with supersymmetry breaking[41] incorporated takes a form which looks remarkably like a slow-roll potential along a particular trajectory in field space. However, the scale of supersymmetry breaking which is required has no particle physics motivation, so it is not yet clear if this can be tied to a particle physics theory of our world.

Another single field model is that of Adams, Ross, and Sarkar[42]. They are interested in the problem of the large quadratic terms which can be present in supergravity theories. They argue that there can be special points where the quadratic terms vanish, and these can be quasi-fixed points in the evolution of the field.

Another paper which addresses the issue of large supergravity corrections is by Gaillard, Murayama, and Olive[43]. They observe that at tree-level, the mass term for the inflaton which occurs generically in supergravity theories is absent if there is a Heisenberg symmetry. Although there is no symmetry reason, it is claimed that gravitational interactions preserve the symmetry (based on a one-loop calculation), so that the potential can be calculated from gauge and superpotential interactions. A more complete realization of this scenario could be interesting.

Stewart[16] also addressed the issue of large supergravity corrections to the inflaton mass. He identified conditions, which when imposed on the super-potential, guarantee the absence of such corrections. These conditions are $W = W_\psi = \phi = 0$ (in the GRS naming convention) during inflation, as well as some conditions on the Kahler potential. He argues that such potentials might arise naturally in superstring theories.

In fact, Copeland, Liddle, Lyth, Stewart, and Wands[14] had initially pointed out that supergravity corrections to the mass can cancel if there is a minimal Kahler term. Linde and Riotto[44] make the assumption that nonminimal terms which would destroy this cancellation are small, and then consider the model with both one-loop and gravitational effects taken into account.

In summary, there are currently many ideas on how to use the naturalness property of supersymmetry to provide candidates for inflaton fields. There are some clever ideas involved in high scale models; however the D-term models generally give too large density fluctuations while the GUT models often lead to the wrong vacuum following inflation. These models might however be incorporated into more complete and realistic models in the future. The low scale inflation models have the advantage that they require no new mass scales aside from that which was already required to give supersymmetry breaking parameters of order the weak scale in a hidden sector scenario. They provide an interesting signature of a spike in the spectrum so they should be subject to experimental verification in the future. Other objections might include the fact that this inflation is relatively late, so this might mean some prior nonstandard evolution (like a previous inflationary phase) is required. These models require mild tunings; presumably even this is not necessary if one will compromise with more complicated scenarios. It is intriguing that new models of particle physics might also lead to new potentials that could provide slow-roll. Exact superpotentials in the strong interaction regime often take nonpolynomial forms which would not have been anticipated on the basis of weakly coupled renormalizable field theories. Other models which fall outside the range of the supersymmetry-motivated field theories we have considered here include dilaton-based inflation [45] and string-theory-motivated-domain walls as the seed for inflation [46].

Although there are many ideas, it should be remembered that there are many requirements for a good inflationary model. Given the indirectness of many cosmological constraints, it is remarkable how constrained models are. Requirements include a sufficiently long period of inflation, a mechanism for ending inflation sufficiently quickly to reheat to temperatures higher than the weak scale (this might be too stringent but in alternative scenarios one needs a mechanism for baryogenesis), a reheat temperature sufficiently low not to over-produce gravitinos, consistency with a spectral index which does not deviate by more than 20% from 1 (the exact constraint is subject to interpretation), and hopefully no ad hoc scales or small numbers. There are only a few models which meet all these criteria. And we have not even addressed the many issues of how inflation fits into a more complete picture of the cosmology of the early universe which includes baryogenesis [47] and a solution to the Polonyi problem [32]. So despite the many recent advances, the field still remains fertile, and it would not be surprising to see new and more compelling models of inflation in the future.

Acknowledgements

I am very grateful to Ewan Stewart and Chris Kolda for discussions and their comments on the manuscript. I also thank Gia Dvali, Marc Kamionkowski, John March-Russell, and Riccardo Rattazzi for their comments. I also thank Princeton University and the Institute for Advanced Study for their hospitality while this work was completed. This work is supported in part by funds provided by the U.S. Department of Energy (D.O.E.) uncer cooperative agreement #DF-FC02-94ER40818.

References

1. A. Albrecht and P. J. Steinhardt, *Phys. Rev. Lett.* **48** (1982) 1220.

2. A. D. Linde, *Phys. Lett.* **B108** (1982) 389.

3. M. S. Turner, Inflation after COBE: Lectures on Inflationary Cosmology, astro-ph/9304012, Boulder TASI 92: 165-234, Cargese Summer School 1992:23 1-398, and references therein.

4. A. D. Linde, Particle physics and inflationary cosmology (harwood Academic, Basel, 1990).

5. M. Sasaki and E. D. Stewart, *Prog. Theor. Phys.* 95, 71 (1996).

6. R. Brandenberger, Topological Defects and the Formation of Structure in the Universe, astro/ph-9604033, Talk given at Pacific Conference on Gravitation and Cosmology, Seoul, Korea, Feb, 1996.

7. B. Allen, R. R. Caldwell, S. Dodelson, L. Knox, E.P.S. Shellard, and A. Stebbins, astro-ph/9704160, U. Pen, U. Seljak, and N. Turok, astro-ph/9704165, N. Turok, U. Pen, and U. Seljak, astro-ph/9706250.

8. A. Albrecht, How to Falsify Scenarios with Primordial Fluctuations from Inflation, astro-ph/9612017, Presented at Conference on Critical Dialogs in Cosmology, Princeton, NJ, 24, June, 1996.

9. A. D. Linde, *Phys. Lett.* **B259** (1991) 38, A. R. Liddle and D. H. Lyth, *Phys. Rep.* D 52 (1995) 6789, A. D. Linde, *Phys. Rev.* **D49** (1994) 748;

10. M. Dine, W. Fischler, and D. Nemeschansky, *Phys. Lett.* **B136** (1984) 169;

11. G. Coughlan, R. Holman, P. Ramond, and G. Ross, *Phys. Lett.* **B140** (1984) 44.

12. M. Dine, L. Randall, and S. Thomas, hep-ph/9503303, *Phys. Rev. Lett.* **75** (1995) 398.

13. R. Barbieri, S. Ferrara, and C. A. Savoy, *Phys. Lett.* **B119** (1982) 343.

14. E. J. Copeland, A.R. Liddle, D.H. Lyth, E.D. Stewart, and D. Wands, *Phys. Rev.* **D49** (1994) 6410, E. Stewart, *Phys. Lett.* **B345** (1995) 414.

15. G. Dvali, Q. Shafi, and R. Schaefer, hep-ph/9406319 *Phys. Rev. Lett.* **73** (1994) 1886.

16. E. Stewart, *Phys. Rev.* **D51** (1995) 6847.

17. E. Halyo, *Phys. Lett.* **B387** (1996) 43; P. Binetruy and G. Dvali, *Phys. Lett.* **B388** (1996) 241.

18. A. de Gouvea, Takeo Moroi, H. Murayama, *Phys. Rev.* **D56** (1997) 1281.

19. J. A. Casas and C. Munoz, *Phys. Lett.* **B216** (1989) 37; J. A. Casas, J. M. Moreno, C. Munoz, and M. Quiros, *Nucl. Phys.* **B328** (1989) 272; E. D. Stewart, *Phys. Rev.* **D51** (1995) 6847; M. Dine, L. Randall, and S. Thomas, *Nucl. Phys.* **B458** (1996) 291.

20. D. H. Lyth, hep-ph/9710347.

21. T. Matsuda, hep-ph/9705448, May, 1997.

22. John March-Russell, private communication.

23. D. H. Lyth and A. Riotto, hep-ph/9707273, June, 1997.

24. S. Dimopoulos, G. Dvali, and R. Rattazzi, hep-ph/9705348, May, 1997.

25. L. Covi, G. Mangano, A. Masiero, and G. Miele, hep-ph/9707405, July, 1997.

26. G. Lazarides, C. Panagiotakopoulos, and N. D. Vlachos, *Phys. Rev.* **D54** (1996) 54.

27. T. Moroi, Effects of the Gravitino on the Inflationary Universe, Ph. D. thesis, and refs. therein.

28. G. Dvali, L. Krauss, and H. Liu, hep-ph/9707456, July, 97.

29. G. Lazarides, R. K. Schaefer, and Q. Shafi, *Phys. Rev.* **D56** (1997) 1324.

30. Gia Dvali, private communication.

31. L. Randall, M. Soljačić, A. Guth, *Nucl. Phys.* **B472** (1996) 377.

32. T. Banks, D. Kaplan, and A. Nelson, *Phys. Rev.* **D49** (1994) 779; L. Randall, S. Thomas, *Nucl. Phys.* **B449** (1995) 229; D. H. Lyth and E. D. Stewart, *Phys. Rev. Lett.* **75** (1995) 201.

33. Z. Berezhiani, C. Csáki, and L. Randall, *Nucl. Phys.* **B444** (1995) 444.

34. J. Garcia-Bellido, A. Linde, and D. Wands *Phys. Rev.* **D54** (1996) 6040.

35. A. Guth, L. Randall, S. Su, in progress.

36. E. Stewart, *Phys. Lett.* **B391** (1997) 34.

37. E. Stewart, *Phys. Rev.* **D56** (1997) 2019.

38. E. Stewart, *Phys. Lett.* **B345** (1995) 414.

39. D. Lyth and E. Stewart, *Phys. Rev.* **D54** (1996) 7186.

40. J. Garcia-Bellido, hep-th/9707059.

41. L. Alvarez-Gaume, J. Distler, C. Kounnas, M. Marino, *Int. J. Mod. Phys. A* 11 (1996) 4745, hep-th/9604004.

42. J. Adams, G. G. Ross, S. Sarkar, *Phys. Lett.* **B391** (1997) 271.

43. M. Gaillard, H. Murayama, and K. Olive, *Phys. Lett.* **B355** (1995) 71.

44. A. Linde and A. Riotto, *Phys. Rev.* **D56** (1997) 1841.

45. R. Brustein, hep-th/9506045, Contribution to International Conference on Unified Symmetry in the Small and in the Large, Coral Gables, FL, Feb, 95 and references therein.

46. T. Banks, M. Berkooz, S. Shenker, G. Moore, P. Steinhardt, *Phys. Rev.* **D52** (1995) 3548.

47. A. Affleck and M. Dine *Nucl. Phys.* **B249** (1985) 361, M. Dine, L. Randall, and S. Thomas *Nucl. Phys.* **B456** (1996) 291 for example.

Z' Physics and Supersymmetry

M. Cvetič and P. Langacker

Department of Physics and Astronomy, University of Pennsylvania,
Philadelphia, PA 19104-6396

We review the status of heavy neutral gauge bosons, Z', with emphasis on constraints that arise in supersymmetric models, especially those motivated from superstring compactifications. After elaborating on the status and (lack of) predictive power for general models with an additional Z', we concentrate on motivations and successes for Z' physics in supersymmetric theories in general and in a class of superstring models in particular. We review phenomenologically viable scenarios with the Z' mass in the electroweak or in the intermediate scale region.

1 Introduction

The existence of heavy neutral (Z') vector gauge bosons are a feature of many extensions of the standard model (SM). In particular, one (or more) additional $U(1)'$ gauge factors provide one of the simplest extensions of the SM. Additional Z' gauge bosons appear in grand unified theories (GUT's), superstring compactifications, and other classes of models.

However, for those models which do not incorporate constraints due to supersymmetry, supergravity, or superstring theory, the masses of additional gauge bosons are generally free parameters which can range from the electroweak scale ($\mathcal{O}(1)$ TeV) to the Planck scale M_{Pl}. In addition, models with extended gauge symmetry generically contain exotic particles, e.g., new heavy quarks or leptons which are non-chiral under $SU(2)_L$, or new SM singlets, with masses that are tied to those of the new Z's, but are otherwise unconstrained. Thus, such models lack predictive power for Z' physics, and much of the phenomenological work in this context is of the "searching under the lamp-post" variety. In particular, there is no strong motivation to think that an extra Z' would actually be light enough to be observed at future colliders. Even within ordinary GUT's, there is no robust prediction for the mass of a Z'. (There *are*, however, concrete predictions for the relative couplings of the ordinary and exotic particles to the Z' in each ordinary or supersymmetric GUT.)

On the other hand $N = 1$ supersymmetric models typically provide more constraints. First, the scalar potential is determined by the superpotential, Kähler potential, D-terms, and soft supersymmetry breaking terms, and is generally more restrictive. In particular, the $U(1)'$ D-term, which is typically of the order of $M_{Z'}^2$, breaks supersymmetry, so the $U(1)'$ breaking scale should not be much larger than the electroweak scale, i.e., no more than a TeV or so[1,3].

The exception is when the $U(1)'$ breaking takes place along a D-flat direction[a]. In this case the breaking can take place at an intermediate scale[b]. Secondly, superstring models often imply constraints on the superpotential, such as the absence of mass terms and the existence of large (order one) Yukawa couplings, which can determine the mechanism and scale of $U(1)'$ breaking.

There are two promising classes of theoretical structures in which the minimal supersymmetric standard model (MSSM) and its extensions are likely to be embedded. One are superstring models which compactify directly to a gauge group consisting of the SM and possibly additional $U(1)'$ factors[4,5,6,7]. Some of the superstring models are based on $E_6 \times E_8$ Calabi-Yau compactifications of the heterotic string, in which E_6 is already broken at the string scale, e.g., to the SM gauge group and an additional $U(1)'$ factor, via the Wilson line (Hosotani) mechanism. Such Z' models, primarily employing the quantum numbers associated with a particular E_6 breaking pattern, were addressed in [8,9,10]. The Z' phenomenology of superstring models with a true GUT symmetry at the string scale has not been explored. Most of the recent work on supersymmetric Z's has been from the point of view of the first type, non-GUT models, and it will be emphasized in this contribution.

In the models studied in [4,5,6,7,9] the $U(1)'$ breaking is radiative. This is analogous to radiative electroweak breaking, in which the (running) Higgs mass-squared terms are positive at the Planck scale, but one of them is driven negative at a low or intermediate scale by the large Yukawa coupling associated with the top quark. Similarly, in radiative $U(1)'$ breaking one or more SM scalars that carry $U(1)'$ charges have positive mass-squared terms at the Planck scale. However, if these scalars have large Yukawa couplings to exotic multiplets or to Higgs doublets their mass-squared terms can be driven negative at lower scales so that the scalar develops a vacuum expectation value and breaks the $U(1)'$ symmetry. Typically, the initial (Planck scale) values of the Higgs and SM scalar mass-squared terms are comparable and given by the scale of soft supersymmetry breaking. In a class of models in which the magnitudes of the Yukawa couplings in the superpotential are motivated to be of $\mathcal{O}(1)$, the radiative breaking can occur, and it depends on the exotic particle content and on the boundary conditions for the soft supersymmetry breaking terms at the large scale. The symmetry breaking can either take place at the electroweak[4,5,6,9] or (when the minimum occurs along a D-flat direction) at

---

[a]It was claimed in [3] that the breaking *must* be along a D-flat direction. In fact, that is only necessary when the $U(1)'$ breaking scale is much larger than the supersymmetry breaking (i.e., TeV) scale.

[b]The actual minimum is typically shifted slightly away from the D-flat direction by soft supersymmetry breaking terms, leading to finite shifts in the sparticle and Higgs masses of order of the soft breaking, even for a large $U(1)'$ breaking scale.

an intermediate scale [11,4,5,7]. In the former case, the Z' mass is comparable to the ordinary Z and to the scale of supersymmetry breaking, and is almost certainly less than a TeV [5,6].

A class of superstring compactifications, based on free fermionic constructions [12,13,14,15,16] [c], contains all of these ingredients, including the general particle content and gauge group of the MSSM, and in general additional non-anomalous $U(1)'$ gauge symmetry factors and vector pairs of exotic chiral supermultiplets [d]. In this class of models there are no bilinear (mass) terms in the superpotential, and the non-vanishing trilinear (Yukawa) terms are of order one. These conditions usually suffice to require and allow radiative $U(1)'$ breaking, respectively [5]. Thus, supersymmetric models (with constraints on couplings from a class of superstring models) *have predictive power* for the masses of Z' and the accompanying exotic particles. In that sense they are superior to general models with extended gauge structures. For these reasons, we would like to advocate that within supersymmetry (and superstring theory constraints), perhaps the best motivated physics for future experiments, next to the Higgs and sparticle searches, are searches for Z' and the accompanying exotic particles. On the other hand, at present we have little theoretical control over the type of dynamically preferred superstring compactification, or of the soft supersymmetry breaking parameters, i.e., the origin of supersymmetry breaking in superstring models; it is therefore hard to make general predictions for the specific $U(1)'$ charges and other details of the models.

In supersymmetric models additional $U(1)'$s would have important theoretical implications. For example, an extra $U(1)'$ breaking at the electroweak scale in a supersymmetric extension of the SM could solve the μ problem [20] by forbidding an elementary μ-parameter, but inducing an effective μ of the order of the electroweak scale by the $U(1)'$ breaking [4,5,6]. There are also implications for baryogenesis; an extra $U(1)'$ might be useful for electroweak baryogenesis, with changes in the scalar sector of such models modifying the nature of the electroweak transition [6,21], or with cosmic strings providing the needed "out of equilibrium" ingredient [22]. However, alternative models of baryogenesis, in which a lepton asymmetry is first created by the out of equilibrium decay of a heavy Majorana neutrino, and then converted to a baryon asymmetry by electroweak effects [23], is forbidden by a light Z' unless the heavy neutrino carries no $U(1)'$ charge [24]. Other implications include the Higgs, scalar partner, chargino,

---

[c] Certain models based on orbifold constructions with Wilson lines [17] also possess the gauge structure and the particle content of the MSSM.

[d] Related classes of models based on higher level Kač-Moody algebra constructions yield [15,18,19] GUT gauge symmetry with adjoint representations. These models have not yet been explored for Z' physics.

and neutralino masses and couplings, and thus the properties of the LSP [25]. Intermediate scale breaking models may, through higher-dimensional operators in the superpotential, yield an attractive mechanism for generating hierarchies of small masses for quarks and charged leptons, and offer possibilities for generating naturally small Dirac neutrino masses without invoking a seesaw [7]. Small Majorana neutrino masses (comparable to the small Dirac neutrino masses), which allow oscillations of ordinary into sterile neutrinos, or large Majorana masses, generating seesaws, are also possible. Non-renormalizable terms may also provide a mechanism for obtaining the μ-parameter.

This review is organized as follows. In Section 2 we describe representative Z' models and introduce notation. In Section 3, we review the features of Z' models with or without supersymmetry and with or without GUT embedding. In Section 4 we review the properties of non-GUT supersymmetric models with additional $U(1)'$ factors, address additional constraints and inputs arising in a class of superstring models, and discuss both the electroweak and the intermediate scale scenarios for $U(1)'$ symmetry breaking.

2 Z' Physics

2.1 Overview of Z' Models

The most commonly studied Z' couplings are GUT, left-right, and string-motivated. In each case, the Z' couples to $g'Q$, with g' the $U(1)'$ gauge coupling and Q the charge. In the GUT-motivated cases $g = \sqrt{5/3}\sin\theta_W G\lambda_g^{1/2}$, with $G \equiv \sqrt{g_L^2 + g_Y^2} = g_L/\cos\theta_W$, where g_L, g_Y are the gauge couplings of $SU(2)_L$ and $U(1)_Y$, and λ_g depends on the symmetry breaking pattern [26]. If the GUT group breaks directly to $SU(3)_C \times SU(2)_L \times U(1)_Y \times U(1)'$, then $\lambda_g = 1$. Standard examples include: (i) Z_χ, which occurs in $SO_{10} \to SU(5) \times U(1)_\chi$; (ii) Z_ψ, from $E_6 \to SO_{10} \times U(1)_\psi$; (iii) $Z_\eta = \sqrt{3/8}Z_\chi - \sqrt{5/8}Z_\psi$, which occurs in Calabi-Yau compactifications of the heterotic string if E_6 breaks directly to a rank 5 group [27] via the Wilson line (Hosotani) mechanism; (iv) the general E_6 boson $Z(\theta_{E_6}) = \cos\theta_{E_6} Z_\chi + \sin\theta_{E_6} Z_\psi$, where $0 \leq \theta_{E_6} < \pi$ is a mixing angle; (v) Z_{LR} occurs in left-right symmetric (LR) models [28], which contain a right-handed charged boson as well as an additional neutral boson. The ratio $\kappa = g_R/g_L$ of the gauge couplings $g_{L,R}$ for $SU(2)_{L,R}$, respectively, parametrizes the whole class of models. $\kappa = 1$ corresponds to manifest or pseudo-manifest left-right symmetry. In this case $\lambda_g = 1$ by construction and $\kappa > 0.55$ for consistency [29,30]. The $U(1)'$ charges for specific models are given in [2].

There are numerous other Z' models. In particular, the superstring models based on the free fermionic constructions yield couplings for the additional Z''s which do not in general correspond to the standard GUT-type models.

We also mention some types of models that have been inspired on phenomenological grounds. The leptophobic models, in which the leptons do not carry any $U(1)'$ charges [31,32,33], were motivated by a reported excess in the $Z \to b\bar{b}$ branching ratio from LEP, which could be accounted for by a small $Z - Z'$ mixing. Effects in leptonic channels such as $e^+e^- \to Z \to \mu^+\mu^-$ would be higher order in the mixing and therefore negligible. Leptophobic models with both heavy (~ 1 TeV) [31] and light (e.g., 150 GeV) Z' [32] were constructed. It was shown that a leptophobic variant of the η model could emerge from an E_6 supersymmetric-GUT with kinetic mixing (see below) [34], and the possible origin of such models from superstring theory was discussed in [35]. Another class of models motivated by $Z \to b\bar{b}$ involve an extra Z' that couples only to the third family [36]. However, more recent LEP and SLD data have significantly weakened the motivation for both classes of models. One can also consider fermiophobic models, which do not couple to ordinary fermions, but which shift the ordinary Z mass due to mixing [37].

2.2 Mass and Kinetic Mixing

The $Z - Z'$ mass matrix takes the form

$$M^2 = \begin{pmatrix} M_Z^2 & \gamma M_Z^2 \\ \gamma M_Z^2 & M_{Z'}^2 \end{pmatrix}, \tag{1}$$

where γ is determined within each model once the Higgs sector is specified. The physical (mass) eigenstates of mass $M_{1,2}$ are then

$$Z_1 = +Z \cos\phi + Z' \sin\phi, \qquad Z_2 = -Z \sin\phi + Z' \cos\phi, \tag{2}$$

where Z_1 is the known boson and $\tan 2\phi = 2\gamma M_Z^2/(M_{Z'}^2 - M_Z^2)$. The mass M_1 is shifted from the SM value M_Z by the mixing, so that $M_Z^2 - M_1^2 = \tan^2\phi(M_2^2 - M_Z^2)$. For $M_Z \ll M_{Z'}$ one has $M_2 \sim M_{Z'}$ and $\phi \sim \gamma M_1^2/M_2^2$.

There may also be a gauge invariant mixing of the $U(1)$ gauge boson kinetic energy terms. This is equivalent, by appropriate redefinition of the gauge fields, to a mixing between the renormalization group equations (RGE's) for the running $U(1)$ gauge couplings. A significant mixing could be induced by field theoretical loop effects [38,39,40,34], which occur in the RGE's when $Tr(Q_Y Q) \neq 0$, with the trace restricted to the light degrees of freedom. This can occur in GUTs, for example, when multiplets are split into light and heavy sectors. Kinetic mixing can also be due to higher genus effects in superstring theory [41]. However, such effects are small for superstring vacua based on the free fermionic construction[e], on the order of at most a %. An important implication of kinetic mixing is that the effective charge of the Z' at low energies may contain a component proportional to the ordinary weak hypercharge[f].

2.3 Precision Electroweak and Collider Limits

Constraints can be placed on the existence of Z''s either indirectly from fits to high precision electroweak data or from direct searches at colliders. The results depend on the chiral couplings of the Z' and on the relation between the Z' mass and the $Z - Z'$ mixing angle. The GUT based Z' models, described at the beginning of the previous Subsection, are fairly representative, with lower limits on the Z' mass in the range 600-1000 GeV, and the mixing restricted to $|\phi| <$ few $\times 10^{-3}$. Larger (a few per cent) mixings are possible in leptophobic and fermiophobic models with a light (e.g., 150 GeV) Z'[31,32,33,37,42].

In the future, the LHC will be sensitive to reference Z's up to several TeV, well above the expected maximum of $\mathcal{O}(1$ TeV) for a class of superstring-motivated models discussed in Section 4. The NLC sensitivity is somewhat less than that of the LHC for $\sqrt{s} = 500$ GeV, but increases rapidly for higher energies. Following the discovery of a heavy Z', one would want to determine its chiral couplings to quarks and leptons. Their ratios would allow one to discriminate between models and to determine the nature of the underlying extended gauge structure. For masses up to $\sim$ 1–2 TeV considerable information should be obtainable from the LHC and NLC, with their capabilities complementary. The range $M_{Z'} < 1$ TeV, expected in a class of models discussed in Section 4, should allow precise determination of a number of useful diagnostics.

Considerably more information about present limits and future prospects can be found on the Web version of this review[2], as well as in [43] and [44].

3 Z's–Theoretical Considerations

Z' models fall into different classes, depending on whether they are embedded into a GUT and whether supersymmetry is included.

---

[e]If one of the $U(1)$'s is associated with a large supersymmetry-breaking D-term in a "hidden" sector, the kinetic mixing could propagate this large scale to the observable sector[41].

[f]The Z' may also contain a component of hypercharge in models which are not based on $U(1)_Y \times U(1)'$, such as the LR.

Z' Models in GUT's without Supersymmetry

In a general class of models with extended gauge structure which do not incorporate constraints of supersymmetry, supergravity or superstring theory, the mass and couplings of additional gauge bosons are free parameters in general. However, one class of models of special interest is based on the GUT gauge structure [45]. At M_U the GUT gauge group is broken to a smaller one which includes the SM gauge group and may also include additional $U(1)'$ factors. As opposed to general models the gauge couplings of the additional Z' are now determined within each GUT model, and the quantum numbers of additional exotic particles are also fixed [2]. However, even within GUT models, there is *no robust prediction for the Z' mass and the mass of the accompanying exotic particles*; without fine-tuning of parameters their masses are likely to be at M_U, while with fine-tuning their masses can be anywhere between M_U and M_Z.

Z' Models in Supersymmetric GUT's

Supersymmetric GUT models possess the advantages of the ordinary GUT models, e.g., gauge coupling unification, and they may provide more constraints on the parameters of the theory. The GUT models with the MSSM below M_U have a gauge coupling unification [46] that is consistent with current experiments. Taking the observed α and weak angle $\sin^2 \theta_W$ as inputs and extrapolating assuming the particle content of the MSSM, one finds [47] that the running $SU(2)_L$ and $U(1)_Y$ couplings meet at a scale $M_U \sim 3 \times 10^{16}$ GeV. One can then predict $\alpha_s(M_Z) \sim 0.130 \pm 0.010$ for the strong coupling, which is roughly consistent with, but slightly above, most experimental determinations, ~ 0.12. To ensure correct electroweak symmetry breaking, fine-tuning of the superpotential parameters or a specific (higher-dimensional) Higgs representation, e.g., the "missing partner" mechanism [48,49], is needed.

Within a symmetry breaking scheme in which the GUT group is broken down to the SM and additional $U(1)'$ factors, the success of gauge coupling unification can be ensured only for certain exotic particle spectra. This in general involves fine-tuning of parameters in the superpotential. (One exception is a special case with complete (light) GUT multiplets.) The parameters should be further constrained to ensure additional symmetry breaking. In particular, the breaking of $U(1)'$ may involve large Higgs representations and/or fine-tuning to ensure $M_{Z'} \ll M_U$ and D- and F-flatness up to $\mathcal{O}(M_{Z'})$. Further fine-tuning of the superpotential parameters would be needed to allow $M_{Z'} \gg M_Z$ [9]. A

---

[9] A somewhat analogous analysis of the symmetry breaking pattern of supersymmetric $SO(10)$ to the left-right and then SM gauge group was addressed in [50].

study of (leptophobic) Z' in supersymmetric E_6 was addressed in [34].

Thus, in spite of the constraining structure of supersymmetric GUT models, the mass of Z' and the accompanying exotic particles again generically tends to be of order M_U. Other mass scales could be achieved by choosing specific Higgs chiral supermultiplets and adjusting the amount of fine-tuning for the superpotential parameters.

Supersymmetric Z' Models without GUT Embedding

Supersymmetric models with additional $U(1)'$ factors, but without GUT embedding, provide a promising class of models with more predictive power, and are a possible *minimal extension of the MSSM*. If mass parameters are absent in the superpotential, the soft supersymmetry breaking masses and the type of the SM singlets responsible for the $U(1)'$ breaking determine the mass scale of the Z' without additional (excessive) fine-tuning.

Such a class of models can be derived within certain classes of superstring compactifications. In these models the massless particle spectrum and couplings in the superpotential are calculable, and there are *no mass parameters in the superpotential*. Thus, supersymmetric Z' models with built in constraints on couplings from superstring models *provide a class of models with a predictive power* for Z' physics and that of the accompanying exotic particles.

In the following Section we review the properties of Z's based on that type of model. It is primarily based on Refs. [5,6,7]. An important related analysis was given in [4] [h]. Earlier work, which addresses Z's in models with softly broken $N = 1$ supergravity with no direct connection to superstring models, but with Z' quantum numbers obtained from E_6 embeddings (i.e., with E_6 broken already at the string scale by the Wilson line (Hosotani) mechanism), was given in [8]. More recent analyses in this context appeared in [9,51].

4 Z's from Superstring Compactifications

A class of superstring compactifications that are based on fermionic constructions [12,13,14,15,18,19,16] provides the necessary ingredients for radiative $U(1)'$ breaking, either at the electroweak or at an intermediate scale. The particle content and the gauge group structure contain, along with the MSSM, additional $U(1)'$ gauge symmetry factors. The massless particle spectrum and the superpotential couplings (which do not have mass terms) are calculable. Importantly, in this class of superstring models the trilinear (Yukawa) couplings in the superpotential are either absent or of order one.

---

[h]Some aspects of Z's in superstring models were also addressed in [35].

320

4.1 Properties of Superstring Vacua

Weakly coupled heterotic string theory, compactified down to $N = 1$ super-symmetric theory in four-dimensions, possesses a number of attractive features, such as a gauge group that contains the SM gauge group and massless chiral supermultiplets, which could play the role of family and Higgs supermultiplets.

The theory also predicts (at tree level) gauge coupling unification at $M_{string} \sim g_U \times 5 \times 10^{17}$ GeV [52], where g_U is the gauge coupling at the string scale. This is to be compared with the MSSM value $M_U \sim 3 \times 10^{16}$ GeV. When properly viewed as predictions for $\ln(M_U/M_Z)$ and $1/\alpha_s$, the gauge unification works to within 10–15% [i].

Using the powerful duality symmetries in superstring theory one can obtain [54] a handle on certain properties of the strongly coupled heterotic string as well. In particular, for the strongly coupled $SO(10)$ heterotic string theory, which is dual to a weakly coupled Type I superstring theory, M_{string}, the scale of the four-dimensional gauge coupling unification, depends on the value of the ten-dimensional dilaton of the Type I superstring theory. M_{string} can be made compatible with M_U for a small enough value of the dilaton (weakly coupled Type I superstring theory). Similarly, for the strongly coupled $E_8 \times E_8$ heterotic string theory, which is dual to eleven dimensional M-theory compactified on S^1/Z_2 (one-dimensional Z_2 orbifold), M_{string} depends on the radius of the circle S^1. M_{string} can be made compatible with M_U for the values of the radius which correspond to the limit in which one of the E_8's is about to become strongly coupled [54]. Unfortunately, explicit constructions of such superstring vacua and their phenomenological analyses are still in their infancy.

There are two major obstacles to connecting the string vacua of the weakly coupled heterotic string to the low energy world: (i) The absence of a fully satisfactory scenario for supersymmetry breaking, either at the level of world-sheet dynamics or of the effective field theory. The supersymmetry breaking is expected to induce soft mass parameters which provide another scale in the theory that can hopefully provide a link between M_{string} and the electroweak scale. (ii) There is a large degeneracy of superstring vacua, i.e., there are by now a large number of superstring models, all dynamically on the same footing.

One may parametrize our ignorance of supersymmetry breaking by introducing soft supersymmetry breaking mass terms. In this manner one looses definitive predictive power, i.e., the quantitative results critically depend on the values of such parameters. However, the explicitly calculable structure of

---

[i] This discrepancy can be remedied [53], e.g., by introducing exotic particles with specific mass ranges. One must, however, be careful because generic exotics can introduce corrections of $\mathcal{O}(1)$. Another possiblity is to embed $U(1)_Y$ with $k < 5/3$.

the superpotential (and other terms in the supersymmetric part of the effective Lagrangian), still allows for useful predictions for the low energy physics.

As for the issue of degeneracy of superstring vacua, the GUT models based on higher level Kač-Moody algebra constructions [15,18,19] provide the needed adjoint representations; however, they possess large additional gauge group factors, and a large number of additional exotic particles. It may therefore be advantageous to study a class of models which at M_{string} already possess the SM gauge group as a part of the observable sector gauge structure, and have a massless particle content that includes three SM families and at least two SM Higgs doublets. A number of such superstring models (not necessarily consistent with gauge coupling unification) were constructed. One class [13,14,15] is based on the free fermionic constructions. A set of such models also constitutes a starting point to address phenomenological issues of Z' physics in supersymmetric models based on superstring compactifications.

4.2   Anomalous and Nonanomalous $U(1)'$ in Superstring Theory

A class of $N = 1$ supersymmetric string models based on free fermionic constructions contain the SM gauge group, three ordinary families, and at least two SM doublets. Such models in general also contain non-Abelian "shadow" sector gauge symmetries and *a number of additional $U(1)'$'s, one of them generically anomalous*. The shadow sector non-Abelian gauge group may play a role in dynamical supersymmetry breaking [56]. In addition, there are a large number of additional matter multiplets, which transform non-trivially under the $U(1)'$ and/or the standard model symmetry[j].

Due to the anomalous $U(1)'$ symmetry, at genus-one there is an additional contribution of $\mathcal{O}(M^2_{string}/16\pi^2)$ [57] to the corresponding D-term in the effective Lagrangian. The numerical suppression factor of $\mathcal{O}(1/16\pi^2)$ renders the scale of the genus-one contribution to be generically smaller than M_{string} by one to two orders of magnitude. The contribution of such a term is canceled [58,57] by giving nonzero vacuum expectation values (VEV's) of $\mathcal{O}(M_{string}/10)$ to certain multiplets in such a way that the D-flatness and F-flatness condition is maintained at genus-one level of the effective string theory, thus providing a mechanism for "restabilizing" the vacuum[k]. The nonzero VEV's typically break a number of additional non-anomalous $U(1)'$'s at M_{string} as well. Thus,

---

[j] We select only those models with the SM embedded into $SU(5)$, to preserve the normalization of the $U(1)_Y$ gauge coupling. The fact that the observable sector gauge group at M_{string} is not $SU(5)$ implies [55] that the theory contains fractionally charged color singlets.

[k] The SM singlets which break the anomalous $U(1)'$ can couple through nonrenormalizable terms to the SM particles, introducing an effective μ-parameter [59] and fermion masses [60,61]. Since the VEV of these singlets is $\mathcal{O}(M_{string}/10)$ the terms must be of high order.

the gauge symmetry and the light exotic particle content of the observable sector is in general drastically reduced. Nevertheless, there often remain one or more *non-anomalous $U(1)'$'s* and associated light exotic matter[l].

In the following we shall concentrate on phenomenological consequences of an additional non-anomalous $U(1)'$ symmetry. Potential additional phenomenological problems[m] will not be addressed.

4.3 $U(1)'$ Symmetry Breaking Scenarios

For the sake of simplicity we consider only one additional $U(1)'$. The symmetry breaking of the additional $U(1)'$ must take place via the Higgs mechanism, in which the scalar components of chiral supermultiplets S_i which carry non-zero charges under the $U(1)'$ acquire non-zero VEV's and spontaneously break the symmetry. The low energy effective action, responsible for symmetry breaking, is specified by the superpotential, Kähler potential, D-terms, and soft supersymmetry breaking terms.

There are two symmetry breaking scenarios: *(i) electroweak scale breaking*[6]. When an additional $U(1)'$ is broken by a *single* SM singlet S, the mass scale of the $U(1)'$ breaking should be[5] in the electroweak range (and not larger than a TeV). The $U(1)'$ breaking may be radiative due to the large (of order one) Yukawa couplings of S to exotic particles. The scale of the symmetry breaking is set by the soft supersymmetry breaking scale, in analogy to the radiative breaking of the electroweak symmetry due to the large top Yukawa coupling. The analysis was generalized[6,4] by including the coupling of the two SM Higgs doublets to the SM singlet S.

(ii) Intermediate scale breaking[7]. This scenario takes place[5] if there are couplings in the renormalizable superpotential of exotic particles to two or more mirror-like singlets S_i with opposite signs of their $U(1)'$ charges. In this case, the potential may have D- and F-flat directions, along which it consists only of the quadratic soft supersymmetry breaking mass terms. If there is a mechanism to drive a particular linear combination negative at $\mu_{RAD} \gg M_Z$, the $U(1)'$ breaking is at *an intermediate scale* of order μ_{RAD}, or in the case of dominant non-renormalizable terms in the superpotential the $U(1)'$ takes place at an intermediate scale governed by these terms.

---

[l]The study of some D-flatness and F-flatness conditions within some models was given, for example, in[14,15]. However a systematic study of these conditions for a large class of such models is needed to classify all possible symmetry breaking scenarios[62].

[m]For example: (i) there may be additional color triplets which could mediate a too fast proton decay; (ii) the mass spectrum of the ordinary fermions may not be realistic; and (iii) the light exotic particle spectrum may not be consistent with gauge coupling unification.

4.4 Electroweak Scale Breaking

The superpotential in this model is of the form:

$$W = h_s \hat{S} \hat{H}_1 \cdot \hat{H}_2 + h_Q \hat{U}_3^c \hat{Q}_3 \cdot \hat{H}_2, \tag{3}$$

where the Higgs doublet superfield $\hat{H}_2$ has only a Yukawa coupling to a single (third) quark family [n]. For simplicity, the Kähler potential is written in a canonical form, providing canonical kinetic energy terms for the matter superfields. Supersymmetry breaking is parametrized with the most general soft mass parameters, with $m^2_{1,2,S,Q,U}$ corresponding to the soft mass-squared terms of the scalar fields H_1, H_2, S, Q_3, U_3^c, respectively, and A and A_Q are the soft mass parameters associated with the first and the second trilinear terms in the superpotential (3).

Gauge symmetry breaking is now driven by the VEV's of the doublets H_1, H_2 and the singlet S, obtained by minimizing the Higgs potential. By an appropriate choice of the global phases of the fields, one can choose the VEV's such that $\langle H_{1,2}^0 \rangle = v_{1,2}/\sqrt{2}$ and $\langle S \rangle = s/\sqrt{2}$ are positive in the true minimum. Whether the obtained local minimum of the potential is acceptable will depend on the location and depth of the other possible minima and of the barrier height and width between the minima [63].

A nonzero value (in the electroweak range) for s renders the first term of the superpotential (3) into an effective μ-parameter, i.e., $\mu \hat{H}_1 \hat{H}_2$, with $\mu \equiv h_s s/\sqrt{2}$ in the electroweak range, thus providing an elegant solution to the μ problem.

The $Z - Z'$ mass-squared matrix is given by (1), where

$$M_Z^2 = \frac{1}{4} G^2 (v_1^2 + v_2^2), \qquad M_{Z'}^2 = g'^2 (v_1^2 Q_1^2 + v_2^2 Q_2^2 + s^2 Q_S^2), \tag{4}$$

$$\gamma = \frac{2g'}{G} \frac{(v_1^2 Q_1 - v_2^2 Q_2)}{v_1^2 + v_2^2}. \tag{5}$$

Here g' is the gauge coupling for $U(1)'$, $G = \sqrt{g_Y^2 + g'^2}$, and $Q_{1,2}$ and Q_S are the $U(1)'$ charges for the $\hat{H}_{1,2}$ and $\hat{S}$ superfields.

Electroweak Scale Conditions

This symmetry breaking scenario can be classified in three different categories according to the value of the singlet s VEV:

---

[n]The masses of other quarks and leptons are obtained in this class of models through non-renormalizable terms [60]. The fermion masses obtained in this way may not be realistic.

324

$s = 0$. In this case the breaking is driven only by the two Higgs doublets (this would be the typical case if the soft mass of the singlet remains positive). The Z' boson would generically acquire mass of the same order as the Z, and some particles (Higgses, charginos and neutralinos) would tend to be dangerously light [o]. There is a possibility of a small $Z - Z'$ mixing due to cancellations, and by considerable fine-tuning one may be able to arrange the parameters to barely satisfy experimental constraints.

$s \sim v_{1,2}$. This case would naturally give $M_{Z'} \gtrsim M_Z$ (if $g'Q_{1,2}$ is not too small) and the effective μ-parameter may be small. Thus, some sparticles may be expected to be light. One requires $Q_1 = Q_2$ to have negligible $Z - Z'$ mixing [p]. A particularly interesting example is the "Large Trilinear Coupling Scenario", in which a large trilinear soft supersymmetry breaking term dominates the symmetry breaking pattern, and relative signs and the magnitudes of the soft mass-squared terms are not important since they contribute negligibly to the location of the minimum. The three Higgs fields assume approximately equal VEV's: $v_1 \sim v_2 \sim s \sim 174$ GeV. In this scenario, the electroweak phase transition may be first order, with potentially interesting cosmological implications.

$s \gg v_{1,2}$. Unless $g'Q_S$ is large, $M_{Z'} \gg M_Z$ requires $s \gg v_{1,2}$ and the effective μ-parameter is naturally large. In this case the breaking of the $U(1)'$ is triggered effectively by the running of the soft mass-squared term m_S^2 towards negative values in the infrared, yielding $s^2 \simeq -(2m_S^2)/(g'^2 Q_S^2)$ and $M_{Z'}^2 \sim -2m_S^2$, with m_S^2 evaluated at s. The presence of this large singlet VEV influences $SU(2)_L \times U(1)_Y$ breaking already at tree level. The hierarchy $M_{Z'} \gg M_Z$ results from a cancellation of different mass terms of order $M_{Z'}$; the fine-tuning involved is roughly given by $M_{Z'}/M_Z$. The $Z - Z'$ mixing is suppressed by the large Z' mass (in addition to any accidental cancellation for particular choices of charges). Excessive fine-tuning would be needed for $M_{Z'} \gg 1$ TeV. More details are given in [6] (see also [4,51]).

String Scale Conditions

A detailed discussion of the relation of the electroweak scale parameters to the boundary conditions at M_{string} is given in [6,2] (see also [5,4,9]), assuming that the non-zero Yukawa couplings are given by $h_s = h_Q = \sqrt{2}g_U$ at M_{string}, as determined in a class of superstring models. It was shown that the symmetry

---

[o]In the $Q_1 + Q_2 = 0$ (which allows an elementary μ-parameter) or large $\tan\beta = v_2/v_1 \gg 1$ cases one of the neutral gauge bosons becomes massless [64]. This does not provide a viable hierarchy since the $W^\pm$ mass is non-zero and related to $v_{1,2}$ and G in the usual way.

[p]Such models are allowed for, e.g., leptophobic couplings [31,32,33,34,42].

breaking scenarios described above can be obtained, but that one requires either nonuniversal soft breaking terms at M_{string}, or additional exotic particles, e.g., additional $SU(3)_C$ triplets $\hat{D}_{1,2}$, which couple to $\hat{S}$ in the superpotential.

The Spectra of Other Particles

The spectrum of physical Higgses after symmetry breaking consists of three neutral CP even scalars (h_i^0, $i = 1, 2, 3$), one CP odd pseudoscalar (A^0) and a pair of charged Higgses ($H^\pm$), i.e., it has one scalar more than in the MSSM (for a detailed analysis see[6], for an earlier discussion see[1,8]). In the neutralino sector, there is an extra $U(1)'$ zino and the higgsino $\tilde{S}$ as well as the four MSSM neutralinos. In these models the LSP is usually mostly $\tilde{B}$. For large gaugino masses, however, the lightest neutralino is the singlino $\tilde{S}$, whose mass is of the order of M_Z. It provides a viable dark matter candidate[25].

Masses for the squarks and sleptons can be obtained directly from the MSSM formulae by setting $\mu \equiv h_s s/\sqrt{2}$ and adding the pertinent D-term diagonal contributions from the $U(1)'$[1,3]. This extra term can produce significant mass deviations with respect to the minimal model and plays an important role in the connection between parameters at the electroweak and string scales.

4.5 Intermediate Scale Breaking

A mechanism to generate intermediate scale $U(1)'$ breaking in supersymmetric theories utilizes the D-flat directions[11], which are in general present in the case of two or more SM singlets with opposite signs of the $U(1)'$ charges. For simplicity, consider *two* chiral multiplets $\hat{S}_{1,2}$ that are singlets under the SM gauge group, with the $U(1)'$ charges Q_{1S} and Q_{2S}, respectively. If these charges have opposite signs ($Q_{1S}Q_{2S} < 0$), there is a D-flat direction:

$$Q_{1S}\langle S_1\rangle^2 + Q_{2S}\langle S_2\rangle^2 = 0. \tag{6}$$

In the absence of the self-coupling of $\hat{S}_1$ and $\hat{S}_2$ in the superpotential there is an F-flat direction in the $S_{1,2}$ scalar field space as well. Then the only contribution to the scalar potential along this D- and F-flat direction is due to the soft mass-squared terms $m_{1S}^2|S_1|^2 + m_{2S}^2|S_2|^2$. For the (real) component along the flat direction $s \equiv (\sqrt{2|Q_{2S}|}\text{Re}S_1 + \sqrt{2|Q_{1S}|}\text{Re}S_2)/\sqrt{|Q_{1S}| + |Q_{2S}|}$ the potential is simply

$$V(s) = \frac{1}{2}m^2 s^2, \quad m^2 \equiv \frac{|Q_{2S}|m_{1S}^2 + |Q_{1S}|m_{2S}^2}{|Q_{1S}| + |Q_{2S}|}, \tag{7}$$

where $m^2_{1S,2S}$ are respectively the $S_{1,2}$ soft mass-squared terms. m^2 is evaluated at the scale s. For m^2 positive at the string scale it can be driven to negative values at the electroweak scale if $\hat{S}_1$ and/or $\hat{S}_2$ have a large Yukawa coupling to other fields, which is in general the case for this class of superstring models. In this case, $V(s)$ develops a minimum along the flat direction and s acquires a VEV.

From the minimization condition for (7) the VEV $\langle s \rangle$ occurs very close to the scale μ_{RAD} at which m^2 crosses zero. This scale is fixed by the RGE evolution of parameters from M_{string} down to the electroweak scale and lies in general at an intermediate scale. It can be achieved with universal boundary conditions at M_{string} as long as there is a large Yukawa coupling of SM singlets to other matter. The precise value depends on the type of couplings of $\hat{S}_{1,2}$ and the particle content of the model. Examples with μ_{RAD} in the range 10^{15} GeV – 10^4 GeV are discussed in detail in[7] (see also[5,4]).

Competition with Non-Renormalizable Operators

The stabilization of the minimum along the D-flat direction can also be due to non-renormalizable terms in the superpotential, which lift the F-flat direction for sufficiently large s. If these terms are important below μ_{RAD}, they will determine $\langle s \rangle$. The relevant non-renormalizable terms are of the form[q]

$$W_{\rm NR} = \left(\frac{\alpha_K}{M_{Pl}}\right)^K \hat{S}^{3+K},\tag{8}$$

where $\hat{S}$ is the (effective) chiral superfield with the (real) scalar component along the D-flat direction, $K = 1, 2, \cdots$, and M_{Pl} is the Planck scale.

Including the F-term from (8), the potential along s is

$$V(s) = \frac{1}{2}m^2 s^2 + \frac{1}{2(K+2)}\left(\frac{s^{2+K}}{M^K}\right)^2,\tag{9}$$

where $M \equiv C_K M_{Pl}/\alpha_K$, and C_K is a coefficient of order unity. The VEV of s is then[r]

$$\langle s \rangle = \left(|m|M^K\right)^{\frac{1}{K+1}} \sim \left(m_{soft}M^K\right)^{\frac{1}{K+1}},\tag{10}$$

---

[q]One can also have terms of the form $\alpha_K^K \hat{S}^{2+K}\hat{\Phi}/M_{Pl}^K$, where Φ is a SM singlet that does not acquire a VEV. These have similar implications as the terms in (8).

[r]For simplicity, soft-terms of the type $(AW_{\rm NR} + {\rm H.c.})$ with $A \sim m_{soft}$ are not included in (9). Such terms do not affect the order of magnitude estimates.

where $m_{soft} = \mathcal{O}(|m|) = \mathcal{O}(M_Z)$ is a typical soft supersymmetry breaking scale. m^2 is evaluated at the scale $\langle s \rangle$ and has to satisfy the necessary condition $m^2(\langle s \rangle) < 0$. If non-renormalizable terms are negligible below μ_{RAD}, no solution to (10) exists and $\langle s \rangle$ is fixed solely by the running m^2.

The coefficients α_K in (8) and thus M in (9) are in principle calculable in a class of superstring models discussed above. Depending on the $U(1)'$ charges and world-sheet symmetries of the superstring models, not all values of K are allowed. It is expected that the world-sheet integrals that determine the couplings of non-renormalizable terms are such that M increases as K increases. For $K = 1$ one obtains (in a class of models): $M \sim 3 \times 10^{17}$ GeV, and for $K = 2$: $M \sim 7 \times 10^{17}$ GeV.

Higgs and Higgsino Mass Spectrum

The mass of the physical field s in the vacuum $\langle s \rangle$ is either $\sim m_{soft}/4\pi$ for pure radiative breaking, or $\sim m_{soft}$ in the case of stabilization by non-renormalizable terms. Thus, the potential is very flat, with possible important cosmological consequences. The physical excitations along the transverse direction have an intermediate mass scale. There remains one massless pseudoscalar, which can acquire a mass from a soft supersymmetry breaking term of the type AW_{NR} or from loop corrections.

The fermionic part of the $Z' - S_1 - S_2$ sector consists of three neutralinos $(\tilde{B}', \tilde{S}_1, \tilde{S}_2)$. One combination is light, with mass of order m_{soft} (if the minimum is fixed by non-renormalizable terms) or of order $m_{soft}/4\pi$ (obtained at one-loop order when the minimum is instead determined by the running of m^2). The two other neutralinos have intermediate scale masses.

Other fields that couple to $\hat{S}_{1,2}$ in the renormalizable superpotential acquire intermediate scale masses, and those that do not remain light.

μ-Parameter

The flat direction S can have a set of non-renormalizable couplings to MSSM states that offer a solution to the μ problem. The non-renormalizable μ-generating terms are of the form,

$$W_\mu \sim \hat{H}_1 \hat{H}_2 \hat{S} \left(\frac{\hat{S}}{M} \right)^P . \tag{11}$$

For breaking due to non-renormalizable terms, $K = P$ yields an effective μ-parameter $\mu \sim m_{soft}$, while for pure radiative breaking $\mu \sim \mu_{RAD}^{P+1}/M^P$, which depends crucially on the value of μ_{RAD}.

Fermion Masses

Non-renormalizable couplings can also yield mass hierarchies between family generations. Generational up, down, and electron mass terms appear, via

$$W_{u_i} \sim \hat{H}_2 \hat{Q}_i \hat{U}_i^c \left(\frac{\hat{S}}{M}\right)^{P'_{u_i}} ; W_{d_i} \sim \hat{H}_1 \hat{Q}_i \hat{D}_i^c \left(\frac{\hat{S}}{M}\right)^{P'_{d_i}} ; W_{e_i} \sim \hat{H}_1 \hat{L}_i \hat{E}_i^c \left(\frac{\hat{S}}{M}\right)^{P'_{e_i}} ,$$

$$(12)$$

with i the family number. K and P'_i can be chosen to yield a realistic hierarchy for the first two generations [7]. Presumably the top mass is associated with a renormalizable coupling ($P'_{u_3} = 0$). The other third family masses do not fit as well; it is possible that m_b and m_τ are associated with some other mechanism, such as non-renormalizable operators involving the VEV of a different singlet.

There may also be non-renormalizable Majorana and Dirac neutrino terms. They can produce small (non-seesaw) Dirac masses. It is also possible to have small Majorana masses, providing an interesting possibility for oscillations of the ordinary into sterile neutrinos [65]. A traditional seesaw can also be obtained, depending on the nature of the non-renormalizable operators [7].

5 Conclusions

Supersymmetric Z' models motivated from a class of superstring models provide a "minimal" extension of the MSSM. The electroweak breaking scenario yields phenomenologically acceptable symmetry breaking patterns, which involve a certain but not excessive amount of fine-tuning. It predicts interesting new phenomena testable at future colliders. The intermediate scale scenario provides a framework in which intermediate scales can naturally occur, with implications for the μ-parameter and the fermion masses.

Acknowledgments

This work was supported by Department of Energy Grant DOE-EY-76-02-3071 and NSF Career Advancement Award. We would like to thank F. del Aguila, D. Demir, G. Cleaver, J. R. Espinosa and L. Everett for fruitful collaboration.

References

1. See C. Kolda and S. Martin, Phys. Rev. **D53**, 3871 (1996), and references theirin, and [2].

2. For a more detailed version of this review, see M. Cvetič and P. Langacker, hep-ph/9707451.

3. J. D. Lykken, hep-ph/9610218.

4. D. Suematsu and Y. Yamagishi, Int. J. Mod. Phys. **A10**, 4521 (1995).

5. M. Cvetič and P. Langacker, Phys. Rev. **D54**, 3570 (1996) and Mod. Phys. Lett. **11A**, 1247 (1996).

6. M. Cvetič, D. A. Demir, J. R. Espinosa, L. Everett and P. Langacker, Phys. Rev. **D56**, 2861 (1997).

7. G. Cleaver, M. Cvetič, J. R. Espinosa, L. Everett and P. Langacker, hep-ph/9705391.

8. J. F. Gunion, L. Roszkowski, and H. E. Haber, Phys. Lett. **B189**, 409 (1987); Phys. Rev. **D38**, 105 (1988).

9. E. Keith, E. Ma, and B. Mukhopadhyaya, Phys. Rev. **D55**, 3111 (1997); E. Keith and E. Ma, Phys. Rev. **D56**, 7155 (1997).

10. For a review, see J. Hewett and T. Rizzo, Phys. Rep. **183**, 193 (1989).

11. M. Dine, V. Kaplunovsky, M. Mangano, C. Nappi and N. Seiberg, Nucl. Phys. **B259**, 549 (1985); M. Mangano, Zeit. Phys. **C28**, 613 (1985); G. Costa, F. Feruglio, F. Gabbiani and F. Zwirner, Nucl. Phys. **B286**, 325 (1987); J. Ellis, K. Enqvist, D. V. Nanopoulos and K. Olive, Phys. Lett. **188B**, 415 (1987); R. Arnowitt and P. Nath, Phys. Rev. Lett. **60**, 1817 (1988).

12. I. Antoniadis, C. Bachas and C. Kounnas, Nucl. Phys. **B289**, 87 (1987); H. Kawai, D. Lewellen and S. H. H. Tye, Phys. Rev. Lett. **57**, 1832 (1986) and Phys. Rev. **D34**, 3794 (1986).

13. I. Antoniadis, J. Ellis, J. Hagelin and D. V. Nanopoulos, Phys. Lett. **B231**, 65 (1989).

14. A. Faraggi, D. V. Nanopoulos and K. Yuan, Nucl. Phys. **B335**, 347 (1990); A. Faraggi, Phys. Lett. **B278**, 131 (1992).

15. S. Chaudhuri, S.-W. Chung, G. Hockney and J. Lykken, Nucl. Phys. **B456**, 89 (1995); S. Chaudhuri, G. Hockney and J. Lykken, Nucl. Phys. **B469**, 357 (1996).

16. For a review, see K. R. Dienes, Phys. Rep. **287**, 447 (1997).

17. L. E. Ibáñez, J. E. Kim, H. P. Nilles, and F. Quevedo, Phys. Lett. **191B**, 282 (1987); A. Font, L. E. Ibáñez, H. P. Nilles and F. Quevedo, Phys. Lett. **210B**, 101 (1988), *erratum, ibid.*, **B213**, 564 (1988).

18. G. Aldazabal, A. Font, L. E. Ibáñez, and A.M. Uranga, Nucl. Phys. **B452**, 3 (1995) and Nucl. Phys. **B465**, 34 (1996).

19. Z. Kakushadze and S. H. H. Tye, Phys. Lett. **B392**, 335 (1997), Phys. Rev. **D55**, 7878 (1997) and Phys. Rev. **D55**, 7896 (1997).

20. J. E. Kim and H. P. Nilles, Phys. Lett. **B138**, 150 (1984).

21. M. Pietroni, Nucl. Phys. **B402**, 27 (1993); J. R. Espinosa, in prep.

22. R. Brandenberger, A.-C. Davis, and M. Rees, Phys. Lett. **B349**, 329 (1995), and references therein.

23. M. Fukugita and T. Yanagida, Phys. Lett. **B174**, 45 (1986); Phys. Rev. **D42**, 1285 (1990); P. Langacker, R.D. Peccei and T. Yanagida, Mod. Phys. Lett. **A1**, 541 (1986); W. Buchmüller and M. Plümacher, Phys. Lett. **B389**, 73 (1996).

24. W. Buchmüller and T. Yanagida, Phys. Lett. **B302**, 240 (1993).

25. B. de Carlos and J. R. Espinosa, Phys. Lett. **B407**, 12 (1997).

26. R. Robinett, Phys. Rev. **D26**, 2388 (1982); R. Robinett and J. Rosner, Phys. Rev. **D25**, 3036 (1982) and **D26**, 2396 (1982).

27. E. Witten, Nucl. Phys. **B258**, 75 (1985).

28. For a review see R.N. Mohapatra, *Unification and Supersymmetry* (Springer, New York, 1986); for constraints on general LR models, see P. Langacker and S. Uma Sankar, Phys. Lett. **D40**, 1569 (1989).

29. U. Amaldi *et al.*, Phys. Rev. **D36**, 1385 (1987).

30. M. Cvetič, B. Kayser, and P. Langacker, Phys. Rev. Lett. **68**, 2871 (1992).

31. P. Chiappetta *et al.*, Phys. Rev. **D54**, 789 (1996); G. Altarelli *et al.*, Phys. Lett. **B375**, 292 (1996).

32. V. Barger, K. Cheung, and P. Langacker, Phys. Lett. **B381**, 226 (1996).

33. K. Agashe *et al.*, Phys. Lett. **B385**, 218 (1996).

34. K. S. Babu, C. Kolda, and J. March-Russell, Phys. Rev. **D54**, 4635 (1996).

35. A. E. Faraggi and M. Masip, Phys. Lett. **B388**, 524 (1996); J. L. Lopez and D. V. Nanopoulos, Phys. Rev. **D55**, 397 (1997).

36. B. Holdom, Phys. Lett. **B339**, 114 (1994), **B351**, 279 (1995); P. Frampton, M. Wise, and B. Wright, Phys. Rev. **D54**, 5820 (1996).

37. A. Donini *et al.*, Nucl. Phys. **B507**, 51 (1997).

38. B. Holdom, Phys. Lett. **B166**, 196 (1986).

39. F. del Aguila *et al.*, Nucl. Phys. **B283**, 50 (1987); F. del Aguila, G. D. Coughlan and M. Quirós; Nucl. Phys. **B307**, 633 (1988), [E-**B312**, 751 (1989)]; F. del Aguila, M. Masip and M. Pérez-Victoria, Nucl. Phys. **B456**, 531 (1995).

40. F. del Aguila, M. Cvetič and P. Langacker, Phys. Rev. **D52**, 37 (1995).

41. K. R. Dienes, C. Kolda and J. March-Russell, Nucl. Phys. **B492**, 104 (1997).

42. See the Appendix of the second paper in [5].

43. For a review see, *e.g.*, M. Cvetič and S. Godfrey, in Proceedings of *Electroweak Symmetry Breaking and Beyond the Standard Model*, eds. T.

Barklow, S Dawson, H. Haber and J. Siegrist (World Scientific 1995), hep-ph/9504216, and references therein.

44. See P. Langacker, p 883 of *Precision Tests of the Standard Electroweak Model*, ed. P. Langacker (World, Singapore, 1995).
45. For a review, see P. Langacker, Phys. Rep. **72**, 185 (1981).
46. S. Dimopoulos, S. Raby and F. Wilczek, Phys. Rev. **D24**, 1681 (1981); L. E. Ibáñez and G. G. Ross, Phys. Lett. **105B**, 439 (1981).
47. For recent discussions, see P. Langacker and N. Polonsky, Phys. Rev. **D52**, 3081 (1985); P. Langacker, in Proceedings of *SUSY-95*, hep-ph/9511207, and references therein.
48. S. Dimopoulos and F. Wilczek, Santa Barbara preprint 81-0600 (1981).
49. For a review, see, e.g., H. P. Nilles, Phys. Rep. **110**, 1 (1984).
50. C. S. Aulakh and R. N. Mohapatra, Phys. Rev. **D28**, 217 (1983).
51. T. Gherghetta, T. A. Kaeding, and G. L. Kane, hep-ph/9701343.
52. V. Kaplunovsky, Nucl. Phys. **B307**, 145 (1988), (*E*, **B383**, 436 (1992)).
53. See e.g., K. R. Dienes and A. E. Faraggi, Phys. Rev. Lett. **75**, 2646 (1995); K. R. Dienes, Ref. [16]; P. Binetruy and P. Langacker, in prep.
54. E. Witten, Nucl. Phys. **B471**, 135 (1996).
55. A. N. Schellekens, Phys. Lett. **B237**, 363 (1990).
56. For recent work, see e.g., P. Binetruy and E. Dudas, Phys. Lett. **B389**, 503 (1996) and references therein.
57. M. Dine, N. Seiberg and E. Witten, Nucl. Phys. **B289**, 585 (1986); J. Atick, L. Dixon and A. Sen, Nucl. Phys. **B292**, 109 (1987).
58. L. Dixon and V. Kaplunovsky, unpublished.
59. V. Jain and R. Shrock, Phys. Lett. **B352**, 83 (1995) and hep-ph/9507238; Y. Nir, Phys. Lett. **B354**, 107 (1995).
60. A. E. Faraggi, Phys. Lett. **B274**, 47 (1992); Nucl. Phys. **B403**, 102 (1993).
61. P. Binetruy, N. Irges, S. Lavignac and P. Ramond, Phys. Lett. **B403**, 38 (1997); P. Binetruy, S. Lavignac and P. Ramond, Nucl. Phys. **B477**, 353 (1996); J. K. Elwood, N. Irges and P. Ramond, Phys. Lett. **B413**, 322 (1997).
62. G. Cleaver, M. Cvetič, J. R. Espinosa, L. Everett and P. Langacker, hep-th/9711178.
63. A. Kusenko, P. Langacker and G. Segrè, Phys. Rev. **D54**, 5824 (1996); A. Kusenko and P. Langacker, Phys. Lett. **B391**, 29 (1997).
64. We thank M.-X. Luo for pointing that out to us.
65. P. Langacker, in preparation. For other mechanisms, see [66].
66. E. Ma, Mod. Phys. Lett. **A11**, 1893 (1996); E. Keith and E. Ma, Phys. Rev. **D54**, 3587 (1996).

PROBING PHYSICS AT SHORT DISTANCES WITH SUPERSYMMETRY

HITOSHI MURAYAMA

Department of Physics, University of California, Berkeley, CA 94720

We discuss the prospect of studying physics at short distances, such as Planck length or GUT scale, using supersymmetry as a probe. Supersymmetry breaking parameters contain information on all physics below the scale where they are induced. We will gain insights into grand unification (or in some cases string theory) and its symmetry breaking pattern combining measurements of gauge coupling constants, gaugino masses and scalar masses. Once the superparticle masses are known, it removes the main uncertainty in the analysis of proton decay, flavor violation and electric dipole moments. We will be able to discuss the consequence of flavor physics at short distances quantitatively.

1 Introduction

The aim of particle physics is very simple: to understand the structure of matter and their interactions at as short distant scale as possible. This is the ultimate form of the reductionist approach of physics. This approach has revealed several layers of distance scales in nature, bulk (1 cm), atomic (10^{-8} cm), nuclear (10^{-13} cm). Understanding physics at shorter distance scales always gave us better understanding of physics at a previously known longer distance scale. Knowing the structure of atoms, we can deduce the chemical properties of atoms and molecules. Knowing the statistics of nuclei, we understand the levels of molecular excitations. The quest continues to understand the origin of the known distance scales, such as the electroweak scale of 10^{-16} cm, and to discover new layer of physics at shorter distance scales.

The main motivation of supersymmetry is to stabilize the electroweak scale against radiative corrections, which tend to make it much shorter (as short as the Planck length) or much longer (no electroweak symmetry breaking). Whenever we speculate about physics at shorter distance scales, we cannot go around the problem of the stability of the electroweak scale. However, once we accept supersymmetry as the stabilization mechanism, we are allowed to speculate physics at much shorter distances, and ask questions about the origin of gauge forces, fermion masses, and even cosmological issues such as baryon asymmetry.

The aim of this short article is to further elaborate on this point. Not only that supersymmetry allows us to speculate physics as the shortest distance scales, it actually provides probes of it. One can even dream about exploring physics at the GUT scale (10^{-30} cm) or the Planck scale (10^{-33} cm) once we see superparticles. We will present various possibilities how we may be able to probe physics at such short distance scales using supersymmetry as a probe (see, *e.g.*, a review on this point[1]). We therefore assume that we will find and can study superparticles at collider experiments.[a]

Of course, such a dream scenario cannot be discussed without certain assumptions. For each examples of such probes in the following sections, we try to make explicit what the underlying assumptions are. One of the main assumptions in any of these discussions is that the layers in distance scales are exponentially apart from each other. This is not an unreasonable assumption from the historic perspective. All the layers of physics came at very disparate distance scales. There appears to be nothing new between the characteristics scales: *deserts*. We do not know if this is the way nature is organized; we can only assume that the shorter distance scales also come with exponential hierarchy and discuss its consequences in our further exploration of physics at yet shorter distance scales.

2 Grand Unification

The simplest example in our approach is the grand unification. Needless to explain, a grand unified theory intends to explain the rather baroque pattern of quark, lepton quantum numbers in the Standard Model by embedding its gauge groups into a simple gauge group. Such a theory would resolve bizarre puzzles in the Standard Model. The fact that the matter is neutral (at the level of at least 10^{-21}) requires a cancellation of electric charges between and electron, two up-quarks and one down-quark. The cancellation of anomalies in the Standard Model appears miraculous. And probably most importantly, the grand unification explains why the strong interaction is stronger than the electromagnetism as a simple consequence of the difference in the size of the gauge groups.

2.1 Gauge Coupling Constants

The supersymmetric grand unification received a strong attention in the past seven years after the precise measurement of the weak mixing angle at LEP in 1991. Many took the agreement of the observed and predicted value of the weak mixing angle as an experimental support for supersymmetry because the

---

[a]Needless to say, it is important to confirm experimentally that the discovered new particles have properties consistent with supersymmetry.[2,3]

334

prediction was quite off if the minimal non-supersymmetric Standard Model was used to predict the weak mixing angle. The situation has not changed qualitatively since. The detailed discussions on the dependence on superparticle spectrum, GUT-scale threshold, and its correlation to the proton decay are all important issues, and we refer to another chapter on gauge unification.

Since our question is what we will learn by using supersymmetry as a probe, let us suppose that we already have found the superparticles. Then the question on grand unification changes dramatically. First of all, we do not need to motivate supersymmetry *assuming* grand unification. The question goes the other way around. Since we know that the supersymmetry is there, we will rather ask if the gauge coupling constants unify given the particle content seen at the electroweak scale. More importantly, we measure all the masses of superparticles, which give us quantitative inputs on the supersymmetry thresholds in the renormalization group analysis. Knowing the particle content at the TeV scale and their masses will completely change the rule of the research. Note, however, that this analysis assumes a desert between the electroweak scale under experimental study and the GUT scale.

If they do unify within a certain accuracy, say within a few percents, we will begin asking what the origin of the small mismatch is (if any). For each of the GUT models we construct, we calculate the GUT-scale threshold corrections and compare them to the data. Such an analysis would certainly exclude parts of the parameter space in each model, and in some cases, the model itself. Especially the correlation to the proton decay becomes important, since the GUT-scale threshold corrections contain information about the mass of the color-triplet Higgsino which mediates the dimension-five proton decay such as $p \to e^+ K^0$ and the mass of the GUT gauge bosons which mediate the dimension-six proton decay such as $p \to e^+ \pi^0$.[4] We will come back to this point in the section on proton decay. Another origin of a small mismatch may be a higher dimension operator in the gauge kinetic function which depends on the GUT-Higgs field, such as $\int d^2\theta \mathrm{Tr}(\frac{\Sigma}{M} W_\alpha W^\alpha)$, where Σ is the adjoint Higgs in $SU(5)$ GUT.[5,6] If we assume that M is the reduced Planck scale and Σ has a VEV at the conventional GUT-scale of order 10^{16} GeV, such an operator gives an order percent correction to the gauge coupling unification. This possibility may unfortunately contaminate the information on GUT-scale threshold. In any case, however, it is clear that the rule of the game changes from motivating supersymmetry using GUT to making selection of GUT models from observed supersymmetry spectrum.

Unfortunately, the fact that the observed gauge couplings appear consistent with the $SU(5)$ unification does not rule out other possibilities, such as intermediate gauge groups with certain matter content.[7,8] Fig. 1 shows two

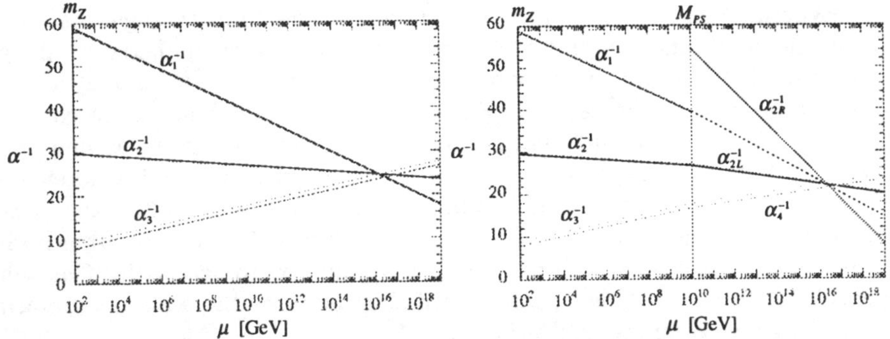

Figure 1: The renormalization group evolution of the gauge coupling constants in two models, $SU(5)$ GUT with grand desert, and $SO(10)$ GUT with an intermediate Pati–Salam symmetry and a particular particle content above the Pati–Salam scale.[8]

patterns of gauge coupling unifications, one with grand-desert $SU(5)$ and the other with intermediate Pati–Salam symmetry. The latter model is intended to be a comparison toy model to make the points clear for the later discussions, how the study of superparticles would help us to sort out the physics at shorter distance scales.

2.2 Gaugino Masses

Another aspect of the grand unification is the unification of superparticle masses. This discussion assumes that the supersymmetry breaking parameters are generated at a scale higher than the GUT-scale, such as the string or Planck scales, and hence respect grand-unified symmetry. Under this assumption, we will see if the superparticle masses unify *at the same scale* as where the gauge coupling constants unify. In fact, the gaugino mass unification,

$$\frac{M_1}{\alpha_1} = \frac{M_2}{\alpha_2} = \frac{M_3}{\alpha_3} \tag{1}$$

holds even when the GUT-group breaks to the Standard Model gauge group in several steps, *i.e.*, with intermediate gauge groups such as Pati–Salam $SU(4) \times SU(2) \times SU(2)$ or its subgroup starting from $SO(10)$ or E_6, as long as the Standard Model gauge groups are embedded in a simple group with a single gaugino mass.[8] At low-energy, the gaugino masses run in the exactly the same way as the gauge coupling constants squared do, which can be read off from Fig. 1 in these two examples. Therefore, the gaugino mass unification tests the idea of grand unification in a highly model-independent manner.

Experimental strategies have been discussed how to disentangle the mixing in the neutralino-chargino sector to measure the supersymmetry breaking masses for $SU(2)$ and $U(1)$ gauginos, M_2 for wino and M_1 for bino, at future e^+e^- linear colliders.[9,10] At the LHC, mass differences can be measured well by identifying the end points in the decay distributions. Upon the assumption that the first and second neutralinos are close to pure bino and wino eigenstates (which may be cross-checked by other analyses of the data), the mass differences also test the gaugino unification.[11] Putting information from both types of experiments, we will have three new numbers to deal with. This will provide us two more independent test if they unify at the scale determined from the gauge coupling unification.

It is an important question if the gaugino mass unification can be spoiled even within GUT models. One possible effect is the threshold correction at the GUT scale. Fortunately, there is no logarithmic threshold corrections unlike to the gauge coupling constants[12] and hence the gaugino mass unification is a better prediction of GUT than the gauge coupling unification. The only way to spoil the gaugino unification from the threshold corrections is to have extremely large representations under the grand unified group with highly non-universal trilinear and bilinear supersymmetry breaking parameters. Another possible effect is a higher dimension operator in the gauge kinetic function which depends on the GUT-Higgs field, such as $\int d^2\theta \text{Tr}(\frac{\Sigma}{M} W_\alpha W^\alpha)$ we discussed before, with an F-component VEV of the Σ field.[9] If we take M at the reduced Planck scale and $F_\Sigma \simeq m_{SUSY} M_{GUT}$, this operator generates an order percent correction to the gaugino mass unification. Note, however, that the size of the F-component VEV tends to be only of m^2_{SUSY} in a wide class of supersymmetry breaking.[13,14]

On the other hand, there is a case where we may be fooled by the apparent gaugino mass unification. In the models of gauge mediated supersymmetry breaking,[15,16,17] the gaugino masses may satisfy the same relation as the case with grand unification even though the supersymmetry breaking gaugino masses have nothing to do with the physics of grand unification. However, this happens only when the messenger fields fall into full $SU(5)$ multiplets and when they acquire masses from the same field which has both A- and F-component VEVs. This is naturally expected in the GUT models, even when the supersymmetry breaking is induced well below the GUT-scale. On the other hand, there is no reason for the messengers to fall into full $SU(5)$ multiplets and acquire masses from the same field if the theory is not grand unified. Even though this remains as a logically possibility that the data can fool us, the apparent gaugino mass unification still strongly suggests grand unification. One case which probably cannot be distinguished on the bases

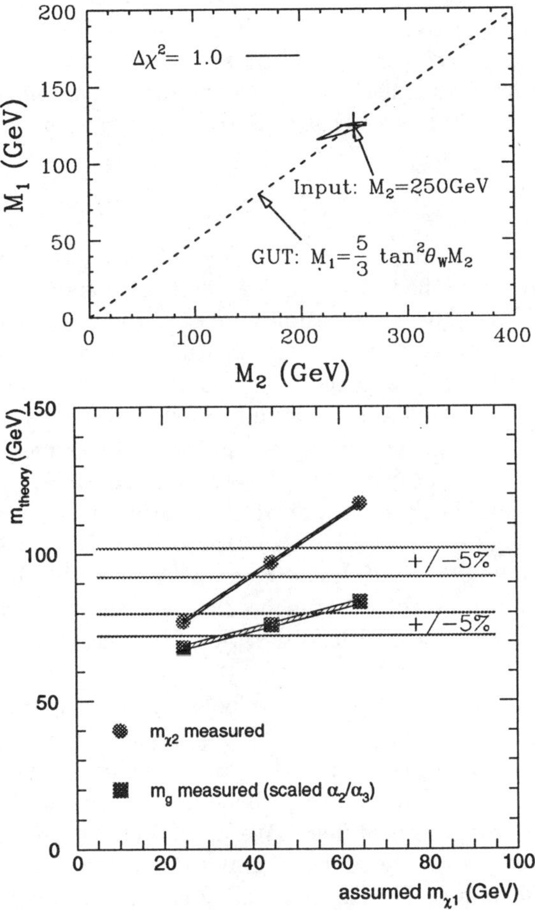

Figure 2: Experimental tests of gaugino mass unification at a future e^+e^- collider[9] and the LHC.[11]

of gaugino masses is the dilaton-dominant supersymmetry breaking in super-string models.[21] In this case, however, there is a specific prediction on the ratio of the scalar masses (universal to all scalars) and gaugino masses (universal to all gauginos) and can be confronted to the data.

There are GUT-like models which do not lead to unified gaugino masses, when the unified group is not simple. One example is flipped $SU(5)$.[18] In this case, we expect to see the unification of M_2 and M_3 at the scale where the gauge coupling constants α_2 and α_3 meet, while M_1 may not. This is an interesting discriminator. The other is the model of dynamical GUT-breaking based on $SU(5) \times SU(3) \times U(1)$.[19,20] Here the gaugino masses do not appear unified at all.

2.3 Scalar Masses

Under $SU(5)$ grand unified group, quarks and leptons belong to either **10** or **5*** multiplets. Under the same assumption that the supersymemtry breaking masses respect grand-unified symmetry, we can extrapolate the observed scalar masses to higher energies and see if they unify at the same scale where the gauge couplings and gaugino masses unify (if they do).

The scalar mass unification will be an independent useful piece of information beyond that from gauge couplings and gaugino masses. One probably very convincing case for grand-desert $SU(5)$ unification is when both the gauge couplings and gaugino masses all unify at the same scale, and also scalars in **10** and **5*** unify there but with different masses. On the other hand, the dilaton-dominated supersymmetry breaking predicts the universal scalar mass, not separate for **10** and **5***, and a definite ratio of the scalar mass to the gaugino mass ratio.

There is a possibility that we get fooled by an apparent unification of gauge couplings and gaugino masses. This happens, for instance, when the supersymmetry breaking is induced by gauge mediation with messengers in full $SU(5)$ multiplets which acquire both supersymmetric and supersymmetry-breaking masses from a single field. We argued in the previous section that such a case already strongly suggests grand unification, but there remains a possibility that it is not. In this case, the scalar masses do not appear grand unified, and provide a way to differentiate the conventional grand-desert $SU(5)$ GUT from gauge-mediated supersymmetry breaking.

Unlike the gaugino masses, the scalar mass spectrum is sensitive to the pattern of GUT symmetry breaking.[8] Many different patterns of scalar masses were discussed from GUT models.[22,23] An important effect is that the scalars which originally resided in the same GUT multiplet may acquire different contributions from the D-term VEV when the rank of the gauge group is reduced.[24] For instance, all quarks and lepton fields live in the same **16** multiplet under $SO(10)$; but the breaking of $SO(10)$ to $SU(5)$ generally splits the **10** and **5*** masses because of the D-term. The D-term contributions are determined solely by the gauge quantum numbers under the broken gauge group and hence generation blind, and are safe from the point of view of flavor-changing effects. A complicated superpotential interactions may modify the scalar masses as well.[13,25,14] An extreme case is when the quarks and leptons in the Standard Model come from different GUT multiplets; then their scalar masses can be totally unrelated.[26] However, the constraint from the flavor-changing neutral currents and smallness of Yukawa couplings for the first, second generation give us a prejudice that the superpotential couplings, which can potentially split the mass of first and second generation scalars in a non-universal manner, are small. Therefore, it is likely that the scalar masses unify according to the patterns of the GUT symmetry breaking, at least for the first and second generations.

By allowing D-term contributions but not F-term contributions motivated by the above argument, one can try to fit the observed scalar mass spectrum as a function of the symmetry breaking scales and original supersymmetry breaking parameters. For more complicated symmetry breaking patterns, there are less relations and hence the model is harder to test. But still in many interesting symmetry breaking patterns, there remain non-trivial relations among scalar masses which can be confronted to data.[8,22,23] The Fig. 3 shows how the scalar masses acquire different patterns at the electroweak scale between the grand-desert $SU(5)$ and the toy Pati–Salam model, which could not be distinguished based on the gauge coupling constants and the gaugino masses. Therefore the role of gaugino mass unification and scalar mass unification are complimentary; the former gives a model-independent test of the grand unification, while the latter selects out particular symmetry breaking patterns and their energy scales.[8]

Even in the case of non-simple GUT groups, the scalar masses still give us useful information. In flipped $SU(5)$, different sets of fields (Q, d^c, ν) belong to **10** and (L, u^c) to **5***, and e^c is a singlet by itself. Therefore, there is still the sclar-mass unification of Q and d^c, and L and u^c separately. In the model of dynamical GUT-breaking based on $SU(5) \times SU(3) \times U(1)$,[19] still all matter fields belong to the ordinary **10**+**5*** multiplets and the pattern of scalar masses is the same as in the grand-desert $SU(5)$.

Experimentally, measurement of scalar masses is also feasible. At an e^+e^- collider, a well-defined kinematics allows a simple kinematic fit to the decay distributions to extract the mass of the parent scalar particle. This comment

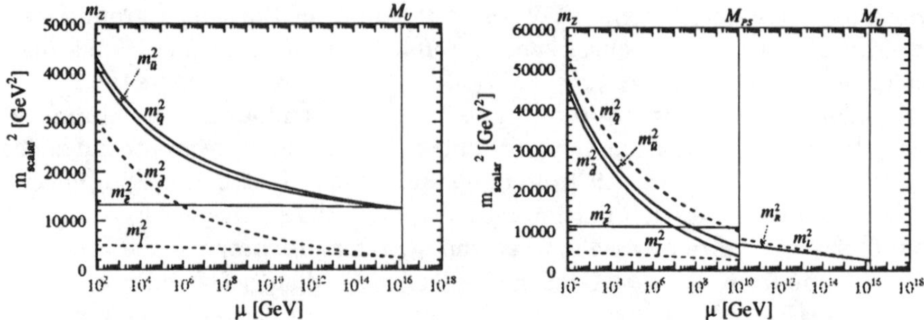

Figure 3: The renormalization group evolution of the scalar masses in two models, $SU(5)$ GUT with grand desert and $SO(10)$ GUT with an intermediate Pati–Salam symmetry used in Fig. 1, which cannot be discriminated based on gauge coupling constants and gaugino masses.

applies both to the sleptons[9,10] and squarks[27] using beam polarizations, as long as they are within the kinematic reach. At the LHC, the mass differences can be measured well as before; especially when the second neutralino decays into on-shell sleptons, one has a high rate and the mass difference between the slepton and the lightest neutralino is measured very well. Many other mass patterns also allow certain mass differences to be measured accurately at a few percent level.[11] It is quite imaginable that the spectroscopy of superparticles will be the main experimental project of the next decade.

3 Proton Decay

Proton decay has been virtually the only direct probe of the physics at the GUT-scale and discussed extensively in the literature. The original idea is that the gauge bosons in $SU(5)$ GUT cause transitions between quarks and leptons in the same $SU(5)$ multiplets and hence allow proton to decay. Assuming that the quarks and leptons of the first generation belong to the same $SU(5)$ multiplets, the exchange of the heavy $SU(5)$ gauge boson generates an operator

$$\mathcal{L} = \frac{1}{M_V^2} uude, \tag{2}$$

which gives rise to a decay $p \to e^+ \pi^0$. The current experimental bound excludes the process for heavy gauge bosons approximately up to 1.5×10^{15} GeV,[29] where we estimated the bound conservatively.[4] Because the operator has a suppression by two powers of a high mass scale, the proton decay rate is suppressed by the fourth power in the mass scale $\Gamma_p \propto m_p^5/M^4$. It is not easy to extend the experimental reach on M. SuperKamiokande will probably extend the limit on the lifetime by a factor of 30 beyond the current one, which translates to a modest improvement by a factor of 2.3 on the GUT-scale. ICARUS may reach the mass scale of the supersymmetric GUT or 10^{16} GeV.

3.1 $D = 5$ operators

The important and novel feature in supersymmetric models is that there are operators of dimension-five which violate baryon and lepton numbers and hence can cause proton decay.[30b] For instance, the following operator is possible in the superpotential:

$$W = \frac{\lambda}{M}(Q_1 Q_1)(Q_2 L_i),$$

(3)

where the subscript refers to the generation, and λ is a coupling constant. The operator involves squarks and sleptons, which need to be converted to quarks or leptons by a loop diagram. The proton decay rate therefore scales as $\Gamma_p \propto m_p^5 \lambda^2/(16\pi^2)^2 M^2 m_{SUSY}^2$ where m_{SUSY} is the mass scale of superparticles. As a result, the reach in the energy scale is drastically improved. The current experimental limit does not allow M below 10^{24} GeV if $\lambda \sim 1$. Therefore, we are sensitive to even Planck-scale suppressed operators which, actually, are excluded with $O(1)$ couplings.

It is interesting that the dimension-five operators necessarily involve quark superfields of different generations (at least two). This is a simple consequence of the Bose symmetry among superfields and the Standard Model gauge invariance. The interesting consequence of this fact is that the proton (or neutron) decay modes preferentially involve kaons in the final state, such as $p \to K^+ \bar{\nu}_\mu$ as predicted to be dominant in the minimal $SU(5)$ GUT. If the dominant proton decay mode will be seen to involve kaons, it is likely to be a consequence of dimension-five operators possible only in supersymmetric theories.

3.2 Minimal SUSY $SU(5)$

In the minimal SUSY $SU(5)$ GUT,[31] the dimension-five operators are generated by the exchange of color-triplet Higgs supermultiplet. The Yukawa couplings of quarks to the Higgs doublets are known from the quark masses, and the couplings to the color-triplet Higgs ($SU(5)$ partner of the doublets) are the those to the Higgs doublets at the GUT-scale because of the $SU(5)$ invariance.

---

[b]In this discussion, we assume that there is no dimension-*four* operators which violate baryon and lepton numbers. Such operators are conveniently forbidden by imposing the R-parity.

Therefore, there is little freedom in this model and the size of the dimension-five operators is completely fixed except possible relative phases which become unobservable below the GUT-scale.[32] The mass of the color-triplet Higgs at first appears to be a free parameter. However, the gauge unification constrains its mass through its threshold correction.[4] At the one-loop level of the renormalization group equations, one obtains

$$(3\alpha_2^{-1} - 2\alpha_3^{-1} - \alpha_1^{-1})(m_Z) = \frac{1}{2\pi}\left\{\frac{12}{5}\ln\frac{M_H}{m_Z} - 2\frac{m_{SUSY}}{m_Z}\right\}, \tag{4}$$

$$(5\alpha_1^{-1} - 3\alpha_2^{-1} - 2\alpha_3^{-1})(m_Z) = \frac{1}{2\pi}\left\{12\ln\frac{M_V^2 M_\Sigma}{m_Z^3} + 8\ln\frac{m_{SUSY}}{m_Z}\right\}. \tag{5}$$

Here, m_{SUSY} stands for some weighted average of the superparticle masses. Once the mass spectrum of the superparticle is measured, one can determine m_{SUSY} in the above formulae, and then extract the mass of the colored Higgs M_{H_C}, and a combination of M_V and M_Σ from the renormalization group equations. In fact, the measured α_s is smaller than the preferred value from the GUT and as a result prefers a low-value of color-triplet Higgs mass. Since various α_s measurements basically converged recently to $\alpha_s(m_Z) = 0.118 \pm 0.003$, the minimal $SU(5)$ model is almost excluded[28,29] unless extreme parameters are chosen for gauginos (preferentially light) and squarks (preferentially heavy). Given the uncertainties in the superparticle spectrum, it is hard to announce the definite exclusion of the model. Once the superparticles are found, however, we will be able to make a final word on the model, assuming the current value of α_s persists and the superKamiokande will not find proton decay.

3.3 Non-minimal SUSY-GUT

There are many good reasons to discuss extensions of the minimal SUSY SU(5) GUT. Among them, there are two points directly relevant to the nucleon decay. (1) The triplet-doublet splitting problem. In minimal SUSY SU(5) GUT, one needs to fine-tune independent parameters at the level of 10^{-14} to keep Higgs doublets light while making the color-triplet Higgs heavy. (2) The wrong fermion mass relations. It predicts $m_s = m_\mu$ and $m_d = m_e$ at the GUT-scale, which are off from the phenomenologically preferred Georgi–Jarlskog relations $m_s = m_\mu/3$, $m_d = 3m_e$.

Solutions to the above-mentioned problems modify the predicted rate and branching ratios of the nucleon decay. One possible attempt to obtain Georgi–Jarlskog relations is to use the SU(5)-adjoint Higgs to construct an effective **45**

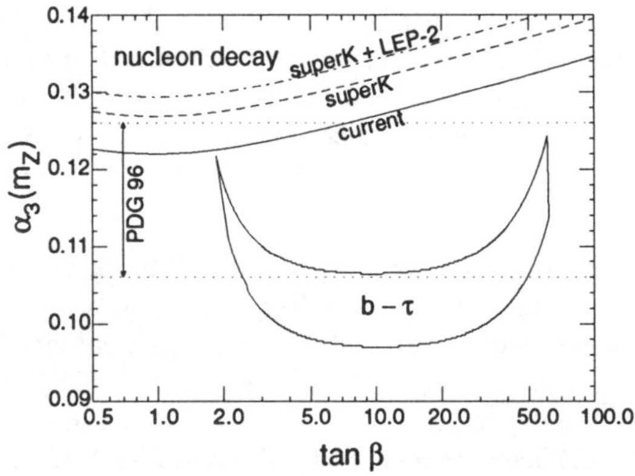

Figure 4: Excluded region on $(\tan\beta, \alpha_s(m_Z))$ plane from nucleon decay based on very conservative assumptions using constraints on superparticle spectrum from LEP-1 only.[29] Expected improvements from superKamiokande and LEP-2 are also shown. The range shown for $\alpha_s(m_Z)$ from PDG96 is two-sigma range. The preferred region from $b - \tau$ unification is also shown for $m_t = 176$ GeV as a crescent-shaped region.

Higgs doublets as composites of ordinary Higgs doublets in **5** and the adjoint. This modification leads to a factor-of-two enhancement in the amplitude; a factor of four in the rate.[36] The relative branching ratios can be also different. It remains true that the $K^{+,0}\bar{\nu}_\mu$ modes are the dominant ones, while the $K^0\mu^+$ mode may be much less suppressed than in the minimal SU(5).[37,38] the proton decay in $SO(10)$ models with realistic fermion mass texture has been also discussed extensively.[39,40]

There are various proposals to solve the triplet-doublet splitting problem, which lead to different nucleon decay phenomenology. I discuss three of them here. (1) The missing partner model,[41] (2) Dimopoulos–Wilczek–Srednicki mechanism,[42] and (3) flipped SU(5) model.[18]

In the missing partner model, one employs **75** representation to break SU(5) instead of the adjoint **24**, and further introduces **50** and **50*** representations which mix with the color-triplet Higgs to make them massive. Since the model involves such large representations, the size of the GUT-scale threshold corrections are significantly larger than that in the minimal model. And the correction changes the determination of the color-triplet Higgs mass as done in Eq. (4), and the measured values of the gauge coupling constants prefer larger M_{H_C} than in the minimal model.[43] In this case the proton decay rates are much more suppressed, by a few orders of magnitudes. One drawback of the model is that it becomes non-perturbative well below the Planck scale due to large representations and one needs to complicate the model further to keep it perturbative.[44] It is worth to recall that the minimal SU(5) model is marginally allowed only with very conservative assumptions. Even though there is an additional suppression to the proton decay rate in this class of models, the decay rate may still well be within the reach of superKamiokande experiment.

The mechanism proposed by Dimopoulos, Wilczek and further by Srednicki employs SO(10) unification with Higgs fields in adjoint and symmetric tensor representations which naturally keep Higgs doublets light. However, their model breaks SO(10) only to $SU(3) \times SU(2)_L \times SU(2)_R$ and has to be extended to achieve the desired symmetry breaking down to the standard model gauge group. One of such extensions by Babu and Barr[45] eliminates $D = 5$ entirely; but it involves rather complicated Higgs sector, and one needs to forbid some allowed interactions in the superpotential arbitrarily. A later attempt to guarantee the special form of the superpotential by symmetries did not eliminate the $D = 5$ operators entirely, but resulted in a weak suppression of the operators. Again in view of the very marginal situation in the minimal model, the decay rate could be within the reach of the superKamiokande.

The flipped SU(5) model solves the triplet-doublet splitting problem in a way that it also eliminates the $D = 5$ operators entirely. A possible problem with this model is that the gauge unification becomes more or less an accident rather than a prediction. On the other hand, the elimination of the $D = 5$ operator is a natural consequence of the structure of the Higgs sector, and is rather a robust prediction of the model except the Planck-scale effects which will be discussed below. An interesting feature of the model is that the GUT-scale is determined by α_2 and α_3 and hence can be *lower* than the scale in the minimal SU(5) which is determined by α_2 and α_1. Since the model does not predict the relation between α_1 and $\alpha_{SU(5)}$, α_1 does not need to meet with the other coupling constants at the same scale. Therefore, the GUT-scale can be as low as $M_{GUT}^{\text{flipped}} = 4\text{--}20 \times 10^{15}$ GeV. If the M_{GUT} is at the low side within this range, the $D = 6$ operator may be observable in the $\pi^0 e^+$ mode,[46] since the superKamiokande is expected to extend the reach by a factor of 30.

Certain models of direct gauge mediation[33] also have $SU(5)$ group broken below the typical GUT-scale and can lead to dimension-six proton decay at a rate observable at superKamiokande. There is also a variant of missing-partner model with dimension-six proton decay within the reach of superKamiokande.[34]

3.4 Planck-scale Operators

Planck-scale physics may generate $D = 5$ operators suppressed by the reduced Planck scale $M_* = 2 \times 10^{18}$ GeV. Even when there is no color-triplet Higgs, such as in string compactifications which breaks the gauge group down to the standard model (with possible U(1) factors) directly, the higher string excitations may give rise to effective non-renormalizable $D = 5$ operators which break baryon- and/or lepton-number symmetries. For $D = 5$ operators which involve first- and second-generation fields, $1/M_*$ suppression is far from enough: one needs a coupling constant of order 10^{-7} to keep the nucleons stable enough as required by experiments.

It is a serious question in supersymmetry phenomenology why the Planck-scale $D = 5$ operators are so much suppressed. One possibility is to forbid them by employing discrete gauge symmetries[47] which are believed to be respected by quantum gravitational effects unlike global symmetries. In this case, there is no baryon-number-violating $D = 5$ operator from Planck-scale physics and we do not have any handle on it. A different type of solution is probably more interesting: the $D = 5$ operators are suppressed because of the same reason why the Yukawa couplings of light generations are suppressed.[35] One way to understand why the Yukawa couplings are so small, such as 10^{-6} for the case of the electron, may be a natural consequence of an approximate flavor symmetry. If a flavor symmetry exists and is only weakly broken to explain smallness of the Yukawa couplings, the same flavor symmetry can well suppress the $D = 5$ operators at the Planck-scale. The $D = 5$ operators with such a flavor origin may have very different flavor structure from those in the GUT models, and may lead to quite different decay modes like $p \to K^0 e^+$.

Suppression of Planck-scale $D = 5$ operators based on certain flavor symmetries were discussed.[48,49] For instance the S_3^3 model[49] explains the hierarchical Yukawa matrices as a consequence of sequential breaking of the flavor symmetry while the symmetry preserves sufficient degeneracy among the squarks and sleptons to suppress flavor-changing neutral currents. It happens that the flavor symmetry in this model also suppresses $D = 5$ operators to the level of about 1/9 of the minimal $SU(5)$ model, so that it can well be within the reach of superKamiokande.[50] What is particularly interesting in this model is that it predicts $p \to K^0 e^+$ as the *dominant* mode over the $K^+ \bar\nu$, while $n \to K^0 \bar\nu_e$ is the dominant mode in neutron decay with a comparable rate. In general, decay modes of proton, if observed, will provide interesting information on the flavor physics.[38,37,40]

4 Flavor Physics

Another interesting topic is how well we will be able to understand the origin of flavor, fermion masses and mixing based on the study of supersymmetry possible at the electroweak scale. Unlike the case of grand unification and proton decay, the answer to this question depends heavily on what the true story is.

4.1 Neutrino Physics

An analogue of proton decay discussed in the previous section is a consequence of flavor physics suppressed by powers of the mass scale, such as the neutrino mass via the seesaw mechanism.[51] The neutrino masses are generated from their Dirac masses m_D with right-handed neutrinos neutral under the Standard Model gauge groups and their Majorana masses M which violate the lepton numbers by two units. The one-generation case is given by a two-by-two mass matrix

$$\mathcal{L}_{mass} = \frac{1}{2}(\nu, N^c) \begin{pmatrix} 0 & m_D \\ m_D & M \end{pmatrix} \begin{pmatrix} \nu \\ N^c \end{pmatrix}, \tag{6}$$

where ν is the Weyl field of the left-handed neutrino, and N the right-handed neutrino. The Lagrangian is written in terms of the charge-conjugated Weyl spinor N^c with the same chirality as the left-handed field ν. After diagonalization of the mass matrix, one obtains a mass for the left-handed neutrino of m_D^2/M, which is power suppressed. This mechanism naturally explains why the neutrino masses are so small, if finite, and leaves imprint of short-distance physics in the pattern of neutrino masses and mixings.

Let us emphasize that this is the area where a dramatic progress is likely to be made in the next few years, from superKamikande (together with neutrino beam from KEK), CHORUS, NOMAD, KARMEN, SNO, BOREXINO, MINOS, and more. Even though supersymmetry does not necessarily help to study the physics at the scale of right-handed neutrinos, many flavor models in supersymmetry predict interesting patterns on neutrino masses.[52] We will certainly be making selections on different flavor models based on neutrino physics if finite neutrino masses and their mixings will be established.

4.2 Flavor-Changing Neutral Currents

As discussed in other chapters, there are severe constraints on the superparticle masses and mixings from the flavor-changing neutral currents (FCNC). There are broadly three categories of models which naturally suppress the FCNC. (1) Flavor symmetry enforces the squarks, sleptons to be degenerate,[53] or aligns

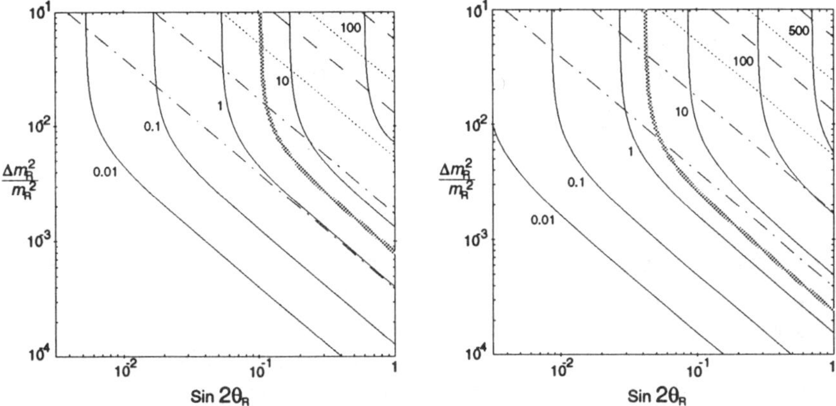

Figure 5: Contours of constant $\sigma(e^+e_R^- \to e^{\pm}\mu^{\mp}\tilde{\chi}^0\tilde{\chi}^0)$ (solid) and $\sigma(e_R^-e_R^- \to e^-\mu^-\tilde{\chi}^0\tilde{\chi}^0)$ (solid) in fb for e^+e^- or e^-e^- linear colliders, with $\sqrt{s} = 500$ GeV, $m_{\tilde{e}_R}, m_{\tilde{\mu}_R} \approx 200$ GeV, and $M_1 = 100$ GeV (solid). The thick gray contour represents the experimental reach in one year. Constant contours of $B(\mu \to e\gamma) = 4.9 \times 10^{-11}$ and 2.5×10^{-12} are also plotted for degenerate left-handed sleptons with mass 120 GeV and $\tilde{l} \equiv -(A + \mu \tan\beta)/\bar{m}_R = 0$ (dotted), 2 (dashed), and 50 (dot-dashed), with left-handed sleptons degenerate at 350 GeV.

their mass basis to that of the down-quark, charged-lepton mass basis.[54] (2) The string theory generates universal scalar mass.[21] (3) The supersymmetry breaking is generated in a flavor-blind fashion below the scale of flavor physics.[15,16,17]

In cases (1) and (2), there may be interesting imprints of flavor physics in the small mixing between squarks and sleptons. The case of GUT also belongs to the category: the large top Yukawa coupling above the GUT-scale may affect the slepton masses-squared with small mixing between, for instance, selectron and smuon.[55,56] The search for rare decays such as $\mu \to e\gamma$,[56,57] K-physics,[58] or electric dipole moments of electron or neutron[59] may reveal the imprints of flavor physics in scalar masses. At present, the main uncertainty in the quantitative analysis is the mass of superparticles. Once they are measured, however, we can try to extract the mixing effects in the scalar mass matrices from the FCNC data.

An interesting case where the flavor-mixing effects in scalar masses can be probed at colliders was discussed.[60] Analogous to neutrino oscillation, a selectron produced from an e^+e^- or e^-e^- collider can oscillate to a smuon as a result of the flavor mixing and is detected as $e\mu$ final state (see Fig. 5).

In the case (3) where the scalar masses are generated in a flavor-blind fashion below the flavor physics scale, such as in the models of low-energy gauge mediation, we unfortunately may not learn about the origin of flavor from the study of flavor signatures at the electroweak scale.

5 Conclusion

Experiments at the electroweak scale will remove the cloud which masks the physics at yet shorter distance scales. If supersymmetry turns out to be the mechanism of stabilizing the electroweak scale, we will have a wealth of new data on superparticle spectroscopy. Combined with data on proton decay, neutrino physics, and FCNC, we will obtain useful information on physics such as grand unification, string, flavor physics. At this point it is just a dream; but we may be able to glimpse the physics at the shortest possible distance scales by this program, which is nothing but the goal of particle physics after all.

Acknowledgments

This work was supported in part by the U.S. Department of Energy under Contracts DE-AC03-76SF00098, in part by the National Science Foundation under grant PHY-95-14797, and also by Alfred P. Sloan Foundation.

1. H. Murayama, Invited Talk at the ICEPP Symposium "From LEP to the Planck World," University of Tokyo, Dec 17–18, 1992. In Proceedings of the ICEPP Symposium "From LEP to the Planck World," eds. K. Kawagoe and T. Kobayashi, UT-ICEPP 93-12, TU-451, 11pp.

2. J.L. Feng, M.E. Peskin, H. Murayama, and X. Tata, *Phys. Rev.* D **52**, 1418 (1995).

3. M. M. Nojiri, K. Fujii, and T.Tsukamoto, *Phys. Rev.* D **54**, 6756 (1996).

4. J. Hisano, H. Murayama, and T. Yanagida, *Phys. Rev. Lett.* **69**, 1014, (1992); *Nucl. Phys.* **B402**, 46 (1993).

5. L. J. Hall and U. Sarid, *Phys. Rev. Lett.* **70**, 2673 (1993).

6. T. Dasgupta, P. Mamales, and P. Nath, *Phys. Rev.* D **52**, 5366 (1995).

7. N.G. Deshpande, E. Keith, and T.G. Rizzo, *Phys. Rev. Lett.* **70**, 3189 (1993).

8. Y. Kawamura, H. Murayama, and M. Yamaguchi, *Phys. Lett.* **B324**, 52 (1994).

9. T. Tsukamoto, K. Fujii, H. Murayama, M. Yamaguchi, and Y. Okada, *Phys. Rev.* D **51**, 3153 (1995).

10. J. L. Feng and M. J. Strassler, *Phys. Rev.* D **51**, 4661 (1995); *Phys. Rev.* D **55**, 1326 (1997); H. Baer, R. Munroe, and X. Tata, *Phys. Rev.* D **54**, 6735 (1996); Erratum-ibid **56**, 4424 (1997).

11. I. Hinchliffe, F.E. Paige, M.D. Shapiro, J. Soderqvist, and W. Yao, *Phys. Rev.* D **55**, 5520 (1997).

12. J. Hisano, H. Murayama, and T. Goto, *Phys. Rev.* **D49**, 1446 (1994).

13. Y. Kawamura, H. Murayama, and M. Yamaguchi, *Phys. Rev.* **D51**, 1337 (1995).

14. A. Pomarol and S. Dimopoulos, *Nucl. Phys.* B **453**, 83 (1995); R. Rattazzi, *Phys. Lett.* B **375**, 181 (1996).

15. M. Dine and A.E. Nelson, *Phys. Rev.* **D48**, 1277 (1993).

16. M. Dine, A.E. Nelson and Y. Shirman, *Phys. Rev.* **D51**, 1362 (1995).

17. M. Dine, A.E. Nelson, Y. Nir and Y. Shirman, *Phys. Rev.* **D53**, 2658 (1996).

18. S. M. Barr, *Phys. Lett.* B **112**, 219 (1982); J.P. Derendinger, J. E. Kim, and D.V. Nanopoulos, *Phys. Lett.* B **139**, 170 (1984); I. Antoniadis, J. Ellis, J.S. Hagelin, and D.V. Nanopoulos, *Phys. Lett.* B **194**, 231 (1987).

19. T. Yanagida, *Phys. Lett.* B **344**, 211 (1995);

20. N. Arkani-Hamed, H.-C. Cheng, and T. Moroi, *Phys. Lett.* B **387**, 529 (1996).

21. V. S. Kaplunovsky and J. Louis, *Phys. Lett.* B **306**, 269 (1993); A. Brignole, L.E. Ibanez, and C. Munoz, *Nucl. Phys.* B **422**, 125 (1994), Erratum-ibid. **436**, 747 (1995); A. Brignole, L.E. Ibanez, C. Munoz, and C. Scheich, *Z. Phys.* C **74**, 157 (1997).

22. C. Kolda, and S. P. Martin, *Phys. Rev.* D **53**, 3871 (1996).

23. H.C. Cheng, and L.J. Hall, *Phys. Rev.* D **51**, 5289 (1995).

24. M. Drees, *Phys. Lett.* **181B**, 279 (1986); J.S. Hagelin and S. Kelley, *Nucl. Phys.* **B342**, 95 (1990); A.E. Faraggi, J.S. Hagelin, S. Kelley, and D.V. Nanopoulos, *Phys. Rev.* **D45** 3272 (1992).

25. H. Murayama, Invited plenary talk given at 4th International Conference on "Physics Beyond the Standard Model," Lake Tahoe, CA, 13-18 Dec 1994. Proceedings, eds. by J. Gunion, T. Han, J. Ohnemus, World Scientific, 1995.

26. S. Dimopoulos and A. Pomarol, *Phys. Lett.* B **353**, 222 (1995).

27. J. L. Feng and D. E. Finnell, *Phys. Rev.* D **49**, 2369 (1994).

28. J. Hisano, T. Moroi, K. Tobe, and T. Yanagida, *Mod. Phys. Lett.* A **10**, 2267 (1995); J. Bagger, K. Matchev, and D. Pierce, *Phys. Lett.* B **348**, 443 (1995).

29. H. Murayama, Invited talk presented at 28th International Conference on High-energy Physics (ICHEP 96), Warsaw, Poland, 25-31 Jul 1996. Published in the Proceedings of the 28th International Conference on High Energy Physics, *eds.*, Z. Ajduk and A. K. Wroblewski, World Scientific, 1997, pp. 1377-1382.

30. N. Sakai and Tsutomu Yanagida, *Nucl. Phys.* B **197**, 533 (1982); S. Weinberg, *Phys. Rev.* D **26**, 287 (1982).

31. S. Dimopoulos and H. Georgi, *Nucl. Phys.* B **193**, 150 (1981); N. Sakai, *Z. Phys.* C **11**, 153 (1981).

32. P. Nath and R. Arnowitt, *Phys. Rev.* D **38**, 1479 (1988).

33. H. Murayama, *Phys. Rev. Lett.* **79**, 18 (1997); S. Dimopoulos, G. Dvali, R. Rattazzi, and G.F. Giudice, CERN-TH/97-98, hep-ph/9705307.

34. J. Hisano, Y. Nomura, and T. Yanagida, KEK-TH-547, hep-ph/9710279.

35. H. Murayama and D. B. Kaplan, *Phys. Lett.* **B336**, 221 (1994).

36. H. Murayama, Invited talk presented at the 22nd INS International Symposium on Physics with High Energy Colliders, Tokyo, Japan, March 8–10, 1994, published in Proceedings of INS Symposium, World Scientific, 1994.

37. A. Antaramian, LBL-36819, Feb 1995, Ph.D. Thesis.

38. K.S. Babu and S.M. Barr, *Phys. Lett.* B **381**, 137 (1996).

39. V. Lucas and S. Raby, *Phys. Rev.* D **55**, 6986 (1997).

40. K.S. Babu, J. C. Pati, and F. Wilczek, IASSNS-HEP-97-136, hep-ph/9712307.

41. A. Masiero, D.V. Nanopoulos, K. Tamvakis, and T. Yanagida, *Phys. Lett.* B **115**, 380 (1982); B. Grinstein, *Nucl. Phys.* B **206**, 387 (1982).

42. S. Dimopoulos and F. Wilczek, NSF-ITP-82-07 (unpublished); M. Srednicki, *Nucl. Phys.* B **202**, 327 (1982).

43. K. Hagiwara and Y. Yamada, *Phys. Rev. Lett.* **70**, 709 (1993); Y. Yamada, *Z. Phys.* C **60**, 83 (1993).

44. J. Hisano, T. Moroi, K. Tobe, and T. Yanagida, *Phys. Lett.* B **342**, 138 (1995).

45. K.S. Babu and S.M. Barr, *Phys. Rev.* D **48**, 5354 (1993).

46. J. Ellis, J. L. Lopez, and D.V. Nanopoulos, *Phys. Lett.* B **371**, 65 (1996).

47. L. E. Ibanez and G. G. Ross, *Nucl. Phys.* **B368**, 3 (1992).

48. V. Ben-Hamo and Y. Nir, *Phys. Lett.* B **339**, 77 (1994).

49. L. J. Hall and H. Murayama, *Phys. Rev. Lett.* **75**, 3985 (1995).

50. C. D. Carone, L. J. Hall, and H. Murayama, *Phys. Rev.* **D53**, 6282 (1996).

51. T. Yanagida, in *Proceedings of Workshop on the Unified Theory and the Baryon Number in the Universe*, Tsukuba, Japan, 1979, edited by A. Sawada and A. Sugamoto (KEK, Tsukuba, 1979), p. 95; M. Gell-Mann, P. Ramond and R. Slansky, in *Supergravity*, proceedings of the Workshop, Stony Brook, New York, 1979, edited by P. Van Nieuwenhuizen and D.Z. Freedman (North-Holland, Amsterdam, 1979), p. 315.

52. Y. Grossman and Y. Nir, *Nucl. Phys.* B **448**, 30 (1995); M. Schmaltz,

Phys. Rev. D **52**, 1643 (1995); C. D. Carone and L. J. Hall, *Phys. Rev.* D **56**, 4198 (1997); P. Binetruy, S. Lavignac, S. Petcov, and P. Ramond, *Nucl. Phys.* B **496**, 3 (1997).

53. M. Dine, A. Kagan, and R. Leigh, *Phys. Rev.* D **48**, 4269 (1993) ; P. Pouliot and N. Seiberg, *Phys. Lett.* B **318**, 169 (1993); D.B. Kaplan and M. Schmaltz, *Phys. Rev.* D **49**, 3741 (1994); A. Pomarol and D. Tommasini, *Nucl. Phys.* B **466**, 3 (1996); R. Barbieri, G. Dvali, and L. J. Hall, *Phys. Lett.* B **377**, 76 (1996); R. Barbieri, L. J. Hall, S. Raby, and A. Romanino, *Nucl. Phys.* B **493**, 3 (1997).

54. Y. Nir and N. Seiberg, *Phys. Lett.* B **309**, 337 (1993).

55. L. J. Hall, V. A. Kostelecky, and S. Raby, *Nucl. Phys.* B **267**, 415 (1986).

56. R. Barbieri and L.J. Hall, *Phys. Lett.* B **338**, 212 (1994).

57. R. Barbieri, L.J. Hall, and A. Strumia, *Nucl. Phys.* B **445**, 219 (1995); J. Hisano, T. Moroi, K. Tobe, M. Yamaguchi, and T. Yanagida, *Phys. Lett.* B **357**, 579 (1995).

58. R. Barbieri, L.J. Hall, and A. Strumia, *Nucl. Phys.* B **449**, 437 (1995).

59. S. Dimopoulos and L.J. Hall, *Phys. Lett.* B **344**, 185 (1995).

60. N. Arkani-Hamed, H.-C. Cheng, J. L. Feng, and L. J. Hall, *Phys. Rev. Lett.* **77**, 1937 (1996).

A HIGGSINO-LSP WORLD

G.L. KANE

Randall Lab of Physics
University of Michigan
Ann Arbor, MI 48109-1120

If the LSP were (mainly) a higgsino the signatures of several superpartners (but not all) at colliders would involve isolated photons. Some possible superpartner candidates at LEP and FNAL are characterized by such photons. If the associated sparticles are indeed light (and if there is a light stop), one expects small effects in $b \to s + \gamma$, R_B, and α_s — the reported data is a little more consistent with what is expected in a higgsino-LSP world than with the SM. The resulting LSP is a good cold dark matter candidate, and the sparticles have properties that allow EW baryogenesis. Each of these phenomena only can be evidence for supersymmetry if $\tan \beta, \mu$ and gaugino masses M_1 and M_2 have certain values, and the whole picture can only be right if the resulting values are the same for all phenomena. The expected behavior of superpartners at colliders is quite different from the usual (Bino-LSP) case studied. There will be a number of tests possible at LEP and FNAL.

In 1986 it was argued [1] that collider events with energetic, isolated γ's were a possible signature of supersymmetry.

In early 1996 we learned about the CDF e"e"$\gamma\gamma\not{E}_T$ event.[2] Analysis [3] showed it had a very low probability of occurring in the SM, and a reasonable probability of occurring from sparticle production. In the interpretation of ref. 3 the photons arose because the lightest neutralino ($\tilde{N}_1$) was mainly higgsino, the next-lightest mainly photino, and the latter predominantly occurred in the decay of the produced superpartners. Such an interpretation was only possible if $\tan \beta, \mu, M_1, M_2$ had certain values.

During 1996, we checked consistency with other phenomenon. It was known that a $\tilde{h}$LSP could give deviations from the SM in $b \to s\gamma$ and in R_b.[4] Both show small 1-1.5σ effects, and both are in the direction and of a size expected from the values of $\tan \beta, \mu, M_1, M_2$ deduced from the CDF event. In addition, if there were an effect in R_b then α_s measured from Γ_Z should be a little larger than the α_s world average without Γ_Z, and it currently is (0.121 $\pm$ 0.003 compared to 0.117 $\pm$ 0.003, with an expected difference of about 0.004-0.005).

It was known that higgsinos did not in general make good cold dark matter candidates, but for the values of $\tan \beta, \mu, M_1, M_2$ here the resulting LSP naturally gave $\Omega h^2 \sim 1/4$.[5]

A number of predictions were made [6] for ways additional superpartners might show up if indeed they were being produced. One prediction was for events from

$$e^+e^- \rightarrow \tilde{N}_2 \left(\rightarrow \gamma\tilde{N}_1 \right) + \tilde{N}_2 \left(\rightarrow \gamma\tilde{N}_1 \right)$$

at LEP. $\tilde{N}_1$ is the LSP and escapes. The missing invariant mass is $\not{M} > 2M_{\tilde{N}_1} \cong 100$ GeV. The minimum photon energy is determined by $M_{\tilde{N}_2} - M_{\tilde{N}_1} \gtrsim 15$ GeV assuming the CDF event was valid, giving $E_\gamma > 5 - 8$ GeV. About six such events were reported[7] at LEP 161+172, combining detectors.

There is background for such events. With the above cuts, and with $|\cos\theta| < 0.85$ since the background from photons rises rapidly near the beam pipe, I estimate (using the results of ref. 8) $\lesssim 1.5$ events background. But the cuts were not chosen a priori, and were different for each detector, so one cannot be precise. Also, there is not yet a well understood calculation of the contribution to the background from events with extra photons that disappear down the beam pipe. Nevertheless, these events are encouraging, though not yet confirmation. If the events are background, then about 2/3 of them should have $|\cos\theta| > 0.85$ and $E_\gamma < 5$ GeV, but only one or two does. Several predictions have been made[6] to test this further in the 1997-1998 runs.

In the $\tilde{h}$LSP world $\tan\beta$ is near 1, and the stop is light. Both of these reduce the value of M_{h°, so it is likely to be found at LEP, and will surely be detectable at FNAL after it turns on in 1999 with upgraded luminosity.

If the stop is light there could be consistency problems at FNAL because of $t \rightarrow \tilde{t}\tilde{N}_1$ decays. One can show[9] that a possible way to satisfy all conditions is production of gluinos, with $\tilde{g} \rightarrow t + \tilde{t}$. Data at FNAL is at least as consistent with this interpretation as with the pure SM.

In a $\tilde{h}$LSP world some SUSY parameters are rather well determined. [3,6] $\tan\beta$ is near one ; $\mu < 0$; at the weak scale $M_W \lesssim M_1 \lesssim M_2; |\mu| < M_2$. These results have implications for the form of the theory at unification(string or GUT).

Acknowledgements

I am very grateful to a number of colleagues for discussions and collaborations: S. Ambrosanio, G. Kribs, G. Mahlon, S. Martin, S. Mrenna, R. Pain, J. Wells, M. Carena, C. Wagner, G. Wilson, D. Stickland, K. Riles, D. Treille. This research was supported in part by the U.S. Department of Energy.

REFERENCES

1. H. Haber, G.L. Kane, and M. Quiros, Phys. Lett. **160B** (1985) 297; H. Komatsu and J. Kuhn, Phys. Lett. **157B** (1985) 90.
2. F. Abe et al., hep-ex/9801019.

354

3. S. Ambrosanio, G.L. Kane, G.D. Kribs, S.P. Martin, and S. Mrenna, Phys. Rev. Lett. **76** (1996) 3498.
4. See the Chapter of P. Chankowski and S. Pokorski.
5. J. Wells and G.L. Kane, Phys. Rev. Lett. **76** (1996) 4458.
6. G.L. Kane and G. Mahlon, Phys. Lett. **B408** (1997) 222.
7. See the talk of G. Wilson, XXXI Rencontres de Moriond, Les Arcs, France, March 1997.
8. See S. Ambrosanio, http://feynman.physics.lsa.umich.edu /ambros/Phys/2Photon+Emiss/; S. Mrenna, "Estimating Two Photon + Missing Energy Backgrounds to SUSY Signals at LEPII", hep-ph/9705441; P. Bain and R. Pain, DELPHI memo, 97-48 PHYS700.
9. G.L. Kane and S. Mrenna, Phys. Rev. Lett. **77** (1996) 3502.

Gauge-Mediated Supersymmetry Breaking

G.F. GIUDICE, R. RATTAZZI

Theory Division, CERN, CH-1211 Genève 23, Switzerland

We review the status of theories with gauge-mediated supersymmetry breaking.

1 Introduction

Theories with gauge-mediated supersymmetry breaking (for a review, see ref.[60]) provide an interesting alternative to the standard scenario in which the information of supersymmetry breaking is communicated to the observable sector by gravity [24,11,67]. They offer the possibility of solving the flavour problem, since the soft terms are generated at a low-mass scale and are not sensitive to the high-energy physics, which is presumably responsible for the breaking of the flavour symmetry.

In the gravity-mediated approach, the soft terms are generated at the Planck mass, and therefore necessarily at a scale larger than or equal to Λ_F, the scale of the unknown dynamics responsible for the observed flavour breaking. Then there is no obvious reason why the supersymmetry-breaking masses for squarks and sleptons should be flavour invariant. Even if at tree level, because of some accidental reason, they are flavour symmetric, still loop corrections from the flavour-violating sector will distort their structure. Even contributions from ordinary grand unified theories (GUTs) lead to significant flavour-breaking effects in the soft terms [68].

These flavour-breaking contributions to the soft terms are very dangerous. The mismatch between the diagonalisation matrices for quarks and squarks (and analogously for leptons and sleptons) lead to flavour-violating gaugino vertices, and eventually to large contributions to flavour-changing neutral-current (FCNC) processes. Studies of $\mu \to e\gamma$, $\bar{K}^0$–K^0 mass difference and similar processes set very stringent bounds on the relative spittings among different generations of squarks and sleptons [66,55].

Of course this does not mean that gravity-mediated scenarios cannot give a realistic theory. It may be that at the level of quantum gravity soft terms are flavour invariant, or some flavour symmetry [45,81] or dynamical mechanism [32] is responsible for an approximate alignment between particles and sparticles. At any rate, it seems unavoidable [13,14] that in these scenarios, flavour violations should be, at best, just at the edge of present bounds and should be soon visible.

On the other hand, in gauge-mediated theories, soft terms are generated at the messenger scale M, which is *a priori* unrelated to Λ_F. If $M \ll \Lambda_F$, the soft terms feel the breaking of flavour only through Yukawa interactions. Yukawa couplings are the only relevant sources of flavour violation, precisely as in the SM. As a consequence, the GIM mechanism is fully operative and it can be generalized to a superGIM mechanism, involving ordinary particles and their supersymmetric partners. Since it is reasonable to expect that Λ_F is as large as the GUT or the Planck scale, in gauge-mediated theories the flavour problem is naturally decoupled, in contrast to the case of supergravity or, in a different context, of technicolour theories [99,97].

Moreover, in gauge-mediated theories, it is possible to describe the essential dynamics without dealing with gravity. This may be viewed as an aesthetic drawback, as it delays the complete unification of forces. However, from a more technical point of view, it is an advantage, because one can solve the model using only field-theoretical tools, without facing our present difficulties in treating quantum gravity. This is particularly interesting in view of the improvements in the understanding of non-perturbative aspects of supersymmetric theories (for reviews, see *e.g.* ref. [70,95,85]).

Finally, as we will illustrate, gauge-mediated theories are quite predictive in the supersymmetric mass spectrum and have distinctive phenomenological features. Future collider experiment can put these predictions fully to test.

2 The Building Blocks of Gauge Mediation

The first ingredient of models with gauge-mediated supersymmetry breaking is an *observable sector* which contains the usual quarks, leptons, and two Higgs doublets, together with their supersymmetric partners. Then the theory contains a sector responsible for supersymmetry breaking. We will refer to it as the *secluded sector*, to distinguish it from the hidden sector of theories where supersymmetry breaking is mediated by gravity. For the moment we will leave the secluded sector unspecified, since it still lacks a standard description. For our purposes, all we need to know is that the Goldstino field overlaps with a chiral superfield X, which acquires a VEV along the scalar and auxiliary components

$$\langle X \rangle = M + \theta^2 F \ . \tag{1}$$

As it will be discussed in detail in the following, the parameters M and $\sqrt{F}$, which are the fundamental mass scales in the theory, can vary between several tens of TeV and somewhat below the GUT scale.

Finally the theory has a *messenger sector*, formed by some new superfields which transform under the gauge group as a real non-trivial representation and couple at tree level with the Goldstino superfield X. This coupling generates a supersymmetric mass of order M for the messenger fields and mass-squared splittings inside the messenger supermultiplets of order F. This sector is also unknown and it is the main source of model dependence. It is fairly reasonable to expect that the secluded and messenger sectors have a common origin, and models in which these two sectors are unified will be discussed in sect. 8.

The simplest messenger sector is described by N_f flavours of chiral superfields Φ_i and $\bar{\Phi}_i$ ($i = 1, ..., N_f$) transforming as the representation $\mathbf{r} + \bar{\mathbf{r}}$ under the gauge group. In order to preserve gauge coupling-constant unification, one usually requires that the messengers form complete GUT multiplets. If this is the case, the presence of messenger fields at an intermediate scale does not modify the value of M_{GUT}, but the inverse gauge coupling strength at the unification scale α_{GUT}^{-1} receives an extra contribution

$$\delta\alpha_{GUT}^{-1} = -\frac{N}{2\pi}\ln\frac{M_{GUT}}{M} , \tag{2}$$

$$N = \sum_{i=1}^{N_f} n_i . \tag{3}$$

Here n_i is twice the Dynkin index of the gauge representation $\mathbf{r}$ with flavour index i; *e.g.* $n = 1$ or 3 for an $SU(5)$ $\mathbf{5}$ or $\mathbf{10}$, respectively. We will refer to N as the *messenger index*, a quantity which will play an important rôle in the phenomenology of gauge-mediated theories. From eq. (2) we infer that perturbativity of gauge interactions up to the scale M_{GUT} implies

$$N \lesssim 150/\ln\frac{M_{GUT}}{M} . \tag{4}$$

If M is as low as 100 TeV, then N can be at most equal to 5. However, this upper bound on N is relaxed for larger values of M. For instance, for $M = 10^{10}$ GeV, eq. (4) shows that N as large as 10 is allowed.

It should also be noticed that, in the minimal $SU(5)$ model, the presence of messenger states at scales of about 100 TeV is inconsistent with proton-decay limits and with b–τ unification[21]. However, these constraints critically depend on the GUT model considered and, after all, minimal $SU(5)$ is not a fully consistent model.

In the case under consideration the interaction between the chiral messenger superfields Φ and $\bar{\Phi}$ and the Goldstino superfield X is given by the superpotential term

$$W = \lambda_{ij}\bar{\Phi}_i X \Phi_j . \tag{5}$$

358

After replacing in eq. (5) the X VEV, see eq. (1), we find that the spinor components of Φ and $\bar{\Phi}$ form Dirac fermions with masses λM, while the scalar components have a squared-mass matrix

$$
\begin{pmatrix} \Phi^\dagger & \bar{\Phi} \end{pmatrix}
\begin{pmatrix} (\lambda M)^\dagger (\lambda M) & (\lambda F)^\dagger \\ (\lambda F) & (\lambda M)(\lambda M)^\dagger \end{pmatrix}
\begin{pmatrix} \Phi \\ \bar{\Phi}^\dagger \end{pmatrix} .
\tag{6}
$$

Here we have dropped flavour indices and, with a standard abuse of notation, we have denote the superfields and their scalar components by the same symbols. If there is a single field X, then the matrices λM and λF can be simultaneously made diagonal and real, and the scalar messengers mass eigenvectors are $(\Phi + \bar{\Phi}^\dagger)/\sqrt{2}$ and $(\bar{\Phi} - \Phi^\dagger)/\sqrt{2}$ with squared-mass eigenvalues $(\lambda M)^2 \pm (\lambda F)$. It is now convenient to absorbe the coupling constant λ in the definition of M and F, $\lambda_{ii} M \to M_i$, $\lambda_{ii} F \to F_i$. In the following we will implicitly assume this redefinition.

3 Physical Mass Spectrum

The mass scale $\sqrt{F}$ is the measure of supersymmetry breaking in the messenger sector. However, we are of course mainly interested in the amount of supersymmetry breaking in the observable sector. Ordinary particle supermultiplets are degenerate at the tree level, since they do not directly couple to X, but splittings arise at the quantum level because of gauge interactions between observable and messenger fields. While vector bosons and matter fermion masses are protected by gauge invariance, gaugino, squarks, and sleptons can acquire masses consistently with the gauge symmetry, once supersymmetry is broken. Gaugino masses are generated at one loop, but squark and slepton masses can only arise at two loops, since the exchange of both gauge and messenger particles is necessary. The corresponding Feynman diagrams have been first evaluated in ref. [3], where it is shown that they lead to an acceptable particle spectrum with positive scalar squared masses. However, it is generally more convenient to employ a simple and systematic method [59] to extract the soft terms for observable fields in theories in which supersymmetry breaking is communicated by renormalizable and perturbative (but not necessarily gauge) interactions.

Working in the leading-log approximation, the supersymmetry-breaking gaugino masses are

$$
\tilde{M}_{\lambda_r}(t) = k_r \frac{\alpha_r(t)}{4\pi} \Lambda_G \qquad (r = 1, 2, 3) ,
\tag{7}
$$

$$\Lambda_G = \sum_{i=1}^{N_f} n_i \frac{F_i}{M_i} \left[1 + \mathcal{O}(F_i^2/M_i^4) \right] , \tag{8}$$

where $k_1 = 5/3$, $k_2 = k_3 = 1$, and the gauge coupling constants are normalized such that $k_r \alpha_r$ ($r = 1, 2, 3$) are all equal at the GUT scale. In the simple case in which there is a single X superfield and therefore the ratio F_i/M_i is independent of the flavour index i, eq. (8) becomes

$$\Lambda_G = N \frac{F}{M} \left[1 + \mathcal{O}(F^2/M^4) \right] . \tag{9}$$

Next-to-leading corrections in α_s can affect the gluino mass prediction by a significant amount. However a complete calculation of these corrections (including ultraviolet threshold effects) is still missing.

Neglecting Yukawa-coupling effects, the supersymmetry-breaking scalar masses at the scale Q ($t = \ln M^2/Q^2$) are

$$m_{\tilde{f}}^2(t) = 2 \sum_{r=1}^{3} C_r^{\tilde{f}} k_r \frac{\alpha_r^2(0)}{(4\pi)^2} \left[\Lambda_S^2 + h_r \Lambda_G^2 \right] , \tag{10}$$

$$h_r = \frac{k_r}{b_r} \left[1 - \frac{\alpha_r^2(t)}{\alpha_r^2(0)} \right] , \tag{11}$$

$$\alpha_r(t) = \alpha_r(0) \left[1 + \frac{\alpha_r(0)}{4\pi} b_r t \right]^{-1} , \tag{12}$$

Here $\alpha_r(0)$ are the gauge coupling constants at the messenger scale M and, in the case of a single X superfield,

$$\Lambda_S^2 = N \frac{F^2}{M^2} \left[1 + \mathcal{O}(F^2/M^4) \right] . \tag{13}$$

It should be clear now that, with no loss of generality, we can use eqs. (9) and (13) also in the case of non-universal values of F_i and M_i, treating them as a definition of the messenger index

$$N \equiv \Lambda_G^2/\Lambda_S^2 . \tag{14}$$

In eq. (10) $C_r^{\tilde{f}}$ depends on the $\tilde{f}$ quantum numbers, $C = \frac{N^2-1}{2N}$ for the N-dimensional representation of $SU(N)$, and $C = Y^2 = (Q - T_3)^2$ for the $U(1)$ factor. Here b_r are the β-function coefficients

$$b_3 = -3, \quad b_2 = 1, \quad b_1 = 11 . \tag{15}$$

The physical scalar mass is obtained by adding to eq. (10) the D-term contribution $M_Z^2 \cos 2\beta(T_3^{\tilde{f}} - Q^{\tilde{f}} \sin^2 \theta_W)$. There is no one-loop messenger contribution to the supersymmetry-breaking trilinear terms. However A terms are generated in the leading-log approximation by RG evolution proportional to gaugino masses.

In deriving these mass formulae, we have assumed $F \ll M^2$. Although this approximation is in most cases justified, as for $F > M^2$ a scalar messenger particle has negative square mass, it is not appropriate whenever $F \simeq M^2$. Notice that in this case the method of ref.[59], used to derive the soft terms, fails, because all higher covariant-derivative operators contribute to supersymmetry-breaking terms at the same order. An explicit Feynman diagram calculation is now necessary. This has been performed in refs.[35,76], with the result

$$\Lambda_G = \sum_{i=1}^{N_f} n_i \frac{F_i}{M_i} g(F_i/M_i^2) \,, \tag{16}$$

$$
\begin{aligned}
g(x) &= \frac{1}{x^2}\left[(1+x)\ln(1+x)\right] + (x \to -x) \\
&= 1 + \frac{x^2}{6} + \frac{x^4}{15} + \frac{x^6}{28} + \mathcal{O}(x^8) \,,
\end{aligned} \tag{17}
$$

$$\Lambda_S^2 = \sum_{i=1}^{N_f} n_i \frac{F_i^2}{M_i^2} f(F_i/M_i^2) \,, \tag{18}$$

$$
\begin{aligned}
f(x) &= \frac{1+x}{x^2}\left[\ln(1+x) - 2\mathrm{Li}_2\left(\frac{x}{1+x}\right) + \frac{1}{2}\mathrm{Li}_2\left(\frac{2x}{1+x}\right)\right] + (x \to -x) \\
&= 1 + \frac{x^2}{36} - \frac{11}{450}x^4 - \frac{319}{11760}x^6 + \mathcal{O}(x^8) \,.
\end{aligned} \tag{19}
$$

The functions $g(x)$ and $f(x)$ represent the corrections with respect to the $F \ll M^2$ case. $g(x)$ is always larger than 1, reaching the maximum value $g(1) = 2\ln 2 = 1.4$. $f(x)$ is approximately equal to 1 for $x < 0.85$, and reaches the minimum value $f(1) = (2\ln 2)(1 + \ln 2) - \pi^2/6 = 0.7$.

The leading contributions to $\Lambda_{G,S}$, proportional to F/M, are universal for the different GUT components of messengers. This is true because, the ratio F/M is a universal quantity, independent of the coupling constant λ between the messenger superfields and the Goldstino superfield X, see eq. (5). On the other hand, the argument of the functions g and f in eqs. (16) and (18) is

F/M^2, a quantity which is not equal for messengers with different SM quantum numbers. Therefore, depending on the particular choice of λ, the inclusion of the correction functions g and f can be relevant for part of the messenger multiplet. This effect gives an uncertainty on the mass prediction which could be significant for gauginos. For instance, gluino masses could be enhanced by up to 40%, giving an apparent violation of gaugino mass unification.

For phenomenological reasons, we have to introduce also the Higgs mixing mass μ, defined by the superpotential interaction $\mu H_1 H_2$. This mass term breaks the Peccei-Quinn symmetry, and therefore it cannot be generated by gauge interactions alone. The origin of the μ parameter [49] is still one of the main problematic issues in theories with gauge mediation. We will discuss this problem in sect. 5. Here we will simply assume that a hard mass parameter μ exists at the messenger scale M.

Although the parameter μ is not determined by the underlying theory, we can compute it from the condition of correct electroweak breaking:

$$\mu^2 = -\frac{M_Z^2}{2} + \frac{1}{\tan^2 \beta - 1} \left(m_{H_1}^2 - \tan^2 \beta \, m_{H_2}^2\right) , \qquad (20)$$

$$B = \frac{\sin 2\beta}{2\mu} \left(m_{H_1}^2 + m_{H_2}^2 + 2\mu^2\right) \qquad (21)$$

Here $m_{H_{1,2}}$ and B are the usual soft-breaking parameters in the Higgs sector. Notice that the low-energy value of B is non-vanishing even if its boundary condition at the messenger scale is $B(0) = 0$. This has motivated [8] several phenomenological studies [37,10,89,16,56] of the very predictive case $B(0) = 0$, in which the low-energy value of B is determined in terms of the gaugino mass. This case is also interesting because the model is automatically free from any dangerous new CP violating parameter, aside from the SM ones, *i.e.* the usual Kobayashi-Maskawa phase and Θ_{QCD}.

Equation (21) determines the low-energy value of B in terms of $\tan \beta$ or, if the condition $B(0) = 0$ is imposed, predicts $\tan \beta$, which turns out to be much larger than one. It is important to stress that, if μ and B are generated radiatively through some new interactions, it is very plausible [49] that also $m_{H_{1,2}}^2$ receive extra corrections. These corrections then modify the values of μ and B extracted from the electroweak-breaking conditions.

4 Properties of the Mass Spectrum

The most important feature of the gauge-mediated mass spectrum is of course flavour universality, which is guaranteed by the symmetry of gauge interactions. This property is maintained if gravity-mediated contributions do not reintroduce large flavour violations in the K^0–$\bar{K}^0$ system or in $\mu \to e\gamma$ transitions. We require therefore that gravity-mediated contributions do not account for more than, say, one permille of the soft squared masses. Since gravity generates soft terms with typical size F/M_P, the flavour criterion gives a rough upper bound on the messenger mass scale

$$M \lesssim \frac{1}{10^{\frac{3}{2}}} \frac{\alpha}{4\pi} M_P \sim 10^{15} \text{ GeV} , \qquad (22)$$

where $M_P = (8\pi G_N)^{-1/2} = 2.4 \times 10^{18}$ GeV is the reduced Planck mass.

Another attractive feature of the gauge-mediated mass formulae is that gaugino mass terms are generated at one loop, and (positive) squark mass terms are generated at two loops. Since fermion and boson bilinears have different canonical dimensions, all supersymmetry-breaking mass parameter have the same scaling property $\tilde{m} \sim (\alpha/\pi)F/M$. However, as we will see in sect. 5, it is not automatic for the μ and B parameters to satisfy this property.

An interesting property of the mass formulae derived here is that they allow a high degree of predictivity. The whole supersymmetric spectrum is determined by the effective supersymmetry breaking scale $\Lambda = F/M$, the messenger index N, the messenger mass M, and $\tan\beta$. The parameter μ is predicted (up to a phase ambiguity) if we assume the absence of any non-minimal contribution to $m^2_{H_{1,2}}$. Comprehensive analyses of the supersymmetric particle masses in gauge-mediated models have been presented in refs.[37,10].

An important feature is the large hierarchy between the strongly-interacting and weakly-interacting particles. For small values of M, this hierarchy is determined by the ratio $\alpha_3(0)/\alpha_2(0)$. For larger values of M, where the RG running is important, the value of N plays a crucial rôle. If N is small, the most important effect is the decrease of the ratio $\alpha_3(0)/\alpha_2(0)$, as M increases, and squarks become lighter and closer in mass to sleptons. On the other hand, if N is large, gaugino masses increase and their effects in the RG evolution dominate the squark and slepton masses. In this case, squarks become heavier as M grows. The ratio between gaugino and scalar masses is determined by the messenger index N. Increasing N, scalars become lighter and the ratio $m_{\tilde{e}_R}/M_1$ can be smaller than one.

Another remarkable success of the gauge-mediated mass spectrum is that eqs. (20)–(21) have an acceptable solution[44], and therefore electroweak-symmetry breaking can be achieved. This happens because the negative contribution to $m^2_{H_2}$ is proportional to the large stop mass. Indeed for small M, $m^2_{H_2}$ is approximately given by

$$m^2_{H_2} = m^2_{\tilde{e}_L} - \frac{3h^2_t}{4\pi^2} m^2_{\tilde{t}} \left(\ln\frac{M}{m_{\tilde{t}}} + \frac{3}{2} \right) . \qquad (23)$$

Therefore, even for moderate values of M, the coefficient in front of the logarithm is large enough to drive $m^2_{H_2}$ negative and trigger electroweak-symmetry breaking. This however requires rather large values of μ to compensate the stop contribution in eq. (20). This rather large value of μ, originating from the hierarchy between strongly- and weakly-interacting particles is at the basis of rather stringent upper bounds on the supersymmetric particle masses obtained from the naturalness criterion. For instance, using the criterion [12] that no independent parameter should be correlated by more than 10%, it is found [25,15,4] that the right-handed selectron is lighter than 100 GeV or less, depending on the parameters. The limit does not significantly change for large N, since the ratio $\mu/m_{\tilde{e}_R}$ only slowly increases as N grows. It was also suggested in ref. [4] that messengers belonging to split GUT multiplets may alleviate the fine-tuning problem present in the minimal model.

With the value of μ extracted from the electroweak-breaking condition, one can compute the spectrum of neutralinos and charginos. Since μ is typically large, it turns out that the lightest neutralino is mainly B-ino, $\chi_1^\pm$ and χ_2^0 form a degenerate W-ino weak triplet, and the nearly higgsinos $\chi_2^\pm$ and $\chi_{3,4}^0$ have masses roughly equal to μ.

The CP-odd neutral Higgs mass is given by

$$m_A^2 = \frac{2B\mu}{\sin 2\beta} \simeq \frac{\mu^2}{\sin^2 \beta} \, . \tag{24}$$

Since m_A is rather large, the theory at low energies is approximately described by a single Higgs doublet, and the Higgs phenomenology at LEP2 should resamble the SM one. The charged Higgs boson mass is given by $m_{H\pm}^2 = M_W^2 + m_A^2$.

The lightest CP-even Higgs boson mass m_h receives important radiative corrections proportional to the top quark Yukawa coupling[83,50,64], and it can be predicted in terms of the fundamental parameters Λ, M, N, and $\tan\beta$[37,10,90].

The largest values of m_h are obtained at large $\tan\beta$. For large $\tan\beta$ and small N, the stop mixing is negligible and the value of m_h corresponds to the value obtained in the usual supersymmetric model with vanishing stop mixing[19]. As we increase M, A_t grows and the typical squark mass decreases; the two effects roughly compensate each other and the value of m_h is not modified. At larger N, the squarks are lighter and the value of m_h is smaller. The effect of increasing M is now more important since A_t receives a large renormalization proportional to the gaugino mass. Including the leading two-loop effects, we obtain $m_h < 110$ GeV (for $M_1 = 100$ GeV) and $m_h < 115$ GeV (for $M_1 = 200$ GeV). A large fraction of the parameter space is within the reach of LEP2. If the Higgs boson is not discovered, the LEP2 search will provide an extremely severe bound on the model.

In conclusion, gauge-mediated models have a very predictive mass spectrum. If supersymmetry is discovered, it seems very likely that these predictions can be used to distinguish these models from the typical spectrum of gravity-mediated models. In presence of unification relations, gaugino masses are the same in both scenarios. However, the scalar masses are different. Even if M is close to the GUT scale, the initial condition of scalar masses is not unified. This is because the gauge bosons are already split and do not form a complete GUT multiplet. From eq. (10), we find

$$\frac{2}{7}\frac{m^2_{\tilde{Q}_L}(0)}{m^2_{\tilde{E}_R}(0)} = \frac{3}{8}\frac{m^2_{\tilde{U}_R}(0)}{m^2_{\tilde{E}_R}(0)} = \frac{3}{7}\frac{m^2_{\tilde{D}_R}(0)}{m^2_{\tilde{E}_R}(0)} = \frac{2}{3}\frac{m^2_{\tilde{L}_L}(0)}{m^2_{\tilde{E}_R}(0)} = 1 \, . \tag{25}$$

However, for large N and large M the mass spectrum is "gaugino dominated" and the signal of gauge-mediation is more difficult to be distinguished. Studies of the comparison between the mass spectrum of gauge-mediated and gravity-mediated models can be found in ref. [37,10,20,96].

5   The Origin of μ and $B\mu$

Usually one refers to the μ-problem as the difficulty in generating the correct mass scale for the Higgs bilinear term in the superpotential

$$W = \mu H_1 H_2 \, , \tag{26}$$

which, for phenomenological reasons, has to be of the order of the weak scale. If this term were present in the limit of exact supersymmetry, then it would have to be of the order of the Planck scale or some other fundamental large scale (like the GUT scale). In the gravity-mediated scenario, however, if this term is absent in the supersymmetric limit and the Kähler potential is general enough, the correct μ is generated after supersymmetry is broken [58]. In particular the two terms

$$K = H_1 H_2 \left(\frac{X^\dagger}{M_P} + \frac{X X^\dagger}{M_P^2} + \dots \right) \, , \tag{27}$$

where X a supersymmetry-breaking chiral spurion generate respectively $\mu \sim F_X/M_P$ and $B\mu \sim (F_X/M_P)^2$. Therefore the size of both μ and $B\mu$ is of the order of the weak scale.

In gauge mediation the problem is somehow more acute. Here μ must be related to the scales X and F_X in the messenger sector, and the natural value is $\mu \sim (1/16\pi^2)F_X/X$. Indeed it is not hard to accomplish that, as we will show below. The real difficulty is to generate *both* μ and B of the same order.

Consider the simplest possibility of a direct coupling $W = \lambda X H_1 H_2$. Even putting $\lambda \sim 1/16\pi^2$ by hand does not accomplish the job since $\mu = \lambda X$ while $B\mu = \lambda F_X$, so that

$$B = \frac{B\mu}{\mu} = \frac{F_X}{X}. \tag{28}$$

The origin of this problem is not the tree level origin of μ, but the fact that μ and $B\mu$ are generated at the same order in the "small" parameter λ.

Equation (28) is the expression of the μ-problem in gauge mediation. When $m^2_{H_{1,2}}$ are given by the usual two-loop gauge-mediated contribution, eq. (28) is inconsistent with electroweak symmetry breaking. This is because the stability condition of the scalar potential, $2|B\mu| < m^2_{H_1} + m^2_{H_2} + 2\mu^2$, cannot be satisfied when μ is determined by electroweak-symmetry breaking as in eq. (20). New contributions to $m^2_{H_1}$ and $m^2_{H_2}$ could make eq. (28) consistent with electroweak breaking, but only at the price of a quite unreasonable fine tuning.

It seems that in order to solve this problem, μ should neither be generated at tree level in W nor via loop corrections to the Kähler potential. This is indeed what happens in the solutions that have been proposed in the literature.

The basic point of the solution outlined in ref.[49] is that μ arises only from higher-derivative operators of the form

$$\int d^4\theta H_1 H_2 D^2 G(X^\dagger, X). \tag{29}$$

Here D_α is the supersymmetric covariant derivative and G a function determined by loop corrections. The crucial point is that a D^2 acting on any function of X and $X^\dagger$ always produces an antichiral superfield. Then, independent of the form of G, no B term is generated along with μ in eq. (29). Higher-order corrections with a different form may then generate B of the right size.

This idea can be implemented[49] in models which involve new singlet messengers, but no new states at low energies. The dynamics also give new two-loop contributions to the Higgs mass parameters $m^2_{H_{1,2}}$, aside from the usual gauge-mediated contributions. Models of this kind can also be embedded in schemes where are mass scales are generated dynamically[39].

A possibile way to dynamically generate μ and $B\mu$ is to add a singlet chiral superfield S to the observable sector and to include the most general superpotential free from mass parameters,

$$W = \lambda S H_1 H_2 + \frac{k}{3} S^3. \tag{30}$$

However, this solution does not work in the case of gauge mediation. This is because the soft terms, which are computable, give rise to a potential which is either inconsistent with the appropriate electroweak-breaking conditions or leads to some unacceptable light particles [44,62].

A more promising direction is to couple directly S to some sizeable source of soft terms. One possibility [44] is to add one light $SU(5)$ flavour $f, \bar{f}$ coupled to S via $W = \lambda_f S f \bar{f}$. If $\lambda_f \sim 1$ then m_S^2 receives a large negative contribution, from the coloured triplets, which is the analogue of the stop contribution to the Higgs mass. Thus a VEV for S of the order of the weak scale can be generated, together with a consistent mass spectrum. However, this scenario implies the existence of exotic, and stable (or very long lived), matter $f, \bar{f}$ with weak scale mass.

Another possibility is to couple S directly to the messengers [59]. This can be done in a theory with at least two messenger flavours and a superpotential

$$W = X(\bar{\Phi}_1 \Phi_1 + \bar{\Phi}_2 \Phi_2) + S \lambda_S \bar{\Phi}_1 \Phi_2 \tag{31}$$

which avoids the radiative generation of the dangerous tadpole mixing $SX^\dagger$. This allows a correct electroweak symmetry breaking, without any approximate R-Goldstone bosons.

Another possibility, suggested in ref. [46], is to invoke a field S with non-renormalizable superpotential couplings $W = S^n H_1 H_2 + S^m$, in units of $1/M_P$. Soft terms, like those from a Kähler potential interaction $SS^\dagger X X^\dagger / M_P^2$, induce a tree level $\mu = \langle S \rangle^n \sim F_X^{n/(m-2)}$, so that for $m = 2n + 2$ one has $\mu \sim \sqrt{F_X}$. This requires further tuning to get the right μ. Anyhow, an implication of a tree level mechanism like this is that B is not generated at tree level along with μ. This solves the supersymmetric CP problem since the only source of A and B terms is the gaugino mass via RG evolution [47]. Moreover the size of the radiatively induced B is small, so that $\tan\beta$ is predicted to be naturally large [8,48,89].

Finally, a different mechanism for generating the μ term was suggested in ref. [102]. It was argued that the cancellation of the cosmological constant in gauge mediation necessarily requires a new mass scale. Indeed the cosmological constant receives two different contributions

$$\langle V \rangle \simeq \langle F \rangle^2 - \frac{\langle W \rangle^2}{3M_P^2} . \tag{32}$$

If supersymmetry is broken at low scales, both the auxiliary fields F and the superpotential W are determined by the same typical energy scale Λ. The cancellation (at the price of a fine tuning) between the two terms can only be achieved if we add to the superpotential a constant term of order $\bar{\Lambda}^3 \sim \Lambda^2 M_P$. It is possible that the scale $\bar{\Lambda}$ is dynamically generated by some new strongly-interacting sector which breaks the continuous R symmetry. If this new sector couples to the ordinary Higgs via non-renormalizable interactions, the μ term is generated by the condensation of a bilinear of new fields,

$$\mu \sim \frac{\bar{\Lambda}^2}{M_P} \sim \left(\frac{\Lambda}{M_P}\right)^{1/3} \Lambda \; . \tag{33}$$

For interesting values of Λ, the ratio $(\Lambda/M_P)^{1/3}$ can mimic the loop factor and give the correct size of the μ term.

6 The Lightest Supersymmetric Particle: the Gravitino

As a result of the spontaneous breakdown of supersymmetry, the physical spectrum contains a massless spin 1/2 fermion, the Goldstino, by definition contained in the chiral superfield X. When the globally supersymmetric theory is coupled to gravity and promoted to a locally supersymmetric theory, the Goldstino provides the longitudinal modes of the spin-3/2 partner of the graviton, the gravitino. As a result of this superhiggs mechanism, the gravitino acquires a supersymmetry-breaking mass which, under the condition of vanishing cosmological constant, is given by [98,28]

$$m_{3/2} = \frac{F_0}{\sqrt{3}M_P} = \frac{F}{k\sqrt{3}M_P} = \frac{1}{k}\left(\frac{\sqrt{F}}{100 \text{ TeV}}\right)^2 2.4 \text{ eV} \; , \tag{34}$$

Here $M_P = (8\pi G_N)^{-1/2} = 2.4 \times 10^{18}$ GeV is the reduced Planck mass. We denote by F_0 the total contribution of the supersymmetry-breaking VEV of the auxillary fields, normalized in such a way that the vacuum energy of the globally supersymmetric theory is $V = F_0^2$. Thus F_0 does not coincide with the definition of F, which appears in the sparticle masses through $\Lambda = F/M$. While F_0 is the fundamental scale of supersymmetry breaking, F is the scale of supersymmetry breaking felt by the messenger particles (*i.e.* the mass splitting inside their supermultiplets). The ratio $k \equiv F/F_0$ depends on how supersymmetry breaking is communicated to the messengers. If the communication occurs via a direct interaction, this ratio is just given by a coupling constant, like the parameter λ in the case described by eq. (5). It can be argued that this coupling should be smaller than one, by requiring perturbativity up to the GUT scale [6]. If the communication occurs radiatively, then k is given by some loop factor, and therefore it is much smaller than one.

In gauge-mediated models, the gravitino is the lightest supersymmetric particle (LSP) for any relevant value of F. Indeed, as argued in sect. 4, a safe solution to the flavour problem requires that gravity-mediated contributions to the sparticle spectrum should be much smaller than gauge-mediated contributions. Since $m_{3/2}$ is exactly the measure of gravity-mediated effects, it is indeed the solution of the flavour problem in gauge mediation, see eq. (22), which implies that the gravitino is the LSP.

If R parity is conserved, all supersymmetric particles decay into the gravitino. In order to compute the decay rate we need to know the interaction Lagrangian at lowest order in the gravitino field. Since, for $\sqrt{F} \ll M_P$, the dominant gravitino interactions come from its spin-1/2 component, the interaction Lagrangian can be computed in the limit of global supersymmetry, using the supersymmetric analogue of the equivalence theorem [51,53,22,23].

For on-shell particles, the Goldstino Lagrangian can be written as a Yukawa interaction to chiral fields and a magnetic moment-like interaction to gauge particles, after use of the equations of motion,

$$\mathcal{L} = \frac{k}{F} \left[(m_\psi^2 - m_\phi^2)\bar{\psi}_L\phi + \frac{M_\lambda}{4\sqrt{2}}\bar{\lambda}^a\sigma^{\nu\rho}F_{\nu\rho}^a \right] \tilde{G} + \text{h.c.} \qquad (35)$$

Notice that both Goldstino interactions are proportional to the ratio between the mass splitting inside the supermultiplet and the scale of supersymmetry breaking [51,53].

For our purposes the interaction Lagrangian in eq. (35) is sufficient to describe the relevant processes involving the Goldstino. Derivations of the complete effective Lagrangian involving multi-Goldstino interactions can be found in ref. [26,57,17,73,18].

7 The Next-to-Lightest Supersymmetric Particle

The next-to-lightest supersymmetric particle (NLSP) plays an important rôle in the phenomenology of gauge mediation. Assuming R-parity conservation, we expect that all supersymmetric particles will proptly decay into cascades leading to the NLSP, with the NLSP then decaying into the gravitino via $1/F$ interactions. Therefore the nature of the NLSP determines the signatures in collider experiments and some cosmological properties of gauge mediation. The NLSP can be, depending on the parameter choice, the neutralino, the stau, or, in a very restricted region of parameters, the sneutrino. Let us review these possibilities.

The NLSP neutralino has, in most cases, a dominant B-ino component, since the ratio μ/M_1 is typically larger than one. An exeption occurs for large N and small M and Λ, as the NLSP neutralino is a non-trivial superposition of different states. Another exeption is given by models where the physics generating the parameter μ also contributes to supersymmetry-breaking Higgs masses.

From eq. (35) we find the following NLSP χ_1^0 decay rates [5,34,10]:

$$\Gamma(\chi_1^0 \to \gamma\tilde{G}) = \frac{k^2 \, \kappa_\gamma m_{\chi_1^0}^5}{16\pi F^2} = k^2 \, \kappa_\gamma \left(\frac{m_{\chi_1^0}}{100 \text{ GeV}}\right)^5 \left(\frac{100 \text{ TeV}}{\sqrt{F}}\right)^4 2 \times 10^{-3} \text{ eV} ,$$

(36)

$$\frac{\Gamma(\chi_1^0 \to Z^0\tilde{G})}{\Gamma(\chi_1^0 \to \gamma\tilde{G})} = \frac{\kappa_Z}{\kappa_\gamma} \left(1 - \frac{M_Z^2}{m_{\chi_1^0}^2}\right)^4 ,$$

(37)

$$\frac{\Gamma(\chi_1^0 \to h^0\tilde{G})}{\Gamma(\chi_1^0 \to \gamma\tilde{G})} = \frac{\kappa_h}{\kappa_\gamma} \left(1 - \frac{m_h^2}{m_{\chi_1^0}^2}\right)^4 ,$$

(38)

$$\kappa_\gamma = |N_{11} \cos\theta_W + N_{12} \sin\theta_W|^2 ,$$

(39)

$$\kappa_Z = |N_{11} \sin\theta_W - N_{12} \cos\theta_W|^2 + \frac{1}{2}|N_{13} \cos\beta - N_{14} \sin\beta|^2 ,$$

(40)

$$\kappa_h = |N_{13} \sin\alpha - N_{14} \cos\alpha|^2 .$$

(41)

Here N_{1i} are the χ_1^0 components in standard notations (see, *e.g.* ref. [63]) and $\tan 2\alpha = \tan 2\beta(m_A^2 + M_Z^2)/(m_A^2 - M_Z^2)$. The decay mode into photon and Goldstino is very likely to dominate. Even if the decay modes into the Z^0 boson or the neutral Higgs boson are kinematically allowed, they are quite suppressed by the β^8 phase-space factor. Moreover, if χ_1^0 is mainly B-ino, $\kappa_Z/\kappa_\gamma = 0.3$ and κ_h/κ_γ is negligible. Complete expressions for the neutralino decay rates into three-body final states can be found in ref. [10].

For roughly

$$N > \frac{66}{5(13\xi_1 - 2)} \quad \xi_1 \equiv \frac{\alpha_1^2(M_1)}{\alpha_1^2(M)} = \left[1 + \frac{11}{4\pi}\alpha_1(M_1) \ln \frac{M_1^2}{M^2}\right]^2 ,$$

(42)

the right-handed slepton is lighter than χ_1^0. The transition occurs at moderate values of N for small M (*e.g.* $N = 1.7$ for $M = 10^5$ GeV), but requires large values of N for large M (*e.g.* $N > 5$ for $M = 10^{12}$ GeV). This is the result of the significant renormalization of $m_{\tilde{E}_R}$ proportional to the gaugino mass in the regime of large N and M.

Among the three generations of right-handed sleptons, $\tilde{\tau}_R$ is the lightest because of mixing effects proportional to m_τ in the stau mass matrix:

$$m_{\tilde{\tau}}^2 = \begin{pmatrix} m_{\tilde{L}_L}^2 + m_\tau^2 - (\frac{1}{2} - \sin^2\theta_W)\cos 2\beta M_Z^2 & m_\tau(A_\tau - \mu\tan\beta) \\ m_\tau(A_\tau - \mu\tan\beta) & m_{\tilde{E}_R}^2 + m_\tau^2 - \sin^2\theta_W \cos 2\beta M_Z^2 \end{pmatrix} .$$

(43)

As the mixing grows with $\tan\beta$, see eq. (43), the stau can become the LSP for values of N quite smaller than what shown in eq. (42). In particular, for extremely large values of $\tan\beta$, the determinant of the matrix in eq. (43) can become negative and destabilize the electromagnetically neutral vacuum [8,89].

The NLSP stau decay rate is

$$\Gamma(\tilde{\tau} \to \tau\tilde{G}) = \frac{k^2 m_{\tilde{\tau}}^5}{16\pi F^2} = k^2 \left(\frac{m_{\tilde{\tau}}}{100\ \text{GeV}}\right)^5 \left(\frac{100\ \text{TeV}}{\sqrt{F}}\right)^4 2 \times 10^{-3}\ \text{eV} \ . \quad (44)$$

If the mixing is small, the lightest stau is mainly right-handed. However, in this case, $\tilde{e}_R$ and $\tilde{\mu}_R$ are so close in mass to $\tilde{\tau}$ that their three-body decays into the NLSP are very much suppressed by phase space. Under these conditions, all three right-handed sleptons directly decay into the corresponding charged lepton and Goldstino. Particles which, in spite of not being the NLSP, have a dominant two-body decay into their supersymmetric partner and a $\tilde{G}$ have been called [6] "co-NLSPs".

For larger values of $\tan\beta$, typically $\tan\beta > 4$–8 depending on the parameter choice, the first two generations of sleptons decay as $\tilde{\ell}_R \to \ell\tau\tilde{\tau}$ and the stau is the "only" NLSP. Another possibility is that χ_1^0, although not the NLSP, is nearly mass degenerate to $\tilde{\tau}$, and therefore dominantly decays into a photon and Goldstino. In this case, $\tilde{\tau}$ and χ_1^0 are again "co-NLSPs".

The possibility that a sneutrino is the NLSP is very marginal. It requires large values of N and values of Λ so low that discovery of supersymmetry should happen quite soon. In this case, the sneutrino decays into a neutrino and a Goldstino.

These various cases corresponds to a quite different phenomenology in high-energy experiments, as reviewed in refs. [60,65].

8 Dynamical Supersymmetry Breaking and Gauge Mediation

When the mass splittings inside supermultiplets arise at tree level, the supertrace sum rule $\text{STr}\mathcal{M}^2 = 0$ holds [54]. This theorem prevents the construction of simple and realistic models in which supersymmetry is broken at tree level in the standard model sector. A generic implication is the existence of sparticles below the mass range of quarks and leptons [30]. The supertrace mass formula, however, does not hold beyond tree level. It was soon realized that when the splittings inside supermultiplets arise from radiative corrections, the sparticles can be all made consistently heavier than the particles. This was a motivation of the first gauge-mediated supersymmetry-breaking models. Indeed the aim of refs. [40,42] was to build a "supersymmetric technicolour" theory, in which the breakdown of supersymmetry is due to some strong gauge dynamics, while its mediation to the SM particles is just due to the usual SM gauge interactions. Those models came early in the understanding of strongly-coupled supersymmetric theories, and many of the dynamical assuptions on which they were based had to be revised in the light of deeper tools of analysis, like the Witten index [101]. The idea of gauge mediation emerged, however, as an independent, and general, aspect of those models. Indeed the paradigm discussed in refs. [29,41,43,3,79,31], consists of an O'Raifertaigh [84] sector coupled to the messengers at the tree level.

In theories with tree-level supersymmetry breaking, the fundamental mass scales are just inputs, whose origin is left unspecified. This implies, in particular, that the small size of the weak scale itself, albeit stable against large quantum corrections, is left unexplained. Undoubtly, a more fundamental theory would be one where the relevant mass scales arise dynamically. The scale of supersymmetry breaking *can* indeed have this nice property [100]. This is because when supersymmetry is unbroken at tree level, it will stay so to all orders in perturbation theory. The reason for that is the non-renormalization theorem, whose range of applicability however does not extend beyond perturbation theory. Indeed there are theories where supersymmetry does break non-perturbatively. Consequently the ratio between the scale of breaking and the fundamental mass scale goes like e^{-1/g^2}, and it is exponentially small at weak coupling. This can explain in a very natural way the smallness of the supersymmetry breaking (or weak) scale with respect to the Planck mass. Notice, indeed the analogy with the case of technicolour theories. There also one uses a *chiral* symmetry, unbroken in perturbation theory, to generate the weak scale hierarchy. A great advantage of supersymmetry, however, is that in many cases the non-perturbative effects that determine its breaking can be studied exactly.

The explanation of the gauge hierarchy is one general motivation for being interested in field theories with dynamical supersymmetry breaking (DSSB). As a matter of fact, in gauge-mediated models the motivation is even stronger that in the gravity-mediated case. This is because in gauge mediation, the whole dynamics of breaking and mediation, takes place at very low scales, where we should be able to describe it field theoretically. In principle this may not be the case in the gravity-mediated scenario, where string theoretic effects may play a crucial rôle. Of course we always have to plead ignorance onto what makes the cosmological constant vanish, in gauge and in gravity mediation.

The first attempts to build realistic models of DSSB with gauge mediation where discussed in ref. [2]. Many of the gauge theories that are known to break supersymmetry have flavour symmetries, which can remain unbroken in the vacuum. A simple strategy [2] is then to consider models with an anomaly-free flavour group $G_F \supset SU(3) \times SU(2) \times U(1)$ (or better $G_F \supset SU(5)$, and consider only complete $SU(5)$ multiplets in order to preserve the unification of gauge couplings) and weakly gauge it. The resulting model still breaks supersymmetry as long as the SM gauge force is weak, since the vacuum energy must be continuous in the gauge coupling constant. Moreover the SM sector "knows" about supersymmetry breaking just via its gauge interactions. In principle this idea seems very promising. In practice, though, there are problems to implement it, some of which are largely present today. The main problem is the loss of asymptotic freedom in the SM gauge factors. Models that both break supersymmetry and have $G_F \supset SU(3) \times SU(2) \times U(1)$ have a large gauge group G_S. From the point of view of the SM groups (or G_F) the different "colours" of G_S are just different flavours. This means that, being G_S large, there are in general many flavours of SM matter in the supersymmetry-breaking (messenger) sector. In practice this makes the gauge couplings of the SM to blow up a few decades above the messenger scale. This is often well below the GUT scale. For instance, consider the class of DSSB of refs. [1,77], based on $G_S = SU(N)$ (N odd) with an antisymmetric tensor A and $N - 4$ antifundamentals $\bar{F}_i$ in the matter sector. The smallest model where an anomaly free $SU(5)$ gauge group can be embedded is based on $SU(15)$, for which there are 15 families of messengers above the supersymmetry-breaking scale. A way out could be to embed only $SU(2)_W \times U(1)$ among the SM group factors in the messenger sector. But this would be problematic: on one side gauge unification would be typically lost and on the other the gluino mass would be tiny as it arises at a high order in perturbation theory.

Another problem of the early attempts was the spontaneous breakdown of the R-symmetry, which is generic in models with spontaneous supersymmetry breaking. The resulting axion couples to the QCD anomaly and is phenomenologically problematic as its scale f_a is typically low $\sim 10 - 100$ TeV. This can however be considered a smaller problem. Indeed there may exist different sources of explicit R breaking, which are small enough not to restore supersymmetry, at least in a nearby vacuum, and large enough to give the axion a mass that renders it phenomenologically harmless. In ref. [80] it was shown that in some models this rôle can be played by $1/M_P$ suppressed dimension-five operators in W. Moreover, in ref. [9] it was pointed out that, when R is broken at the same scale as supersymmetry, the cancellation of the cosmological constant, by adding a constant term to W in supergravity, implies an amount of explicit R breaking that gives the axion an acceptably large mass. Finally, there are now models, where R is broken at a higher scale than supersymmetry. This scale can consistently be in the astrophysically allowed window $10^9 - 10^{12}$ GeV, so that the R axion could indeed solve the strong CP problem.

The revival of DSSB and gauge mediation, that was started by Dine, Nelson and collaborators[44,46,47] in 1993, had the main goal of tackling the difficulties due to the loss of asymptotic freedom. In order to build realistic models, the approach was to assume the DSSB sector is completely neutral under the gauge forces of the SM. The basic structure of this class of models consist of three sectors, one to break supersymmetry, one to mediated it and finally one containing the ordinary particles. These sectors only communicate with each other via weak gauge forces. Supersymmetry is first broken in sector I, and the news of it is transferred to sector II (the messengers) via an abelian gauge group $U(1)_m$, denoted as messenger hypercharge. Sector III, the observable sector, is neutral under $U(1)_m$, so that the breakdown of supersymmetry is felt there only at the next step, via the usual gauge interactions. This structure seems fairly complicated and *ad hoc*, nonetheless models in this class provided the first examples of a realistic theory with calculable, dynamical, and universal soft terms.

In the last few years, there has been an active research towards simple theories of DSSB. An interesting attempt to circumvent the problem of Landau poles was suggested in refs.[87,7]. The idea is to generate models in which $X \gg \sqrt{F_X}$ in a natural way. The larger value of the messenger mass X, can displace the Landau pole above the GUT scale, even for a large number of messenger families. One considers DSSB models, like those introduced in ref.[86], which have classically flat directions that are only lifted by non-renormalizable operators. Unfortunately these kind of models have, quite generically, negative squark mass squared. There have also been attempts[72,71,69] to consider gauge theories with quantum moduli spaces[91], which do not break supersymmetry, and couple them to singlets to obtain effective O'Raifertaigh models. A characterization of this class of models was given in refs.[78,38,94]. The basic idea is to introduce some gauge interactions for the chiral field which couples to the strongly-interacting mesons on the quantum moduli space. This allows to create a stable minimum far away from the origin, along some flat direction. This mechanism for stabilizing a tree-level flat potential is just Witten's inverse hierarchy[100]. Other stabilization mechanisms have also been discussed in refs.[74,75]. Finally, attempts to avoid intermediate sectors involving singlet fields were presented in refs.[88,93,27].

In conclusion, there are at present many different ways gauge mediation can be realized in the context of dynamical supersymmetry breaking. Future research will hopefully clarify if any of these models is relevant for the description of Nature.

References

1. I. Affleck, M. Dine and N. Seiberg, *Phys. Lett.* B **137**, 187 (1984).
2. I. Affleck, M. Dine and N. Seiberg, *Nucl. Phys.* B **256**, 557 (1985).
3. L. Alvarez-Gaumé, M. Claudson, and M. Wise, *Nucl. Phys.* B **207**, 96 (1982).
4. K. Agashe and M. Graesser, preprint hep-ph/9704206.
5. S. Ambrosanio, G.L. Kane, G.D. Kribs, S.P. Martin, and S. Mrenna, *Phys. Rev.* D **54**, 5395 (1996).
6. S. Ambrosanio, G.D. Kribs, and S.P. Martin, *Phys. Rev.* D **56**, 1761 (1997).
7. N. Arkani-Hamed, J. March-Russell, and H. Murayama, preprint hep-ph/9701286.
8. K.S. Babu, C. Kolda, F. Wilczek, *Phys. Rev. Lett.* **77**, 3070 (1996).
9. J. Bagger, E. Poppitz and L. Randall, *Nucl. Phys.* B **426**, 3 (1994).
10. J.A. Bagger, K. Matchev, D.M. Pierce, and R.J. Zhang, *Phys. Rev.* D **55**, 3188 (1997).
11. R. Barbieri, S. Ferrara, and C.A. Savoy, *Phys. Lett.* B **119**, 343 (1982).
12. R. Barbieri and G.F. Giudice, *Nucl. Phys.* B **306**, 63 (1988).
13. R. Barbieri and L.J. Hall, *Phys. Lett.* B **338**, 212 (1994).
14. R. Barbieri, L.J. Hall, and A. Strumia, *Nucl. Phys.* B **445**, 219 (1995).
15. G. Bhattacharyya and A. Romanino, *Phys. Rev.* D **55**, 7015 (1997).
16. F.M. Borzumati, preprint hep-ph/9702307.
17. A. Brignole, F. Feruglio, and F. Zwirner, *Nucl. Phys.* B **501**, 332 (1997).
18. A. Brignole, F. Feruglio, and F. Zwirner, preprint hep-ph/9709111.
19. M. Carena, J.R. Espinosa, M. Quiros, and C.E.M. Wagner, *Phys. Lett.* B **355**, 209 (1995).
20. M. Carena, P. Chankowski, M. Olechowski, S. Pokorski, and C.E.M. Wagner, *Nucl. Phys.* B **491**, 103 (1997).
21. C.D. Carone and H. Murayama, *Phys. Rev.* D **53**, 1658 (1996).
22. R. Casalbuoni, S. De Curtis, D. Dominici, F. Feruglio, and R. Gatto, *Phys. Lett.* B **215**, 313 (1988).
23. R. Casalbuoni, S. De Curtis, D. Dominici, F. Feruglio, and R. Gatto, *Phys. Rev.* D **39**, 2281 (1989).
24. A.H. Chamseddine, R. Arnowitt, and P. Nath, *Phys. Rev. Lett.* **49**, 970 (1982).
25. P. Ciafaloni and A. Strumia, *Nucl. Phys.* B **494**, 41 (1997).
26. T.E. Clark and S.T. Love, *Phys. Rev.* D **54**, 5723 (1996).
27. C. Csaki, L. Randall, and W. Skiba, preprint hep-ph/9707386.
28. S. Deser and B. Zumino, *Phys. Rev. Lett.* **38**, 1433 (1977).

29. S. Dimopoulos and S. Raby, *Nucl. Phys.* B **192**, 1981 (353).

30. S. Dimopoulos and H. Georgi, *Nucl. Phys.* B **193**, 150 (1981).

31. S. Dimopoulos and S. Raby, *Nucl. Phys.* B **219**, 479 (1983).

32. S. Dimopoulos, G.F. Giudice, and N. Tetradis, *Nucl. Phys.* B **454**, 59 (1995).

33. S. Dimopoulos and G.F. Giudice, *Phys. Lett.* B **379**, 105 (1996).

34. S. Dimopoulos, S. Thomas, and J.D. Wells, *Phys. Rev.* D **54**, 3283 (1996).

35. S. Dimopoulos, G.F. Giudice, and A. Pomarol, *Phys. Lett.* B **389**, 37 (1996).

36. S. Dimopoulos and G.F. Giudice, *Phys. Lett.* B **393**, 72 (1997).

37. S. Dimopoulos, S. Thomas, and J.D. Wells, *Nucl. Phys.* B **488**, 39 (1997).

38. S. Dimopoulos, G. Dvali, R. Rattazzi, and G.F. Giudice preprint hep-ph/9705307.

39. S. Dimopoulos, G. Dvali and R. Rattazzi, preprint hep-ph/9707537.

40. M. Dine, W. Fischler, and M. Srednicki, *Nucl. Phys.* B **189**, 575 (1981).

41. M. Dine and W. Fischler, *Phys. Lett.* B **110**, 227 (1982).

42. M. Dine and M. Srednicki, *Nucl. Phys.* B **202**, 238 (1982).

43. M. Dine and W. Fischler, *Nucl. Phys.* B **204**, 346 (1982).

44. M. Dine and A.E. Nelson, *Phys. Rev.* D **48**, 1277 (1993).

45. M. Dine, R. Leigh, and A. Kagan, *Phys. Rev.* D **48**, 2214 (1993).

46. M. Dine, A.E. Nelson, and Y. Shirman, *Phys. Rev.* D **51**, 1362 (1995).

47. M. Dine, A.E. Nelson, Y. Nir, and Y. Shirman, *Phys. Rev.* D **53**, 2658 (1996).

48. M. Dine, Y. Nir, and Y. Shirman, *Phys. Rev.* D **55**, 1501 (1997).

49. G. Dvali, G.F. Giudice, and A. Pomarol, *Nucl. Phys.* B **478**, 31 (1996).

50. J. Ellis, G. Ridolfi, and F. Zwirner, *Phys. Lett.* B **257**, 83 (1991).

51. P. Fayet, *Phys. Lett.* B **70**, 461 (1977).

52. P. Fayet, *Phys. Lett.* B **84**, 416 (1979).

53. P. Fayet, *Phys. Lett.* B **86**, 272 (1979).

54. S. Ferrara, L. Girardello, and F. Palumbo, *Phys. Rev.* D **20**, 403 (1979).

55. F. Gabbiani, E. Gabrielli, A. Masiero, and L. Silvestrini, *Nucl. Phys.* B **477**, 321 (1996)

56. E. Gabrielli and U. Sarid, preprint hep-ph/9707546.

57. T. Gherghetta, *Nucl. Phys.* B **485**, 25 (1997).

58. G. F. Giudice and A. Masiero, *Phys. Lett.* B **206**, 480 (1988).

59. G.F. Giudice and R. Rattazzi, preprint hep-ph/9706540.

60. G.F. Giudice and R. Rattazzi, preprint hep-ph/9801271.

61. A. de Gouvêa, T. Moroi, and H. Murayama, *Phys. Rev.* D **56**, 1281

(1997).

62. A. de Gouvêa, A. Friedland, and H. Murayama, preprint hep-ph/9711264.

63. H.E. Haber and G.L. Kane, *Phys. Reports* **117**, 75 (1985).

64. H.E. Haber and R. Hempfling, *Phys. Rev. Lett.* **66**, 1815 (1991).

65. H.E. Haber and J. Gunion, this volume.

66. J.S. Hagelin, S. Kelley, and T. Tanaka, *Nucl. Phys.* B **415**, 293 (1994).

67. L.J. Hall, J. Lykken, and S. Weinberg, *Phys. Rev.* D **27**, 2359 (1983).

68. L.J. Hall, V.A. Kostelecky, and S. Raby, *Nucl. Phys.* B **267**, 415 (1986).

69. T. Hotta, K.-I. Izawa and T. Yanagida, *Phys. Rev.* D **55**, 415 (1997).

70. K. Intriligator and N. Seiberg, *Nucl. Phys. Proc. Suppl.* **45BC** (1996) 1.

71. K. Intriligator and S. Thomas, *Nucl. Phys.* B **473**, 121 (1996).

72. K.-I. Izawa and T. Yanagida, *Prog. Theor. Phys.* **95**, 829 (1996).

73. M.A. Luty and E. Ponton, preprint hep-ph/9706268.

74. M.A. Luty, preprint hep-ph/9706554.

75. M.A. Luty and J. Terning, preprint hep-ph/9709306.

76. S.P. Martin, *Phys. Rev.* D **55**, 3177 (1997).

77. Y. Meurice and G. Veneziano, *Phys. Lett.* B **141**, 69 (1984).

78. H. Murayama, *Phys. Rev. Lett.* **79**, 18 (1997).

79. C.R. Nappi and B.A. Ovrut, *Phys. Lett.* B **113**, 175 (1982).

80. A.E. Nelson and N. Seiberg, *Nucl. Phys.* B **416**, 46 (1994).

81. Y. Nir and N. Seiberg, *Phys. Lett.* B **309**, 337 (1993).

82. Y. Nir, *Phys. Lett.* B **354**, 107 (1995)

83. Y. Okada, M. Yamaguchi, and T. Yanagida, *Prog. Theor. Phys.* **85**, 1 (1991).

84. L. O'Raifertaigh, *Nucl. Phys.* B **96**, 331 (1975).

85. M.E. Peskin, Proc. Theoretical Advanced Study Institute in Elementary Particle Physics (TASI 96), Fields, Strings, and Duality, Boulder, CO, USA, preprint hep-th/9702094.

86. E. Poppitz, Y. Shadmi and S. Trivedi *Phys. Rev. Lett.* **388**, 561 (1996).

87. E. Poppitz and S.P. Trivedi, *Phys. Rev.* D **55**, 5508 (1997).

88. L. Randall, *Nucl. Phys.* B **495**, 37 (1997).

89. R. Rattazzi and U. Sarid, *Nucl. Phys.* B **501**, 297 (1997).

90. A. Riotto, O. Tornkvist, R.N. Mohapatra, *Phys. Lett.* B **388**, 599 (1996).

91. N. Seiberg, *Phys. Lett.* B **318**, 469 (1993).

92. N. Seiberg, *Phys. Rev.* D **49**, 6857 (1994).

93. Y. Shadmi, *Phys. Lett.* B **405**, 99 (1997).

94. Y. Shirman, preprint hep-ph/9709383.

95. M.A. Shifman, *Prog. Part. Nucl. Phys.* **39**, 1 (1997).

96. A. Strumia, *Phys. Lett.* B **409**, 213 (1997).
97. L. Susskind, *Phys. Rev.* D **20**, 2619 (1979).
98. D.V. Volkov and V.A. Soroka, *JETP Lett.* **18**, 312 (1973).
99. S. Weinberg, *Phys. Rev.* D **19**, 1277 (1978).
100. E. Witten, *Phys. Lett.* B **105**, 267 (1981).
101. E. Witten, *Nucl. Phys.* B **202**, 253 (1982).
102. T. Yanagida, *Phys. Lett.* B **400**, 109 (1997).

CHARGE AND COLOR BREAKING

J. ALBERTO CASAS

*Instituto de Estructura de la Materia, Serrano 123 (CSIC),
28006 Madrid, Spain*

The presence of scalar fields with color and electric charge in supersymmetric theories makes feasible the existence of dangerous charge and color breaking (CCB) minima and unbounded from below directions (UFB) in the effective potential, which would make the standard vacuum unstable. The avoidance of these occurrences imposes severe constraints on the supersymmetric parameter space. We give here a comprehensive and updated account of this topic.

1   Introduction

Experimental observation tells us that color and electric charge are gauge quantum numbers preserved in nature. From the thoretical point of view, in the Standard Model they are certainly conserved in an automatical way since the only fundamental scalar field is the Higgs boson, a colorless electroweak doublet. The Higgs potential has a continuum of degenerate minima, but these are all physically equivalent and one can always define the unbroken $U(1)$ generator to be the electric charge. In supersymmetric (SUSY) extensions of the Standard Model things become more complicated. First, the Higgs sector must contain for consistency at least two Higgs doublets H_1, H_2 (plus perhaps some singlets or triplets). Hence, one has to check that the minimum of the Higgs potential $V(H_1, H_2)$ still occurs for values of H_1, H_2 which are apropriately aligned in order to preserve the electric charge; otherwise the whole electroweak symmetry becomes spontaneously broken. Second, the supersymmetric theory has a large number of additional charged and colored scalar fields, namely all the sleptons and squarks, say $\tilde{l}_i$, $\tilde{q}_i$. Consequently one has to verify that the minimum of the whole potential $V(H_1, H_2, \tilde{q}_i, \tilde{l}_i)$ still occurs at a point in the field space, which we will call "realistic minimum" in what follows, where $\tilde{q}_i, \tilde{l}_i = 0$, thus preserving color and electric charge.

The generic situation is that the scalar potential does not present just a single minimum, and, besides the realistic minimum, there is a number of additional charge and color breaking (CCB) minima. Then, a reasonable requirement is that the realistic minimum is the deepest one, i.e the global minimum of the theory. This is certainly the usual constraint imposed in the literature and represents the most conservative attitude in order to be safe. Nevertheless, a situation with CCB minima deeper than the realistic minimum could still be acceptable if the cosmology leads the universe to the latter and this is stable enough. This issue will be discussed in the last section of this chapter.

CCB minima are not the only disease that the supersymmetric scalar potential can present. It may also happen that the field space contains directions along which the potential becomes unbounded form below (UFB), which is obviously undesirable. Both issues, CCB and UFB, are closely related, as we will see throughout the chapter.

In order to introduce some notation and to illustrate some relevant aspects and warnings concerning CCB, let us briefly review the CCB condition which has been most extensively used in the literature, namely the "traditional" bound, first studied by Frere et al. and subsequently by others [1,2]. The *tree-level* scalar potential, V_0, in the minimal supersymmetric standard model (MSSM) is given by

$$V_0 = V_F + V_D + V_{\text{soft}} \ , \tag{1}$$

with

$$V_F = \sum_\alpha \left| \frac{\partial W}{\partial \phi_\alpha} \right|^2 \ , \tag{2a}$$

$$V_D = \frac{1}{2} \sum_a g_a^2 \left(\sum_\alpha \phi_\alpha^\dagger T^a \phi_\alpha \right)^2 \ , \tag{2b}$$

$$
\begin{aligned}
V_{\text{soft}} \ = \ & \sum_\alpha m_{\phi_\alpha}^2 |\phi_\alpha|^2 + \sum_{i \equiv generations} \{ A_{u_i} \lambda_{u_i} Q_i H_2 u_i + A_{d_i} \lambda_{d_i} Q_i H_1 d_i \\
+ \ & A_{e_i} \lambda_{e_i} L_i H_1 e_i + \text{h.c.} \} + (B\mu H_1 H_2 + \text{h.c.}) \ ,
\end{aligned} \tag{2c}
$$

where W is the MSSM superpotential

$$W = \sum_{i \equiv generations} \{ \lambda_{u_i} Q_i H_2 u_i + \lambda_{d_i} Q_i H_1 d_i + \lambda_{e_i} L_i H_1 e_i \} + \mu H_1 H_2 \ , \tag{3}$$

ϕ_α runs over all the scalar components of the chiral superfields and a, i are gauge group and generation indices respectively. Q_i (L_i) are the scalar partners of the quark (lepton) $SU(2)_L$ doublets and u_i, d_i (e_i) are the scalar partners of the quark (lepton) $SU(2)_L$ singlets. In our notation $Q_i \equiv (u_L, \ d_L)_i$, $L_i \equiv (\nu_L, \ e_L)_i$, $u_i \equiv u_{Ri}$, $d_i \equiv d_{Ri}$, $e_i \equiv e_{Ri}$. Finally, $H_{1,2}$ are the two SUSY Higgs doublets. The first observation is that the previous potential is extremely involved since it has a large number of independent fields. Furthermore, even assuming universality of the soft breaking terms at the unification scale, M_X, it contains a large number of independent parameters: m, M, A, B, μ, i.e. the universal scalar and gaugino masses, the universal coefficients of the trilinear and bilinear scalar terms, and the Higgs mixing mass, respectively. In addition, there are the gauge (g) and Yukawa (λ) couplings which are constrained by the experimental data. Notice that M does not appear explicitly in V_0, but it does through the renormalization group equations (RGEs) of all the remaining parameters.

The complexity of V has made that until recently only particular directions in the field-space have been explored. The best-known example of this is the "traditional" bound, first studied by Frere et al. and subsequently by others[1,2]. These authors considered just the three fields present in a particular trilinear scalar coupling, e.g. $\lambda_u A_u Q_u H_2 u$, assuming equal vacuum expectation values (VEVs) for them:

$$|Q_u| = |H_2| = |u| \ , \tag{4}$$

where only the u_L-component of Q_u takes a VEV in order to cancel the D–terms. The phases of the three fields are taken in such way that the trilinear scalar term in the potential gets negative sign. Then, the potential (1) gets extremely simplified and it is easy to show that a very deep CCB minimum appears *unless* the famous constraint

$$|A_u|^2 \leq 3 \left(m_{Q_u}^2 + m_u^2 + m_2^2 \right) \tag{5}$$

is satisfied. In the previous equation $m_{Q_u}^2, m_u^2, m_2^2$ are the mass parameters of Q_u, u, H_2. Notice from eq.(1) that m_2^2 is the sum of the H_2 squared soft mass, $m_{H_2}^2$, plus μ^2. Similar constraints for the other trilinear terms can straightforwardly be written. These "traditional" bounds have extensively been used in the literature. Notice that the trilinear coefficient, A, plays a crucial role for the appearance of a CCB minimum. This is logical since the scalar trilinear terms are essentially negative contributions to the scalar potential (they are negative for a certain combination of the phases of the fields). However, we will see in sect.4 that they are irrelevant for UFB directions.

From the previous bound we can extract two important lessons. First, many ordinary CCB bounds (as the one of eq.(5)) come from the analysis of particular directions in the field-space, thus corresponding to *necessary but not sufficient* conditions to avoid dangerous CCB minima. Consequently a complete analysis requires a more exhaustive exploration of the field space. Second, the bound of eq.(5) has been obtained from the analysis of the tree-level potential V_0. Hence, the *radiative corrections* should be incorporated in some way. With regard to this point a usual practice has been to consider the tree-level scalar potential improved by one-loop RGEs, so that all the parameters appearing in it (see eq.(1)) are running with the renormalization scale, Q. Then it is demanded that the previous CCB constraints, i.e. eq.(5) and others, are satisfied at any scale between M_X and M_Z. As we will see in sect.2 this procedure is not correct and leads to an overestimate of the restrictive power of the bounds. Therefore a more careful treatment of the radiative corrections is necessary when analyzing CCB bounds.

The chapter is organized as follows. Section 2 is devoted to analyze and give prescriptions to handle the above-mentioned issue of the radiative corrections. Section 3 deals with the Higgs part of the potential, which is a requirement for subsequent analyses. In sections 4 and 5 a complete analysis of the UFB and UFB directions of the MSSM field space is performed, giving a complete set of optimized bounds. Special attention will be paid to the most powerful one, the so-called UFB-3 bound. The effective restrictive power of these bounds is examined in section 6. Section 7 is devoted to the bounds that CCB pose on flavour mixing couplings, which turn out to be surprisingly strong. Finally, the cosmological considerations are left for section 8.

2 The role of the radiative corrections

As has been mentioned in sect.1, in the CCB analysis the scalar potential is usually considered at tree-level, improved by one-loop RGEs, so that all the parameters appearing in it (see eq.(1)) are running with the renormalization scale, Q. The two questions that arise are:

- What is the appropriate scale, say $Q = \hat{Q}$, to evaluate V_0?

- How important are the radiative corrections that are being ignored?

These two questions are intimately related. To understand this it is important to recall that the *exact* effective potential

$$V(Q, \lambda_\alpha(Q),\ m_\beta(Q), \phi(Q)) \tag{6}$$

(in short $V(Q,\phi)$), where $\lambda_\alpha(Q), m_\beta(Q)$ are running parameters and masses and $\phi(Q)$ are the generic classical fields, is scale-independent, i.e.

$$\frac{dV}{dQ} = 0 \ . \tag{7}$$

This property allows in principle any choice of Q, and in particular a different one for each value of the classical fields, i.e. $Q = f(\phi)$. When analyzing CCB bounds, one is interested in possible CCB minima , so one has to minimize the scalar potential. Denoting by $\langle \phi \rangle$ the VEVs of the ϕ–fields obtained from

382

the minimization of V, it is clear from (7) that the two following minimization conditions

$$\frac{\partial V(Q = f(\phi), \phi)}{\partial \phi} = 0 \tag{8}$$

$$\left.\frac{\partial V(Q, \phi)}{\partial \phi}\right|_{Q=f(\phi)} = 0 \tag{9}$$

yield equivalent results for $\langle \phi \rangle$ (for a more detailed discussion see refs.[3,4]).

The previous results apply exactly *only* to the exact effective potential. In practice, however, we can only know V with a certain degree of accuracy in a perturbative expansion. In particular, at one-loop level

$$V_1 = V_0(Q, \phi) + \Delta V_1(Q, \phi) \tag{10}$$

where V_0 is the (one-loop improved) tree-level potential and ΔV_1 is the one-loop radiative correction to the effective potential

$$\Delta V_1 = \sum_\alpha \frac{n_\alpha}{64\pi^2} M_\alpha^4 \left[\log \frac{M_\alpha^2}{Q^2} - \frac{3}{2}\right] . \tag{11}$$

Here $M_\alpha^2(Q)$ are the improved tree-level squared mass eigenstates and $n_\alpha = (-1)^{2s_\alpha}(2s_\alpha + 1)$, where s_α is the spin of the corresponding particle. It is important to notice that $M_\alpha^2(Q)$ are in general field–dependent quantities since they are the eigenvalues of the $(\partial^2 V_0/\partial\phi_i\partial\phi_j)$ matrix. Hence, the values of $M_\alpha^2(Q)$ depend on the values of the fields and thus on which direction in the field space is being analyzed. $V_1(Q, \phi)$ does *not* obey eq.(7) for all values of Q. However, in the region of Q of the order of the most significant masses appearing in (11), the logarithms involved in the radiative corrections, and the radiative corrections themselves (i.e. ΔV_1), are minimized, thus improving the perturbative expansion. So we expect V to be well approximated by V_1 and it is not surprising that in that region of Q, V_1 is approximately scale-independent [5,6], i.e. eq.(7) is nearly satisfied. On the other hand, due to the smallness of ΔV_1, V_1 and V_0 are, in this region, very similar. Consequently (always in this region of Q) we can safely approximate V by V_1 or even[a] V_0, and minimize by using either eq.(8) or eq.(9), although of course eq.(9) is much easier to handle. This statement can be numerically confirmed, see e.g. refs.[4,7].

In conclusion, the radiative corrections are reasonably well incorporated by using the tree-level potential $V_0(\phi, \hat{Q})$, where the renormalization scale $\hat{Q}$

---

[a]More precisely, for a choice of Q such that $\partial \Delta V_1/\partial \phi = 0$ the results from V_0 and V_1 are the same. In practice this precise condition is quite involved and such a degree of precision is not necessary.

is of the order of the most significant mass, normally $\hat{Q} \sim \phi$. The application of these recipes to our task of determining the CCB minima and extract the corresponding CCB bounds will be shown in sects.4,5.

3 The Higgs potential and the realistic minimum

The Higgs part of the MSSM potential can be extracted (at tree level) from eq.(1). It reads

$$
\begin{aligned}
V_{\text{Higgs}} &= m_1^2|H_1|^2 + m_2^2|H_2|^2 - m_3^2\left(\epsilon_{ij}H_1^i H_2^j + \text{h.c.}\right) - \frac{1}{2}g_2^2\left|\epsilon_{ij}H_1^i H_2^j\right|^2 \\
&+ \frac{1}{8}(g_2^2 + g'^2)(|H_2|^4 + |H_1|^4) + \frac{1}{8}(g_2^2 - g'^2)|H_2|^2|H_1|^2,
\end{aligned} \tag{12}
$$

where $H_1 \equiv (H_1^0,\ H_1^-)$, $H_2 \equiv (H_2^+,\ H_2^0)$, $m_1^2 \equiv m_{H_1}^2 + \mu^2$, $m_2^2 \equiv m_{H_2}^2 + \mu^2$, $m_3^2 \equiv -\mu B$ and g_2, g' are the gauge couplings of $SU(2) \times SU(1)_Y$. All these parameters are understood to be running parameters evaluated at some renormalization scale Q.

Our first interest in V_{Higgs} comes from the fact that V_{Higgs} depends not only on the neutral components of H_1, H_2, but also on the charged ones, i.e. H_1^-, H_2^+. Hence, one should check that $\langle H_1^- \rangle$, $\langle H_2^+ \rangle$ remain vanishing when V_{Higgs} is minimized (one of them, say $\langle H_2^+ \rangle$, can always be chosen as vanishing through an $SU(2)$ rotation). Fortunately, it is easy to show from (12) that the minimum of V_{Higgs} always lies at $H_2^+ = H_1^- = 0$. So the MSSM is safe from this point of view. It is worth remarking that non-minimal supersymmetric extensions of the standard model do not have this nice property, at least in such an automatic way. (This is e.g. the case of the so-called next-to-minimal supersymmetric standard model (NMSSM), which contains an extra singlet in the Higgs sector [8].) Therefore we can set $H_2^+ = H_1^- = 0$ and focuss our attention on the neutral part of V_{Higgs}, which reads

$$
V_{\text{Higgs}} = m_1^2|H_1|^2 + m_2^2|H_2|^2 - 2|m_3^2||H_1||H_2| + \frac{1}{8}(g'^2 + g_2^2)(|H_2|^2 - |H_1|^2)^2. \tag{13}
$$

Notice that, since we are interested in the minimization of the potential, we have implicitly chosen in (13) a phase of H_1, H_2 such that the mixing term $\propto (\epsilon_{ij}H_1^i H_2^j + \text{h.c.})$ gets negative.

The second aspect of V_{Higgs} which interests us is that V_{Higgs} should develope a minimum at $|H_1^0| = v_1$, $|H_2^0| = v_2$, such that $SU(2) \times U(1)_Y$ is broken in the correct way, i.e. $v_1^2 + v_2^2 = 2M_W^2/g_2^2 \simeq (175\ rmGeV)^2$. This is the realistic minimum that corresponds to the standard vacuum. This requirement fixes one of the five independent parameters (m, M, A, B, μ) of the MSSM, say

μ, in terms of the others. Actually, for some choices of the four remaining parameters (m, M, A, B), there is no value of μ capable of producing the correct electroweak breaking. Therefore, this requirement restricts the parameter space further, as is illustrated in Fig.1 (darked region) with a representative example (which will be discussed in detail in sect.6). In addition, the actual value of the potential at the realistic minimum, say $V_{\text{real min}}$, is important for the CCB analysis since the possible CCB vacua are dangerous as long as they deeper than $V_{\text{real min}}$. From (13) it is straightforward to get $V_{\text{real min}}$

$$V_{\text{real min}} = -\frac{1}{8}(g'^2 + g_2^2)(v_2^2 - v_1^2)^2 = -\frac{\left\{\left[(m_1^2 + m_2^2)^2 - 4|m_3|^4\right]^{1/2} - m_1^2 + m_2^2\right\}^2}{2\,(g'^2 + g_2^2)} \tag{14}$$

Note that this is the result obtained by minimizing just the tree-level part of (13). As explained in sect.2 this procedure is correct if the minimization is performed at some sensible scale Q, which should be of the order of the most relevant mass entering ΔV_1, see eq.(11). Since we are dealing here with the Higgs-dependent part of the potential, that mass is necessarily of the order of the largest Higgs-dependent mass, namely the largest stop mass. From now on we will denote this scale by $M_S{}^b$.

Finally, to be considered as realistic, the previous minimum must be really a minimum in the *whole* field-space. This simply implies that all the scalar squared mass eigenvalues (squarks and sleptons) must be positive. Actually, we should go further and demand that all the not yet observed particles, i.e. charginos, squarks, etc., have masses compatible with the experimental bounds.

4 Unbounded from below (UFB) constraints

In this section we analyze the constraints that arise from directions in the field-space along which the (tree-level) potential can become unbounded from below (UFB). It is in fact possible to give a *complete* clasification of the potentially dangerous UFB directions and the corresponding constraints in the MSSM. In order to understand what are the dangerous directions and the form of the corresponding bounds it is useful to notice the following two general properties about UFB in the MSSM:

1 Contrary to what happens to the CCB minima (see sect.1), the trilinear scalar terms cannot play a significant role along an UFB direction

---

[b]A more precise estimate of M_S was given in [7], but for our purposes this is accurate enough.

since for large enough values of the fields the corresponding quartic (and positive) F–terms become unavoidably larger.

2 Since all the physical masses must be positive at $Q = M_S$, the only negative terms in the (tree-level) potential that can play a relevant role along an UFB direction are[c]

$$m_2^2|H_2|^2 \ , \quad -2|m_3^2||H_1||H_2| \quad . \tag{15}$$

Therefore, any UFB direction must involve, H_2 and, *perhaps*, H_1. Furthermore, since the previous terms are cuadratic, all the quartic (positive) terms coming from F– and D–terms must be vanishing or kept under control along an UFB direction. This means that, in any case, besides H_2 some additional field(s) are required for that purpose. In all the instances, the preferred additional fields are H_1 and/or sleptons since they normally have smaller soft masses and therefore amount to a less positive contribution to the potential.

Using the previous general properties we can completely clasify the possible UFB directions in the MSSM. Special attention should be paid to the UFB–3 bound, which is the strongest one:

UFB-1

The first possibility is to play just with H_1 and H_2. Then, the relevant terms of the potential are those written in eq.(13). Obviously, the only possible UFB direction corresponds to choose $H_1 = H_2$ (up to $O(m_i)$ differences which are negligible for large enough values of the fields), so that the quartic D–term is cancelled. Thus, the (tree-level) potential along the UFB-1 direction is

$$V_{\text{UFB}-1} = (m_1^2 + m_2^2 - 2|m_3^2|)|H_2|^2 \ . \tag{16}$$

The constraint to be imposed is that, for *any* value of $|H_2| < M_X$,

$$V_{\text{UFB}-1}(Q = \hat{Q}) > V_{\text{real min}}(Q = M_S) \ , \tag{17}$$

where $V_{\text{real min}}$ is the value of the realistic minimum, given by eq.(14), and $V_{\text{UFB}-1}$ is evaluated at an appropriate scale $\hat{Q}$ (see sect.2). $\hat{Q}$ must be of the same order as the most significant mass along this UFB-1 direction,

---

[c]The only possible exception are the stop soft mass terms $m_{Q_t}^2|Q_t|^2 + m_t^2|t|^2$ since the stop masses are given by $\sim (m_{Q_t,t}^2 + M_{top}^2 \pm \text{mixing})$, but this possibility is barely consistent with the present bounds on squark masses.

which is obviously of order H_2. More precisely $\hat{Q} \sim \text{Max}(g_2|H_2|,\ \lambda_{top}|H_2|,\ M_S)$. Consequently, from (16) the bound (17) is accurately equivalent to the well-known condition

$$m_1^2 + m_2^2 \geq 2|m_3^2|. \tag{18}$$

From the previous discussion, it is clear that the bound (18) must be satisfied at any $Q > M_S$ and, in particular, at $Q = M_X$.

UFB-2

If, besides H_2, H_1, we consider additional fields in the game, it is easy to check by simple inspection (see property 2 above) that the best possible choice is a slepton L_i (along the ν_L direction), since it has the lightest mass without contributing to further quartic terms in V. Consequently, from eq.(1), the relevant potential reads

$$
\begin{aligned}
V &= m_1^2|H_1|^2 + m_2^2|H_2|^2 - 2|m_3^2||H_1||H_2| + m_{L_i}^2|L_i|^2 \\
&+ \frac{1}{8}(g'^2 + g_2^2)(|H_2|^2 - |H_1|^2 - |L_i|^2)^2.
\end{aligned} \tag{19}
$$

By minimizing V with respect to H_1, L_i, it is possible to write these two fields in terms of H_2. This step leads to non-trivial results provided that $|m_3^2| < \mu^2$, $|H_2|^2 > 4m_{L_i}^2/(g'^2 + g_2^2)\left[1 - \frac{|m_3|^4}{\mu^4}\right]$; otherwise the optimum value for L_i is $L_i = 0$, and we come back to the direction UFB-1. Then, the potential along the UFB-2 direction reads[d]

$$V_{\text{UFB}-2} = \left[m_2^2 + m_{L_i}^2 - \frac{|m_3|^4}{\mu^2}\right]|H_2|^2 - \frac{2m_{L_i}^4}{g'^2 + g_2^2}\ . \tag{20}$$

From (20) it might seem that the potential is unbounded from below unless $m_2^2 + m_{L_i}^2 - \frac{|m_3|^4}{\mu^2} \geq 0$. However, strictly, the UFB-2 constraint reads

$$V_{\text{UFB}-2}(Q = \hat{Q}) > V_{\text{real min}}(Q = M_S)\ , \tag{21}$$

where $V_{\text{real min}}$ is the value of the realistic minimum, given by eq.(14), and $V_{\text{UFB}-2}$ is evaluated at an appropriate scale $\hat{Q}$. Again $\hat{Q} \sim \text{Max}(g_2|H_2|, \lambda_{top}|H_2|,\ M_S)$.

---

[d]Eq.(20) relies on the equality $m_1^2 - m_{L_i}^2 = \mu^2$, which only holds under the assumption of degenerate soft scalar masses for H_1 and L_i at M_X and in the approximation of neglecting the bottom and tau Yukawa couplings in the RGEs. Otherwise, one simply must replace μ^2 by $m_1^2 - m_{L_i}^2$ in eq.(20).

UFB-3

The only remaining possibility is to take $H_1 = 0$. Then, the H_1 F–term can be cancelled with the help of the VEVs of sleptons of a particular generation, say e_{L_j}, e_{R_j}, without contributing to further quartic terms. More precisely

$$\left| \frac{\partial W}{\partial H_1} \right|^2 = \left| \mu H_2 + \lambda_{e_j} e_{L_j} e_{R_j} \right|^2 = 0 \ , \tag{22}$$

where λ_{e_j} is the corresponding Yukawa coupling. It is important to note that this trick is not useful if $H_1 \neq 0$, as it happens in the UFB–2 direction, since then the e_{L_j}, e_{R_j} F–terms would eventually dominate. Now, in order to cancel (or keep under control) the $SU(2)_L$ and $U(1)_Y$ D–terms we need the VEV of some additional field, which cannot be H_1 for the above mentioned reason. Once again the optimum choice is a slepton L_i (with $i \neq j$) along the ν_L direction, as in the UFB–2 case. Denoting $|e_{L_j}| = |e_{R_j}| \equiv |e| = \sqrt{\frac{|\mu|}{\lambda_{e_j}}|H_2|}$, the relevant potential reads

$$\begin{aligned}
V &= (m_2^2 - \mu^2)|H_2|^2 + (m_{L_j}^2 + m_{e_j}^2)|e|^2 + m_{L_i}^2|L_i|^2 \\
&+ \frac{1}{8}(g'^2 + g_2^2)(|H_2|^2 + |e|^2 - |L_i|^2)^2.
\end{aligned} \tag{23}$$

Now, the value of L_i can be written, by simple minimization, in terms of H_2, namely $|L_i|^2 = -\frac{4m_{L_i}^2}{g'^2 + g_2^2} + (|H_2|^2 + |e|^2)$. It turns out that for any value of $|H_2| < M_X$ satisfying

$$|H_2| > \sqrt{\frac{\mu^2}{4\lambda_{e_j}^2} + \frac{4m_{L_i}^2}{g'^2 + g_2^2}} - \frac{|\mu|}{2\lambda_{e_j}} \ , \tag{24}$$

the value of the potential along the UFB–3 direction is simply given by

$$V_{\text{UFB}-3} = \left[m_2^2 - \mu^2 + m_{L_i}^2 \right] |H_2|^2 + \frac{|\mu|}{\lambda_{e_j}} \left[m_{L_j}^2 + m_{e_j}^2 + m_{L_i}^2 \right] |H_2| - \frac{2m_{L_i}^4}{g'^2 + g_2^2} \ , \tag{25}$$

Otherwise

$$\begin{aligned}
V_{\text{UFB}-3} &= \left[m_2^2 - \mu^2 \right] |H_2|^2 + \frac{|\mu|}{\lambda_{e_j}} \left[m_{L_j}^2 + m_{e_j}^2 \right] |H_2| \\
&+ \frac{1}{8}(g'^2 + g_2^2) \left[|H_2|^2 + \frac{|\mu|}{\lambda_{e_j}}|H_2| \right]^2 .
\end{aligned} \tag{26}$$

Then, the UFB-3 condition reads

$$V_{\text{UFB}-3}(Q = \hat{Q}) > V_{\text{real min}}(Q = M_S) \,, \qquad (27)$$

where $V_{\text{real min}}$ is given by eq.(14), $\hat{Q} \sim \text{Max}(g_2|e|, \lambda_{top}|H_2|, g_2|L_i|, M_S)$. It is interesting to mention that the previous constraint (27) with the replacements $e \to d$, $\lambda_{e_j} \to \lambda_{d_j}$, $L_j \to Q_j$, must also be imposed. Now i may be equal to j (the optimum choice is d_j = sbottom) and $\hat{Q} \sim \text{Max}(\lambda_{top}|H_2|, g_3|d|, \lambda_{u_j}|d|, g_2|L_i|, M_S)$.

Anyway, the optimum condition is the one with the sleptons (note e.g. that the second term in eqs.(25, 26) is proportional to the slepton masses and thus smaller) and indeed represents, as we will see in sect.6, the *strongest* one of *all* the UFB and CCB constraints in the parameter space of the MSSM.

5 Charge and color breaking (CCB) constraints

These constraints arise from the existence of CCB minima in the potential deeper than the realistic minimum. We have already mentioned the "traditional" CCB constraint [1] of eq.(5). Other particular CCB constraints have been explored in the literature [9,10,11]. In this section we will perform a *complete* analysis of the CCB minima, obtaining a set of analytic constraints that represent the *necessary and sufficient* conditions to avoid the dangerous ones. As we will see, for certain values of the initial parameters, the CCB constraints "degenerate" into the previously found UFB constraints since the minima become unbounded from below directions. In this sense, the following CCB constraints comprise the UFB bounds of the previous section, which can be considered as special (but extremely important) limits of the former.

In order to gain intuition about CCB, let us enumerate a number of general properties which are relevant when one is looking for CCB constraints in the MSSM. (Formal proofs of the following statements can be found in ref.[7].)

1 The most dangerous, i.e. the deepest, CCB directions in the MSSM potential involve only one particular trilinear soft term of one generation (see eq.(2c)). This can be either of the leptonic type (i.e. $A_{e_i}\lambda_{e_i}L_iH_1e_i$) or the hadronic type (i.e. $A_{u_i}\lambda_{u_i}Q_iH_2u_i$ or $A_{d_i}\lambda_{d_i}Q_iH_1d_i$). Along one of these CCB directions the remaining trilinear terms are vanishing or negligible. This is because the presence of a non-vanishing trilinear term in the potential gives a net negative contribution only in a region of the field space where the relevant fields are of order A/λ with λ and A the corresponding Yukawa coupling and soft trilinear coefficient; otherwise either

the (positive) mass terms or the (positive) quartic F–terms associated with these fields dominate the potential. In consequence two trilinear couplings with different values of λ cannot efficiently "cooperate" in any region of the field space to deepen the potential. Accordingly, to any optimized CCB constraint there corresponds a unique relevant trilinear coupling, which makes the analysis much easier.

2 If the trilinear term under consideration has a Yukawa coupling $\lambda^2 \ll g^2$, where g represents a generic gauge coupling constant, then along the corresponding deepest CCB direction the D-term must be vanishing or negligible. This occurs essentially in all the cases except for the top, and simplifies enormously the analysis.

From the previous properties it can be checked that for a given trilinear coupling under consideration there are *two* different relevant directions to explore. Next, we illustrate them taking the trilinear coupling of the first generation, $A_u \lambda_u Q_u H_2 u_R$, as a guiding example.

Direction (a)

It exploits the trick expounded in the direction UFB-3. Namely, if $H_1 = 0$, then one can take two d-type squarks d_{L_j}, d_{R_j} (or sleptons e_{L_j}, e_{R_j}) such that $\lambda_{d_j} \gg \lambda_u$ (or $\lambda_{e_j} \gg \lambda_u$), so that their VEVs cancel the H_1 F–term, i.e.

$$\left| \frac{\partial W}{\partial H_1} \right|^2 = \left| \mu H_2 + \lambda_{d_j} d_{L_j} d_{R_j} \right|^2 = 0 \ . \tag{28}$$

Notice that H_1 must be very small or vanishing, otherwise the (positive) d_{L_j} and d_{R_j} F–terms, $\lambda_{d_j}^2 \left\{ |H_1 d_{R_j}|^2 + |d_{L_j} H_1|^2 \right\}$, would clearly dominate the potential (this is also in agreement with the property 1 above). Since $|d_{L_j}|^2, |d_{R_j}|^2 \ll |H_2|^2, |Q_u|^2, |u_R|^2$, the d_{L_j}, d_{R_j} mass terms are negligible and the net effect of eq.(28) is to decrease the H_2 squared mass from m_2^2 to^e $m_2^2 - \mu^2$. Furthermore, in addition to $H_2, Q_u, u_R, d_{L_j}, d_{R_j}$, other fields could take extra non-vanishing VEVs. As in the above-explained UFB-2 direction (see sect.4) and for similar reasons, it turns out that the optimum choice is $L_i \neq 0$, with the VEV along the ν_L direction. Therefore, along the direction (a)

$$H_2, Q_u, u_R \neq 0 \ , \quad \text{Possibly} \quad L_i \neq 0 \ , \tag{29a}$$

$$|d_{L_j}|^2 = |d_{R_j}|^2 \ ; \quad d_{L_j} d_{R_j} = -\frac{\mu}{\lambda_{d_j}} H_2 \tag{29b}$$

---

^{e}Recall that $m_2^2 - \mu^2 = m_{H_2}^2$, i.e. the H_2 soft mass, see sect.3.

Direction (b)

If we allow for $H_1 \neq 0$, then we cannot play the trick of eq.(28) to cancel the H_1 F–term. Therefore, along this alternative direction

$$H_2, Q_u, u_R, H_1 \neq 0 \,, \quad \text{Possibly} \quad L_i \neq 0 \,, \tag{30}$$

Let us now write the potential along the directions (a), (b). It is useful for this task to express the various VEVs in terms of the H_2 one, using the following notation [10]

$$\begin{aligned}
|Q_u| &= \alpha|H_2| \,, \quad |u_R| = \beta|H_2| \,, \\
|H_1| &= \gamma|H_2| \,, \quad |L_i| = \gamma_L|H_2| \,.
\end{aligned} \tag{31}$$

E.g. the "traditional" direction, eq.(4), is recovered for the particular values $\alpha = \beta = 1$, $\gamma = \gamma_L = 0$. Now, the basic expression for the scalar potential (see eq.(1)) is

$$V = \lambda_u^2 F(\alpha,\beta,\gamma,\gamma_L)\alpha^2\beta^2|H_2|^4 - 2\lambda_u \hat{A}(\gamma)\alpha\beta|H_2|^3 + \hat{m}^2(\alpha,\beta,\gamma,\gamma_L)|H_2|^2 \,, \tag{32}$$

where

$$\begin{aligned}
F(\alpha,\beta,\gamma,\gamma_L) &= 1 + \frac{1}{\alpha^2} + \frac{1}{\beta^2} + \frac{f(\alpha,\beta,\gamma,\gamma_L)}{\alpha^2\beta^2} \,, \\
f(\alpha,\beta,\gamma,\gamma_L) &= \frac{1}{\lambda_u^2}\left\{ \frac{1}{8}g_2^2\left(1 - \alpha^2 - \gamma^2 - \gamma_L^2\right)^2 \right. \\
&\quad \left. + \frac{1}{8}g'^2\left(1 + \frac{1}{3}\alpha^2 - \frac{4}{3}\beta^2 - \gamma^2 - \gamma_L^2\right)^2 + \frac{1}{6}g_3^2\left(\alpha^2 - \beta^2\right)^2 \right\} \,, \\
\hat{A}(\gamma) &= |A_u| + |\mu|\gamma \,, \\
\hat{m}^2(\alpha,\beta,\gamma,\gamma_L) &= m_2^2 + m_{Q_u}^2\alpha^2 + m_u^2\beta^2 + m_1^2\gamma^2 + m_{L_i}^2\gamma_L^2 - 2|m_3^2|\gamma \,. \tag{33}
\end{aligned}$$

For the (a)–direction eqs.(32,33) hold replacing $\gamma = 0$, $m_2^2 \to m_2^2 - \mu^2$ in eq.(33). For the (b)–direction, when $\text{sign}(A_u) = \text{sign}(B)$, it is not possible to choose the phases of the fields in such a way that the trilineal scalar coupling ($\propto A_u\lambda_u Q_u H_2 u_R$), the cross term in the H_2 F–term ($\propto \mu\lambda_u Q_u H_1 u_R$) and the Higgs mixing term ($\propto \mu B H_1 H_2$) become negative at the same time [7]. Correspondingly, one (any) of the three terms $\{|A_u|, |\mu|\gamma, -2|m_3^2|\gamma\}$ in eq.(33) must flip the sign.

Minimizing V with respect to $|H_2|$ for fixed values of $\alpha,\beta,\gamma,\gamma_L$, we find, besides the $|H_2| = 0$ extremal (all VEVs vanishing), the following CCB solution

$$|H_2|_{ext} = |H_2(\alpha,\beta,\gamma,\gamma_L)|_{ext} = \frac{3\hat{A}}{4\lambda_u\alpha\beta F}\left\{1 + \sqrt{1 - \frac{8\hat{m}^2 F}{9\hat{A}^2}}\right\} \,. \tag{34}$$

$$V_{\text{CCB min}} = -\frac{1}{2}\alpha\beta|H_2|^2_{ext}\left(\hat{A}\lambda_u|H_2|_{ext} - \frac{\hat{m}^2}{\alpha\beta}\right) . \tag{35}$$

Notice that, as was stated above (see property 1), the typical VEVs at a CCB minimum are indeed of order A/λ. The previous CCB minimum will be negative[f] (and much deeper than the realistic minimum) unless

$$\hat{A}^2 \leq F\hat{m}^2 \tag{36}$$

This is in fact the most general form of a CCB constraint.

The previous CCB bound takes a more handy form if we realize that since $\lambda_u^2 \ll 1$ (see property 3 above) the D–terms should vanish. This implies $\alpha^2 - \beta^2 = 0$, $1 - \alpha^2 - \gamma^2 - \gamma_L^2 = 0$. As a consequence $f(\alpha, \beta, \gamma, \gamma_L)$ becomes vanishing and $F = 1 + \frac{2}{\alpha^2}$.

Now, we can write the explicit form of the bounds for the directions (a, b):

CCB-1

This bound arises by considering the direction (a) and thus the general condition (36) takes the form

$$|A_u|^2 \leq \left(1 + \frac{2}{\alpha^2}\right)\left[m_2^2 - \mu^2 + (m_{Q_u}^2 + m_u^2)\alpha^2 + m_{L_i}^2\gamma_L^2\right] , \tag{37}$$

where α^2 is arbitrary and γ_L^2 is given by $\gamma_L^2 = 1 - \alpha^2$. More precisely

1. If $m_2^2 - \mu^2 + m_{L_i}^2 > 0$ and $3m_{L_i}^2 - (m_{Q_u}^2 + m_u^2) + 2(m_2^2 - \mu^2) > 0$, then the optimized CCB-1 occurs for $\alpha = 1$, i.e.

$$|A_u|^2 \leq 3\left[m_2^2 - \mu^2 + m_{Q_u}^2 + m_u^2\right] \tag{38}$$

2. If $m_2^2 - \mu^2 + m_{L_i}^2 > 0$ and $3m_{L_i}^2 - (m_{Q_u}^2 + m_u^2) + 2(m_2^2 - \mu^2) < 0$, then the optimized CCB-1 bound is

$$|A_u|^2 \leq \left(1 + \frac{2}{\alpha^2}\right)\left[m_2^2 - \mu^2 + (m_{Q_u}^2 + m_u^2)\alpha^2 + m_{L_i}^2(1 - \alpha^2)\right] \tag{39}$$

with $\alpha^2 = \sqrt{\dfrac{2(m_{L_i}^2 + m_2^2 - \mu^2)}{m_{Q_u}^2 + m_u^2 - m_{L_i}^2}}.$

---

[f] The mere existence of a CCB minimum is discarded by demanding $\hat{A}^2 < (8/9)F\hat{m}^2$, see eq.(34).

3. If $m_2^2 - \mu^2 + m_{L_i}^2 < 0$, then the CCB-1 bound is automatically violated. In fact the minimization of the potential in this case gives $\alpha^2 \to 0$, and we are exactly led to the UFB-3 direction shown above, which represents the correct analysis in this instance.

CCB-2

This bound arises by considering the direction (b). Then the general condition (36) takes the form

$$
\begin{aligned}
(|A_u| + |\mu|\gamma)^2 \;\; \leq \;\; & \left(1 + \frac{2}{\alpha^2}\right)\left[m_2^2 + (m_{Q_u}^2 + m_u^2)\alpha^2\right. \\
& + \;\; \left. m_1^2\gamma^2 + m_{L_i}^2\gamma_L^2 - 2|m_3^2|\gamma\right]
\end{aligned} \tag{40}
$$

where α^2, γ^2 are arbitrary and γ_L^2 is given by $\gamma_L^2 = 1 - \alpha^2 - \gamma^2$. Rules to handle this bound in an efficient way (i.e. to take the values of α^2, γ^2 that make the bound as strong as possible) can be found in ref.[7].

If $\mathrm{sign}(A_u) = \mathrm{sign}(B)$, the sign of one of the three terms $\{|A_u|, |\mu|\gamma, -2|m_3^2|\gamma\}$ in (40) must be flipped (see comments after eq.(33)). Notice that, due to the form of (40) flipping the sign of $|A_u|$ or the sign of $|\mu|\gamma$ leads to the same result. Therefore, there are only two choices to examine.

Concerning the renormalization scale at which the previous CCB-1, CCB-2 constraints must be evaluated, a sensible choice is $\hat{Q} \sim \mathrm{Max}(M_S, g_3\frac{A_u}{4\lambda_u}, \lambda_t\frac{A_u}{4\lambda_u})$, since $H_2 \sim \frac{A_u}{4\lambda_u}$, see eq.(34).

The previous CCB-1, CCB-2 bounds are straightforwardly generalized to all the couplings with coupling constant $\lambda \ll 1$. This essentialy includes all the couplings apart from the top. The generalization to the top Yukawa coupling case is more involved since $\lambda_{top} = O(1)$, so the D-terms should not be assumed to vanish anymore. Furthermore, the CCB-1 bounds are not longer applicable due to the absence of d–type squarks such that $\lambda_{d_j} \gg \lambda_{top}$. Finally, the associated CCB minima have in many cases a similar size to the realistic one. So, it is important to examine explicitly the condition $V_{\mathrm{CCB\ min}} > V_{\mathrm{real\ min}}$.

For more details, the interested reader is referred to ref.[7].

6 Constraints on the SUSY parameter space

In sections 4–6 a complete analysis of all the potentially dangerous unbounded from below (UFB) and charge and color breaking (CCB) directions has been carried out. Now, we wish to show explicitly, through a numerical analysis, the restrictive power of the constraints on the MSSM parameter space. We will see that this is certainly remarkable.

We will consider the whole parameter space of the MSSM, m, M, A, B, μ, with the only assumption of universality[9]. Actually, universality of the soft SUSY-breaking terms at M_X is a desirable property not only to reduce the number of independent parameters, but also for phenomenological reasons, particularly to avoid flavour-changing neutral currents (see, e.g. ref.[12]). As discussed in sect.3, the requirement of correct electroweak breaking fixes one of the five independent parameters of the MSSM, say μ, so we are left with only four parameters (m, M, A, B). In order to present the results in a clear way we will start by considering the particular case $m = 100$ GeV and $B = A - m$ (i.e. the well–known minimal SUGRA relation [13]), and later we will let B to vary freely.

Fig.1a shows the region excluded by the "traditional" CCB bounds of the type of eq.(5), evaluated at an *appropriate* scale (see sect.2). Clearly, the "traditional" bounds, when correctly evaluated, turn out to be very weak. In fact, only the leptonic (circles) and the d–type (diamonds) terms do restrict, very modestly, the parameter space. Let us recall here that it has been a common (incorrect) practice in the literature to evaluate these traditional bounds at all the scales between M_X and M_W, thus obtaining very important (and of course overestimated) restrictions in the parameter space. Fig.1b shows the region excluded by the "improved" CCB constraints obtained in sect.5. Clearly, the excluded region becomes dramatically increased. Notice also that all the trilinear couplings (except the top one in this case) give restrictions, producing areas constrained by different types of bounds simultaneously. The restrictions coming from the UFB constraints, obtained in sect.4, are shown in Fig.1c. By far, the most restrictive bound is the UFB–3 one (small filled squares). Indeed, the UFB–3 constraint is the *strongest* one of *all* the UFB and CCB constraints, excluding extensive areas of the parameter space. This is a most remarkable result. Finally, in Fig.1d we summarize all the constraints plotting also the excluded region due to the experimental bounds on SUSY particle masses (filled diamonds). The finally allowed region (white) is quite small.

How do these results evolve when we vary the values of m and B? The results indicate that the smaller the value of m the more restrictive the bounds become (an explanation of this behavior will be given below). More precisely for $m < 50$ GeV the whole parameter space becomes forbidden (for any value of the remaining parameters). So, from UFB and CCB constraints we can

---

[9]Let us remark, however, that the constraints found in previous sections are general and they can also be applied to the non-universal case.

conclude

$$m \geq 50 \text{ GeV} . \tag{41}$$

Obviously, the limiting case $m = 0$ is excluded. This is very relevant for no-scale models, since $m = 0$ is a typical prediction in that kind of scenarios. Concerning the remaing parameter, B, the results indicate that the larger the value of B, the more restrictive the bounds. In general, for $m \lesssim 500$ GeV[h], B has to satisfy the bound

$$|B| \lesssim 3.5 \, m . \tag{42}$$

Figures illustrating eqs.(41,42) can be found in refs.[7,15].

So far, we have just presented the numerical results in the figs.1a–d and eqs.(41,42) with no attempt to explain the physical reasons underlying them. It is, however, very instructive to examine this question. The first thing to note is that, due to their structure, the CCB bounds on A/m (see eqs.(37,40)) are essentially m–invariant and B–invariant. The numerical analysis confirms this fact [7,15]. On the other hand, the UFB–3, which is the strongest (CCB and UFB) bound, becomes more stringent as $m_{H_2}^2 = m_2^2 - \mu^2$ (i.e. the H_2 soft mass) becomes more negative. This is clear from eqs.(25–27). The precise value of $m_{H_2}^2$ at low energy depends on its initial value at M_X, i.e. m, and on the RG running that, due to the effect of λ_{top}, brings $m_{H_2}^2$ to negative values. Consequently, the smaller m and the larger λ_{top}, the stronger the UFB–3 bound becomes. Concerning m this result is certainly well reflected in eq.(41). Concerning λ_{top}, since $m_{top} \sim \lambda_{top}\langle H_2 \rangle$, where $\langle H_2 \rangle = 2M_W^2 \sin \beta / g_2^2$, it is clear that the smaller $\tan \beta$, the larger λ_{top} and therefore the stronger the UFB–3 bound. But $\tan \beta$ decreases as B increases, thus the form of eq.(42). On the other hand, values of $\tan \beta$ too close to 1 demand a value of λ_{top} at low energy higher than the infrared fixed point value, which is impossible to get from the running from high energies. This fact also contributes to the upper bound on $|B|$, eq.(42).

---

[h]Larger values of m start to conflict clearly the naturality bounds for electroweak breaking [14,6], so they are not realistic.

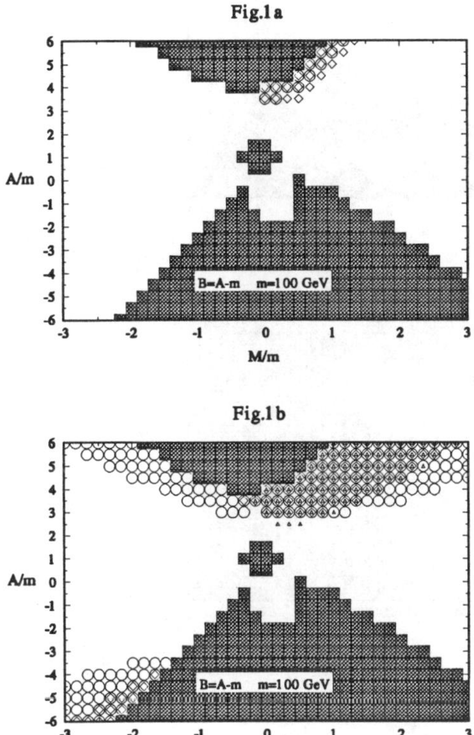

Figure 1: Excluded regions in the parameter space of the MSSM, with $B = A - m$, $m = 100$ GeV and $M_{\text{top}}^{\text{phys}} = 174$ GeV. The darked region is excluded because there is no solution for μ capable of producing the correct electroweak breaking. (a) The circles and diamonds indicate regions excluded by the "traditional" CCB constraints associated with the e and d-type trilinear terms respectively. (b) The same as (a) but using the "improved" CCB constraints. The triangles correspond to the u-type trilinear terms. (c) The crosses, squares and small filled squares indicate regions excluded by the UFB-1,2,3 constraints respectively. d) The previous excluded regions together with the one arising from the experimental lower bounds on supersymmetric particle masses (filled diamonds).

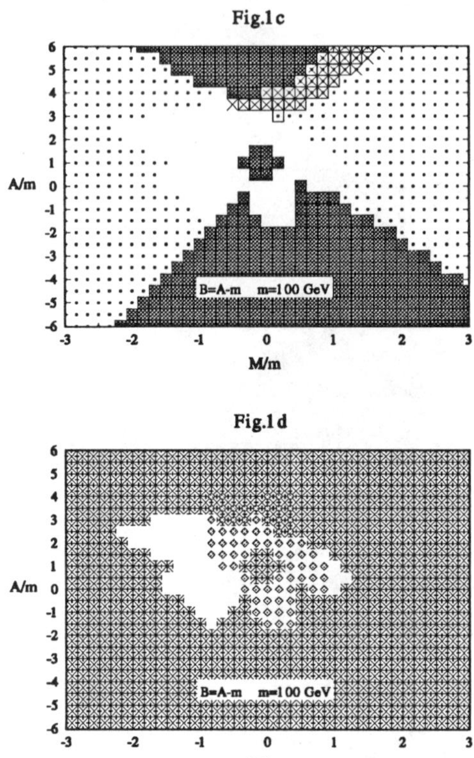

Figure 2: – Continued.

To summarize, the UFB and CCB bounds, specially the UFB-3 bound, put important constraints on the MSSM parameter space. Contrary to a common believe, the bounds affect not only the trilinear parameter, A, but also the values of the universal scalar mass, m, the bilinear term parameter, B, and the universal gaugino mass, M. This can be noted from the figures 1a–d and eqs.(41, 42). Also, the frequently used constraint $|A| \leq 3m$ is not in general a good approximation. The actual bounds on A depend on the values of the other SUSY parameters (m, M, B).

The application of the UFB and CCB bounds to particular SUSY scenarios has been considered in some works. It is worth–mentioning that the string-inspired dilaton-dominated scenario is completely excluded on these grounds[16] (as the above-mentioned no-scale scenarios). The infrared fixed point scenario is also severely constrained [15] (in particular it requires $M < 1.1\, m$).

7 CCB constraints on flavor-mixing couplings

Supersymmetry has sources of flavor violation which are not present in the Standard Model [17]. These arise from the possible presence of non-diagonal terms in the squark and slepton mass matrices, coming from the soft-breaking potential (see[i] eq.(2c))

$$
\begin{aligned}
V_{\text{soft}} \;=\; & \left(m_L^2\right)_{ij} \bar{L}_{L_i} L_{L_j} \;+\; \left(m_{e_R}^2\right)_{ij} \bar{e}_{R_i} e_{R_j} \\
+\; & \left(m_Q^2\right)_{ij} \bar{Q}_{L_i} Q_{L_j} \;+\; \left(m_{u_R}^2\right)_{ij} \bar{u}_{R_i} u_{R_j} \;+\; \left(m_{d_R}^2\right)_{ij} \bar{d}_{R_i} d_{R_j} \\
+\; & \left[A_{ij}^l \bar{L}_{L_i} H_1 e_{R_j} + A_{ij}^u \bar{Q}_{L_i} H_2 u_{R_j} + A_{ij}^d \bar{Q}_{L_i} H_1 d_{R_j} + \text{h.c.} \right] + ..(43)
\end{aligned}
$$

where $i, j = 1, 2, 3$ are generation indices. A usual simplifying assumption of the MSSM is that m_{ij}^2 is diagonal and universal and A_{ij} is proportional to the corresponding Yukawa matrix. Actually, we have implicitly used this assumption in all the previous sections. However, there is no compelling theoretical argument for these hypotheses[j].

The size of the off-diagonal entries in m_{ij}^2 and A_{ij} is strongly restricted by FCNC experimental data [17,18,19,20]. Here, we will focus our attention on the $A_{ij}^{(f)}$ terms; a summary of the corresponding FCNC bounds is given in the second column of Table 1 [18,20]. The $\left(\delta_{LR}^{(f)}\right)_{ij}$ parameters used in the table are defined as

$$
\left(\delta_{LR}^{(f)}\right)_{ij} \equiv \frac{\left(\Delta M_{LR}^{2\,(f)}\right)_{ij}}{M_{av}^{2\,(f)}} \,, \tag{44}
$$

where $f = u, d, l$; $M_{av}^{2\,(f)}$ is the average of the squared sfermion ($\tilde{f}_L$ and $\tilde{f}_R$) masses and $\left(\Delta M_{LR}^{2\,(f)}\right)_{ij} = A_{ij}^{(f)} \langle H_f^0 \rangle$, with $H_u^0 \equiv H_2^0$, $H_{d,l}^0 \equiv H_1^0$, are the off-diagonal entries in the sfermion mass matrices. It is remarkable that the $A_{ij}^{(f)}$ terms are also restricted on completely different grounds, namely from the requirement of the absence of dangerous charge and color breaking (CCB) minima or unbounded from below (UFB) directions. These bounds are in general *stronger than the FCNC ones*. Other properties of these bounds are the following:

---

[i] We work in a basis for the superfields where the Yukawa coupling matrices are diagonal.

[j] This is why, contrary to eq.(1), we have not factorized the Yukawa couplings, λ, in the trilinear terms in eq.(43)

i) Some of the bounds, particularly the UFB ones, are genuine effects of the non-diagonal $A_{ij}^{(f)}$ structure, i.e. they do not have a "diagonal counterpart".

ii) Contrary to the FCNC bounds, the strength of the CCB and UFB bounds does not decrease as the scale of supersymmetry breaking increases.

There is no room here to review in detail how these bounds arise, although the philosophy is similar to that explained in sects.4, 5 (for further details see ref.[21]). Let us write however the final form of the constraints

CCB bounds

$$\left|A_{ij}^{(u)}\right|^2 \leq \lambda_{u_k}^2 \left(m_{u_{L_i}}^2 + m_{u_{R_j}}^2 + m_2^2\right), \qquad k = \text{Max}\,(i,j)$$

$$\left|A_{ij}^{(d)}\right|^2 \leq \lambda_{d_k}^2 \left(m_{d_{L_i}}^2 + m_{d_{R_j}}^2 + m_1^2\right), \qquad k = \text{Max}\,(i,j)$$

$$\left|A_{ij}^{(l)}\right|^2 \leq \lambda_{e_k}^2 \left(m_{e_{L_i}}^2 + m_{e_{R_j}}^2 + m_1^2\right), \qquad k = \text{Max}\,(i,j) \qquad (45)$$

UFB bounds

$$\left|A_{ij}^{(u)}\right|^2 \leq \lambda_{u_k}^2 \left(m_{u_{L_i}}^2 + m_{u_{R_j}}^2 + m_{e_{L_p}}^2 + m_{e_{R_q}}^2\right), \qquad k = \text{Max}\,(i,j),\ p \neq q\,.$$

$$\left|A_{ij}^{(d)}\right|^2 \leq \lambda_{d_k}^2 \left(m_{d_{L_i}}^2 + m_{d_{R_j}}^2 + m_{\nu_m}^2\right), \qquad k = \text{Max}\,(i,j)$$

$$\left|A_{ij}^{(l)}\right|^2 \leq \lambda_{e_k}^2 \left(m_{e_{L_i}}^2 + m_{e_{R_j}}^2 + m_{\nu_m}^2\right), \qquad k = \text{Max}\,(i,j),\ m \neq i,j. \qquad (46)$$

The CCB bounds must be evaluated at a renormalization scale $Q \sim 2A_{ij}^{(f)}/\lambda_{f_k}^2$, while the UFB bounds must be imposed at any $Q^2 \gg (m/\lambda_{f_k})^2$. This can be relevant in many instances. For example, for universal gaugino and scalar masses ($M_{1/2}$ and m respectively) satisfying $M_{1/2} \gtrsim m$, the UFB bounds are more restrictive at M_X than at low energies (especially the hadronic ones). This trend gets stronger as the ratio $M_{1/2}/m$ increases.

The previous CCB and UFB bounds can be expressed in terms of the $\left(\delta_{LR}^{(f)}\right)_{ij}$ parameters defined in eq.(44) and compared with the corresponding FCNC bounds. It turns out that the former are almost always stronger. This is illustrated in Table 1 for the particular case $M_{av}^{(f)} = 500$ GeV. The only exception is $\left(\delta_{LR}^{(l)}\right)_{12}$, which is experimentally constrained by the $\mu \to e, \gamma$ process. As the scale of supersymmetry breaking increases the FCNC bounds are easily satisfied whereas the CCB and UFB bounds continue to strongly constrain the theory.

Another case in which the FCNC constraints are satisfied is when approximate "infrared universality" emerges from the RG equations [22,18,19]. Again, the CCB and UFB bounds continue to impose strong constraints on such theories. This is because, as argued before, these bounds have to be evaluated at different large scales and do not benefit from RG running.

Table 1: FCNC bounds versus CCB and UFB bounds on $(\delta_{LR}^{(f)})_{ij}$ for $M_{av}^{(f)} = 500$ GeV. The bounds have been obtained from ref.[20] taking $x = (m_{\mathrm{gaugino}}/M_{av}^{(f)})^2 = 1$.

	FCNC	CCB and UFB
$\left(\delta_{LR}^{(d)}\right)_{12}$	4.4×10^{-3}	2.9×10^{-4}
$\left(\delta_{LR}^{(d)}\right)_{13}$	3.3×10^{-2}	10^{-2}
$\left(\delta_{LR}^{(d)}\right)_{23}$	1.6×10^{-2}	10^{-2}
$\left(\delta_{LR}^{(u)}\right)_{12}$	3.1×10^{-2}	2.3×10^{-3}
$\left(\delta_{LR}^{(l)}\right)_{12}$	8.5×10^{-6}	3.6×10^{-4}
$\left(\delta_{LR}^{(l)}\right)_{13}$	5.5×10^{-1}	6.1×10^{-3}
$\left(\delta_{LR}^{(l)}\right)_{23}$	10^{-1}	6.1×10^{-3}

8 Cosmological considerations and final comments

As has been mentioned in sect.1, the CCB and UFB bounds presented here are conservative; they correspond to sufficient, but not necessary, conditions for the viability of the standard vacuum. It is possible that we live in a metastable vacuum [2,23], whose lifetime is longer than the age of the universe. This certainly softens the constraints obtained here.

The first study on CCB-metastability bounds was performed by Claudson et al [2]. They showed that only the top-Yukawa CCB bounds are dangerous from this point of view. In other words, among the various CCB minima (see sect.5), the one associated with the top-Yukawa coupling is the only one to which the realistic minimum has a substantial probability to decay during the universe life-time. The remaining CCB minima, although deeper, present too high barriers for an efficient tunnelling. That analysis has been re-done by Kusenko et al. [23], taking into account some subtleties when analyzing the transition probabilities. Their results are qualitatively similar to those of ref.[2]. Quantitatively, they obtain a bound similar to the CCB-1 bound (see eq.(38)), empirically modified as $|A_t|^2 \lesssim 3\left[m_2^2 - \mu^2 + 2.5(m_{Q_t}^2 + m_t^2)\right]$, which of course is weaker than the pure stability bound. On the other hand, the UFB bounds (in particular the UFB-3 bound, which is the strongest of all the CCB and UFB bounds) have not been analyzed yet from this point of view.

It is important to keep in mind that the metastability bounds represent necessary but perhaps not sufficient conditions to be safe (in the same sense that the stability bounds presented in sects.4–5 represent sufficient but perhaps not necessary conditions). The reason is that for the applicability of the metastability bounds, the universe should be driven by some mechanism into the realistic (but local and metastable) minimum. This problem has been treated in several papers [24,23]. Of course, a definite answer (not based in an anthropic principle) requires the consideration of a particular cosmological scenario in order to determine the initial values of the relevant fields at early times. Apparently, the realistic minimum is indeed favoured in many cosmological scenarios. Namely, if the initial conditions are dicted by thermal effects, the universe tends to fall into the realistic minimum since it is the closest one to the origin. This can also be the case in some inflationary scenarios. However, a more systematic analysis of these issues would be welcome.

Finally, from a more philosophic point of view, it is conceptually difficult to understand how the cosmological constant is vanishing precisely in a local "interim" vacuum (especially from an inflationary point of view). It is also interesting that many of the (tree-level) UFB directions presented here, particularly the ones associated with flavour violating couplings, are really unbounded from below (after radiative corrections) and, if present, make the theory ill-defined, at least until Planckean physics comes to the rescue. These issues, however, enter the realm of still unknown pieces of fundamental physics.

References

1. J.M. Frere, D.R.T. Jones and S. Raby, *Nucl. Phys.* **B222** (1983) 11; L. Alvarez-Gaumé, J. Polchinski and M. Wise, *Nucl. Phys.* **B221** (1983) 495; J.P. Derendinger and C.A. Savoy, *Nucl. Phys.* **B237** (1984)

307; C. Kounnas, A.B. Lahanas, D.V. Nanopoulos and M. Quirós, *Nucl. Phys.* **B236** (1984) 438.

2. M. Claudson, L.J. Hall and I.Hinchliffe, *Nucl. Phys.* **B228** (1983) 501.
3. C. Ford, D.R.T. Jones, P.W. Stephenson and M.B. Einhorn, *Nucl. Phys.* **B395** (1993) 17.
4. J.A. Casas, J.R. Espinosa, M. Quirós and A. Riotto, *Nucl. Phys.* **B436** (1995) 3.
5. G. Gamberini, G. Ridolfi and F. Zwirner, *Nucl. Phys.* **B331** (1990) 331.
6. B. de Carlos and J.A. Casas, *Phys. Lett.* **B309** (1993) 320.
7. J.A. Casas, A. Lleyda and C. Muñoz, *Nucl. Phys.* **B471** (1996) 3.
8. J. Ellis, J.F. Gunion, H.E. Haber, L. Roszkowski and F. Zwirner, *Phys. Rev.* **D39** (1989) 844.
9. M. Drees, M. Glück and K. Grassie, *Phys. Lett.* **B157** (1985) 164.
10. J.F. Gunion, H.E. Haber and M. Sher, *Nucl. Phys.* **B306** (1988) 1.
11. H. Komatsu, *Phys. Lett.* **B215** (1988) 323.
12. G.G. Ross, "Grand Unified Theories", Benjamin Publishing Co. (1985).
13. R. Barbieri, S. Ferrara and C.A. Savoy, *Phys. Lett.* **119B** (1982) 343; L. Hall, J. Lykken and S. Weinberg, *Phys. Rev.* **D27** (1983) 2359.
14. R. Barbieri and G.F. Giudice, *Nucl. Phys.* **B306** (1988) 63.
15. J.A. Casas, A. Lleyda and C. Muñoz, *Phys. Lett.* **B389** (1996) 305.
16. J.A. Casas, A. Lleyda and C. Muñoz, *Phys. Lett.* **B380** (1996) 59.
17. S. Dimopoulos and H. Georgi, *Nucl. Phys.* **B193** (1981) 150; F. Gabbiani and A. Masiero *Nucl. Phys.* **B322** (1989) 235; J.S. Hagelin, S. Kelley and T. Tanaka *Nucl. Phys.* **B415** (1994) 293; R. Barbieri and L.J. Hall, *Phys. Lett.* **B338** (1994) 212; S. Dimopoulos and D. Sutter *Nucl. Phys.* **B452** (1995) 49.
18. D. Choudhury, F. Eberlein, A. Konig, J. Louis and S. Pokorski, *Phys. Lett.* **B342** (1995) 180
19. B. de Carlos, J.A. Casas and J.M. Moreno, *Phys. Rev.* **D53** (1996) 6398.
20. F. Gabbiani, E. Gabrielli, A. Masiero and L. Silvestrini, *Nucl. Phys.* **B477** (1996) 321
21. J.A. Casas and S. Dimopoulos, *Phys. Lett.* **B387** (1996) 107.
22. M. Dine, A. Kagan and S. Samuel, *Phys. Lett.* **B243** (1990) 250
23. A. Kusenko, P. Langacker and G. Segre, *Phys. Rev.* **D54** (1996) 5824.
24. T. Falk, K. Olive, L. Roszkowski and M. Srednicki, *Phys. Lett.* **B367** (1996) 183; A. Riotto and E. Roulet, *Phys. Lett.* **B377** (1996) 60; A. Strumia, *Nucl. Phys.* **B482** (1996) 24.

SUPERSYMMETRIC LOOP EFFECTS

PIOTR H. CHANKOWSKI[a]

Santa Cruz Institute for Particle Physics
University of California, Santa Cruz, CA 95064, U.S.A.

STEFAN POKORSKI

Inst. of Theoretical Physics, Warsaw University, Hoża 69, 00-681 Warsaw, Poland
and
Laboratoire de Physique Theorique, Univ. de Paris-Sud 91400, Orsay, France.

We review the loop effects of the low energy supersymmetry. The global success of the Standard Model rises two related questions: how strongly the mass scales of the superpartners are constrained and can they be, nevertheless, indirectly seen in precision measurements. The bulk of the electroweak data is well screened from supersymmetric loop effects, due to the structure of the theory, even with superpartners generically light, $\mathcal{O}(M_Z)$. The only exception are the left-handed squarks of the third generation which have to be $\gtrsim \mathcal{O}(300\text{ GeV})$ to maintain the success of the SM. The other superpartners can still be light, at their present experimental mass limits, and would manifest themselves through virtual corrections to the small number of observables such as R_b, $b \to s\gamma$, $K^0\text{-}\bar{K}^0$ and $B^0\text{-}\bar{B}^0$ mixing and a few more for large $\tan\beta$. Those effects require still higher experimental precision to be detectable.

1 Introduction

The masses of the weak bosons, $W^\pm$ and Z^0, as well as fermion masses originate from the mechanism of spontaneous symmetry breaking. This mechanism requires the presence of elementary or composite scalar modes, the so-called Higgs modes which develop nonvanishing vacuum expectation values. In the Standard Model (SM), viewed as an effective low energy theory, the Higgs potential looks very unnatural and the theory faces the well known hierarchy problem. In brief, scalar potential is generically unstable with respect to quantum corrections from any new physics (the mass squared parameter of the potential receives loop corrections proportional to masses squared of the heavy new particles). Thus, the structure of the vacuum in the SM is strongly suggestive of the existence of new scale of fundamental interactions, the physical cut-off to the SM, close to the electroweak scale.

Supersymmetry offers an interesting solution to the hierarchy puzzle and, moreover, has several other theoretical and phenomenological (gauge coupling

---

[a]On leave of absence from the Institute of Theoretical Physics, Warsaw University, Hoża 69, 00-681 Warsaw, Poland.

unification) virtues. The new scale, the mentioned earlier cut-off to the SM, is the scale of soft supersymmetry breaking. In other words, this is the scale (often it can be defined only in some average sense) of the mass spectrum of the superpartners to the particles of the SM. Two immediate and most important remarks about the superpartner spectrum are the following ones: if supersymmetry is to cure the hierarchy problem that scale is expected to be not much above the electroweak scale. On the other hand, it is totally unknown in detail, as we do not have at present any realistic model of supersymmetry breaking. Therefore, the minimal supersymmetric extension of the SM, the so-called Minimal Supersymmetric Standard Model (MSSM) is a very well defined theoretical framework but contains many free parameters: superpartner soft masses and their dimensionful couplings.

Lack of any detailed knowledge about the superpartner masses has obvious implications for the direct search for superpartners which can only be based on systematic exploration of the higher and higher energy scales. It is, therefore, very interesting to discuss the question to what extent the superpartner spectrum can manifest itself through virtual (loop) effects on the electroweak observables. Do very high precision measurements of the electroweak observables provide us with a tool to see supersymmetric effects indirectly or, at least, to put stronger limits on its spectrum? We remember the important rôle played by precision measurements in seeing, indirectly, some evidence for the top quark long ago its direct discovery and with the mass quite close to its measured mass. Also, the present level of precision makes the electroweak measurements to some extent sensitive even to the Higgs boson mass, although the dependence is only logarithmic. With supersymmetric corrections the situation is different. The dependence on the top quark (and Higgs) mass in the SM is due to *nondecoupling* of heavy particles which get their masses through the mechanism of spontaneous symmetry breaking. The soft SUSY breaking is explicit and the Appelquist-Carazzone theorem [1] applies to the superpartner spectrum. Thus, supersymmetric virtual effects dissapear at least as $\mathcal{O}(1/M_{SUSY})$. Nevertheless, several interesting questions can be discussed and this is the content of this Chapter. First, we discuss the impact of the general succes of the SM in describing the precision data on the existence of new physics and on supersymmetry in particular. The resulting constraints on the SUSY spectrum are reviewed with emphasize on the existence of the room for very light, $\mathcal{O}(M_Z)$, particles. Indeed, most of the superpartners effectively decouple from most of the electroweak observables much faster than $\mathcal{O}(1/M_{SUSY})$. This high degree of screening follows from the basic structure of the theory. There are only few exceptions to this general rule: effects of light, $\mathcal{O}(M_Z)$, charginos, stops and the charged Higgs boson can be substantial in some specific observables like $R_b \equiv \Gamma(Z^0 \to \bar{b}b)/\Gamma(Z^0 \to hadrons)$ and some flavour changing neutral current (FCNC) processes. In addition, for large $\tan \beta$ sizeable loop corrections to the Yukawa couplings from those particles are also possible. Those effects are discussed in subsequent sections. The Chapter ends with a brief summary of the overall prospects.

2 Supersymmetry and the electroweak precision data

The bulk of the electroweak precision measurements (M_W, Z^0-pole observables, νe, ep scattering data, etc.) shows that the global comparison of the SM predictions with the data is impressive. Both, the experiment and the theory have at present similar accuracy, typically $\mathcal{O}(1 \; \%_{oo})$! The predictions of the SM are usually given in terms of the very precisely known parameters G_μ, α_{EM}, M_Z and the other three parameters $\alpha_s(M_Z)$, m_t, M_h. The top quark mass and the strong coupling constant are now also reported from independent experiments with considerable precision: $m_t = (175.6 \pm 5.5)$ GeV and $\alpha_s(M_Z) = 0.118 \pm 0.003$, but those measurements are difficult and it is safer to take α_s, m_t, M_h as parameters of an overall fit. Such fits give values of m_t and α_s very well consistent with the above values [2,3,4,5].

The theoretical uncertainties in the SM predictions (for fixed m_t, M_h, α_s) come mainly from the RG evolution of $\alpha_{EM} \equiv \alpha(0) \to \alpha(M_Z)$ which depends on the hadronic contribution to the photon vacuum polarization $\alpha(s) = \alpha(0)/(1 - \Delta\alpha(s))$ where $\Delta\alpha(s) = \Delta\alpha_{hadr} + \ldots$ and $\Delta\alpha_{hadr} = 0.0280 \pm 0.0007$ [6]. The present error in the hadronic vacuum polarization propagates as $\mathcal{O}(1 \; \%_{oo})$ errors in the final predictions. The other uncertainties come from the neglected higher order corrections and manifest themselves as renormalization scheme dependence, higher order arbitrariness in resummation formulae etc. Those effects have been estimated to be smaller than $\mathcal{O}(1 \; \%_{oo})$, hence the conclusion is that the theory and experiment agree with each other at the level of $\mathcal{O}(1 \; \%_{oo})$ accuracy. In particular, the genuine weak loop corrections are now tested at $\mathcal{O}(5\sigma)$ level and the precision is already high enough to see some sensitivity to the Higgs boson mass.

The electroweak observables depend only logaritmically on the Higgs boson mass (whereas the dependence on the top quark mass is quadratic). Global fits to the present data give $M_h \approx 130^{+130}_{-70}$ GeV and the 95% C.L. upper bound is around 470 GeV [7,5]. Thus, the data give some indication for a light Higgs boson. (It is worth noting that $M_h = 1$ TeV is more than $\gtrsim 3\sigma$ away from the best fit). The direct experimental lower limit on the SM Higgs boson mass M_h is ~ 70 GeV.

The overall global succes of the SM is a bit overshadowed by a couple of (quite relevant!) scattered clouds. The results for R_b are still preliminary and differ by about 2σ between different experiments. The world average is only 1.8σ away from the SM prediction but with the error which is only slightly smaller than the maximal possible enhancement of R_b in the MSSM (see later). The effective Weinberg angle is now reported with a very high precision. However, this result comes from averaging over the SLD and LEP results which are more then 3σ apart. Hoping for further experimental clarification of those few points it is interesting to discuss already now the impact of the general succes of the SM on the existence of new physics. The simplest interpretation of the success of the SM within the MSSM is that the superpartners are heavy enough to decouple from the electroweak observables. Explicit calculations (with the same precision as in the SM) show that this happens if the common supersymmetry breaking scale is $\geq \mathcal{O}(300-400)$ GeV[8,9,10,11]. This is very important as such a scale of supersymmetry breaking is still low enough for supersymmetry to cure the hierarchy problem. However, in this case the only supersymmetric signature at the electroweak scale and just above it is the Higgs sector with a light, $M_h \leq \mathcal{O}(150)$ GeV, Higgs boson. This prediction is consistent with the SM fits discussed earlier[7,5]. We can, therefore, conclude at this point that the supersymmetric extension of the SM, with all superpartners $\geq \mathcal{O}(300)$ GeV, is phenomenologically as succesful as the SM itself and has the virtue of solving the hierarchy problem. Discovery of a light Higgs boson is the crucial test for such an extension.

The relatively heavy superpartners discussed in the previous paragraph are sufficient for explaining the success of the SM. But is it necessary that all of them are that heavy? Is there a room for some light superpartners with masses $\mathcal{O}(M_Z)$ or even below? This question is of great importance for LEP2. Indeed, a closer look at the electroweak observables shows that the answer to this question is positive. The dominant quantum corrections to the electroweak observables are the so-called "oblique" corrections to the gauge boson self-energies. They are economically summarized in terms of the S, T, U[12,3] parameters

$$\alpha S \sim \Pi'_{3Y}(0) = \Pi'_{L3,R3} + \Pi'_{L3,B-L} \tag{1}$$

(the last decomposition is labelled by the $SU_L(2) \times SU_R(2) \times U_{B-L}(1)$ quantum numbers[13])

$$\alpha T \equiv \Delta\rho \sim \Pi_{11}(0) - \Pi_{33}(0) \tag{2}$$

$$\alpha U \sim \Pi'_{11}(0) - \Pi'_{33}(0) \tag{3}$$

where $\Pi_{ij}(0)$ $(\Pi'_{ij}(0))$ are the (i,j) left-handed gauge boson self-energies at the zero momentum (their derivatives) and the self-energy correction to the S parameter mixes W^3_μ and B_μ gauge bosons. It is clear from their definitions that the parameters S, T, U have important symmetry properties: T and U vanish when quantum corrections to the left-handed gauge boson self-energies leave unbroken "custodial" $SU_V(2)$ symmetry. The parameter S vanishes if $SU_L(2)$ remains an exact symmetry (notice that, since $\mathbf{3_L \otimes 3_R = 1 \oplus 5}$ under $SU_V(2)$, exact $SU_V(2)$ is not sufficient for vanishing of S [13]).

In terms of the parameters S, T and U the "new physics" contribution to the basic electroweak observables can be approximately written as [b]

$$\delta M_W = \frac{M_W}{2}\frac{\alpha}{c_W^2 - s_W^2}\left(c_W^2 T^{new} - \frac{1}{2}S^{new} + \frac{c_W^2 - s_W^2}{4s_W^2}U^{new}\right) \tag{4}$$

$$\delta \sin^2 \theta^{eff}_{lept} = -\frac{s_W^2 c_W^2}{c_W^2 - s_W^2}\left(\alpha T^{new} - \frac{\alpha}{4s_W^2 c_W^2}S^{new}\right) \tag{5}$$

$$\delta\Gamma(Z^0 \to \overline{f}f) = \frac{\alpha M_Z}{12s_W^2 c_W^2}\left[\left(g_V^2 + g_A^2\right)\left(\alpha T^{new}\right)\right.$$
$$\left. - 4g_V Q^f \frac{s_W^2 c_W^2}{c_W^2 - s_W^2}\left(-\alpha T^{new} - \frac{\alpha}{4s_W^2 c_W^2}S^{new}\right)\right] \tag{6}$$

where the parameters M_W, $c_W \equiv M_W/M_Z$, s_W, $g_V = -I_3^f + 2Q^f \sin^2 \theta^{eff}_{lept}$, $g_A = -I_3^f$ are computed in the SM (taking into account loop corrections) and with some reference values of m_t and M_h. S^{new}, T^{new}, U^{new} contain only the contributions from physics beyond the SM.

The success of the SM means that it has just the right amount of the $SU_V(2)$ breaking (and of the $SU_L(2)$ breaking), encoded mainly in the top quark-bottom quark mass splitting. Any extension of the SM, to be consistent with the precision data, should not introduce additional sources of large $SU_V(2)$ breaking in sectors which couple to the left-handed gauge bosons. In the MSSM, the main potential origin of new $SU_V(2)$ breaking effects in the left-handed sector is the splitting between the left-handed stop and sbottom

---

[b]We assume that the supersymmetric contributions of order M_Z^2/M_{SUSY}^2 to those observables are small and can be neglected. If it is not the case, the parameters S, T and U should be defined at non-zero momentum transfer [14] and one should take into account also the "new physics" contributions through additional parameters like $\Delta \alpha = \Pi'_{\gamma\gamma}(M_Z) - \Pi'_\gamma(0)$ and $e_5 = M_Z^2 F'_{ZZ}(M_Z^2)$ [15].

masses [16,14]:

$$M_{\tilde{t}_L}^2 = m_Q^2 + m_t^2 - \frac{1}{6}\cos 2\beta(M_Z^2 - 4M_W^2)$$

$$M_{\tilde{b}_L}^2 = m_Q^2 + m_b^2 - \frac{1}{6}\cos 2\beta(M_Z^2 + 2M_W^2) \tag{7}$$

The $SU_V(2)$ breaking is small if the common soft mass m_Q^2 is large enough. So, from the bulk of the precision data one gets a lower bound on the masses of the left-handed squarks of the third generation[c]. However, the right-handed top and bottom squarks can be very light, at their experimental lower bounds ~ 70 and ~ 150 GeV, respectively.

The other possible source of $SU_V(2)$ violation is in the slepton sector. For small values of m_L^2 the splitting between the left-handed slepton and sneutrino masses

$$M_{\tilde{\nu}}^2 = m_L^2 + \frac{1}{2}\cos 2\beta M_Z^2$$

$$M_{\tilde{l}}^2 = m_L^2 + m_l^2 + \frac{1}{2}\cos 2\beta(M_Z^2 - 2M_W^2) \tag{8}$$

becomes non-negligible. Since this mass splitting may be of similar magnitude for all three generations of sleptons this effect should also be considered[14]. It is also worth noting different behaviour of the slepton and squark contributions to the $SU_V(2)$ breaking: the former vanishes in the limit $\tan\beta \to 1$ whereas the latter is maximal in this limit and slightly decreases as $\tan\beta \to \infty$.

In Fig. 1 we show the lower bound on the mass of the heavier top squark as a function of the mass of the lighter stop, which follows from the requirement that a fit in the MSSM is at most by $\Delta\chi^2 = 2$ worse than the analogous fit in the SM. From the analysis of the SUSY contributions to the parameters T and S it follows that in the MSSM T^{new} is always positive whereas S^{new} is always negative[d]. Therefore, the full fits to the electroweak observables give more restrictive limits on the MSSM parameter space than do e.g. bounds on the $\Delta\rho(0)$ parameter alone because, as follows from eqs (4-6), the effects in the T and S always add up. In the context of Fig. 1 there is one more interesting observation to be made. In the low $\tan\beta$ region, for a given $M_{\tilde{t}_1}$, an absolute

---

[c]Additional source of the $SU_V(2)$ breaking is also in the A-terms. In principle, there can be cancellations between the soft mass term and the A-term contribution, such that another solution with small $SU_V(2)$ breaking exists with a large inverse hierarchy $m_U^2 \gg m_Q^2$. This is very unnatural from the point of view of the GUT boundary conditions and here we assume $m_Q^2 > m_U^2$.

[d]S^{new} could be positive only in the small window of the chargino parameter space which is already excluded by the unsuccesfull LEP search.

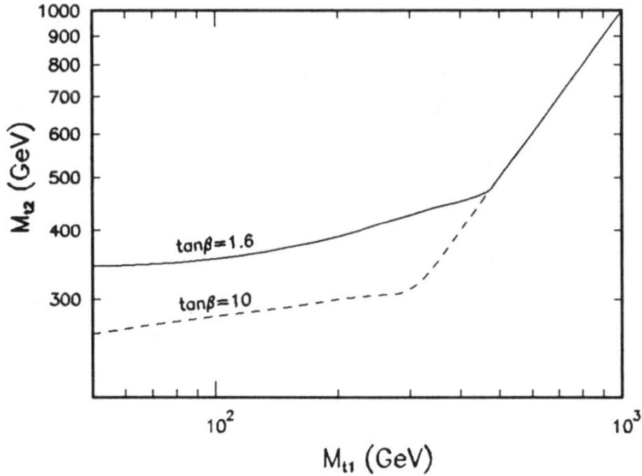

Figure 1: Lower bounds on the heavier stop mass $M_{\tilde{t}_2}$, as a function of $M_{\tilde{t}_1}$ for $\tan\beta = 1.6$ (solid line) and $\tan\beta = 10$ (dashed line). A scan over the top quark mass and the top squarks mixing angle $\theta_{\tilde{t}}$ has been performed.

lower bound on $M_{\tilde{t}_2}$ is set by the (conservative) experimental lower bound on the lightest supersymmetric Higgs boson mass, $M_h \gtrsim 60$ GeV. For low $\tan\beta$, the tree level Higgs boson mass is close to zero and radiative corrections are very important. They depend logarithmically on the product $M_{\tilde{t}_1} M_{\tilde{t}_2}$. Also, since the best fit in the SM requires $M_h \approx 130$ GeV, too small values of $M_{\tilde{t}_2}$ (leading to too small value of M_h) are disfavoured by the MSSM fit. The limit shown in Fig. 1 take both effects into account. They explain the difference between the limits for small and large $\tan\beta$ cases. The absolute limits on the stop masses obtained from the bound on M_h (which apply only for low $\tan\beta$) turns out to be slightly weaker than the limits from the fit shown in Fig. 1.

The important rôle played in the fit by the precise result for $\sin^2\theta_{eff}^{lept}$ is illustrated in Fig. 2a. The world average value (used in obtaining the bounds shown in Fig. 1) is obtained in the SM model with $m_t = (175 \pm 6)$ GeV and $M_h \sim (120 - 150)$ GeV, with little room for additional supersymmetric contribution. Hence, the relevant superpartners ($\tilde{t}_L$ and $\tilde{b}_L$) have to be heavy. With lighter superpartners, one obtains the band (solid lines) shown in Fig. 2a. We see that the SLD result for $\sin^2\theta_{eff}^{lept}$ leaves much more room for light superpartners. Thus, settling the SLD/LEP dispute is very relevant for new physics. Similar dependence for M_W is shown in Fig. 2b. The experimental

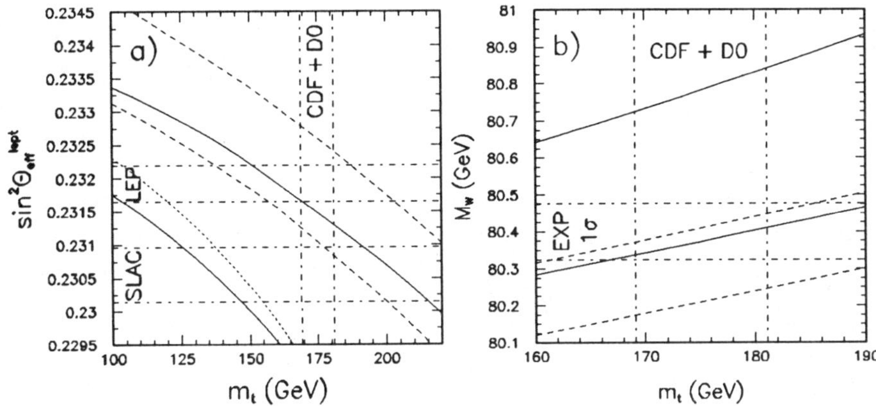

Figure 2: Predictions for $\sin^2 \theta_{eff}^{lept}$ (a) and M_W (b) in the SM (the band bounded by the dashed lines) and in the MSSM (solid lines) as functions of the top-quark mass. The bands are obtained by scanning over the MSSM parameters (M_h in the SM) respecting all available experimental limits. The SLC and the (average) LEP measurements for $\sin^2 \theta_{eff}^{lept}$ and 1σ experimental range for M_W are marked by horizontal dash-dotted lines. Dotted line in (a) shows the lower limit for $\sin^2 \theta_{eff}^{lept}$ in the MSSM if all sparticles are heavier than Z^0.

result, $M_W = 80.401 \pm 0.076$ GeV, is the average of M_W measured by UA2, LEP and Tevatron [17].

All squarks of the first two generations as well as sleptons can be still at their present lower experimental limits, and the success of the SM in the description of the precision electroweak data is still maintained. The same applies to the gaugino/higgsino sectors, since they do not give any strong $SU_V(2)$ breaking effects. In conclusion, most of the superpartners decouple from most of the electroweak observables, even if very light, $\mathcal{O}(M_Z)$. This high degree of screening follows from the basic structure of the model. The remarkable exception is the famous R_b [22].

3 R_b in the MSSM

As already mentioned, although the new ALEPH measurement [19] of R_b is in perfect agreement with its value predicted in the SM, the average of all measurements still deviates from the SM value by $\sim 1.8\sigma$ [5]. In view of this fact and because the R_b value in the MSSM is sensitive to different set of parameters than the bulk of the other electroweak observables, it is interesting

410

to discuss this observable in more detail.

In the MSSM there are new contributions to the $Z^0 b\bar{b}$ vertex, namely, Higgs bosons exchange in the loops, neutralino-sbottom and chargino-stop loops [20]. For low and intermediate values of $\tan\beta$, the first give negative contribution to R_b and to minimize this effect the pseudoscalar mass M_A has to be sufficiently large, say, $M_A > \mathcal{O}(300 \text{ GeV})$ whereas the neutralino-sbottom contribution is negligible. The chargino-stop loops can be realized in two ways: with stop coupled to Z^0 and with charginos coupled to Z^0. In both cases the lighter the stop and chargino the larger is the positive contribution.

The Dirac charginos are defined as

$$C_i^- = \begin{pmatrix} \lambda_i^- \\ \bar{\lambda}_i^+ \end{pmatrix} \qquad i = 1, 2 \tag{9}$$

where λ_i^- (λ_i^+) are linear combinations of the negatively (positively) charged $SU(2)$ gauginos and down-(up-)type higgsinos [e] The $b\tilde{t}_1 C^-$ coupling is enhanced for a right-handed stop (it is then proportional to the top quark Yukawa coupling). Then, however, the stop coupling to Z^0 is suppressed (it is proportional to $g\sin^2\theta_W$). Therefore, significant contribution can only come from the diagrams in which charginos are coupled to Z^0. Their actual magnitude depend on the interplay of the couplings in the $C_i^- \tilde{t}_1 b$ vertex and the $Z^0 C_i^- C_j^-$ vertex. The first one is large only for charginos with large up-higgsino component, the second - for charginos with large gaugino component in at least one of its two-component spinors. It has been observed [23,24] that, the situation in which both couplings are simultaneously large never happens for $\mu > 0$. Large R_b can then only be achieved at the expense of extremly light C_j^- and $\tilde{t}_1$. In addition, for fixed m_{C_1} and $M_{\tilde{t}_1}$, R_b is larger for $r \equiv M_2/|\mu| > 1$ i.e. for higgsino-like chargino as the enhancement of the $C_1^- \tilde{t}_1 b$ coupling is more important than of the $Z^0 C_1^- C_1^-$ coupling.

For $\mu < 0$ the situation is different. In the range $0.5 \lesssim r \lesssim 1.5$ a light chargino can be a strongly mixed state with a large up-higgsino and gaugino components (the higgsino-gaugino mixing comes from the chargino mass matrix). Large couplings in both vertices of the diagram with charginos coupled to Z^0 give significant increase in R_b even for the lighter chargino as heavy as $80 - 90$ GeV (similar increase in R_b for $\mu > 0$ requires $m_{C_1} \approx 50$ GeV and $M_{\tilde{t}_1} \approx 50$ GeV). This is illustrated in Fig. 3a where we show the contours of constatnt δR_b in the (M_2, μ) plane for fixed parameters of the stop sector.

The chargino-stop contributions do not change the value of the left-right asymmetry in b quarks $\mathcal{A}_b \equiv (g_L^2 - g_R^2)/(g_L^2 + g_R^2)$ where g_L (g_R) are the effective

---

[e] Up- and down- type higgsinos are superpartners of the Higgs boson doublets giving masses to the up- and down-type quarks, respectively.

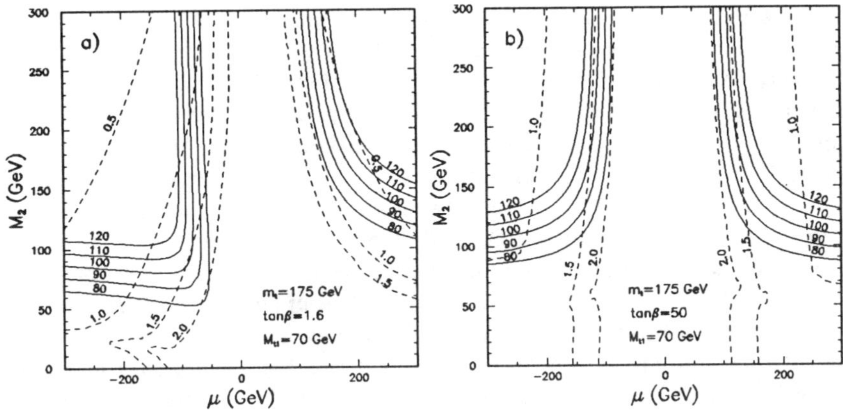

Figure 3: Contours of constant lighter chargino masses $m_{C_1^\pm}$ =80 − 120 GeV (solid lines) and of $\delta R_b \times 10^3$ =2.0, 1.5, 1.0, 0.5 (dashed lines) in the (μ, M_2) plane for $m_t = 175$ GeV, $M_{\tilde{t}_1} = 70$ GeV. a) $\tan\beta = 1.6$, b) $\tan\beta = 50$ and $M_A = 60$ GeV. Chargino masses $m_{C_1^\pm} \lesssim 83$ GeV are already excluded by LEP2.

couplings of left-handed (right-handed) b quarks to Z^0 (measured at SLD [29]) as they mostly modify only the left-handed effective coupling g_L [20]. In this case we get $\delta \mathcal{A}_b \approx 5.84 \times (1 - \mathcal{A}_b^{SM}) \times \delta R_b$ i.e. a very small, positive shift [18].

An enhancement of R_b is also possible for large $\tan\beta$ values, $\tan\beta \approx m_t/m_b$ [20,21]. In this case, in addition to the stop-chargino contribution (and neutralino-sbottom contribution enhanced by large $\tan\beta$) there can be even larger positive contribution from the h^0, H^0 and A^0 exchanges in the loops, provided those particles are sufficiently light (in this range of $\tan\beta$, $M_h \approx M_A$) and non-negligible sbottom-neutralino loop contributions. The main difference with the low $\tan\beta$ case is the approximate independence of the results on the sign of μ (which can be traced back to the approximate symmetry of the chargino masses and mixings under $\mu \rightarrow -\mu$). Also, the effects are always maximal for $M_2/|\mu| \gg 1$ i.e. for higgsino-like chargino. With present experimental constraints ($M_A \gtrsim 60$ GeV [27]) values of R_b up to ~ 0.2178 can still be obtained [23,25]. This is illustrated in Fig. 3b. Significant enhancement of R_b in the large $\tan\beta$ regime is, however, rather unlikely as it requires very precise cancellation of the SUSY contributions to obtain acceptable rate for $b \rightarrow s\gamma$ [23] (see later).

We conclude that additional supersymmetric contributions to the $Z^0 b\bar{b}$ vertex, from the chargino-right-handed stop loop (and from a light CP-odd Higgs boson for large $\tan\beta$), can be non-negligible when both are light, $\sim \mathcal{O}(M_Z)$. However, even with the chargino and stop at their present experimental mass limit, the prediction for R_b in the MSSM depends strongly on the chargino composition (see Fig. 3) and on the stop mixing angle. The values ranging from 0.2158 (the SM prediction) up to ~ 0.2178 for both small and large $\tan\beta$ values can be obtained (given all the experimental constraints) [23,24]. (Only marginal modification of the SM result for R_c is possible, though [26].) Those predictions hold with or without R-parity conservation [28] and with or without the GUT relation for the gaugino masses. The upper bound is reachable for chargino masses up to ~ 90 GeV provided they are mixed gaugino-higgsino states ($M_2/|\mu| \sim 1$) with $\mu < 0$ for low $\tan\beta$ and higgsino-like for large $\tan\beta$. In the same chargino mass range $\delta R_b \approx 0$ in the deep higgsino and gaugino regions for low $\tan\beta$ and gaugino region for large $\tan\beta$.

In conclusion, the new values of R_b and R_c are good news for supersymmetry. At the same time, one should face the fact that, unfortunately, in the MSSM

$$\delta R_b^{max} \sim \mathcal{O}(1\ \sigma^{exp})$$

so much better experimental precision is needed for a meaningful discussion.

4 FCNC with light superpartners

Another important class of processes, where light superpartners could manifest themselves through virtual corrections, are the Flavour Changing Neutral Current (FCNC) transitions. Gauge invariance, renormalizability and particle content of the SM imply the absence (in the lepton sector) or strong suppression (in the quark sector) of such processes. This prediction of the SM is in beautiful agreement with the presently available experimental data. In the MSSM there are two kinds of new contributions to the FCNC transitions. First of all, they may originate from flavour mixing in the sfermion mass matrices [30]. The strong suppression of the FCNC transisions observed in Nature puts severe constraints on flavour changing elements in the sfermion mass matrices [31]. Even in the absence of such effects the other kind of new contributions to FCNC processes arise through the ordinary K-M mixings due to additional exchanges of (light) supersymmetric particles in loops.

The present section is devoted to discussing such a scenario. The only extra MSSM contributions to the FCNC processes we consider are the charged Higgs boson-top and chargino-stop loops. The third generation sfermions and chargino(s) are indeed expected to be among the lightest superpartners. It is reasonable to assume (as follows from the analysis of the renormalization group equations for soft supersymmetry breaking parameters) that the first two generations of sfermions are heavier and degenerate in mass. The relevant parameter space is then identical to the one tested in corrections to R_b discussed earlier. Following the results on the precision tests we will assume that the heavier stop is heavy enough to decouple and that the lighter one is dominantly right-handed, i.e. that stop left-right mixing angle $\theta_{\tilde t}$ is relatively small (of order 10°).

Within this scenario, sizeable effects can still occur in the neutral meson mixing (K^0-$\bar K^0$ and B^0-$\bar B^0$). Supersymmetric contributions to other FCNC processes are either small or screened by long-distance QCD effects. A remarkable exception is the inclusive weak radiative B meson decay, $B \to X_s\gamma$, to which light superpartners can contribute significantly, and where strong interaction effects are under control. In the following, we shall focus on the $B \to X_s\gamma$ decay only and summarize the main points. More extensive discussion is given in ref. [32].

In the first approximation the $B \to X_s\gamma$ decay rate is given by simple one-loop graphs. However, the strong interaction corrections to these one-loop diagrams are enhanced by the large logarithms $\ln(M_W^2/m_b^2)$ and, in the SM, they increase the decay rate by a factor of order 2. Thus, resumming these large QCD logarithms up to (at least) next-to-leading order (NLO) is necessary to acquire sufficient accuracy [33]. This is conveniently done in three steps of which only the first one depends on the presence of "new physics" (supersymmetry) close to the electroweak scale.

In the first step one integrates out at the scale $Q = M_W$ all heavy fields and introduces the effective Hamiltonian

$$H_{eff} = -\frac{4G_F}{\sqrt{2}} V_{ts}^* V_{tb} \sum_{i=1}^{8} C_i(Q)\mathcal{O}_i(Q) \tag{10}$$

where $\mathcal{O}_i$ are the operators and $C_i(Q)$ are their Wilson coefficients. The relevant for $B \to X_s\gamma$ operators are

$$\mathcal{O}_7 = \frac{e}{16\pi^2}m_b(\bar s_L \sigma^{\mu\nu} b_R)F_{\mu\nu} \tag{11}$$

$$\mathcal{O}_8 = \frac{g_s}{16\pi^2}m_b(\bar s_L \sigma^{\mu\nu} T^a b_R)G_{\mu\nu}^a \tag{12}$$

where $F_{\mu\nu}$ and $G_{\mu\nu}^a$ are the photonic and gluonic field strength tensors, respectively. The leading-order SM [34] and MSSM [35] contributions to the coefficients $C_7(M_W)$ and $C_8(M_W)$ are well known. The next-to-leading corrections to $C_7(M_W)$ have been computed fully only in the case case of the SM [36]. In the supersymmetric case, only contributions proportional to logarithms of super-partner masses are known[37].

In the next step, resummation of large logarithms $\ln(M_W^2/m_b^2)$ is achieved by evolving the coefficients $C_i(Q)$ from $Q \sim M_W$ to $Q \sim m_b$ according to the renormalization group equations. The necessary for the complete NLO evolution coefficients of the RGE have been computed only recently[38].

Finally, the Feynman rules derived from the effective Hamiltonian at the scale $Q \sim m_b$ are used to calculate the b-quark decay rate $\Gamma(b \to X_s\gamma)$ which is a good approximation to the corresponding B-meson decay rate[41]. Radiative corrections to this computation, necessary to achieve NLO precision of the whole procedure, have also been computed recently[39].

We stress again that all these calculations are identical in the SM and MSSM except for the initial numerical values of the Wilson coefficients C_7 and C_8 at $Q \sim M_W$ which contain the information about "new physics". Another important remark is that even in the NLO computation, the theoretical prediction for $\Gamma(b \to X_s\gamma)$ still has an uncertainy (shown in Fig 4a) of order 15%[38] which is always taken into account in the bounds on sparticle masses presented below.

Fig. 4a shows the lower limits on the mass of the CP-odd MSSM Higgs boson mass[f], M_A, as a function of $\tan\beta$. Solid lines correspond to the case when all the superpartner masses are very large (above 1 TeV). In this case, the MSSM results are the same as in the Two Higgs Doublet Model (2HDM). Dashed (dotted) lines in Fig. 4a show the same limits in the presence of chargino and stop with masses $m_{C_1} = M_{\tilde{t}_1} = 500$ (250) GeV (with all other sparticles heavy) obtained by scanning over the values of $r = M_2/\mu$ and $\theta_{\tilde{t}}$. In the presence of light stop and chargino limits on M_A are significantly weaker and totally disappear for large values of $\tan\beta$ for which the chargino-stop contribution can be very large. The price however is a high degree of fine tuning (see the next Section).

The existing measurement of $BR(b \to s\gamma)$ imposes already significant constraints on the MSSM parameter space. To understand them, it is important to remember that the charged Higgs contribution to the $b \to s\gamma$ amplitude has always the same sign as the SM one whereas the chargino-stop contribution to this amplitude may have opposite sign. Since the actually measured value of $BR(b \to s\gamma)$ is close to the SM prediction, SUSY and charged Higgs contributions must either be small by themselves or cancel each other to a large extent. There exists, however, a third possibility where negative chargino-stop contribution overcomes the SM and charged Higgs ones yielding the correct absolute

---

[f]The charged Higgs boson mass is in one-to-one correspondence with M_A: $M_{H^\pm}^2 = M_A^2 + M_W^2$ (up to small radiative corrections[43]).

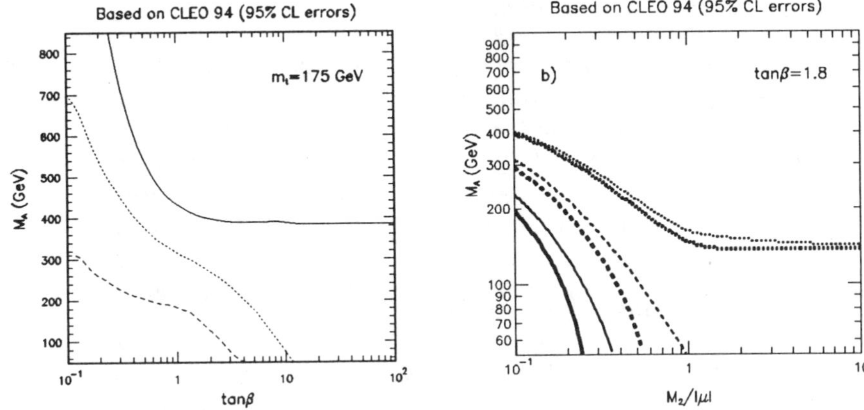

Figure 4: a) Lower limits on M_A from $b \to s\gamma$ as a function of $\tan\beta$. Solid line correspond to very heavy, $> \mathcal{O}(1\,\text{TeV})$ sparticles. Dashed (dotted) line show the limit for $m_{C_1} = M_{\tilde{t}_1} = 250$ (500) GeV. b) Lower limits on M_A as a function of $M_2/|\mu|$, based on CLEO $BR(B \to X_s\gamma)$ measurement. Thick lines show limits for $\mu > 0$, thin lines for $\mu < 0$. Solid, dashed and dotted lines show limits for lighter stop and chargino masses $M_{\tilde{t}_1} = m_{C_1^\pm} = 90$, 150 and 300 GeV, respectively.

magnitude of the total amplitude but with the opposite sign compared to the SM case. In paticular, it is worth stressing that the supersymmetric contribution, coming from a light chargino and stop, can provide a natural mechanism for lowering the $b \to s\gamma$ rate compared to the SM value, in agreement with the trend seen in the present data [35,23].

Another important observation is that, large chargino-stop contribution to $b \to s\gamma$ amplitude arise when the chargino is higgsino-like rather then gaugino-like i.e. when $M_2/|\mu| > 1$. In addition, the size of the chargino-stop contribution can be modified by changing the stop mixing angle $\theta_{\tilde{t}}$.

Fig. 4b (taken from ref. [32]) shows the lower limit on the allowed pseudoscalar Higgs boson mass M_A as a function of $r = M_2/|\mu|$ for three different values of the lighter chargino and lighter stop masses. It shows that for small $M_2/|\mu|$, i.e. for gaugino-like lighter chargino (when the chargino-stop contribution to $BR(b \to s\gamma)$ is suppressed) the resulting limits on M_A are quite strong even for very light chargino and stop e.g. $M_A \geq \mathcal{O}(200\,\text{GeV})$ for $M_{\tilde{t}_1} = m_{C_1^\pm} = 90$ GeV (we take 95% errors of CLEO measurement). The limits decrease when $M_2/|\mu|$ increases and approximately saturate for $M_2/|\mu| \geq 1$.

416

5 Large effects for $\tan\beta \sim m_t/m_b$

Large values of $\tan\beta$, $\tan\beta \sim m_t/m_b$, have been frequently advocated in the literature as a possible dynamical explanation of the large top to bottom quark mass ratio [44]. In this scenario the Yukawa couplings of the down-type quarks and leptons are enhanced leading in some cases to large loop effects.

Particularily large in this regime are the chargino-stop corrections to the $b \to s\gamma$ amplitude. Indeed, in the limit of higgsino-like lighter chargino we get

$$C_7^{C_1\tilde{t}_1}(M_W) \approx -\frac{m_t^2}{2m_{C_1}^2}\cos^2\theta_{\tilde{t}_1}f_\gamma^{(1)}(x) \pm \tan\beta\frac{m_t}{2m_{C_1}}\sin\theta_{\tilde{t}}\cos\theta_{\tilde{t}}f_\gamma^{(3)}(x) \quad (13)$$

where $x \equiv (M_{\tilde{t}_1}/m_{C_1})^2$, functions $f_\gamma^{(i)}(x)$ are defined in the second paper of ref. [35] and the sign in the second term is the same as the sign of the μ parameter. It is clear that for large $\tan\beta$, $m_{C_1} \sim M_W$ and the stop mixing angle not too small the second term dominates and is much larger than the SM $W^\pm$-t contribution

$$C_7^{tW^\pm} = \frac{3}{2}\frac{m_t^2}{M_W^2}f_\gamma^{(1)}\left(\frac{m_t^2}{M_W^2}\right) \quad (14)$$

Moreover, the higgsino-like chargino-stop contribution vanishes only as $1/m_{C_1}$ and remains noneglibible up to relatively large m_{C_1}. For the gaugino-like higgsino, instead, we get:

$$C_7^{C_1\tilde{t}_1}(M_W) \approx -\frac{M_W^2}{m_{C_1}^2}\cos^2\theta_{\tilde{t}}f_\gamma^{(1)}(x) \quad (15)$$

which is much smaller and vanishes as $1/m_{C_1}^2$. Large contribution from light higgsino-like chargino and stop can be cancelled by the charged Higgs boson loop. However, this requires a high degree of fine-tuning [23]. The allowed region in the plane $(\theta_{\tilde{t}}, M_{\tilde{t}_1})$ consists of two very narrow bands (corresponding to two different signs of the total amplitude). Outside these bands the generic prediction for the $BR(b \to s\gamma)$ is one-two orders of magnitude larger than the value measured by CLEO.

For similar reasons, the existence of very light, $\mathcal{O}(M_Z)$, pseudoscalar and charged Higgs bosons (and consequently significant enhancement of R_b) in the large $\tan\beta$ regime is rather unlikely. Indeed, charged Higgs contribution by itself would then give too high a rate for $b \to s\gamma$. It can be compensated by the chargino-stop loop but, again, at the expense of strong fine-tuning (due to the $\tan\beta$ enhancement factor present in eq. (13)) Nothing, of course, prevents the existence of light gaugino-like chargino and stop in the large $\tan\beta$

regime. For heavy enough pseudoscalar A and chargino or stop (or both) the contribution of eq. (13) may have, however, interesting consequences. Indeed, for negative values of $A_t \times \mu$ the rate can be easily smaller than the one of the SM [52] in agreement with the trend of the data. (This scenario can be realized in supergravity models with non-universal soft terms [53] and in the gauge mediated models with $B = 0$ [54]).

Another class of interesting effects in the large $\tan\beta$ regime originate from finite corrections to the down-type quark and lepton Yukawa couplings [45]. For sparticle masses $\gtrsim M_Z$ these effects can be concisely described by the effective lagrangian [47]

$$
\begin{aligned}
\mathcal{L}_{Yuk} \;=\; & -Y^d_{1ab} H_1 \bar{d}_{Ra} q_{Lb} - Y^d_{2ab} H^\dagger_2 \bar{d}_{Ra} q_{Lb} \\
& - Y^u_{2ab} H_2 \bar{u}_{Ra} q_{Lb} - Y^u_{1ab} H^\dagger_1 \bar{u}_{Ra} q_{Lb} + h.c.
\end{aligned}
\tag{16}
$$

(we suppressed the terms with lepton interactions) describing Yukawa interactions arising after integrating out (heavy) sparticles [46]. At the tree level, terms with Y^d_{2ab} and Y^u_{1ab} are absent in the MSSM (and in the SM). They are, however, generated by triangle diagrams (with helicity flips on fermion lines) with squarks and either gauginos or higgsinos circulating in loops [g]. Most interesting effects are due to the new term Y^d_{2ab} which reads

$$
\begin{aligned}
Y^d_{2ab} \;=\; & Y^d_{1ae} \left(\delta_{eb} \frac{2}{3} \frac{\alpha_s}{\pi} \mu m_{\tilde{g}} I(M^2_{\tilde{q}_L}, M^2_{\tilde{d}_R}, m^2_{\tilde{g}}) \right. \\
& \left. + \frac{1}{(4\pi)^2} (Y^{u\dagger}_2 Y^u_2)_{eb} \mu A_{\tilde{u}} I(M^2_{\tilde{q}_L}, M^2_{\tilde{u}_R}, \mu^2) \right) + \text{smaller terms} \\
\;=\; & Y^d_{1ae} \Delta_{eb}
\end{aligned}
\tag{17}
$$

where $A_{\tilde{u}}$ are the trilinear soft supersymmetry breaking terms (for simplicity we assume that the soft SUSY breaking matrices $M^2_{\tilde{q}_L}$, $M^2_{\tilde{d}_L}$ and $A_{\tilde{u}}$ are all proportional to the unit matrix) and

$$
I(x, y, z) \equiv \frac{xy \log(x/y) + yz \log(y/z) + zx \log(z/x)}{(x-y)(y-z)(z-x)}
$$

The presence of Y^d_{2ab} modifies the value of Y^d_{1ab} and (neglecting nondiagonal terms in Yukawa couplings) we get

$$
Y_b \equiv Y^d_{1bb} = \frac{e}{\sqrt{2}s_W} \frac{m_b}{M_W} \frac{\sqrt{1 + \tan^2\beta}}{1 + \tan\beta \Delta_{bb}}
\tag{18}
$$

---

[g] These diagrams are finite due to the so-called non-renormalization theorems which in the case of unbroken supersymmetry would also force these corrections to vanish.

It is clear that sizeable effects appear for large values of $\tan\beta$ and light sparticles involved. Moreover, these effects vanish only as $1/M_{soft}$ and persist therefore even for relatively heavy sparticles. The corrections affect all processes where the bottom quark Yukawa coupling is involved. In particular, they modify the well known limits on the $(\tan\beta, M_{H\pm})$ plane [48] derived from the experimental result for $b \to c\tau\bar{\nu}_\tau$ [49]. Qualitatively, with these corrections included, Y_b becomes larger for $\mu < 0$ and enhaces the contribution of the H^+ Higgs boson to the process $b \to c\tau\bar{\nu}_\tau$ strenghtening thus the limits on the $(\tan\beta, M_{H\pm})$ plane. For $\mu > 0$, instead, the bound is weakened and can even dissapear. For detailed discussion, see the ref. [49].

The same effective lagrangian (16) describes also the dominant part of the (large) supersymmetric corrections to the processes $H^+ \to tb$ [50] and $t \to H^+b$ which in the case of large $\tan\beta$ and sufficiently light H^+ (e.g. when R_b is enhanced) competes with the standard decay $t \to W^+b$ [51].

These corrections, interpreted as supersymmetric threshold corrections, are also very important in context of the GUT models and unification of the Yukawa couplings. In particular, they significantly lower the values of $\tan\beta$ and m_t predicted from the bottom-tau Yukawa coupling unification (for details see ref. [45,52]) and also, when their full generation dependence is taken into account, they significantly modify the naive predictions of the GUT models for CKM mixing angles [55].

6 Summary

There is an apparent contradiction between the hierarchy problem (which suggest new physics to be close to the electroweak scale) and the striking success of the Standard Model in describing the electroweak data. The supersymmetric extension of the SM offers an interesting solution to this puzzle. The bulk of the electroweak data is well screened from supersymmetric loop effects, due to the structure of the theory, even with superpartners generically light, $\mathcal{O}(M_Z)$. The only exception are the left-handed squarks of the third generation which have to be $\gtrsim \mathcal{O}(300~\text{GeV})$ to maintain the success of the SM. The other superpartners can still be light, at their present experimental mass limits, and would manifest themselves through virtual corrections to the small number of observables such as R_b, $b \to s\gamma$, K^0-$\bar{K}^0$ and B^0-$\bar{B}^0$ mixing and a few more for large $\tan\beta$. Those effects are very interesting but require still higher experimental precision to be detectable.

Our goal here was to study unconstrained minimal supersymmetric model, with arbitrary soft supersymmetry breaking parameters. Under stronger assumption, e.g. of universal soft terms at the GUT scale [56,45,52], one can get from virtual effects stronger constraints on the superpartner spectrum (for recent studies see e.g. refs. [57]).

Acknowledgments

We would like to thank M. Misiak and J. Rosiek for their contribution to this review.

This work was supported by Polish State Commitee for Scientific Research under grant 2 P03B 040 12 (for 1997-98). The work of PHCh was also partly supported by the U.S.-Polish Maria Skłodowska-Curie Joint Fund II (MEN/DOE-96-264).

1. T. Appelquist, J. Carazzone *Phys. Rev.* **D11** (1976) 2856.
2. J. Ellis, G. Fogli, E. Lisi *Phys. Lett.* **292B** (1992) 427, **318B** (1993) 375, **333B** (1994) 118, P.H. Chankowski, S. Pokorski *Phys. Lett.* **356B** (1995) 307.
3. G. Altarelli, R. Barbieri *Phys. Lett.* **253B** (1991) 161; G. Altarelli, R. Barbieri, S. Jadach *Nucl. Phys.* **B369** (1993) 3; G. Altarelli, R. Barbieri, F. Caravaglios *Phys. Lett.* **349** (1995) 145, *Nucl. Phys.* **B405** (1993) 3.
4. A. Blondel in *Proceedings of the XXVIII Int. Conf. on High Energy Physics* Warsaw, 1996, Z. Ajduk, A.K. Wróblewski eds., World Scientific Singapore.
5. The LEP Electroweak Working Group, CERN preprint LEPEWWG/97-01.
6. S. Eidelman, F. Jegerlehner, *Z. Phys.* **C67** (1995) 585.
7. J. Ellis, G. Fogli, E. Lisi *Phys. Lett.* **389B** (1996) 321.
8. G. Altarelli, R. Barbieri, F. Caravaglios *Phys. Lett.* **314B** (1993) 357.
9. J. Ellis, G. Fogli, E. Lisi *Phys. Lett.* **324B** (1994) 173.
10. P.H. Chankowski, S. Pokorski *Phys. Lett.* **366B** (1996) 188.
11. W. de Boer, A. Dabelstein, W.F.L. Hollik, W. Mösle preprint IEKP-KA-96-08 (hep-ph/9609209).
12. M.E. Peskin, T. Takeuchi *Phys. Rev. Lett.* **65** (1990) 964, *Phys. Rev.* **D46** (1992) 381;
13. T. Inami, C.S. Lim, A. Yamada *Mod. Phys. Lett.* **A7** (1992) 2789.
14. D. Garcia, J. Solà *Mod. Phys. Lett.* **A9** (1994) 211; P.H. Chankowski et al. *Nucl. Phys.* **B417** (1994) 101.
15. R. Barbieri, M. Frigeni, F. Caravaglios *Phys. Lett.* **279B** (1992) 169.
16. R. Barbieri, M. Frigeni, F. Giuliani, H.E. Haber *Nucl. Phys.* **B341** (1990) 309; H.E. Haber in *Proc. of the Workshop on the INFN Eloisotron Project* (hep-ph/9305248).

17. J. L. Rosner preprint EFI-97-18 (hep-ph/9704331).
18. P. Bamert et al. *Phys. Rev.* **D54** (1996) 4275.
19. I. Tomalin (ALEPH Collaboration) in *Proceedings of the XXVIII Int. Conf. on High Energy Physics* Warsaw, 1996, Z. Ajduk, A.K. Wróblewski eds., World Scientific Singapore.
20. M. Boulware, D. Finnell *Phys. Rev.* **D44** (1991) 2054.
21. J. Rosiek *Phys. Lett.* **252B** (1990) 135; A. Denner et al. *Z. Phys.* **C51** (1991) 695.
22. G.L. Kane C. Kolda J.D. Wells *Phys. Lett.* **338B** (1994) 219; P.H. Chankowski, S. Pokorski in Proceedings of the Beyond the Standard Model IV conference, Lake Tahoe C.A., eds. J.F. Gunion, T. Han, J. Ohnemus (1994) p. 233; D. Garcia, R. Jiménez J. Solà *Phys. Lett.* **347** (1995) 309, 321, E **351B** (1995) 602; D. Garcia, J. Solà *Phys. Lett.* **354B** (1995) 335, **357B** (1995) 349; G.L. Kane R.G. Stuart, J.D. Wells *Phys. Lett.* **354B** (1995) 350; P.H. Chankowski, S. Pokorski *Phys. Lett.* **366B** (1996) 188; G.L. Kane, J.D. Wells *Phys. Rev. Lett.* **76** (1996) 869; J. Ellis, J.L. Lopez, D.V. Nanopoulos *Phys. Lett.* **372B** (1996) 95, **397B** (1997) 88.
23. P.H. Chankowski, S. Pokorski *Nucl. Phys.* **B475** (1996) 3.
24. M. Drees et al. *Phys. Rev.* **D54** (1996) 5598.
25. J. Solà, *Phys. Lett.* **357B** (1996) 349.
26. D. Garcia, J. Solà, *Phys. Lett.* **354B** (1995) 335.
27. The ALEPH Collaboration preprint CERN PPE/97-071
28. P.H. Chankowski, D. Choudhury, S. Pokorski *Phys. Lett.* **389B** (1996) 677; P.H. Chankowski, S. Pokorski *Acta Phys. Pol.* **B27** (1996) 1719.
29. P. Rowson (SLD Collaboration) to appear in *Proceedings of the XXXII Rencontres de Moriond, Electroweak Session*, J. Trân Van Than ed., Edition Frontieres.
30. M.J. Duncan *Nucl. Phys.* **B221** (1983) 285; J.F. Donoghue, H.-P. Nilles, D. Wyler *Phys. Lett.* **128B** (1983) 55; A. Boquet, J. Kaplan, C.A. Savoy *Phys. Lett.* **148B** (1984) 69.
31. A. Masiero, F. Gabbiani *Nucl. Phys.* **B322** (1989) 235; J.S. Hagelin, S. Kelley, T. Tanaka *Nucl. Phys.* **B415** (1994) 293; A. Masiero, F. Gabbiani, L. Silvestrini *Nucl. Phys.* **B477** (1996) 321.
32. M. Misiak, S. Pokorski, J. Rosiek preprint IFT 3/97 (hep-ph/9703442), to appear in "Heavy Flavours II" eds. A.J. Buras, M. Lindner, Advanced Series on Directions in High Energy Physics, World Scientific, Singapore.
33. A. J. Buras, M. Misiak, M. Münz, S. Pokorski *Nucl. Phys.* **B424** (1994) 374.
34. T. Inami, C.S. Lim *Prog. Theor. Phys.* **65** (1981) 297, (E.) *ibid.* **65**

(1981) 1772.

35. S. Bertolini, F. Borzumati, A. Masiero, G. Ridolfi *Nucl. Phys.* **B353** (1991) 591; R. Barbieri, G.-F. Giudice *Phys. Lett.* **309B** (1993) 86; P. Cho, M. Misiak, D. Wyler, *Phys. Rev.* **D54** (1996) 3329.

36. K. Adel, Y.P. Yao *Phys. Rev.* **D49** (1994) 4945.

37. H. Anlauf *Nucl. Phys.* **B430** (1994) 245.

38. K. Chetyrkin, M. Misiak, M. Münz *Phys. Lett.* **400B** (1997) 206.

39. C. Greub, T. Hurth, D. Wyler *Phys. Lett.* **380B** (1996) 385, *Phys. Rev.* **D54** (1996) 3350.

40. A. Ali, C. Greub *Phys. Lett.* **361B** (1995) 146.

41. A.F. Falk, M. Luke, M. Savage *Phys. Rev.* **D49** (1994) 3367; M. Neubert *Phys. Rev.* **D49** (1994) 4623.

42. M.S. Alam et al. *Phys. Rev. Lett.* **74** (1995) 2885.

43. P.H. Chankowski, S. Pokorski, J. Rosiek *Phys. Lett.* **274B** (1992) 191; A. Brignole *Phys. Lett.* **277B** (1992) 313.

44. T. Banks *Nucl. Phys.* **B303** (1988) 172; M. Olechowski, S. Pokorski *Phys. Lett.* **214B** (1988) 393; S. Pokorski in *Proc. XII Int. Workshop on Weak Interactions and Neutrinos*, Ginosar, Israel *Nucl. Phys. Proc. Suppl.* **B13** (1990) 606; S. Dimopoulos, L.J. Hall, S. Raby *Phys. Rev. Lett.* **68** (1992) 1984, *Phys. Rev.* **D45** (1992) 4192; G. W. Anderson, S. Raby, S. Dimopoulos, L.J. Hall *Phys. Rev.* **D49** (1994) 3660; H.-P. Nilles in Proc. of the 1990 Theoretical Advanced Study Institute in Elementary Particle Physics, eds. M. Cvetič, P. Langacker, World Scientific, Singapore, p. 633; W. Majerotto, B. Mösslacher *Z. Phys.* **C48** (1990) 273; M. Drees, M.M. Nojiri *Nucl. Phys.* **B369** (1992) 54; B. Ananhtarayan, G. Lazarides, Q. Shafi *Phys. Lett.* **300B** (1993) 245.

45. L.J. Hall, R. Ratazzi, U. Sarid *Phys. Rev.* **D50** (1994) 7048.

46. R. Hempfling *Phys. Rev.* **D49** (1994) 6168.

47. R. Hempfling private comunication.

48. P. Krawczyk, S. Pokorski *Phys. Rev. Lett.* **60** (1988) 182; G. Isidori *Phys. Lett.* **298B** (1993) 409; Y. Grossman, H.E. Haber, Y. Nir *Phys. Lett.* **357B** (1995) 630.

49. J.A. Coarasa, R.A. Jiménez, J. Solà preprint UAB-FT-407 (hep-ph/9701392).

50. R.A. Jiménez, J. Solà *Phys. Lett.* **389B** (1996) 53.

51. J.A. Coarasa, D. Garcia, J. Guasch, R.A. Jiménez, J. Solà preprint UAB-FT-397 (hep-ph/9607485).

52. M. Carena, M. Olechowski, S. Pokorski, C.E.M. Wagner *Nucl. Phys.* **B426** (1994) 269; R. Hempfling *Z. Phys.* **C63** (1994) 309;

53. F. Borzumati, M. Olechowski, S. Pokorski *Phys. Lett.* **349B** (1995) 311.

54. R. Rattazzi, U. Sarid *Nucl. Phys.* **B501** (1997) 297.

55. T. Blažek, S. Pokorski, S. Raby *Phys. Rev.* **D52** (1995) 4151.

56. G.L. Kane, C. Kolda, L. Roszkowski, J.D. Wells *Phys. Rev.* **D49** (1994) 6173, **D50** (1994) 3498; M. Olechowski, S. Pokorski *Nucl. Phys.* **B404** (1993) 590; M. Carena, M. Olechowski, S. Pokorski, C.E.M. Wagner *Nucl. Phys.* **B419** (1994) 213;

57. M. Carena, C.E.M. Wagner *Nucl. Phys.* **B452** (1995); T. Blažek, M. Carena, S. Raby, C.E.M. Wagner *Nucl. Phys. Proc. Suppl.* **52A** (1997) 133; J.D. Wells, C. Kolda, G.L. Kane *Phys. Lett.* **338B** (1994) 219; D. Pierce, J. Bagger, K. Matchev, R. Zhang *Nucl. Phys.* **B491** (1996) 3; H. Baer, M. Brhlik *Phys. Rev.* **D55** (1997) 3201; N.G. Deshpande, B. Dutta, S. Oh *Phys. Rev.* **D56** (1997) 519.

CP Violation in Low-Energy SUSY

A. MASIERO

SISSA – ISAS, Trieste and Dip. di Fisica, Università di Perugia and INFN, Sezione di Perugia, Via Pascoli, I-06100 Perugia, Italy

L. SILVESTRINI

Physik Department, Technische Universität München, D-85748 Garching, Germany

We discuss CP violation in the context of the minimal SUSY extension of the standard model as well as in a generic low-energy SUSY model. After analyzing the SUSY contributions to the observed CP violation in the kaon system, we emphasize the prospects for disentangling SUSY loop contributions from pure SM effects in B physics.

1 Introduction

CP violation has major potentialities to exhibit manifestations of new physics beyond the Standard Model (SM). Indeed, it is quite a general feature that new physics possesses new CP violating phases in addition to the Cabibbo-Kobayashi-Maskawa (CKM) phase (δ_{CKM}) or, even in those cases where this does not occur, δ_{CKM} shows up in interactions of the new particles, hence with potential departures from the SM expectations. Moreover, although the SM is able to account for the observed CP violation in the kaon system, we cannot say that we have tested so far the SM predictions for CP violation. The detection of CP violation in B physics will constitute a crucial test of the standard CKM picture within the SM. Again, on general grounds, we expect new physics to provide departures from the SM CKM scenario for CP violation in B physics. A final remark on reasons that make us optimistic in having new physics playing a major role in CP violation concerns the matter-antimatter asymmetry in the Universe. Starting from a baryon-antibaryon symmetric Universe, the SM is unable to account for the observed baryon asymmetry. The presence of new CP-violating contributions when one goes beyond the SM looks crucial to produce an efficient mechanism for the generation of a satisfactory ΔB asymmetry.

The above considerations apply well to the new physics represented by low-energy supersymmetric extensions of the SM. Indeed, as we will see below, supersymmetry (SUSY) introduces CP violating phases in addition to δ_{CKM} and, even if one envisages particular situations where such extra-phases vanish, the phase δ_{CKM} itself leads to new CP-violating contributions in processes where SUSY particles are exchanged. CP violation in B decays has all potentialities to exhibit departures from the SM CKM picture in low-energy SUSY extensions, although, as we will discuss, the detectability of such deviations strongly depends on the regions of the SUSY parameter space under consideration.

The discussion will proceed in two steps. In the first part we will briefly review the status of CP violation in the context of the minimal SUSY extension of the SM (MSSM). As interesting as this particular SUSY extension is, in terms of economy of parameters and, hence, predictivity, it is becoming clearer and clearer that the MSSM is based upon drastic assumptions of simplification which do not find so far a strong theoretical motivation. Therefore we will devote much of our attention to the analysis of a generic low-energy SUSY extension of the SM without any specific assumption on the large SUSY parameter space. In this context the main point is the identification of the most promising processes in B physics for a manifestation of the SUSY CP violation compatibly with all the stringent constraints which come from the existing data on FCNC phenomena.

2 CP violation in the MSSM

In the MSSM two new "genuine" SUSY CP violating phases are present. They originate from the SUSY parameters μ, M, A and B. The first of these parameters is the dimensionful coefficient of the $H_u H_d$ term of the superpotential. The remaining three parameters are present in the sector that softly breaks the N=1 global SUSY. M denotes the common value of the gaugino masses, A is the trilinear scalar coupling, while B denotes the bilinear scalar coupling. In our notation all these three parameters are dimensionful. The simplest way to see which combinations of the phases of these four parameters are physical[1] is to notice that for vanishing values of μ, M, A and B the theory possesses two additional symmetries.[2] Indeed, letting B and μ vanish, a $U(1)$ Peccei-Quinn symmetry originates, which in particular rotates H_u and H_d. If M, A and B are set to zero, the Lagrangian acquires a continuous $U(1)$ R symmetry. Then we can consider μ, M, A and B as spurions which break the $U(1)_{PQ}$ and $U(1)_R$ symmetries. In this way the question concerning the number and nature of the meaningful phases translates into the problem of finding the independent combinations of the four parameters which are invariant under $U(1)_{PQ}$ and $U(1)_R$ and determining their independent phases. There are three such independent combinations, but only two of their phases are independent. We use here the commonly adopted choice:

$$\Phi_A = \arg\left(A^* M\right), \qquad \Phi_B = \arg\left(B^* M\right). \tag{1}$$

The main constraints on Φ_A and Φ_B come from their contribution to the electric dipole moments of the neutron and of the electron. For instance, the effect of Φ_A and Φ_B on the electric and chromoelectric dipole moments of the light quarks (u, d, s) lead to a contribution to d_N^e of order [3]

$$d_N^e \sim 2 \left(\frac{100 \text{GeV}}{\tilde{m}} \right)^2 \sin \Phi_{A,B} \times 10^{-23} \text{e cm}, \tag{2}$$

where $\tilde{m}$ here denotes a common mass for squarks and gluinos. The present experimental bound, $d_N^e < 1.1 \times 10^{-25}$ e cm, implies that $\Phi_{A,B}$ should be $< 10^{-2}$, unless one pushes SUSY masses up to O(1 TeV). A possible caveat to such an argument calling for a fine-tuning of $\Phi_{A,B}$ is that uncertainties in the estimate of the hadronic matrix elements could relax the severe bound in eq. (2).[4]

In view of the previous considerations most authors dealing with the MSSM prefer to simply put Φ_A and Φ_B equal to zero. Actually, one may argue in favour of this choice by considering the soft breaking sector of the MSSM as resulting from SUSY breaking mechanisms which force Φ_A and Φ_B to vanish. For instance, it is conceivable that both A and M originate from one same source of $U(1)_R$ breaking. Since Φ_A "measures" the relative phase of A and M, in this case it would "naturally" vanish. In some specific models it has been shown [5] that through an analogous mechanism also Φ_B may vanish.

If $\Phi_A = \Phi_B = 0$, then the novelty of SUSY in CP violating contributions merely arises from the presence of the CKM phase in loops where SUSY particles run.[6] The crucial point is that the usual GIM suppression, which plays a major role in evaluating ε and ε' in the SM, in the MSSM case is replaced by a super-GIM cancellation which has the same "power" of suppression as the original GIM.[7] Again, also in the MSSM, as it is the case in the SM, the smallness of ε and ε' is guaranteed not by the smallness of δ_{CKM}, but rather by the small CKM angles and/or small Yukawa couplings. By the same token, we do not expect any significant departure of the MSSM from the SM predictions also concerning CP violation in B physics. As a matter of fact, given the large lower bounds on squark and gluino masses, one expects relatively tiny contributions of the SUSY loops in ε or ε' in comparison with the normal W loops of the SM. Let us be more detailed on this point.

In SUSY, FCNC CP-violating phenomena arise at one-loop [a] through the exchange of: i)W and quarks (SM contribution); ii)charginos and squarks; iii) charged Higgs and quarks and iv) gluinos (or, more generally, neutralinos) and squarks. Here we are considering the low-energy minimal SUSY extension of

---

[a]In SUSY models with broken R-parity even tree-level contributions may be present.

the SM where universality of the soft-breaking terms is assumed. This implies that at the tree level no flavour mixing in the squark mass matrices is present when one goes to the physical quark basis. Hence, no tree-level flavour changing $q - \tilde{q} - \tilde{g}$ vertices exist. However, radiative corrections or, equivalently, the running from the large supergravity scale down to the electroweak scale induces a mismatch in the simultaneous diagonalization of q and $\tilde{q}$ mass matrices, hence producing flavour changes at the $q - \tilde{q} - \tilde{g}$ vertices. As detailed in the chapter on FCNC in SUSY, the amount of this mismatch in the MSSM is again proportional to a very good approximation to the CKM angles and the Yukawa couplings. Due to the large gluino and squark masses (their lower bound is almost 200 GeV), in spite of the presence of the strong coupling, it has been shown[8] that the gluino exchange (or, for that matter, any neutralino exchange) is subleading with respect to chargino ($\chi^{\pm}$) and charged Higgs ($H^{\pm}$) exchanges. Hence, when dealing with CP violation in the MSSM with exact universality and $\Phi_A = \Phi_B = 0$, one can confine the analysis to $\chi^{\pm}$ and $H^{\pm}$ loops. If one takes all squarks to be degenerate in mass and heavier than ~ 200 GeV, then $\chi^{\pm} - \tilde{q}$ loops are obviously severely penalized with respect to the SM $W - q$ loops (remember that at the vertices the same CKM angles occur in both cases).

The only chance for the MSSM to produce some sizeable departure from the SM situation in CP violation is in the particular region of the parameter space where one has light $\tilde{q}$, $\chi^{\pm}$ and/or $H^{\pm}$. We have already seen in the previous chapters that the best candidate (indeed the only one unless $\tan\beta \sim m_t/m_b$) for a light $\tilde{q}$ is the stop. Hence one can ask the following question: can the MSSM present some novelties in CP-violating phenomena when we consider $\chi^{+} - \tilde{t}$ loops with light $\tilde{t}$, χ^{+} and/or H^{+}?

Several analyses in the literature tackle the above question or, to be more precise, the more general problem of the effect of light $\tilde{t}$ and χ^{+} on FCNC processes.[9-10] A first important observation concerns the relative sign of the $W - t$ loop with respect to the $\chi^{+} - \tilde{t}$ and $H^{+} - t$ contributions. As it is well known, the latter contribution always interferes positively with the SM one. Interestingly enough, in the region of the MSSM parameter space that we consider here, also the $\chi^{+} - \tilde{t}$ contribution constructively interferes with the SM contribution. The second point regards the composition of the lightest chargino, i.e. whether the gaugino or higgsino component prevails. This is crucial since the light stop is predominantly $\tilde{t}_R$ and, hence, if the lightest chargino is mainly a wino then it couples to $\tilde{t}_R$ mostly through the LR mixing in the stop sector. Consequently, a suppression in the contribution to box diagrams going as $\sin^4 \theta_{LR}$ is present (θ_{LR} denotes the mixing angle between the lighter and heavier stops). On the other hand, if the lightest chargino is predominantly a higgsino (i.e. $M_2 \gg \mu$ in the chargino mass matrix), then the χ^+–lighter $\tilde{t}$ contribution grows. In this case contributions $\propto \theta_{LR}$ become negligible and, moreover, it can be shown that they are independent on the sign of μ. A detailed study is provided in ref. [10] For instance, for $M_2/\mu = 10$ they find that the inclusion of the SUSY contribution to the box diagrams doubles the usual SM contribution for values of the lighter $\tilde{t}$ mass up to $100-120$ GeV, using $\tan\beta = 1.8$, $M_{H+} = 100$ TeV, $m_\chi = 90$ GeV and the mass of the heavier $\tilde{t}$ of 250 GeV. However, if m_χ is pushed up to 300 GeV, the $\chi^+ - \tilde{t}$ loop yields a contribution which is roughly 3 times less than in the case $m_\chi = 90$ GeV, hence leading to negligible departures from the SM expectation. In the cases where the SUSY contributions are sizeable, one obtains relevant restrictions on the ρ and η parameters of the CKM matrix by making a fit of the parameters A, ρ and η of the CKM matrix and of the total loop contribution to the experimental values of ε_K and ΔM_{B_d}. For instance, in the above-mentioned case in which the SUSY loop contribution equals the SM $W - t$ loop, hence giving a total loop contribution which is twice as large as in the pure SM case, combining the ε_K and ΔM_{B_d} constraints leads to a region in the $\rho - \eta$ plane with $0.15 < \rho < 0.40$ and $0.18 < \eta < 0.32$, excluding negative values of ρ.

In conclusion, the situation concerning CP violation in the MSSM case with $\phi_A = \phi_B = 0$ and exact universality in the soft-breaking sector can be summarized in the following way: the MSSM does not lead to any significant deviation from the SM expectation for CP-violating phenomena as d_N^e, ε, ε' and CP violation in B physics; the only exception to this statement concerns a small portion of the MSSM parameter space where a very light $\tilde{t}$ ($m_{\tilde{t}} < 100$ GeV) and χ^+ ($m_\chi \sim 90$ GeV) are present. In this latter particular situation sizeable SUSY contributions to ε_K are possible and, consequently, major restrictions in the $\rho - \eta$ plane can be inferred. Obviously, CP violation in B physics becomes a crucial test for this MSSM case with very light $\tilde{t}$ and χ^+. Interestingly enough, such low values of SUSY masses are at the border of the detectability region at LEP II.

3 CP violation in general SUSY extensions of the SM

We now move to the more interesting case where one envisages a generic SUSY extension of the SM without making the quite drastic assumptions entering the construction of the MSSM. In previous chapters we have already seen that FCNC processes severely limit the amount of non-universality in the $\tilde{q}$ sector. The questions we tackle in this chapter concerning a general SUSY extension of the SM are: i) what are the constraints imposed on $\tilde{q}$ mass matrices by ε, ε' and d_N^e? ii) is the SUSY contribution to CP violation in kaon physics of the

428

superweak or milliweak kind? iii) given the constraints from CP conserving FCNC processes involving B mesons, can we still hope that low-energy SUSY manifests itself through sizeable departures from the SM expectations in CP violating B decays? Obviously the latter question, in particular, seems of great relevance in a moment when new B-factories and B-dedicated programs at existing accelerators are going to become operative very soon.

To answer the above questions we use a model-independent parameterization of FCNC and CP violating contributions which makes use of the so-called super-CKM basis [11] where $q - \tilde{q} - \tilde{g}$ couplings are diagonal in flavour and all the flavour changing effects are due to the non-diagonality of the $\tilde{q}$ mass matrices. As long as the ratio of the off-diagonal entries over an average $\tilde{q}$ mass remains a small parameter, the first term of the expansion which is obtained by an off-diagonal mass insertion in the $\tilde{q}$ propagators represents a suitable approximation. This method avoids the specific knowledge of the $\tilde{f}$ mass matrices and allows to account for the FCNC $\tilde{g}$ and $\chi^{\pm}$ loops. Notice that, for a given process, the mass insertions refer to squarks of different electric charge when $\tilde{g}$ and $\chi^{\pm}$ are exchanged in the loops and that the relative size and sign of such different mass insertions is undetermined as long as one keeps with generic SUSY extensions of the SM.[b] Hence one cannot be conclusive about the interplay of $\tilde{g}$ and $\chi^{\pm}$ exchanges to a given FCNC or CP violating process, but one should consider all the various possibilities taking into account the full set of processes which constrain such mass insertions. For the sake of our discussion here it is enough to concentrate on one of the two contributions, for instance the SUSY loops with $\tilde{g}$ running inside. In the approach to a generic SUSY extension we expect similar contributions from $\tilde{g}$ and $\chi^{\pm}$ exchanges, unless one invokes particular regions of the SUSY parameter space where $\chi^{\pm}$ are very light. Indeed, here, differently from the very constrained case of the MSSM, we generally obtain severe constraints on the FC mass insertions even for conspicuous values of the $\tilde{g}$, $\chi^{\pm}$ and $\tilde{q}$ masses. Gluino exchange is enhanced with respect to $\chi^{\pm}$ exchange by powers of (α_s / α_W) couplings, but, on the other hand, on quite general grounds we expect $\chi^{\pm}$ to be lighter than $\tilde{g}$ and hence a factor $(m_{\chi}/m_{\tilde{g}})^2$ can penalize the $\tilde{g}$ exchange contribution. With that in mind, and barring significant cancellations between the $\tilde{g}$ and $\chi^{\pm}$ contributions, it makes sense to focus just on the $\tilde{g}$ contribution to CP violating processes.

We briefly recall here the main ingredients for our analysis in the framework of the mass-insertion approximation in the super-CKM basis (this is useful also to establish our notation). There exist four different Δ mass insertions connecting flavours i and j along a sfermion propagator: $(\Delta_{ij})_{LL}$, $(\Delta_{ij})_{RR}$, $(\Delta_{ij})_{LR}$ and $(\Delta_{ij})_{RL}$. The indices L and R refer to the helicity of the fermion

---

[b]Except for LL mass insertions (see ref. [10] for details on this point).

Table 1: Limits on $\mathrm{Im}\left(\delta_{12}^d\right)_{AB}\left(\delta_{12}^d\right)_{CD}$, with $A,B,C,D = (L,R)$, for an average squark mass $m_{\tilde q} = 500\,\mathrm{GeV}$ and for different values of $x = m_{\tilde g}^2/m_{\tilde q}^2$. For different values of $m_{\tilde q}$, the limits can be obtained multiplying the ones in the table by $m_{\tilde q}(\mathrm{GeV})/500$.

x	$\sqrt{\left|\mathrm{Im}\left(\delta_{12}^{\mathrm d}\right)_{\mathrm{LL}}^2\right|}$	$\sqrt{\left|\mathrm{Im}\left(\delta_{12}^{\mathrm d}\right)_{\mathrm{LR}}^2\right|}$	$\sqrt{\left|\mathrm{Im}\left(\delta_{12}^{\mathrm d}\right)_{\mathrm{LL}}\left(\delta_{12}^{\mathrm d}\right)_{\mathrm{RR}}\right|}$
0.3	1.5×10^{-3}	6.3×10^{-4}	2.0×10^{-4}
1.0	3.2×10^{-3}	3.5×10^{-4}	2.2×10^{-4}
4.0	7.5×10^{-3}	4.2×10^{-4}	3.2×10^{-4}

partners. The size of these Δ's can be quite different. For instance, it is well known that in the MSSM case only the LL mass insertion can change flavour, while all the other three above mass insertions are flavour conserving, i.e. they have $i = j$. In this case to realize a LR or RL flavour change one needs a double mass insertion with the flavour changed solely in a LL mass insertion and a subsequent flavour-conserving LR mass insertion. Even worse is the case of a FC RR transition: in the MSSM this can be accomplished only through a laborious set of three mass insertions, two flavour-conserving LR transitions and an LL FC insertion. Notice also that generally the Δ_{LR} quantity does not necessarily coincide with Δ_{RL}. For instance, in the MSSM and in several other cases, one flavour-conserving mass insertion is proportional to the mass of the corresponding right-handed fermion. Hence , $(\Delta_{ij})_{LR}$ and $(\Delta_{ij})_{RL}$ are proportional to the mass of the i-th and j-th fermion, respectively. Instead of the dimensional quantities Δ it is more useful to provide bounds making use of dimensionless quantities, δ, that are obtained dividing the mass insertions by an average sfermion mass.

We start by considering CP violation in the kaon system, i.e. ε and ε'. In reference [12] we provide a very detailed description of how to calculate the effective hamiltonian for $\Delta S = 2$ and $\Delta S = 1$ processes as well as a determination of the hadronic matrix elements. Asking for each $\tilde g$ exchange contribution not to exceed the experimental value $\varepsilon = 2.268 \times 10^{-3}$ we obtain bounds on the quantities $\left|\mathrm{Im}\left(\delta_{12}^d\right)_{LL}^2\right|^{1/2}$, $\left|\mathrm{Im}\left(\delta_{12}^d\right)_{LR}^2\right|^{1/2}$ and $\left|\mathrm{Im}\left(\delta_{12}^d\right)_{LL}\left(\delta_{12}^d\right)_{RR}\right|^{1/2}$ as a function of the average squark mass $m_{\tilde q}$, and of the ratio $x = \left(m_{\tilde g}/m_{\tilde q}\right)^2$. In table 1 we report such limits taking $x = 0.3$, 1 and 4, for $m_{\tilde q} = 500$ GeV. For different values of $m_{\tilde q}$, the limits can be obtained multiplying the values in table 1 by $m_{\tilde q}$ (GeV)/500. In fig. 1 we plot the bound on $\left|\mathrm{Im}\left(\delta_{12}^d\right)_{LL}^2\right|^{1/2}$ as a function of x for $m_{\tilde q} = 500\,\mathrm{GeV}$. It should be noticed that the bounds derived from ε on the imaginary parts of products of δ's are one order of mag-

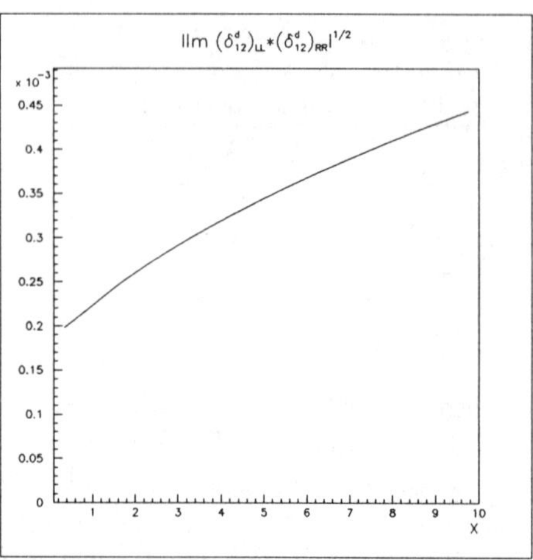

Figure 1: The $\sqrt{\left|\mathrm{Im}\left(\delta^d_{12}\right)^2_{LL}\right|}$ as a function of $x = m^2_{\tilde{g}}/m^2_{\tilde{q}}$, for an average squark mass $m_{\tilde{q}} = 500$ GeV.

nitude more stringent than the corresponding limits on the real parts which are obtained from ΔM_K.

Coming to $\Delta S = 1$ processes, both superpenguin and superboxes contribute to ε'. It was only very recently[12-13] that it was realized that superboxes are at least as important as superpenguin diagrams in contributions which proceed through a $\left(\delta^d_{12}\right)_{LL}$ insertion. In fig. 2 we report the bound on $\mathrm{Im}\left(\delta^d_{12}\right)_{LL}$ as a function of x for $m_{\tilde{q}} = 500$ GeV which comes from the conservative demand that $\varepsilon'/\varepsilon < 2.7 \times 10^{-3}$. The contribution of box and penguin diagrams to the LL terms have opposite signs and a sizeable cancellation occurs for x close to one, where the two contribution are of comparable size (this explains the peak around $x = 1$ in the plot of fig. 2). A much more stringent limit is obtained for $\left(\delta^d_{12}\right)_{LR}$ (fig. 3). For the LR contribution only superpenguins play a relevant role. Speaking of superpenguins, it is interesting to notice that, differently from the SM case, the SUSY contributions are negligibly affected by electroweak penguins, i.e. gluino-mediated Z^0- or γ-penguins are strongly suppressed with respect to gluino-mediated gluon penguins.[12]

In table 2 we summarize the bounds on $\mathrm{Im}\left(\delta^d_{12}\right)_{LL}$ and $\mathrm{Im}\left(\delta^d_{12}\right)_{LR}$ coming from $\varepsilon'/\varepsilon < 2.7 \times 10^{-3}$ for the same values of SUSY masses chosen in table 1. The comparison of the two tables leads to the following two conclusions:

1. if we consider a SUSY extension of the SM where the LR insertions are

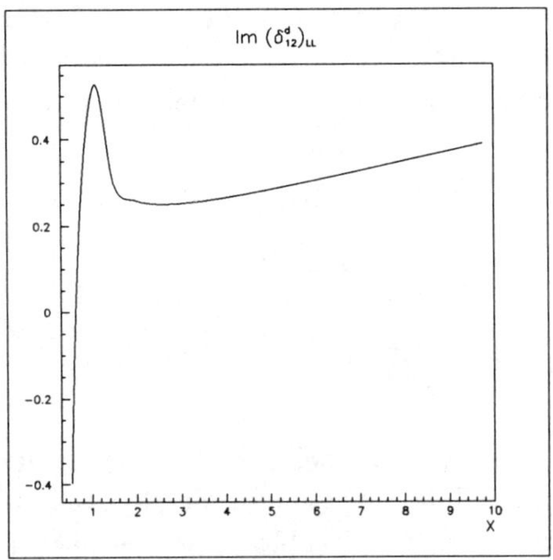

Figure 2: The Im $\left(\delta_{12}^d\right)_{LL}$ as a function of $x = m_{\tilde{g}}^2/m_{\tilde{q}}^2$, for an average squark mass $m_{\tilde{q}} = 500$ GeV.

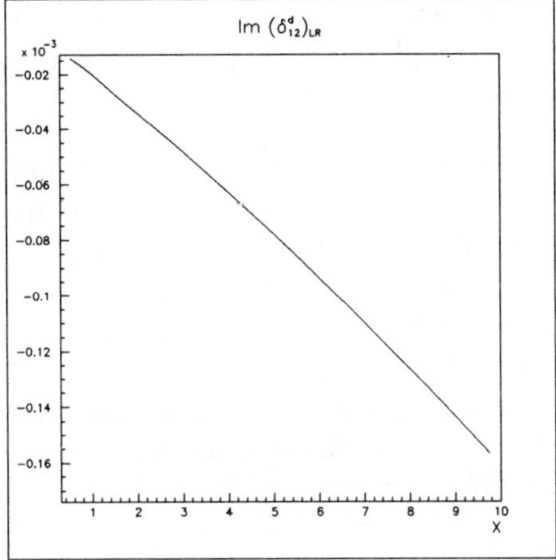

Figure 3: The Im $\left(\delta_{12}^d\right)_{LR}$ as a function of $x = m_{\tilde{g}}^2/m_{\tilde{q}}^2$, for an average squark mass $m_{\tilde{q}} = 500$ GeV.

Table 2: Limits from $\varepsilon'/\varepsilon < 2.7 \times 10^{-3}$ on $\mathrm{Im}\left(\delta_{12}^{d}\right)$, for an average squark mass $m_{\tilde{q}} = 500\,\mathrm{GeV}$ and for different values of $x = m_{\tilde{g}}^2/m_{\tilde{q}}^2$. For different values of $m_{\tilde{q}}$, the limits can be obtained multiplying the ones in the table by $\left(m_{\tilde{q}}(\mathrm{GeV})/500\right)^2$.

x	$\left|\mathrm{Im}\left(\delta_{12}^{d}\right)_{LL}\right|$	$\left|\mathrm{Im}\left(\delta_{12}^{d}\right)_{LR}\right|$
0.3	1.0×10^{-1}	1.1×10^{-5}
1.0	4.8×10^{-1}	2.0×10^{-5}
4.0	2.6×10^{-1}	6.3×10^{-5}

much smaller than the LL ones (this is what occurs in the MSSM, for instance), then fulfilling the bound coming from ε implies that $\mathrm{Im}\left(\delta_{12}^{d}\right)_{LL}$ is too small to provide a sizeable contribution to ε' unless $\left(\delta_{12}^{d}\right)_{LL}$ is almost purely imaginary (remember that ε bounds $\mathrm{Im}\left(\delta_{12}^{d}\right)_{LL}^{2}$). Hence, in this case the SUSY contribution would be of superweak nature;

2. if, on the contrary, we have a SUSY model with sizeable LR $\Delta S = 1$ mass insertions, then it is possible to respect the bound from ε, while obtaining a large contribution to ε'/ε. In this case we would have a SUSY milliweak contribution to CP violation. For this to occur, we need a SUSY model where $\left(\delta_{12}^{d}\right)_{LR}$ is no longer proportional to m_s, but rather to some much larger mass.

The above latter remark would lead us to the natural conclusion that to have sizeable SUSY contributions to ε'/ε one needs a SUSY extension of the SM where $\tilde{q}_L - \tilde{q}_R$ transitions are no longer proportional to m_q (we remind the reader that in the MSSM a mass term $\tilde{q}_L \tilde{q}_R^*$ receives two contributions, one proportional to the parameter A and the other to μ, but both of them are proportional to m_q). However, if this enhancement occurs also for flavour-conserving $\tilde{q}_L - \tilde{q}_R$ transitions, one may envisage some problem with the very stringent bound on the d_N^e. Indeed, imposing this latter constraint yields the following limits on $\mathrm{Im}\left(\delta_{11}^{d}\right)_{LR}$ for $m_{\tilde{q}} = 500$ GeV:

$$
\begin{array}{cc}
x & \mathrm{Im}\left(\delta_{11}^{d}\right)_{LR} \\
0.3 & 2.4 \times 10^{-6} \\
1.0 & 3.0 \times 10^{-6} \\
4.0 & 5.6 \times 10^{-6}
\end{array}
\tag{3}
$$

Table 3: Limits on $\mathrm{Re}\,(\delta_{13})_{AB}\,(\delta_{13})_{CD}$, with $A,B,C,D = (L,R)$, for an average squark mass $m_{\tilde{q}} = 500\,\mathrm{GeV}$ and for different values of $x = m_{\tilde{g}}^2/m_{\tilde{q}}^2$. For different values of $m_{\tilde{q}}$, the limits can be obtained multiplying the ones in the table by $m_{\tilde{q}}(\mathrm{GeV})/500$.

x	$\sqrt{\left\|\mathrm{Re}\left(\delta_{13}^{d}\right)_{\mathrm{LL}}^{2}\right\|}$	$\sqrt{\left\|\mathrm{Re}\left(\delta_{13}^{d}\right)_{\mathrm{LR}}^{2}\right\|}$	$\sqrt{\left\|\mathrm{Re}\left(\delta_{13}^{d}\right)_{\mathrm{LL}}\left(\delta_{13}^{d}\right)_{\mathrm{RR}}\right\|}$
0.3	4.6×10^{-2}	5.6×10^{-2}	1.6×10^{-2}
1.0	9.8×10^{-2}	3.3×10^{-2}	1.8×10^{-2}
4.0	2.3×10^{-1}	3.6×10^{-2}	2.5×10^{-2}

A quick comparison of the above numbers with the bounds on $\mathrm{Im}\left(\delta_{12}^{d}\right)_{LR}$ from ε'/ε reveals that to get a sizeable SUSY contribution to ε' we need values of $\mathrm{Im}\left(\delta_{12}^{d}\right)_{LR}$ which exceed the bound on $\mathrm{Im}\left(\delta_{11}^{d}\right)_{LR}$ arising from d_N^e. Obviously, strictly speaking it is not forbidden for $\mathrm{Im}\left(\delta_{12}^{d}\right)_{LR}$ to be $\geq \mathrm{Im}\left(\delta_{11}^{d}\right)_{LR}$, but certainly such a probability does not look straightforward. In conclusion, although technically it is conceivable that some SUSY extension may provide a large ε'/ε, it is rather difficult to imagine how to reconcile such a large enhancement of $\mathrm{Im}\left(\delta_{12}^{d}\right)_{LR}$ with the very strong constraint on the flavour-conserving $\mathrm{Im}\left(\delta_{11}^{d}\right)_{LR}$ from d_N^e.

4   CP violation in B physics

We now move to the next frontier for testing the unitarity triangle in general and in particular CP violation in the SM and its SUSY extensions: B physics. We have seen above that the transitions between 1st and 2nd generation in the down sector put severe constraints on $\mathrm{Re}\delta_{12}^{d}$ and $\mathrm{Im}\delta_{12}^{d}$ quantities. To be sure, the bounds derived from ε and ε' are stronger than the corresponding bounds from ΔM_K. If the same pattern repeats itself in the transition between 3rd and 1st or 3rd and 2nd generation in the down sector we may expect that the constraints inferred from $B_d - \bar{B}_d$ oscillations or $b \to s\gamma$ do not prevent conspicuous new contributions also in CP violating processes in B physics. We are going to see below that this is indeed the case ad we will argue that measurements of CP asymmetries in several B-decay channels may allow to disentangle SM and SUSY contributions to the CP decay phase.

First, we consider the constraints on δ_{13}^{d} and δ_{23}^{d} from $B_d - \bar{B}_d$ and $b \to s\gamma$, respectively. From the former process we obtain the bounds on $\mathrm{Re}\left(\delta_{13}^{d}\right)_{\mathrm{LL}}^{2}$, $\mathrm{Re}\left(\delta_{13}^{d}\right)_{\mathrm{LR}}^{2}$ and $\mathrm{Re}\left(\delta_{13}^{d}\right)_{\mathrm{LL}}\left(\delta_{13}^{d}\right)_{\mathrm{RR}}$ which are reported in table 3 for the usual three typical values of x ($x = 0.3$, 1 and 4) and for $m_{\tilde{q}} = 500$ GeV (the results scale with $m_{\tilde{q}}(\mathrm{GeV})/500$ for different values of $m_{\tilde{q}}$).

The radiative decay $b \to s\gamma$ constraints only the $\left|\left(\delta^d_{23}\right)_{LR}\right|$ quantity in a significant way. For $m_{\tilde{q}} = 500$ GeV we obtain bounds on $\left|\left(\delta^d_{23}\right)_{LR}\right|$ in the range $(1.3 \div 3) \times 10^{-2}$ for x varying from 0.3 to 4, respectively (the bound scales as $m_{\tilde{q}}^2$). On the other hand, $b \to s\gamma$ does not limit $\left|\left(\delta^d_{23}\right)_{LL}\right|$. In the following, we will take $\left|\left(\delta^d_{23}\right)_{LL}\right| = 1$ (corresponding to $x_s = (\Delta M/\Gamma)_{B_s} > 70$ for $m_{\tilde{q}} = 500$ GeV).

New physics can modify the SM predictions on CP asymmetries in B decays[14] by changing the phase of the B_d–$\bar{B}_d$ mixing and the phase and absolute value of the decay amplitude. The general SUSY extension of the SM that we discuss here affects both these quantities.

The remaining part of this chapter tackles the following question of crucial relevance in the next few years: where and how can one possibly distinguish SUSY contributions to CP violation in B decays? [15] As we said before we want our answer to be as general as possible, i.e. without any commitment to particular SUSY models. Obviously, a preliminary condition to properly answer the above question is to estimate the amount of the uncertainties of the SM predictions for CP asymmetries in B decays.

To discuss the latter above-mentioned point, we choose to work in the theoretical framework of ref. [16] We use the effective Hamiltonian ($\mathcal{H}_{eff}$) formalism, including LO QCD corrections; in the numerical analysis, we use the LO SM Wilson coefficients evaluated at $\mu = 5$ GeV, as given in ref. [17] In most of the cases, by choosing different scales (within a resonable range) or by using NLO Wilson coefficients, the results vary by about $20 - 30\%$. This is true with the exception of some particular channels where uncertainties are larger. The matrix elements of the operators of $\mathcal{H}_{eff}$ are given in terms of the following Wick contractions between hadronic states: Disconnected Emission (DE), Connected Emission (CE), Disconnected Annihilation (DA), Connected Annihilation (CA), Disconnected Penguin (DP) and Connected Penguin (CP) (either for left-left (LL) or for left-right (LR) current-current operators). Following ref. [18], where a detailed discussion can be found, instead of adopting a specific model for estimating the different diagrams, we let them vary within reasonable ranges. In order to illustrate the relative strength and variation of the different contributions, in table 4 we only show, for six different cases, results obtained by taking the extreme values of these ranges. In the first column only $DE = DE_{LL} = DE_{LR}$ are assumed to be different from zero. For simplicity, unless stated otherwise, the same numerical values are used for diagrams corresponding to the insertion of LL or LR operators, i.e. $DE = DE_{LL} = DE_{LR}$, $CE = CE_{LL} = CE_{LR}$, etc. We then consider, in addition to DE, the CE contribution by taking $CE = DE/3$. Annihilation diagrams are included in the third column, where we use $DA = 0$ and

$CA = 1/2DE$.[18] Inspired by kaon decays, we allow for some enhancement of the matrix elements of left-right (LR) operators and choose $DE_{LR} = 2DE_{LL}$ and $CE_{LR} = 2CE_{LL}$ (fourth column). Penguin contractions, CP and DP, can be interpreted as long-distance penguin contributions to the matrix elements and play an important role: if we take $CP_{LL} = CE$ and $DP_{LL} = DE$ (fifth column), in some decays these terms dominate the amplitude. Finally, in the sixth column, we allow for long distance effects which might differentiate penguin contractions with up and charm quarks in the loop, giving rise to incomplete GIM cancellations (we assume $\overline{DP} = DP(c) - DP(u) = DE/3$ and $\overline{CP} = CP(c) - CP(u) = CE/3$).

In addition to the ratios of the different SM contributions to the decay amplitudes given in table 4, obtained letting the matrix elements vary in the broad range defined above, we also give, in table 5, the branching ratios for the channels of interest to us. These branching ratios are obtained following the approach of ref.[19] We use QCD sum rules form factors[20] to compute the factorizable DE contribution, then fit CE using the available data on $b \to c$ two-body decays; CP and DP are extracted from the measured $B \to K\pi$ branching ratios, CA is varied between 0 and 0.5 and DA, $\overline{DP}$ and $\overline{CP}$ are set to zero. The range of values in table 5 corresponds to the variation of the CKM angles in the presently allowed range and to the inclusion of the contributions proportional to CP and DP (see ref.[19] for further details).

Coming to the SUSY contributions, we make use of the Wilson coefficients for the gluino contribution (see eq. (12) of ref.[12]) and parameterize the matrix elements as we did before for the SM case. We obtain the ratios of the SUSY to the SM amplitudes as reported in table 4 for $\tilde{q}$ and $\tilde{g}$ masses of 250 GeV and 500 GeV (second and third row, respectively). From the table, one concludes that the inclusion of the various terms in the amplitudes, DE, DA, etc., can modify the ratio r of SUSY to SM contributions up to one order of magnitude.

In terms of the decay amplitude A, the CP asymmetry reads

$$\mathcal{A}(t) = \frac{(1 - |\lambda|^2)\cos(\Delta M_d t) - 2\mathrm{Im}\lambda \sin(\Delta M_d t)}{1 + |\lambda|^2} \tag{4}$$

with $\lambda = e^{-2i\phi^M} \bar{A}/A$. In order to be able to discuss the results model-independently, we have labelled as ϕ^M the generic mixing phase. The ideal case occurs when one decay amplitude only appears in (or dominates) a decay process: the CP violating asymmetry is then determined by the total phase $\phi^T = \phi^M + \phi^D$, where ϕ^D is the weak phase of the decay. This ideal situation is spoiled by the presence of several interfering amplitudes. If the ratios r in table 4 are small, then the uncertainty on the sine of the CP phase is $< r$, while if r is O(1) ϕ^T receives, in general, large corrections.

Table 4: Ratios of amplitudes for exclusive B decays. For each channel, whenever two terms with different CP phases contribute in the SM, we give the ratio r of the two amplitudes. For each channel, the second and third lines, where present, contain the ratios of SUSY to SM contributions for SUSY masses of 250 and 500 GeV respectively.

Process	DE	$DE+CE$	$DE+CE$ $+CA$	$DE+CE+$ $CA+LR$	$DE+CE+$ $DP+CP$	$DE+CE+$ $\overline{DP}+\overline{CP}$
	$-$	$-$	$-$	$-$	$-$	$-$
$B_d^0 \to J/\psi K_S$	-0.03	0.1	0.1	0.1	0.1	0.1
	-0.008	0.02	0.02	0.04	0.02	0.02
	$-$	$-$	$-$	$-$	$-$	$-$
$B_d^0 \to \phi K_S$	0.7	0.7	0.7	0.6	0.4	0.4
	0.2	0.2	0.2	0.1	0.1	0.09
	0.08	-0.06	-0.05	-0.02	-0.009	-0.01
$B_d^0 \to K_S \pi^0$	0.7	0.7	0.6	0.6	0.4	0.4
	0.2	0.2	0.2	0.1	0.1	0.09
$B_d^0 \to D_{CP}^0 \pi^0$	0.02	0.02	0.02	0.02	0.02	0.02
	-0.6	0.9	-0.7	-2.	6.	4.
$B_d^0 \to \pi^0 \pi^0$	0.3	-0.07	0.4	-0.4	-0.07	-0.06
	0.06	-0.02	0.09	-0.1	-0.02	-0.02
	-0.09	-0.1	-0.1	-0.3	-0.9	-0.8
$B_d^0 \to \pi^+ \pi^-$	0.02	0.02	0.03	0.09	0.8	0.4
	0.005	0.006	0.008	0.02	0.2	0.1
	0.03	0.04	0.05	0.1	0.3	0.2
$B_d^0 \to D^+ D^-$	-0.007	-0.008	-0.01	-0.02	-0.02	-0.02
	-0.002	-0.002	-0.002	-0.005	-0.006	-0.005
	0	0	0	0	0.	0.07
$B_d^0 \to K^0 \bar{K}^0$	-0.2	-0.2	-0.2	-0.2	-0.09	-0.08
	-0.06	-0.05	-0.05	-0.04	-0.02	-0.02
	$-$	$-$	-0.2	-0.4	$-$	$-$
$B_d^0 \to K^+ K^-$	$-$	$-$	0.04	0.1	$-$	$-$
	$-$	$-$	0.01	0.03	$-$	$-$
	$-$	$-$	$-$	$-$	$-$	$-$
$B_d^0 \to D^0 \bar{D}^0$	$-$	$-$	-0.01	-0.03	$-$	$-$
	$-$	$-$	-0.003	-0.006	$-$	$-$
	-0.04	0.1	0.1	0.3	0.1	0.1
$B_d^0 \to J/\psi \pi^0$	0.007	-0.02	-0.02	-0.03	-0.02	-0.02
	0.002	-0.005	-0.005	-0.008	-0.005	-0.005
	$-$	$-$	$-$	$-$	$-$	$-$
$B_d^0 \to \phi \pi^0$	-0.06	-0.1	-0.1	-0.1	-0.1	-0.1
	-0.01	-0.03	-0.03	-0.03	-0.03	-0.03

Table 5: Branching ratios for B decays.

Channel	BR $\times 10^5$
$B \to J/\psi K_S$	40
$B \to \phi K_S$	$0.6 - 2$
$B \to \pi^0 K_S$	$0.02 - 0.4$
$B \to D^0_{CP}\pi^0$	16
$B \to D^+ D^-$	$30 - 50$
$B \to J/\psi \pi^0$	2
$B \to \phi \pi^0$	$1 - 4 \times 10^{-4}$
$B \to K^0 K^0$	$0.007 - 0.3$
$B \to \pi^+ \pi^-$	$0.2 - 2$
$B \to \pi^0 \pi^0$	$0.003 - 0.09$
$B \to K^+ K^-$	< 0.5
$B \to D^0 D^0$	< 20

The results of our analysis are summarized in tables 5 and 6 which collect the branching ratios and CP phases for the relevant B decays of table 4. Φ^D_{SM} denotes the decay phase in the SM; for each channel, when two amplitudes with different weak phases are present, we indicate the SM phase of the Penguin (P) and Tree-level (T) decay amplitudes. The range of variation of r in the SM (r_{SM}) is deduced from table 4. For $B \to K_S\pi^0$ the penguin contributions (with a vanishing phase) dominate over the tree-level amplitude because the latter is Cabibbo suppressed. For the channel $b \to s\bar{s}d$ only penguin operators or penguin contractions of current-current operators contribute. The phase γ is present in the penguin contractions of the $(\bar{b}u)(\bar{u}d)$ operator, denoted as u-P γ in table 6.[21] $\bar{b}d \to \bar{q}q$ indicates processes occurring via annihilation diagrams which can be measured from the last two channels of table 6. In the case $B \to K^+K^-$ both current-current and penguin operators contribute. In $B \to D^0 \bar{D}^0$ the contributions from the $(\bar{b}u)(\bar{u}d)$ and the $(\bar{b}c)(\bar{c}d)$ current-current operators (proportional to the phase γ) tend to cancel out.

SUSY contributes to the decay amplitudes with phases induced by δ_{13} and δ_{23} which we denote as ϕ_{13} and ϕ_{23}. The ratios of A_{SUSY}/A_{SM} for SUSY masses of 250 and 500 GeV as obtained from table 4 are reported in the r_{250} and r_{500} columns of table 6.

We now draw some conclusions from the results of table 6. In the SM, the first six decays measure directly the mixing phase β, up to corrections which, in most of the cases, are expected to be small. These corrections, due to the presence of two amplitudes contributing with different phases, produce

Table 6: CP phases for B decays. ϕ^D_{SM} denotes the decay phase in the SM; T and P denote Tree and Penguin, respectively; for each channel, when two amplitudes with different weak phases are present, one is given in the first row, the other in the last one and the ratio of the two in the r_{SM} column. ϕ^D_{SUSY} denotes the phase of the SUSY amplitude, and the ratio of the SUSY to SM contributions is given in the r_{250} and r_{500} columns for the corresponding SUSY masses.

Incl.	Excl.	ϕ^D_{SM}	r_{SM}	ϕ^D_{SUSY}	r_{250}	r_{500}
$b \to c\bar{c}s$	$B \to J/\psi K_S$	0	–	ϕ_{23}	$0.03 - 0.1$	$0.008 - 0.04$
$b \to s\bar{s}s$	$B \to \phi K_S$	0	–	ϕ_{23}	$0.4 - 0.7$	$0.09 - 0.2$
$b \to u\bar{u}s$		P 0				
	$B \to \pi^0 K_S$		$0.01 - 0.08$	ϕ_{23}	$0.4 - 0.7$	$0.09 - 0.2$
$b \to d\bar{d}s$		T γ				
$b \to c\bar{u}d$		0				
	$B \to D^0_{CP}\pi^0$		0.02	–	–	–
$b \to u\bar{c}d$		γ				
	$B \to D^+D^-$	T 0	$0.03 - 0.3$		$0.007 - 0.02$	$0.002 - 0.006$
$b \to c\bar{c}d$				ϕ_{13}		
	$B \to J/\psi\pi^0$	P β	$0.04 - 0.3$		$0.007 - 0.03$	$0.002 - 0.008$
	$B \to \phi\pi^0$	P β	–		$0.06 - 0.1$	$0.01 - 0.03$
$b \to s\bar{s}d$				ϕ_{13}		
	$B \to K^0\bar{K}^0$	u-P γ	$0 - 0.07$		$0.08 - 0.2$	$0.02 - 0.06$
$b \to u\bar{u}d$	$B \to \pi^+\pi^-$	T γ	$0.09 - 0.9$	ϕ_{13}	$0.02 - 0.8$	$0.005 - 0.2$
$b \to d\bar{d}d$	$B \to \pi^0\pi^0$	P β	$0.6 - 6$	ϕ_{13}	$0.06 - 0.4$	$0.02 - 0.1$
	$B \to K^+K^-$	T γ	$0.2 - 0.4$		$0.04 - 0.1$	$0.01 - 0.03$
$b\bar{d} \to q\bar{q}$				ϕ_{13}		
	$B \to D^0\bar{D}^0$	P β	only β		$0.01 - 0.03$	$0.003 - 0.006$

uncertainties of $\sim 10\%$ in $B \to K_S \pi^0$, and of $\sim 30\%$ in $B \to D^+D^-$ and $B \to J/\psi \pi^0$. In spite of the uncertainties, however, there are cases where the SUSY contribution gives rise to significant changes. For example, for SUSY masses of $O(250)$ GeV, SUSY corrections can shift the measured value of the sine of the phase in $B \to \phi K_S$ and in $B \to K_S \pi^0$ decays by an amount of about 70%. For these decays SUSY effects are sizeable even for masses of 500 GeV. In $B \to J/\psi K_S$ and $B \to \phi \pi^0$ decays, SUSY effects are only about 10% but SM uncertainties are negligible. In $B \to K^0 \bar{K}^0$ the larger effect, $\sim 20\%$, is partially covered by the indetermination of about 10% already existing in the SM. Moreover the rate for this channel is expected to be rather small. In $B \to D^+D^-$ and $B \to K^+K^-$, SUSY effects are completely obscured by the errors in the estimates of the SM amplitudes. In $B^0 \to D^0_{CP} \pi^0$ the asymmetry is sensitive to the mixing angle ϕ_M only because the decay amplitude is unaffected by SUSY. This result can be used in connection with $B^0 \to K_s \pi^0$, since a difference in the measure of the phase is a manifestation of SUSY effects.

Turning to $B \to \pi\pi$ decays, both the uncertainties in the SM and the SUSY contributions are very large. Here we witness the presence of three independent amplitudes with different phases and of comparable size. The observation of SUSY effects in the $\pi^0 \pi^0$ case is hopeless. The possibility of separating SM and SUSY contributions by using the isospin analysis remains an open possibility which deserves further investigation. For a thorough discussion of the SM uncertainties in $B \to \pi\pi$ see ref. [18]

In conclusion, our analysis shows that measurements of CP asymmetries in several channels may allow the extraction of the CP mixing phase and to disentangle SM and SUSY contributions to the CP decay phase. The golden-plated decays in this respect are $B \to \phi K_S$ and $B \to K_S \pi^0$ channels. The size of the SUSY effects is clearly controlled by the the non-diagonal SUSY mass insertions δ_{ij}, which for illustration we have assumed to have the maximal value compatible with the present experimental limits on B^0_d–$\bar{B}^0_d$ mixing.

References

1. M. Dugan, B. Grinstein and L. Hall, *Nucl. Phys.* B **255**, 413 (1985).
2. S. Dimopoulos and S. Thomas, *Nucl. Phys.* B **465**, 23 (1996).
3. W. Buchmuller and D. Wyler, *Phys. Lett.* B **121**, 321 (1983);
   J. Polchinski and M. Wise, *Phys. Lett.* B **125**, 393 (1983);
   W. Fischler, S. Paban and S. Thomas, *Phys. Lett.* B **289**, 373 (1992).
4. J. Ellis and R. Flores, *Phys. Lett.* B **377**, 83 (1996).
5. M. Dine, A. Nelson and Y. Shirman, *Phys. Rev.* D **51**, 1362 (1994);
   M. Dine, Y. Nir and Y. Shirman, *Phys. Rev.* D **55**, 1501 (1997).

6. M. J. Duncan and J. Trampetic, *Phys. Lett.* B **134**, 439 (1984);
E. Franco and M. Mangano, *Phys. Lett.* B **135**, 445 (1984);
J.M. Gerard, W. Grimus, A. Raychaudhuri and G. Zoupanos, *Phys. Lett.* B **140**, 349 (1984);
J. M. Gerard, W. Grimus, A. Masiero, D. V. Nanopoulos and A. Raychaudhuri, *Phys. Lett.* B **141**, 79 (1984); *Nucl. Phys.* B **253**, 93 (1985);
P. Langacker and R. Sathiapalan, *Phys. Lett.* B **144**, 401 (1984);
M. Dugan, B. Grinstein and L. Hall, in ref.[1]

7. For details on this point, see the chapter on FCNC in this book.

8. S. Bertolini, F. Borzumati, A. Masiero and G. Ridolfi, *Nucl. Phys.* B **353**, 591 (1991).

9. G.C. Branco, G.C. Cho, Y. Kizukuri and N. Oshimo, *Phys. Lett.* B **337**, 316 (1994), *Nucl. Phys.* B **449**, 483 (1995);
G.C. Branco, W. Grimus and L. Lavoura, *Phys. Lett.* B **380**, 119 (1996);
A. Brignole, F. Feruglio and F. Zwirner, *Z. Phys.* C **71**, 679 (1996);
J. Rosiek, talk given at the ICHEP96 conference, Warsaw, July 1996.

10. M. Misiak, S. Pokorski and J. Rosiek, hep-ph/9703442.

11. L.J. Hall, V.A. Kostelecky and S. Raby, *Nucl. Phys.* B **267**, 415 (1986);
F. Gabbiani and A. Masiero, *Nucl. Phys.* B **322**, 235 (1989);
J.S. Hagelin, S. Kelley and T. Tanaka, *Nucl. Phys.* B **415**, 293 (1994).

12. F. Gabbiani, E. Gabrielli, A. Masiero and L. Silvestrini, *Nucl. Phys.* B **477**, 321 (1996).

13. E. Gabrielli, A. Masiero and L. Silvestrini, *Phys. Lett.* B **374**, 80 (1996).

14. For a general discussion on CP violation and new physics, see the recent work of Y. Grossman, Y. Nir and R. Rattazzi, hep-ph/9701231 and references therein.

15. For some recent discussions tackling this question, see:
N. Deshpande, B. Dutta and S. Oh, *Phys. Rev. Lett.* **77**, 4499 (1996);
J. Silva and L. Wolfenstein, *Phys. Rev.* D **53**, 5331 (1997);
A. Cohen, D. Kaplan, F. Leipentre and A. Nelson, *Phys. Rev. Lett.* **78**, 2300 (1997);
Y. Grossman and M. Worah, *Phys. Lett.* B **395**, 241 (1997);
M. Ciuchini, E. Franco, G. Martinelli, A. Masiero and L. Silvestrini, preprint CERN-TH/97-47, April 1997, hep-ph/9704274;
Y. Grossman, Y. Nir and M. Worah, hep-ph/9704287;
R. Barbieri and A. Strumia, hep-ph/9704402.

16. M. Ciuchini, E. Franco, G. Martinelli, A. Masiero and L. Silvestrini, in ref.[15]

17. M. Ciuchini, E. Franco, G. Martinelli, L. Reina and L. Silvestrini, *Z. Phys.* C **68**, 239 (1995).

18. M. Ciuchini, E. Franco, G. Martinelli and L. Silvestrini, Preprint CERN-TH/97-30, March 1997, hep-ph/9703353.
19. M. Ciuchini, R. Contino, E. Franco, G. Martinelli and L. Silvestrini, Preprint TUM-HEP-279/97, in preparation.
20. V.M. Belayev et al., *Phys. Rev.* D **51**, 6177 (1995);
    P. Ball, *Phys. Rev.* D **48**, 3190 (1993);
    P. Ball and V.M. Braun, *Phys. Rev.* D **54**, 2182 (1996).
21. R. Fleischer, *Phys. Lett.* B **341**, 379 (1995).

SUPERGRAVITY UNIFIED MODELS

R. ARNOWITT

Center for Theoretical Physics, Department of Physics, Texas A&M University
College Station, TX 77843-4242

PRAN NATH

Department of Physics, Northeastern University
Boston, MA 02115

The development of supergravity unified models and their implications for current and future experiment are discussed.

1 Introduction

Supersymmetry (SUSY) was initially introduced as a global symmetry [1] on purely theoretical grounds that nature should be symmetric between bosons and fermions. It was soon discovered, however, that models of this type had a number of remarkable properties [2]. Thus the bose-fermi symmetry led to the cancellation of a number of the infinities of conventional field theories, in particular the quadratic divergences in the scalar Higgs sector of the Standard Model (SM). Thus SUSY could resolve the gauge hierarchy problem that plagued the SM. Further, the hierarchy problem associated with grand unified models (GUT) [3], where without SUSY, loop corrections gave all particles GUT size masses [4] was also resolved. In addition, SUSY GUT models with minimal particle spectrum raised the value for the scale of grand unification, M_G, to $M_G \cong 2 \times 10^{16}$ GeV, so that the predicted proton decay rate [5,6] was consistent with existing experimental bounds. Thus in spite of the lack of any direct experimental evidence for the existence of SUSY particles, supersymmetry became a highly active field among particle theorists.

However, by about 1980, it became apparent that global supersymmetry was unsatisfactory in that a phenomenologically acceptable picture of spontaneous breaking of supersymmetry did not exist. Thus the success of the SUSY grand unification program discussed above was in a sense spurious in that the needed SUSY threshold M_S (below which the SM held) could not be theoretically constructed. In order to get a phenomenologically viable model, one needed "soft breaking" masses (i.e. supersymmetry breaking terms of dimenison ≤ 3 which maintain the gauge hierarchy [7]) and these had to be introduced in an ad hoc fashion by hand. In the minimal SUSY model, the MSSM [8], where the particle spectrum is just that of the supersymmeterized

SM, one could introduce as many as 105 additional parameters (62 new masses and mixing angles and 43 new phases) leaving one with a theory with little predictive power.

A resolution of the problem of how to break supersymmetry spontaneously was achieved by promoting supersymmetry to a local symmetry i.e. supergravity[9]. Here gravity is included into the dynamics. One can then construct supergravity [SUGRA] grand unified models[10,11] where the spontaneous breaking of supersymmetry occurs in a "hidden" sector via supergravity interactions in a fashion that maintains the gauge hierarchy. In such theories there remains, however, the question of at what scale does supersymmetry break, and what is the "messenger" that communicates this breaking from the hidden to the physical sector. In this chapter we consider models where supersymmetry breaks at a scale $Q > M_G$ with gravity being the messenger[12]. Such models are economical in that both the messenger field and the agency of supersymmetry breaking are supersymmetrized versions of fields and interactions that already exist in nature (i.e. gravity). In addition, models of this type may turn out to be consequences of string theory.

The strongest direct evidence supporting supergravity GUT models is the apparent experimental grand unification of the three gauge coupling constants[13]. This result is non-trivial not only because three lines do not ordinarily intersect at one point, but also because there is only a narrow acceptable window for M_G. Thus one requires $M_G \gtrsim 5 \times 10^{15}\ GeV$ so as not to violate current experimental bounds on proton decay for the $p \to e^+ + \pi^0$ channel (which occurs in almost all GUT models) and one requires $M_G \lesssim 5 \times 10^{17}$ GeV $\cong M_{string}$ (the string scale) so that gravitational effects do not become large invalidating the analysis. Further, assuming an MSSM type of particle spectrum between the electroweak scale M_Z and M_G, acceptable grand unification occurs only with one pair of Higgs doublets and at most four generations. Finally, naturalness requires that SUSY thresholds be at $M_S \lesssim 1$ TeV which turns out to be the case. Thus the possibility of grand unification is tightly constrained.

At present, grand unification in SUGRA GUTs can be obtained to within about 2-3 std.[14,15] However, the closeness of M_G to the Planck scale, $M_{P\ell} = (\hbar c/8\pi G_N)^{1/2} \cong 2.4 \times 10^{18}$ GeV, suggests the possibility that there are $O(M_G/M_{P\ell})$ corrections to these models. One might, in fact, expect such structures to arise in string theory as nonrenormalizable operators (NROs) obtained upon integrating out the tower of Planck mass states. Such terms would produce $\approx 1\%$ corrections at M_G which might grow to $\approx 5\%$ corrections at M_Z. Indeed, as will be seen in Sec.2, it is just such NRO terms involving the hidden sector fields that give rise to the soft breaking masses, and so it would not be surprising to find such structures in the physical sector as well. Thus SUGRA GUTs should be viewed as an effective theory and, as will be discussed in Sec.4, with small deviations between theory and experiment perhaps opening a window to Planck scale physics.

One of the fundamental aspects of the SM, not explained by that theory, is the origin of the spontaneous breaking of SU(2) x U(1). SUGRA GUTS offers an explanation of this due to the existence of soft SUSY breaking masses at M_G. Thus as long as at least one of the soft breaking terms are present at M_G, breaking of SU(2) x U(1) can occur at a lower energy[10,16], providing a natural Higgs mechanism. Further, radiative breaking occurs at the electroweak scale provided the top quark is heavy ie. 100 GeV $\stackrel{<}{\sim} m_t \stackrel{<}{\sim}$ 200 GeV. The minimal SUGRA model [10,17-19] (MSGM), which assumes universal soft breaking terms, requires only four additional parameters and one sign to describe all the interactions and masses of the 32 SUSY particles. Thus the MSGM is a reasonably predictive model, and for that reason is the model used in much of the phenomenological analysis of the past decade. However, we will see in Sec. 2 that there are reasons to consider non-universal extensions of the MSGM, and discuss some of the experimental consequences they produce in Sec.4.

2 Soft Breaking Masses

Supergravity interactions with chiral matter fields, $\{\chi_i(x), \phi_i(x)\}$ (where $\chi_i(x)$ are left (L) Weyl spinors and $\phi_i(x)$ are complex scalar fields) depend upon three functions of the scalar fields: the superpotential $W(\phi_i)$, the gauge kinetic function $f_{\alpha\beta}(\phi_i, \phi_i^\dagger)$ (which enters in the Lagrangian as $f_{\alpha\beta}F_{\mu\nu}^\alpha F^{\mu\nu\beta}$ with α, β = gauge indices) and the Kahler potential $K(\phi_i, \phi_i^\dagger)$ (which appears in the scalar kinetic energy as $K_j^i \partial_\mu \phi_i \partial^\mu \phi_j^\dagger$, $K_j^i \equiv \partial^2 K^2/\partial\phi_i \partial\phi_j^\dagger$ and elsewhere). W and K enter only in the combination

$$G(\phi_i, \phi_i^\dagger) = \kappa^2 K(\phi_i, \phi_i^+) + \ell n[\kappa^6 \mid W(\phi_i) \mid^2] \tag{1}$$

where $\kappa = 1/M_{P\ell}$. Writing $\{\phi_i\} = \{\phi_a, z\}$ where ϕ_a are physical sector fields (squarks, sleptons, higgs) and z are the hidden sector fields whose VEVs $\langle z \rangle = \mathcal{O}(M_{P\ell})$ break supersymmetry, one assumes that the superpotential decomposes into a physical and a hidden part,

$$W(\phi_i) = W_{phy}(\phi_a, \kappa z) + W_{hid}(z) \tag{2}$$

Supersymmetry breaking is scaled by requiring $\kappa^2 W_{hid} = \mathcal{O}(M_S)\tilde{W}_{hid}(\kappa z)$ and the gauge hierarchy is then guaranteed by the additive nature of the terms in

Eq.(2). Thus only gravitational interactions remain to transmit SUSY breaking from the hidden sector to the physical sector.

A priori, the functions W, K and $f_{\alpha\beta}$ are arbitrary. However, they are greatly constrained by the conditions that the model correctly reduce to the SM at low energies, and that non-renormalizable corrections be scaled by κ (as would be expected if they were the low energy residue of string physics of the Planck scale). Thus one can expand these functions in polynomials of the physical fields ϕ_a

$$f_{\alpha\beta}(\phi_i, \phi_i^\dagger) = c_{\alpha\beta} + \kappa d_{\alpha\beta}^a(x,y)\phi_a + \cdots \tag{3}$$

$$W_{phys}(\phi_i) = \frac{1}{6}\lambda^{abc}(x)\phi_a\phi_b\phi_c + \frac{1}{24}\kappa\lambda^{abcd}(x)\phi_a\phi_b\phi_c\phi_d + \cdots \tag{4}$$

$$\begin{aligned}
K(\phi_i, \phi_i^\dagger) &= \kappa^{-2}c(x,y) + c_b^a(x,y)\phi_a\phi_b^\dagger + (c^{ab}(x,y)\phi_a\phi_b + h.c.) \\
&\quad + \kappa(c_{bc}^a\phi_a\phi_b^\dagger\phi_c^\dagger + h.c.) + \cdots
\end{aligned} \tag{5}$$

x=κz and y =$\kappa z^\dagger$, so that $\langle x \rangle$, $\langle y \rangle = \mathcal{O}(1)$. The scaling hypothesis for the NRO's imply then that the VEVs of the coefficients $c_{\alpha\beta}$, $c_{\alpha\beta}^a$, λ^{abc}, c, c_b^a, c^{ab} etc. are all $\mathcal{O}(1)$.

The holomorphic terms in K labeled by c^{ab} can be transformed from K to W by a Kahler transformation, $K \to K - (c^{ab}\phi_a\phi_b + h.c.)$ and

$$W \to W exp[\kappa^2 c^{ab}\ \phi_a\phi_b] = W + \tilde{\mu}^{ab}\phi_a\phi_b + \cdots \tag{6}$$

where $\tilde{\mu}^{ab}(x,y) = \kappa^2 W c^{ab}$. Hence $\langle \tilde{\mu}^{ab} \rangle = \mathcal{O}(\mathcal{M}_S)$, and one obtains a μ-term with the right order of magnitude after SUSY breaking provided only that c^{ab} is not zero [20]. The cubic terms in W are just the Yukawa couplings with $\langle \lambda^{abc}(x) \rangle$ being the Yukawa coupling constants. Also $\langle c_{\alpha\beta} \rangle = \delta_{\alpha\beta}$, $\langle c_b^a \rangle = \delta_b^a$ and $\langle c_{xy} \rangle = 1$ ($c_x \equiv \partial c/\partial x$ etc.) so that the field kinetic energies have canonical normalization.

The breaking of SUSY in the hiddden sector leads to the generation of a series of soft breaking terms [10,11,17-21]. We consider here the case where $\langle x \rangle = \langle y \rangle$ (i.e. the hidden sector SUSY breaking does not generate any CP violation) and state the leading terms. Gauginos gain a soft breaking mass term at M_G of $(m_{1/2})_{\alpha\beta}\lambda^\alpha\gamma^0\lambda^\beta$ (λ^α = gaugino Majorana field) where

$$(m_{1/2})_{\alpha\beta} = \kappa^{-2}\langle G^i(K^{-1})_j^i Re f_{\alpha\beta j}^\dagger \rangle m_{3/2} \tag{7}$$

Here $G^i \equiv \partial G/\partial \phi_i$, $(K^{-1})^i_j$ is the matrix universe of K^i_j, $f_{\alpha\beta j} = \partial f_{\alpha\beta}/\partial \phi^\dagger_j$ and $m_{3/2}$ is the gravitino mass: $m_{3/2} = \kappa^{-1}\langle exp[G/2]\rangle$. In terms of the expansion of Eqs.(3-5) one finds

$$(m_{1/2})_{\alpha\beta} = [c + \ell n(W_{hid})]_x Re\, c^*_{\alpha\beta y} m_{3/2} \tag{8}$$

and $m_{3/2} = (exp\, \tfrac{1}{2}c)\kappa^2 W_{hid}$ where it is understood from now on that x is to be replaced by its VEV in all functions (e.g. c(x)$\to$ $c(\langle x\rangle) = \mathcal{O}(1)$) so that $m_{3/2} = \mathcal{O}(M_S)$. One notes the following about Eq.(8): (i) For a simple GUT group, gauge invariance implies that $c_{\alpha\beta} \sim \delta_{\alpha\beta}$ and so gaugino masses are universal at mass scales above M_G. (ii) While $m_{1/2}$ is scaled by $m_{3/2} = \mathcal{O}(M_S)$, it can differ from it by a significant amount. (iii) From Eq.(8) one sees that it is the NRO such as $\kappa z m_{3/2}\lambda^\alpha\gamma^0\lambda^\alpha$ that gives rise to $m_{1/2}$. Below M_G, where the GUT group is broken, the second term of Eq.(3) would also contribute yielding a NRO of size $\kappa d^a_{\alpha\beta}\phi_a m_{3/2}\, \lambda^\alpha\gamma^0\lambda^\beta \sim (M_G/M_{P\ell})\, m_{3/2}\lambda\gamma^0\lambda$ [23] for fields with VEV $\langle\phi_a\rangle = \mathcal{O}(M_G)$ which break the GUT group. Such terms give small corrections to the universality of the gaugino masses and effect grand unification. They are discussed in Sec.4.

The effective potential for the scalar components of chiral multiplets is given by [10,22]

$$V = exp[\kappa^2 K]\,[(K^{-1})^j_i(W^i + \kappa^2 K^i W)\,(W^j + \kappa^2 K^j W)^\dagger$$

$$-3\kappa^2 \mid W \mid^2] + V_D \tag{9}$$

where $W^i = \partial W/\partial \phi_i$ etc., and

$$V_D = \frac{1}{2}g_\alpha g_\beta\,(Ref^{-1})_{\alpha\beta}\,(K^i(T^\alpha)_{ij}\phi_j)\,(K^k(T^\beta)_{k\ell}\phi_\ell) \tag{10}$$

where g_α are the gauge coupling constants. Eqs.(2-6) then lead to the following soft breaking terms at M_G:

$$V_{soft} = (m^2_0)^a_b\,\phi_a\phi^\dagger_b + \left[\frac{1}{3}\tilde{A}^{abc}\phi_a\phi_b\phi_c + \frac{1}{2}\tilde{B}^{ab}\phi_a\phi_b + h.c.\right] \tag{11}$$

In the following, we impose the condition that the cosmological constant vanish after SUSY breaking, i.e. $\langle V\rangle = 0$. [This is a fine tuning of $\mathcal{O}(M^2_S M^2_{P\ell})$.] Then one finds for D-flat breaking

$$(m^2_0)^a_b = [3(c^a_{cx}\, c^c_{by} - c^a_{b\,xy}) + \delta^a_b]m^2_{3/2} \tag{12}$$

$$\tilde{A}^{abc} = \pm\sqrt{3}\left[\left\{c_x h^{abc} - (c^a_{dx} h^{dac} + c^b_{dx} h^{dac} + c^c_{dx} h^{abd})\right\} + h^{abc}_x\right] m_{3/2} \qquad (13)$$

$$\tilde{B}^{ab} = \left[\left\{(\pm\sqrt{3}c_x - 1)\mu^{ab} \mp \sqrt{3}(c^a_{dx}\mu^{db} + c^b_{dx}\mu^{ad})\right\} \pm \sqrt{3}\mu^{ab}_x\right] m_{3/2} \qquad (14)$$

where the sign ambiguity arises from the two roots of the equation $\langle V \rangle = 0$. In Eqs.(12-14), $h^{abc} = \langle exp(\frac{1}{2}c)\lambda^{abc}\rangle$ are the Yukawa coupling constants and $\mu^{ab} = \langle exp(\frac{1}{2}c)\,\tilde{\mu}^{ab}\rangle$ is the μ-parameter, $h^{abc}_x \equiv \langle exp(\frac{1}{2}c)\,\partial\lambda^{abc}/\partial x\rangle$ and $\mu^{ab}_x \equiv \langle exp(\frac{1}{2}c)\partial\tilde{\mu}^{ab}/\partial x\rangle$. Since K is hermitian, c^a_b (defined in Eq.(5)) obeys $c^a_b = c^{b*}_a$.

One notes the following about Eqs.(12-14): (i) The scalar soft breaking masses $(m^2_0)^a_b$ are in general not universal unless the Kahler metric is flat, i.e. $c^a_b = \delta^a_b$ [10,20], or more generally unless the fields z which break SUSY couple universally to the physical sector i.e. $c^a_b = \tilde{c}(x,y)\delta^a_b$. (ii) $\tilde{A}^{abc}$ can contain a term, λ^{abc}_x which is not scaled by the Yukawa coupling constants h^{abc} if λ^{abc} is a function of the z fields. This possibility may occur in string theory, where the role of z is possibly played by the dilaton and moduli fields [21]. In the following we will neglect the λ^{abc}_x term (and similarly the μ^{ab}_x term in Eq.(14)).

The soft breaking masses in Eqs.(12-14) depend upon two hermitian matrices c^a_{bx} and c^a_{bxy} and the parameter c_x. The number of new SUSY parameters this implies depends upon the GUT group. We consider here the example of SU(5) supergravity GUT with R parity invariance. The superpotential is given by

$$\begin{aligned} W_{phys} &= h^{(1)}_{ij} M^{XY}_i \overline{M}_{Xj} H_{1Y} + \frac{1}{4} h^{(2)}_{ij} \epsilon_{XYZWU} M^{XY}_i M^{ZW}_j H^U_2 \\ &\quad + \mu H^X_2 \overline{H}_{1X} + W_{GUT} \end{aligned} \qquad (15)$$

where $X,Y,\cdots = 1\cdots 5$ are SU(5) indices, $i,j = 1,2,3$ are generation indices, M^{XY}_i and $\overline{M}_{Xi}$ are 10 and $\bar{5}$ matter fields and H^X_2 and $\overline{H}_{1X}$ are 5 and $\bar{5}$ Higgs fields. W_{GUT} contains additional Higgs fields which break SU(5) at M_G. H_2 and $\overline{H}_1$ contain the pair of light Higgs doublets below M_G. The soft breaking parameters then lead to 42 additional parameters for this model (25 new masses and mixing angles and 17 new phases), considerably fewer than the 105 additional parameters of the MSSM. While this is still a formidable number, many are experimentally uninteresting (e.g. scaled by the small Yukawa coupling constants), and we will see in Sec.4 that the remaining parameters are few enough (at least for the CP conserving sectors of the theory) to obtain interesting experimental predictions.

3 Radiative Breaking and the Low Energy Theory

In Sec.2, the SUGRA GUT model above the GUT scale i.e. at $Q > M_G$ was discussed. Below M_G the GUT group is spontaneously broken, and we will assume here that the SM group, SU(3) x SU(2) xU(1), holds for $Q < M_G$ [24]. Contact with accelerator physics at low energy can then be achieved using the renormalization group equations (RGE) running from M_G to the electroweak scale M_Z. As one proceeds downward from M_G, the coupling constants and masses evolve, and provided at least one soft breaking parameter and also the μ parameter at M_G is not zero, the large top quark Yukawa can turn the H_2 running (mass)^2, $m^2_{H_2}(Q)$, negative at the electroweak scale [16]. Thus the spontaneous breaking of supersymmetry at M_G triggers the spontaneous breaking of SU(2) xU(1) at the electroweak scale. In this fashion all the masses and coupling constants at the electroweak scale can be determined in terms of the fundamental parameters (Yukawa coupling constants and soft breaking parameters) at the GUT scale, and the theory can be subjected to experimental tests.

The conditions for electroweak symmetry breaking arise from minimizing the effective potential V at the electroweak scale with respect to the Higgs VEVs $v_{1,2} = \langle H_{1,2} \rangle$. This leads to the equations [16]

$$\mu^2 = \frac{\mu_1^2 - \mu_2^2 tan^2\beta}{tan^2\beta - 1} - \frac{1}{2}M_Z^2; \quad sin^2\beta = \frac{-2B\mu}{2\mu^2 + \mu_1^2 + \mu_2^2} \tag{16}$$

where $tan\beta = v_2/v_1$, B is the quadratic soft breaking parameter ($V^B_{soft} = B\mu H_1 H_2$), $\mu_i = m^2_{H_i} + \Sigma_i$, and Σ_i are loop corrections [26]. All parameters are running parameters at the electroweak scale which we take for convenience to be $Q = M_Z$. Eq.(16) then determines the μ parameter and allows the elimination of B in terms of $tan\beta$. This determination of μ greatly enhances the predictive power of the model.

We parameterize the Higgs soft breaking masses at M_G by

$$m^2_{H_1} = m_0^2(1 + \delta_1); \quad m^2_{H_2} = m_0^2(1 + \delta_2) \tag{17}$$

and the third generation of sfermion masses by

$$\begin{aligned} m^2_{q_L} &= m_0^2(1 + \delta_3); \quad m^2_{u_R} = m_0^2(1 + \delta_4); \quad m^2_{e_R} = m_0^2(1 + \delta_5) \\ m^2_{d_R} &= m_0^2(1 + \delta_6); \quad m^2_{\ell_L} = m_0^2(1 + \delta_7); \end{aligned} \tag{18}$$

where $q_L = (u_L, d_L)$, $\ell_L = (\nu_L, e_L)$ etc. The δ_i thus represent non-universal deviations from the reference mass m_0. For GUT groups which contain an

SU(5) subgroup (e.g. SU(N), N$\geq$5; SO(N), N$\geq$10, E_6 etc.) with matter embedded in the 10 and $\bar{5}$ representations, one has

$$\delta_3 = \delta_4 = \delta_5; \quad \delta_6 = \delta_7 \tag{19}$$

In the following we will choose m_0 to be the soft breaking mass of the first two generations (assumed universal to suppress FCNC effects[5]). We will also keep only the large top Yukawa coupling constant h_t and its associated cubic soft SUSY breaking parameter A_t (an approximation valid for $tan\beta \stackrel{<}{\sim}$ 20). This allows an analytic solution of the 1-loop RGE and enables one to understand physically the predictions of the model. One finds[27]

$$\begin{aligned}
\mu^2 &= \frac{t^2}{t^2-1}\left[\left(\frac{1-3D_0}{2}+\frac{1}{t^2}\right)+\left(\frac{1-D_0}{2}(\delta_3+\delta_4)-\frac{1+D_0}{2}\delta_2\right.\right. \\
&\quad \left.\left. +\frac{1}{t^2}\delta_1\right)\right]m_0^2 + \frac{t^2}{t^2-1}\left[\frac{1}{2}(1-D_0)\frac{A_R^2}{D_0}+C_\mu m_{1/2}^2\right]-\frac{1}{2}M_Z^2 \\
&\quad + \frac{1}{22}\frac{t^2+1}{t^2-1}S_0\left(1-\frac{\alpha_1(Q)}{\alpha_G}\right)
\end{aligned} \tag{20}$$

where $C_\mu = \frac{1}{2}D_0(1-D_0)(H_3/F)^2 + e - g/t^2$, $t \equiv tan\beta$, and $D_0 \cong 1 - m_t^2/(200sin\beta)^2$. D_0 vanishes at the t-quark Landau pole (for $m_t = 175$ GeV, $D_0 \leq 0.23$) and $A_R = A_t - m_{1/2}(H_2 - H_3/F)$ is the residue at the Landau pole ($A_R \cong A_t - 0.61(\alpha_3/\alpha_G)m_{1/2}$), i.e. $A_0 = A_R/D_0 - (H_3/F)m_{1/2}$ where A_0 is the cubic soft SUSY breaking parameter at M_G[28]. α_G is the GUT coupling constant ($\alpha_G \simeq 1/24$) and

$$S_0 = Tr(Ym^2) = (m_{H_2}^2 - m_{H_1}^2) + \sum_{i=1}^{ng}(m_{q_L}^2 - 2m_{u_R}^2 + m_{d_R}^2 - m_{\ell_L}^2 + m_{e_R}^2) \tag{21}$$

In Eq.(20), the form factors e.g. H_2, H_3, F are given in Ibañez et al[16], n_g is the number of generations, and the (mass)^2 in Eq.(21) are all to be taken at M_G. We note that S_0 vanishes for universal soft breaking masses, and reduces to $S_0 = m_{H_2}^2 - m_{H_1}^2$ for three generation GUT models obeying Eq.(19). (The coefficient $(1-\alpha_1(M_Z)/\alpha_G)/22 \cong 0.0268$ and so this term is generally small.)

The renormalization group equations evolve the universal gaugino mass $m_{1/2}$ at M_G to separate masses for SU(3), SU(2) and U(1) at M_Z:

$$\tilde{m}_i = (\alpha_i(M_Z)/\alpha_G)m_{1/2}; \quad i = 1,2,3 \tag{22}$$

450

where at 1-loop, the gluino mass $m_{\tilde{g}} = \tilde{m}_3$ [29]. The breaking of SU(2) x U(1) causes a mixing of the gaugino-higgsino states giving rise to two charginos $\chi_i^{\pm}$, i=1,2 and four neutralinos χ_i^0, $i = 1 \cdots 4$. In the $\tilde{W}^{\pm} - \tilde{H}^{\pm}$ basis, the chargino mass matrix is

$$M_{\chi}^{\pm} = \begin{pmatrix} \tilde{m}_2 & \sqrt{2}M_W sin\beta \\ \sqrt{2}M_W cos\beta & \mu \end{pmatrix} \tag{23}$$

and in the $(\tilde{W}_3, \tilde{B}, \tilde{H}_1, \tilde{H}_2)$ basis and the neutralino mass matrix is

$$M_{\chi}^0 = \begin{pmatrix} \tilde{m}_2 & o & a & b \\ o & \tilde{m}_1 & c & d \\ a & c & o & -\mu \\ b & d & -\mu & o \end{pmatrix} \tag{24}$$

where $a = M_Z cos\theta_W cos\beta$, $c = -tan\theta_W a$, $b = -M_Z cos\theta_W sin\beta$, d=-$tan\theta_W b$ and θ_W is the weak mixing angle. In the domain where $\mu^2/M_Z^2 >> 1$ one has

$$2m_{\chi_1}^0 \simeq m_{\chi_1}^{\pm} \simeq m_{\chi_2}^0 \simeq \frac{1}{3}m_{\tilde{g}}$$
$$m_{\chi_3}^0 \simeq m_{\chi_4}^0 \simeq m_{\chi_2}^{\pm} \simeq \mu >> m_{\chi_1}^0 \tag{25}$$

i.e. the light states are mostly gauginos and the heavy states mostly higgsinos.

The large top mass also causes L-R mixing in the stop mass matrix

$$M_{\tilde{t}}^2 = \begin{pmatrix} m_{t_L}^2 & -m_t(A_t + \mu ctn\beta) \\ -m_t(A_t + \mu ctn\beta) & m_{t_R}^2 \end{pmatrix} \tag{26}$$

where

$$m_{t_L}^2 = m_{Q_L}^2 + m_t^2 + \left(\frac{1}{2} - \frac{2}{3}sin^2\theta_W\right) M_Z^2 cos2\beta \tag{27}$$

$$m_{t_R}^2 = m_U^2 + m_t^2 + \frac{2}{3}\sin^2\theta_W M_Z^2 \cos 2\beta \qquad (28)$$

and

$$m_{Q_L}^2 = \left[\left(\frac{1+D_0}{2}\right) + \frac{5+D_0}{6}\delta_3 - \frac{1-D_0}{6}(\delta_2 + \delta_4)\right]m_0^2$$
$$- \frac{1}{6}(1-D_0)\frac{A_R^2}{D_0} + C_Q m_{1/2}^2 - \frac{1}{66}S_0(1-\alpha_1(Q)/\alpha_G) \qquad (29)$$

$$m_U^2 = \left[D_0 + \frac{2+D_0}{3}\delta_4 - \frac{1-D_0}{6}(\delta_2 + \delta_3)\right]m_0^2$$
$$- \frac{1}{3}(1-D_0)\frac{A_R^2}{D_0} + C_U m_{1/2}^2 + (2/33)S_0(1-\alpha_1(Q)/\alpha_G) \qquad (30)$$

The coefficients C_Q, C_U are given by

$$C_Q = \frac{1}{3}\left[-\frac{1}{2}D_0(1-D_0)(H_3/F)^2 + e\right] + \frac{\alpha_G}{4\pi}\left(\frac{8}{3}f_3 + f_2 - \frac{1}{15}f_1\right) \qquad (31)$$

$$C_U = \frac{2}{3}\left[-\frac{1}{2}D_0(1-D_0)(H_3/F)^2 + e\right] + \frac{\alpha_G}{4\pi}\left(\frac{8}{3}f_3 - f_2 + \frac{1}{3}f_1\right) \qquad (32)$$

the form factors f_i being given in Ibañez et al. [16].

The simplest model is the one with universal soft breaking masses, i.e. $\delta_i=0$. This model depends on only the four SUSY parameters and one sign

$$m_0, \quad m_{1/2}, \quad A_0, \quad B_0, \quad \text{sign}(\mu) \qquad (33)$$

Alternately one may choose m_0 , $m_{\tilde{g}}$, A_t , $\tan\beta$, and $\text{sign}(\mu)$ as the independent parameters. This case has been extensively discussed in the literature[30]. As can be seen above, however, the deviations from universality can significantly effect the values of μ^2 and the stop masses, and these parameters play a crucial role in predictions of the theory. Thus more recently, efforts to include non-universalities into calculations have been made, and some of these effects will be discussed in the context of dark matter in Sec.4.

4 Phenomenological Implications of Supergravity Models

We shall discuss the implications of supergravity models mostly in the context of the parameter space of Eq.(33) although many of the results of the analysis are valid for a broader class of supergravity models which include non-universalities as discussed in Secs.2 and 3. We shall analyse the effect of non-universalities discussed in Sec.3 in some detail in the context of dark matter detectors. There are many phenomenological implications of supergravity unified models. One of the important implications concerns the signatures of supersymmetry in collider experiments. Such signatures depend crucially on the stability of the lowest supersymmetric particle (LSP). A stable LSP can be achieved by the imposition of an R parity symmetry. This is an attractive possibility as it also provides a candidate for cold dark matter. In fact using the renormalization group analysis discussed earlier, one finds that over most of the parameter space of supergravity unified models the lightest neutralino is the LSP[31]. In this circumstance the decay of the supersymmetric particles will have missing energy signals and at least one of the carriers of missing energy will be the neutralino. Signals of this type were studied early on after the advent of supergravity models in the supersymmetric decays of the W and Z bosons[32] and such analyses have since been extended to the decays of all of the supersymmetric particles. Using these decay patterns one finds a variety of supersymmetric signals for SUSY particles at colliders where SUSY particles are expected to be pair produced when sufficient energies in the center of mass system are achieved. One signal of special interest in the search for supersymmetry is the trileptonic signal[33]. For example in $p\bar{p}$ collisions one can have $p\bar{p} \rightarrow \tilde{\chi}_1^{\pm} + \tilde{\chi}_2^0 + X \rightarrow (l_1\bar{\nu}_1\tilde{\chi}_1^0) + (l_2\bar{l}_2\tilde{\chi}_1^0) + X$ which gives a signal of three leptons and missing energy. Using this signal a chargino of mass up to 230 GeV can be explored at the Tevatron with an integrated luminosity of[34] $10fb^{-1}$.

One of the most accurate probes of new physics beyond the Standard Model (SM) is provided by (g-2) for the muon and for the electron[35]. Currently the experimental value of $a_\mu \equiv (g_\mu/2-1)$ is $a_\mu^{exp} = 11659230(84) \times 10^{-10}$ where as $a_\mu^{theory}(SM) = 11659172(15.4) \times 10^{-10}$. Here $a_\mu^{theory}(SM)$ contains the $O(\alpha^5)$ Q.E.D. contributions, $O(\alpha^2)$ hadronic corrections and up to two loop corrections from the SM electro-weak sector. Essentially all of the error in the theoretical computations comes from the hadronic contributions which give $a_\mu^{hadron}(SM) = 687.0(15.4) \times 10^{-10}$ while the contribution from the electro-weak sector has a very small error[36], i.e., $a_\mu^{EW}(SM) = 15.1(0.4) \times 10^{-10}$. It is expected that the new experiment underway at Brookhaven will reduce the experimental uncertainty by a factor of about 20. Further, a new analysis by

Alemany et.al. [37] using the τ data at LEP indicates a significant reduction in the hadronic error, i.e., $a_\mu^{hadron}(SM) = 701.4(9.4) \times 10^{-10}$. The hadronic error is expected to reduce further perhaps by another factor of two when new data from VEPP-2M, DAΦNE, and BEPC experiments comes in. Thus the improved determination of $g_\mu - 2$ in the Brookhaven experiment as well as the expected reduction in the hadronic error will allow one to test the electro-weak SM contribution.

It was pointed out in ref. [38] that any experiment which tests the SM electro-weak contribution to $g_\mu - 2$ can also test the supersymmetric contribution. This conclusion is supported by the recent precision analyses of $g_\mu - 2$ using the renormalization group [39,40]. Thus the Brookhaven experiment is an important probe of supergravity grand unification.

Flavor changing neutral current processes also provide an important constraint on supergravity unified models. A process of great interest here is the decay $b \to s + \gamma$ which has been observed by the CLEO collaboration [41] with a branching ratio of BR($b \to s+\gamma$) $=(2.32\pm0.67)\times10^{-4}$. In the SM this decay proceeds at the loop level with the exchange of W and Z bosons and the most recent analyses [42] of the branching ratio including leading and most of the next to leading order QCD corrections yield BR($b \to s+\gamma$) $=(3.48\pm0.31)\times10^{-4}$ for $m_t = 176$ GeV. In supersymmetry there are additional diagrams which contribute to this process [43]. Thus in SUGRA unification one has contributions from the exchange of the charged Higgs, the charginos, the neutralinos and from the gluino. Interestingly one finds that while the contribution from the charged Higgs exchange is always positive [44] over much of the parameter space its effects are small since $H^\pm$ is heavy. The contribution from the exchange of the other SUSY particles can be either positive or negative with the contribution of the charginos being usually the dominant one [45]. Thus supergravity can accomodate a value of the BR($b \to s+\gamma$) which is lower than the SM value [45,46]. If the more accurate experimental determination of BR($b \to s + \gamma$) continues to show a trend with a value of BR($b \to s+\gamma$) lower than what the SM predicts, the results would provide a strong hint for a low lying chargino and a low lying third generation squark. The $b \to s + \gamma$ experiment also imposes an important constraint on dark matter analyses and we discuss this below.

As mentioned earlier one of the remarkable results of supergravity grand unification with R parity invariance is the prediction that the lightest neutralino χ_1^0 is the LSP over most of the parameter space. In this part of the parameter space the χ_1^0 is a candidate for cold dark matter (CDM). We discuss now the relic density of χ_1 within the framework of the Big Bang Cosmology. The quantity that is computed theoretically is $\Omega_{\chi_1} h^2$ where $\Omega_{\chi_1} = \rho_{\chi_1}/\rho_c$, ρ_{χ_1} is the neutralino relic density and ρ_c is the critical relic density needed to close the universe, $\rho_c = 3H^2/8\pi G_N$, and $H = h100km/sMpc$ is the Hubble constant. The most conservative constraint on $\Omega_{\chi_1}h^2$ is $\Omega_{\chi_1}h^2 < 1$. A variety of more constraining possibilities have also been discussed in the literature. One of the important elements in the computation of the relic density concerns the correct thermal averaging of the quantity (σv) where σ is the neutralino annihilation cross section in the early universe and v is the relative neutralino velocity. Normally the thermal average is calculated by first making the approximation $\sigma v = a + bv^2$ and then evaluating its thermal average[47]. However, it is known that such an approximation breaks down in the vicinity of thresholds and poles[48]. Precisely such a situation exists for the case of the annihilation of the neutralino through the Z and Higgs poles. An accurate analysis of the neutralino relic density in the presence of Z and Higgs poles was given in ref.[49] and similar analyses have also been carried out since by other authors[50,51].

There are a number of possibilites for the detection of dark matter both direct and indirect[52,51]. We discuss here the direct method which involves the scattering of incident neutralino dark matter in the Milky Way from nuclei in terrestial targets. The event rates consist of two parts[53]: one involves an axial interaction and the other a scalar interaction. The axial (spin dependent) part R_{SD} falls off as $R_{SD} \sim 1/M_N$ for large M_N where M_N is the mass of the target nucleus, while the scalar (spin independent) part behaves as $R_{SI} \sim M_N$ and increases with M_N. Thus for heavy target nuclei the spin independent part R_{SI} dominates over most of the parameter space of the model. Analysis of event rates for the scattering of neutralinos off targets such as Ge and Xe indicates that the event rates can lie over a wide range, from $O(1)$ event/kg d to $O(10^{-(5-6)})$ event/kg d[54]. Inclusion of non-universalities is seen to produce definite signatures in the event rate analysis[27,55]. Thus the minimum event rates in the region $m_{\chi_1} < 65$ GeV can be enhanced (for the case $\delta_1 = -1 = -\delta_2$) or reduced (for the case $\delta_1 = 1 = -\delta_2$) by a factor of $O(10\text{-}100)$ due to the presence of non-universalities[27], while the effect of non-universalities is generally small for $m_{\chi_1} > 65\ GeV$. To illustrate how this comes about we see that from Eq.(20) the non-universality correction to μ is given by

$$\Delta\mu^2 = \frac{t^2}{t^2 - 1}\left[\left(\frac{1 - D_0}{2}(\delta_3 + \delta_4) - \frac{1 + D_0}{2}\delta_2 + \frac{1}{t^2}\delta_1\right)\right]m_0^2 \qquad (34)$$

μ^2 influences detection event rates in that an increase (decrease) in μ^2 reduces (increases) R_{SI}. Thus for $\delta_1 = -1 = -\delta_2$ one finds that $\Delta\mu^2$ is negative and tends to raise R_{SI}. In addition, this could drive μ^2 negative, eliminating such points from the parameter space of the theory. These effects are more significant for small $\tan\beta$ values and thus drive the minimum event rates higher in the region $m_{\chi_1} < 65$ GeV. For $\delta_1 = -1 = -\delta_2$ one has $\Delta\mu^2 > 0$ and the effect here is opposite to that for the previous case. (Note also since D_0 is small, δ_3 and δ_4 can cancel or enhance the effects of δ_1 and δ_2). Further, it is found that the event rates for $\mu < 0$ are much suppressed (by a factor ≈ 100) compared to the case $\mu > 0$ [54]. This feature persists in the presence of non-universalities [27].

The detection sensitivity of current detectors is typically of order of a few events/kg d [56] which is as yet insufficent to probe a significant part of the parameter space of SUGRA models. One needs more sensitive detectors with sensitvities 2-3 orders of magnitude better to sample a significant part of the parameter space of SUGRA models. (See in this context ref. [57]).

As mentioned already the minimal supergravity grand unification gives a result for $\alpha_3(M_Z)$ which is about 2-3 std larger than the current world average [14,15]. However, because of the proximity of the GUT scale to the Planck scale one can expect corrections of size $O(M_G/M_{Pl})$ where M_{Pl} is the Planck mass. For example, Planck scale corrections can modify the gauge kinetic energy function so that one has for the gauge kinetic energy term - $(1/4)f_{\alpha\beta}F_\alpha^{\mu\nu}F_{\beta\mu\nu}$. For the minimal SU(5) theory, Eq.(3) can give terms $f_{\alpha\beta} = \delta_{\alpha\beta} + (c/2M_{Pl})d_{\alpha\beta\gamma}\Sigma^\gamma$ where Σ is the the scalar field in the 24 plet of SU(5). After the spontaneous breaking of SU(5) and a re-diagonalization of the gauge kinetic energy function, one finds a splitting of the $SU(3) \times SU(2) \times U(1)$ gauge coupling constants at the GUT scale. These splittings can easily generate the 2 std correction to $\alpha_s(M_Z)$. In fact, using the LEP data one can put constraints on c. One finds that [15] $-1 \leq c \leq 3$. The Planck scale correction also helps relax the stringent constraint on $tan\beta$ imposed by $b - \tau$ unification. Thus in the absence of Planck scale correction one has that $b - \tau$ unification requires $tan\beta$ to lie in two rather sharply defined corridors [58]. One of these corresponds to a small value of $tan\beta$, i.e., $tan\beta \sim 2$ and the second a large value $tan\beta \sim 50$. This stringent constraint is somewhat relaxed by the inclusion of Planck scale corrections [15].

SUSY grand unified models contain many sources of proton instability. Thus in addition to the p decay occuring via the exchange of superheavy vector lepto-quarks, one has the possibility of p dacay from dimension (dim) 4 (dim 3 in the superpotential) and dim 5 (dim 4 in the superpotential) operators [59]. The lepto-quarks exchange would produce $p \to e^+\pi^0$ as its dominant mode with an expected lifetime [60] of $\sim 1 \times 10^{35\pm1}[M_X/10^{16}]^4 y$ where for SU(5) $M_X \cong 1.1 \times 10^{16}$ while the sensitivity of Super Kamiokande (Super-K) to this mode [61,62] is 1×10^{34}. Thus the $e^+\pi^0$ mode may be at the edge of being accessible at Super-K. Proton decay from dim 4 operators is much too rapid but is easily forbidden by the imposition of R parity invariance. The p decay from dim 5 operators is more involved. It depends on both the GUT physics as well as on the low energy physics such as the masses of the squarks and of the gauginos. Analysis in supergravity unified models [63-65] shows that one can make concrete predictions of the p decay modes within these models. Thus one can show that with the Higgs triplet mass $M_{H_3} < 10 M_G$, the parameter space of minimal SU(5) SUGRA will come close to being exhausted for $m_0 < 1 TeV$ and $m_{\tilde{g}} < 1 TeV$ if Super-K[62] and Icarus[66] reach the sensitivity for $\bar{\nu} K^+$ mode of 2×10^{34} y [64,65].

Precision determination of soft SUSY breaking parameters can be utilized as a vehicle for the test of the predictions of supergravity grand unification. It was recently proposed that precision measurement of the soft breaking parameters can also act as a test of physics at the post GUT and string scales [67]. Thus, for example, if one has a concrete model of the soft breaking parameters at the string scale then these parameters can be evolved down to the grand unification scale leading to a predicted set of non-universalities there. If the SUSY particle spectra and their interactions are known with precision at the electro-weak scale, then this data can be utilized to test a specific model at the post GUT or string scales. Future colliders such as the LHC [68] and the NLC [69] will allow one to make mass measurements with significant accuracy. Thus accuracies of up to a few percent in the mass measurements will be possible at these colliders allowing a test of post GUT and string physics up to an accuracy of $\sim 10\%$ [67].

5 Conclusion

Supergravity grand unification provides a framework for the supersymmetric unification of the electro-weak and the strong interactions where supersymmetry is broken spontaneously by a super Higgs effect in the hidden sector and the breaking communicated to the visible sector via gravitational interactions. The minimal version of the model based on a generation independent Kahler potential contains only four additional arbitrary parameters and one sign in terms of which all the SUSY mass spectrum and all the SUSY interaction structure is determined. This model is thus very predictive. A brief summary of the predictions and the phenomenological implications of the model was given. Many of the predictions of the model can be tested at current collider energies and at energies that would be achievable at future colliders. Other predictions of the model can be tested in non-accelerator experiments such as at Super Kamiokande and at Icarus, and by dark matter detectors. We also discussed here extensions of the minimal supergravity model to include nonunversalities in the soft SUSY breaking parameters. Some of the implications of these non-universalities on predictions of the model were discussed. Future experiments should be able to see if the predictions of supergravity unification are indeed verified in nature.

Acknowledgements

This work was supported in part by NSF grant numbers PHY-9411543 and PHY-96020274.

References

1. Yu A. Golfand and E.P. Likhtman, JETP Lett. **13**, 452 (1971); D. Volkov and
   V.P. Akulov, JETP Lett. **16**, 438 (1972).
2. J. Wess and B. Zumino, Nucl. Phys. **B78**, 1 (1974).
3. H. Georgi and S.L. Glashow, Phys. Rev. Lett. **32**, 438 (1974).
4. E. Gildener, Phys. Rev. **D14**, 1667 (1976).
5. S. Dimopoulos and H. Georgi, Nucl. Phys. **B193**, 150 (1981).
6. S. Dimopoulos, S. Raby and F. Wilczek, Phys. Rev. **D24**, 1681 (1981); N. Sakai, Z. Phys. **C11**, 153 (1981).
7. L. Giradello and M.T. Grisaru, Nucl. Phys. **B194**, 65 (1982).
8. For a review of the properties of the MSSM see H. Haber and G. Kane, Phys. Rep. **117**, 75 (1985).
9. D.Z. Freedman, P. van Nieuwenhuizen and S. Ferrara, Phys. Rev. **D14**, 912 (1976); S. Deser and B. Zumino, Phys. Lett. **B62**, 335 (1976).
10. A.H. Chamseddine, R. Arnowitt and P. Nath, Phys. Rev. Lett. **49**, 970 (1982).
11. For reviews see P. Nath, R. Arnowitt and A.H. Chamseddine, *Applied N =1 Supergravity*(World Scientific, Singapore, 1984); H.P. Nilles, Phys. Rep. **110**, 1 (1984); R. Arnowitt and P. Nath, Proc. of VII J.A. Swieca Summer School ed. E. Eboli (World Scientific, Singapore, 1994).
12. Alternate possibilities are considered elsewhere in this book.
13. P. Langacker, Proc. PASCOS 90-Symposium, Eds. P. Nath and S. Reucroft (World Scientific, Singapore 1990); J. Ellis, S. Kelley and D.V. Nanopoulos, Phys. Lett. **B249**, 441 (1990); **B260**, 131 (1991); U. Amaldi, W. de Boer and H. Furstenau, Phys. Lett. **B260**, 447 (1991); F. Anselmo, L. Cifarelli, A. Peterman and A. Zichichi, Nuov. Cim. **104A**, 1817 (1991); **115A**, 581 (1992).

14. P.H. Chankowski, Z. Pluciennik, and S. Pokorski, Nucl. Phys. **B439**, 23 (1995); J. Bagger, K. Matchev and D. Pierce, Phys. Lett **B348**, 443 (1995); R.Barbieri and L.J. Hall, Phys. Rev. Lett. **68**, 752 (1992); L.J. Hall and U. Sarid, Phys. Rev. Lett. **70**, 2673 (1993); P. Langacker and N. Polonsky, Phys. Rev. **D47**, 4028 (1993).

15. T. Dasgupta, P. Mamales and P. Nath, Phys. Rev. **D52**, 5366 (1995); D. Ring, S. Urano and R. Arnowitt, Phys. Rev. **D52**, 6623 (1995); S. Urano, D. Ring and R. Arnowitt, Phys. Rev. Lett. **76**, 3663 (1996); P. Nath, Phys. Rev. Lett. **76**, 2218 (1996).

16. K. Inoue et al., Prog. Theor. Phys. **68**, 927 (1982); L. Ibañez and G.G. Ross, Phys. Lett. **B110**, 227 (1982); L. Alvarez-Gaumé, J. Polchinski and M.B. Wise, Nucl. Phys. **B221**, 495 (1983); J. Ellis, J. Hagelin, D.V. Nanopoulos and K. Tamvakis, Phys. Lett. **B125**, 2275 (1983); L. E. Ibañez and C. Lopez, Phys. Lett. **B128**, 54 (1983); Nucl. Phys. **B233**, 545 (1984); L.E. Ibañez, C. Lopez and C. Muñoz, Nucl. Phys. **B256**, 218 (1985).

17. R. Barbieri, S. Ferrara and C.A. Savoy, Phys. Lett. **B119**, 343 (1982).

18. L. Hall, J. Lykken and S. Weinberg, Phys. Rev. **D27**, 2359 (1983).

19. P. Nath, R. Arnowitt and A.H. Chamseddine, Nucl. Phys. **B227**, 121 (1983).

20. S. Soni and A. Weldon, Phys. Lett. **B126**, 215 (1983).

21. V.S. Kaplunovsky and J. Louis, Phys. Lett. **B306**, 268 (1993).

22. E. Cremmer, S. Ferrara, L.Girardello and A. van Proeyen, Phys. Lett. **116B**, 231(1982).

23. C.T. Hill, Phys. Lett. **B135**, 47 (1984); Q. Shafi and C. Wetterich, Phys. Rev. Lett. **52**, 875 (1984).

24. SO(10) models exist with an intermediate breaking scale. However, such models may have additional difficulties in satisfying grand unification and proton decay constraints[25].

25. S. Urano and R. Arnowitt, hep-ph/9611389.

26. G. Gamberini, G. Ridolfi and F. Zwirner, Nucl. Phys. **B331**, 331 (1990); R. Arnowitt and P. Nath, Phys. Rev. **D46**, 3981 (1992). as this

27. P. Nath and R. Arnowitt, hep-ph/9701301, (to be pub. Phys.Rev.D)

28. P. Nath, J. Wu and R. Arnowitt, Phys. Rev. **D52**, 4169 (1995).

29. QCD corrections to this result are discussed in S.P. Martin and M.T. Vaughn, Phys. Lett. **B318**, 331 (1993); D. Pierce and A. Papadopoulos, Nucl. Phys. **B430**, 278 (1994).

30. G. Ross and R.G. Roberts, Nucl. Phys. **B377**, 571 (1992); R. Arnowitt and P. Nath, Phys. Rev. Lett. **69**, 725 (1992); M. Drees and M.M. Nojiri, Nucl. Phys. **B369**, 54 (1993); S. Kelley *et. al.*, Nucl. Phys.

B398, 3 (1993); M. Olechowski and S. Pokorski, Nucl. Phys. **B404**, 590 (1993); G. Kane, C. Kolda, L. Roszkowski and J. Wells, Phys. Rev. **D49**, 6173 (1994); D.J. Castaño, E. Piard and P. Ramond, Phys. Rev. **D49**, 4882 (1994); W. de Boer, R. Ehret and D. Kazakov, Z.Phys. **C67**, 647 (1995); V. Barger, M.S. Berger, and P. Ohmann, Phys. Rev. **D49**, 4908 (1994); H. Baer, M. Drees, C. Kao, M. Nojiri and X. Tata, Phys. Rev. **D50**, 2148 (1994); H. Baer, C.-H. Chen, R. Munroe, F. Paige and X. Tata, Phys. Rev. **D51**, 1046 (1995).

31. R. Arnowitt and P. Nath, Phys. Rev. Lett. **69**, 725(1992); P. Nath and R. Arnowitt, Phys. Lett. **B289**, 368(1992).

32. S. Weinberg, Phys. Rev. Lett. **50**, 387(1983); R. Arnowitt, A.H. Chamseddine and P. Nath, Phys. Rev. Lett. **50**, 232(1983); Phys. Lett. **129B**, 445(1983); D.A. Dicus, S. Nandi, W.W. Repko and X.Tata, Phys. Lett. **129B**, 451(1983); J. Ellis, J.S. Hagelin, D.V. Nanopoulos, and M. Srednicki, Phys. Rev. Lett. **127B**, 233(1983).

33. P. Nath and R. Arnowitt, Mod. Phys. Lett. **A2**, 331(1987); R. Barbieri, F. Caravaglio, M. Frigeni, and M. Mangano, Nucl. Phys. **B367**, 28(1991); H. Baer and X. Tata, Phys. Rev. **D47**, 2739(1992); J.L. Lopez, D.V. Nanopoulos, X. Wang and A. Zichichi, Phys. Rev. **D48**, 2062(1993); H. Baer, C. Kao and X. Tata, Phys. Rev. **D48**, 5175(1993).

34. D. Amidie and R. Brock, "Report of the tev-2000 Study Group", FERMILAB-PUB-96/082.

35. T. Kinoshita and W. J. Marciano, in *Quantum Electrodynamics*, edited by T. Kinoshita (World Scientific, Singapore, 1990).pp. 419-478.

36. A. Czarnicki, B. Krauss and W. Marciano, Phys.Rev. **D52**, 2619(1995)

37. R. Alemany, M. Davier and A. Hocker, hep-ph/9703220.

38. T. C. Yuan, R. Arnowitt, A.H. Chamseddine and P. Nath, Z. Phys. **C26**, 407(1984); D. A. Kosower, L. M. Krauss, N. Sakai, Phys. Lett. **133B**, 305(1983).

39. J. Lopez, D.V. Nanopoulos, and X. Wang, Phys. Rev. **D49**, 366(1994).

40. U. Chattopadhyay and P. Nath, Phys. Rev. **D53**, 1648(1996); M. Carena, G.F. Giudice and C.E.M. Wagner, CERN-TH/96-271.

41. M. S. Alam et. al. (CLEO Collaboration), Phys. Rev. Lett.**74**,2885(1995).

42. A. J. Buras, M. Misiak, M. Munz and S. Pokorski, Nucl. Phys. **B424**, 374(1994); M. Ciuchini et. al., Phys. Lett. **B316**, 127(1993); K. Chetyrkin, M. Misiak, and M. Munz, Phys.Lett. **B400**, 206 (1997); A. J. Buras, A. Kwiatkowski and N. Pott, hep-ph/9707482.

43. S. Bertolini, F. Borzumati and A. Masiero, Phys. Rev. Lett. **59**, 180(1987); R. Barbieri and G. Giudice, Phys. Lett. **B309**, 86(1993).

44. J.L. Hewett, Phys. Rev. Lett. **70**, 1045(1993); V. Barger, M. Berger, P. Ohmann. and R.J.N. Phillips, Phys. Rev. Lett. **70**, 1368(1993).

45. M. Diaz, Phys. Lett. **B304**, 278(1993); J. Lopez, D.V. Nanopoulos, and G. Park, Phys. Rev. **D48**, 974(1993); R. Garisto and J.N. Ng, Phys. Lett. **B315**, 372(1993); J. Wu, R. Arnowitt and P. Nath, Phys. Rev. **D51**, 1371(1995); V. Barger, M. Berger, P. Ohman and R.J.N. Phillips, Phys. Rev. **D51**, 2438(1995); H. Baer and M. Brhlick, hep-ph/9610224.

46. P. Nath and R. Arnowitt, Phys. Lett. **B336**, 395(1994); F. Borzumati, M. Drees, and M.M. Nojiri, Phys. Rev.**D51**, 341(1995).

47. For a review see, E. W. Kolb and M. S. Turner, *The Early Universe* (Addison- Wesley, Redwood City, CA, 1989).

48. K. Greist and D. Seckel, Phys. Rev. **D43**, 3191 (1991); P. Gondolo and G. Gelmini, Nucl. Phys. **B360**, 145(1991).

49. R. Arnowitt and P. Nath, Phys. Lett. **B299**, 58(1993); **303**, 403(E)(1993); P. Nath and R. Arnowitt, Phys. Rev. **70**, 3696(1993).

50. H. Baer and M. Brhlick, Phys. Rev. **D53**, 597(1996); V. Barger and C. Kao, hep-ph/970443.

51. For a review see, P. Nath and R. Arnowitt, "Supersymmetric Dark Matter", in proceedings of the Heidelberg Workshop on "Aspects of Dark Matter in Astro- and Particle Physics", Heidelberg, September 16-20,1996.

52. For a review see, G. Jungman, M. Kamionkowski and K. Greist, Phys. Rep. **267**,195(1995).

53. M.W. Goodman and E. Witten, Phys. Rev. **D31**, 3059(1983); K.Greist, Phys. Rev. **D38**, (1988)2357; **D39**,3802(1989)(E); J.Ellis and R.Flores, Phys. Lett. **B300**,175(1993); R. Barbieri, M. Frigeni and G.F. Giudice, Nucl. Phys. **B313**,725(1989); M.Srednicki and R.Watkins, Phys. Lett. **B225**,140(1989); R.Flores, K.Olive and M.Srednicki, Phys. Lett. **B237**,72(1990); A. Bottino etal, Astro. Part. Phys. **1**, 61 (1992); **2**, 77 (1994); V.A. Bednyakov, H. V. Klapdor-Kleingrothaus and S.G. Kovalenko, Phys. Rev. **D50**,7128(1994).

54. R.Arnowitt and P.Nath, Mod. Phys. Lett. **A 10**,1257(1995); P.Nath and R.Arnowitt, Phys. Rev. Lett.**74**,4592(1995); R.Arnowitt and P.Nath, Phys. Rev. **D54**,2374(1996); E.Diehl, G.Kane, C.Kolda and J.Wells, Phys. Rev.**D52**,4223(1995); V. A. Bednyakov, S.G. Kovalenko, H.V. Klapdor-Kleingrothaus, and Y. Ramachers, Z. Phys. **A357**, 339(1997); L. Bergstrom and P.Gondolo, Astropart.Phys.**5**, 263(1996); J.D. Vergados, J. Phys. **G22**, 253 (1996);

55. V. Berezinsky, A. Bottino, J. Ellis, N. Forrengo, G. Mignola, and S. Scopel, Astropart.Phys.**5**, 1(1996).

56. R. Bernabei et. al., Phys. Lett. **B389**, 757(1996); A. Bottino, F. Donato, G. Mignoa, S. Scopel, P.Belli, and A. Incicchitti, Phys.Lett.**B402**, 113(1997).

57. D.Cline, Nucl.Phys.B (Proc.Suppl.) **51B**, 304 (1996); P.Benetti et al, Nucl. Inst. and Method for Particle Physics Research, **A307**,203 (1993).

58. V. Barger, M.S. Berger, and P. Ohman, Phys. Lett. **B314**, 351(1993); W. Bardeen, M. Carena, S. Pokorski, and C.E.M. Wagner, Phys. Lett. **B320**, 110(1994).

59. S.Weinberg, Phys.Rev.**D26**,287(1982); N.Sakai and T.Yanagida, Nucl.Ph
ys.**B197**, 533(1982); S.Dimopoulos, S.Raby and F.Wilcek, Phys.Lett. **112B**, 133(1982); J.Ellis, D.V.Nanopoulos and S.Rudaz, Nucl.Phys. **B202**,43(1982); B.A.Campbell, J.Ellis and D.V.Nanopoulos, Phys.Lett.**141B**,299(1984); S.Chadha, G.D.Coughlan, M.Daniel and G.G.Ross, Phys.Lett.**149B**,47(1984).

60. W.J. Marciano, talk at SUSY-97, Philadelphia, May 1997.

61. Particle Data Group, Phys.Rev. **D50**,1173(1994).

62. Y.Totsuka, Proc. XXIV Conf. on High Energy Physics, Munich, 1988,Eds. R.Kotthaus and J.H. Kuhn (Springer Verlag, Berlin, Heidelberg,1989).

63. R.Arnowitt, A.H.Chamseddine and P.Nath, Phys.Lett. **156B**,215(1985); P.Nath, R.Arnowitt and A.H.Chamseddine, Phys.Rev.**32D**,2348(1985); J.Hisano, H.Murayama and T. Yanagida, Nucl.Phys. **B402**,46(1993).

64. R. Arnowitt and P. Nath, Phys. Rev. **D49**, 1479(1994).

65. P. Nath and R. Arnowitt, Proc. of the Workshop "Future Prospects of Baryon Instability Search in p-Decay and $n\bar{n}$ Oscillation Experiments", Oak Ridge, Tennesse, U.S.A., March 28-30, 1996, ed: S.J. Ball and Y.A. Kamyshkov, ORNL-6910, p. 59.

66. Icarus Detector Group, Int. Symposium on Neutrino Astrophysics, Takayama. 1992.

67. R. Arnowitt and P. Nath, hep-ph/9701325 (to be pub. Phys. Rev. D).

68. H. Baer, C. Chen, F. Paige, and X. Tata, Phys. Rev. **DD52**, 2746(1995); I. Hinchliffe, F.E. Paige, M.D. Shapiro, J. Soderqvist and W. Yao, Phys. Rev. **D55**, 5520(1997).

69. T. Tsukamoto, K. Fujii, H. Murayama, M. Yamaguchi, and Y. Okada, Phys. Rev. **D. 51**, 3153(1995); J.L. Feng, M.E. Peskin, H. Murayama, and X. Tata, Phys. Rev. **D52**, 1418(1992); S. Kuhlman et. al., "Physics and Technology of the NLC: Snowmass 96", hep-ex/9605011.

AN INTRODUCTION TO EXPLICIT R-PARITY VIOLATION

HERBI DREINER

Rutherford Appleton Laboratory, Chilton, Didcot, Oxon OX11 0QX, UK

I discuss the theoretical motivations for R-parity violation, review the experimental bounds and outline the main changes in collider phenomenology compared to conserved R-parity. I briefly comment on the effects of R-parity violation on cosmology.

1   Introduction

Until recently, R-parity violation ($\not{R}_p$) has been considered an unlikely component of the supersymmetric extension of the Standard Model (SM). In the past two years, it has motivated potentially favoured solutions to experimentally observed discrepancies (*e.g.* R_b, R_c, ALEPH four-jet events, HERA high Q^2 excess). It is the purpose of this chapter to present $\not{R}_p$ as an equally well motivated supersymmetric extension of the SM and provide an introductory guide. I start out with the definition of R_p and the most serious problem of proton decay. Then I discuss the various motivations for $\not{R}_p$, contrasting them with the R_p-conserving MSSM. Afterwards, I give an overview of the phenomenology of $\not{R}_p$. I finish with a discussion on cosmological effects.

2   What is R-parity?

R-parity (R_p) is a discrete multiplicative symmetry. It can be written as [1]

$$R_p = (-\mathbf{1})^{3B+L+2S}. \tag{1}$$

Here B denotes the baryon number, L the lepton number and S the spin of a particle. The electron has $R_p = +1$ and the selectron has $R_p = -1$. In fact, for all superfields of the supersymmetric SM, the SM field has $R_p = +1$ and its superpartner has [a] $R_p = -1$. R_p is conserved in the MSSM, superpartners can only be produced in pairs (all initial states at colliders are R_p even) and the LSP is stable. When extending the SM with supersymmetry one doubles the particle content to accomodate the superpartners and adds an additional Higgs doublet superfield. The minimal symmetries required to construct the

---

[a] In general symmetries for which the anticommuting parameters, θ, transform nontrivially (and thus superpartners differently) are denoted R-symmetries. They can be discrete (R_p), global continuous, or even gauged [2,3]. R-symmetries can be broken without supersymmetry being broken.

Lagrangian are the gauge symmetry of the SM: $G_{SM} = SU(3)_c \times SU(2)_L \times U(1)_Y$ and supersymmetry (including Lorentz invariance). The most general superpotential with these symmetries and this particle content (*cf.* Ch. 1) is[4]

$$W = W_{MSSM} + W_{\not{R}_p}, \tag{2}$$

$$W_{MSSM} = h^e_{ij} L_i H_1 \bar{E}_j + h^d_{ij} Q_i H_1 \bar{D}_j + h^u_{ij} Q_i H_2 \bar{U}_j + \mu H_1 H_2, \tag{3}$$

$$W_{\not{R}_p} = \frac{1}{2}\lambda_{ijk} L_i L_j \bar{E}_k + \lambda'_{ijk} L_i Q_j \bar{D}_k + \frac{1}{2}\lambda''_{ijk} \bar{U}_i \bar{D}_j \bar{D}_k + \kappa_i L_i H_2. \tag{4}$$

$i,j = 1,2,3$ are generation indices and a summation is implied. L_i (Q_i) are the lepton (quark) $SU(2)_L$ doublet superfields. $\bar{E}_j$ $(\bar{D}_j, \bar{U}_j)$ are the electron (down- and up-quark) $SU(2)_L$ singlet superfields. λ, λ', and λ'' are Yukawa couplings. The κ_i are dimensionful mass parameters. The $SU(2)_L$ and $SU(3)_C$ indices have been suppressed. When including them we see that the first term in $W_{\not{R}_p}$ is anti-symmetric in $\{i,j\}$ and the third term is anti-symmetric in $\{j,k\}$. Therefore $i \neq j$ in $L_i L_j \bar{E}_k$ and $j \neq k$ in $\bar{U}_i \bar{D}_j \bar{D}_k$. Eq.(4) thus contains $9 + 27 + 9 + 3 = 48$ new terms beyond those of the MSSM.

The last term in Eq.(4), $L_i H_2$, mixes the lepton and the Higgs superfields. In supersymmetry L_i and H_1 have the same gauge and Lorentz quantum numbers and we can redefine them by a rotation in (H_1, L_i). The terms $\kappa_i L_i H_2$ can then be rotated to zero in the superpotential[5]. If the corresponding soft supersymmetry breaking parameters B_i are aligned with the κ_i they are simultaneously rotated away[5,6]. However, the alignment of the superpotential terms with the soft breaking terms is not stable under the renormalization group equations[7]. Assuming an alignment at the unification scale, the resulting effects are small[7] except for neutrino masses[7,8]. The effects can be further suppressed by a horizontal symmetry. Throughout the rest of this Chapter, I will assume the $L_i H_2$ terms have been rotated away[b]

$$W_{\not{R}_p} = \lambda_{ijk} L_i L_j \bar{E}_k + \lambda'_{ijk} L_i Q_j \bar{D}_k + \lambda''_{ijk} \bar{U}_i \bar{D}_j \bar{D}_k. \tag{5}$$

Expanding for example the $LL\bar{E}$ term into the Yukawa couplings yields

$$\mathcal{L}_{LL\bar{E}} = \lambda_{ijk} \left[\tilde{\nu}^i_L \bar{e}^k_R e^j_L + \tilde{e}^j_L \bar{e}^k_R \nu^i_L + (\tilde{e}^k_R)^* (\bar{\nu}^i_L)^c e^j_L - (i \leftrightarrow j) \right] + h.c. \tag{6}$$

The tilde denotes the scalar fermion superpartners. These terms thus violate lepton-number. The $LQ\bar{D}$ terms also violate lepton number and the $\bar{U}\bar{D}\bar{D}$ terms violate baryon number. The entire superpotential (5) violates R_p.

---

[b]The ambiguity on bounds due to rotations in (L_i, H_1) space has been discussed in[9].

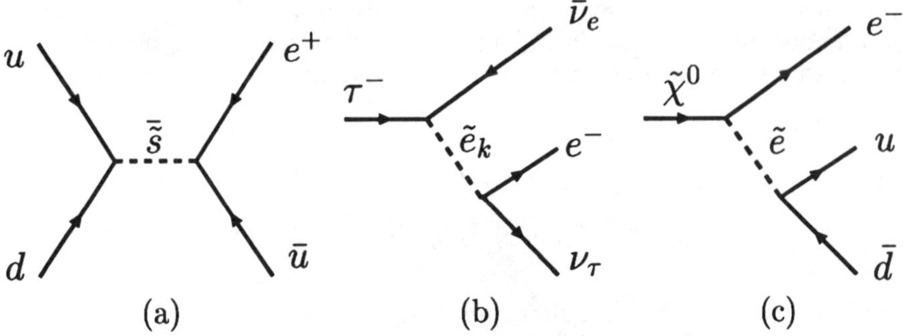

Figure 1: (a) Proton decay via $\bar{U}_1\bar{D}_1\bar{D}_2$ and $L_1Q_1\bar{D}_2$, (b) Tau decay via two $L_1L_3\bar{E}_k$ insertions, (c) Neutralino decay via $L_1Q_1\bar{D}_1$.

3   Proton Decay and Discrete Symmetries

The combination of lepton- and baryon-number violating operators in the Lagrangian can possibly lead to rapid proton decay. For example the two operators $L_1Q_1\bar{D}_k$ and $\bar{U}_1\bar{D}_1\bar{D}_k$ ($k \neq 1$) can contribute to proton decay via the interaction shown in Figure 1a. On dimensional grounds we estimate

$$\Gamma(P \to e^+\pi^0) \approx \frac{\alpha(\lambda'_{11k})\alpha(\lambda''_{11k})}{\tilde{m}^4_{dk}} M^5_{proton}. \tag{7}$$

Here $\alpha(\lambda) = \lambda^2/(4\pi)$. Given that [10] $\tau(P \to e\pi) > 10^{32}\, yr$, we obtain

$$\lambda'_{11k} \cdot \lambda''_{11k} \lesssim 2 \cdot 10^{-27} \left(\frac{\tilde{m}_{dk}}{100\, GeV}\right)^2. \tag{8}$$

For a more detailed calculation see [11]. This bound is so strict that the only natural explanation is for at least one of the couplings to be zero. Thus the simplest supersymmetric extension of the SM is excluded: an extra symmetry is required to protect the proton.

In the MSSM, R_p is imposed by hand. This forbids all the terms in $W_{\cancel{R}_p}$ and thus protects the proton. An alternative discrete symmetry with the same physical result is matter parity

$$(L_i, \bar{E}_i, Q_i, \bar{U}_i, \bar{D}_i) \to -(L_i, \bar{E}_i, Q_i, \bar{U}_i, \bar{D}_i), \quad (H_1, H_2) \to (H_1, H_2). \tag{9}$$

This forbids all terms with an odd power of matter fields and thus forbids all the terms in $W_{\cancel{R}_p}$. However, there are other solutions, which protect the proton equally well. If baryon number is conserved the proton can not decay. Thus forbidding just the interactions $\bar{U}_i\bar{D}_j\bar{D}_k$ is sufficient. This can be achieved by baryon-parity

$$(Q_i, \bar{U}_i, \bar{D}_i) \to -(Q_i, \bar{U}_i, \bar{D}_i), \quad (L_i, \bar{E}_i, H_1, H_2) \to (L_i, \bar{E}_i, H_1, H_2). \tag{10}$$

This symmetry thus protects the proton but allows for $\not{R}_p$ via the $L_iL_j\bar{E}_k$ and $L_iQ_j\bar{D}_k$ operators. If only the the interactions $\bar{U}_i\bar{D}_j\bar{D}_k$ are allowed *and* the proton is lighter than the LSP the proton is stable as well. This can be achieved by lepton parity

$$(L_i, \bar{E}_i) \rightarrow -(L_i, \bar{E}_i), \quad (Q_i, \bar{U}_i, \bar{D}_i, H_1, H_2) \rightarrow (Q_i, \bar{U}_i, \bar{D}_i, H_1, H_2). \quad (11)$$

Baryon-parity and lepton parity are two possible solutions to maintain a stable proton *and* allow for $\not{R}_p$. There is a large number of discrete symmetries which can achieve this [12].

4 Motivation

The symmetries discussed in the previous section were all imposed *ad hoc* with no deeper motivation than to ensure the stability of the proton. On this purely phenomenological level there is no reason to prefer the models with conserved R_p versus those with $\not{R}_p$. However, this is not a satisfactory view of the weak-scale picture. Hopefully, the correct structure will emerge from a simpler theory at a higher energy predicting either R_p-conservation or $\not{R}_p$.

Grand Unified Theories: In GUTs quarks and leptons are typically in common multiplets and thus have the same quantum numbers. The discrete symmetries protecting the proton and resulting in $\not{R}_p$ typically treat quarks and leptons differently and thus seem incompatible with a GUT. All the same, several GUT models have been constructed [5,13,14,15] which have low-energy $\not{R}_p$. This is typically achieved by non-renormalisable GUT scale operators involving Higgs fields. These operators become renormalisable $\not{R}_p$-operators after the GUT symmetry has been broken. Such models have been constructed for the GUT gauge groups $SU(5)$ [5,14], $SO(10)$ [14], and $SU(5) \times U(1)$ [13,14]. They have been constructed such that the only set of low-energy operators is $LL\bar{E}$ or $LQ\bar{D}$, or $\bar{U}\bar{D}\bar{D}$, respectively. There is thus no problem with proton decay. In order to ensure that only the required set of non-renormalisable operators are allowed, additional symmetries are required beyond the GUT gauge group. This is true for both R_p-conservation and $\not{R}_p$. Thus from a grand unified point of view there is no preference for either R_p-conservation or $\not{R}_p$.

String Theory: In string theories unification can be achieved without a simple gauge group. There is thus no difficulty in having distinct quantum numbers for quarks and lepton superfields. Indeed R_p-conserving and $\not{R}_p$ string theories have been constructed [16]. At present, there does not seem to be a preference at the string level for either of the two.

In both string theory and in GUTs, there is no generic prediction for the size of the $\not{R}_p$-Yukawa couplings. This is analogous to the fermion mass problem.

Discrete Gauge Symmetries: There has been a further attack on this problem from a slightly different angle. If a discrete symmetry is a remnant of a broken gauge symmetry it is called a discrete gauge symmetry. It has been argued that quantum gravity effects maximally violate all discrete symmetries unless they are discrete gauge symmetries [17]. The condition that the underlying gauge symmetry be anomaly-free can be translated into conditions on the discrete symmetry. A systematic analysis of all $\mathcal{Z}_N$ symmetries [18] has been performed. The result was that only two symmetries were discrete gauge anomaly-free: R_p and baryon-parity (10). Baryon-parity was slightly favoured since in addition it prohibited dimension-5 proton-decay operators. It has since been shown [19] that the non-linear constraints in [18] are model dependent thus possibly allowing an even larger set of discrete symmetries.

Given the quantum gravity argument it is more appealing to determine the low-energy structure directly from gauge symmetries instead of discrete symmetries. This can possibly even be connected with the fermion-mass or flavour problem. This is an on-going field of research and it is too early to draw any conclusions. I just point out that gauged models with $\not{R}_p$ have been constructed [3,6].

In conclusion, from the theoretical understanding of unification, there is no clear preference between R_p and $\not{R}_p$. In light of the very distinct phenomenology which we discuss below, it is thus mandatory to experimentally search for both possibilities. R_p-conservation and $\not{R}_p$ have the same minimal particle content. They also in principle have the same kind of symmetries, as we have just argued: G_{SM} plus an additional symmetry to protect the proton. They should thus both be considered as different versions of the MSSM. We shall denote the R_p conserving version of the MSSM as R_p-MSSM and the R_p-violating version as $\not{R}_p$-MSSM.

5 Indirect Bounds

The $\not{R}_p$ interactions can contribute to various (low-energy) processes through the virtual exchange of supersymmetric particles [20]. To date, all data are in good agreement with the SM. This leads directly to bounds on the $\not{R}_p$ operators. When determining such limits one must make some simplifying assumptions due to the large number of operators. In the following, we shall assume that one $\not{R}_p$ operator at a time is dominant while the others are negligible. We thus do not include the sometimes very strict bounds on products of operators, for example from $\mu \to e\gamma$ [21]. This is an important assumption but not unreasonable. It holds for the SM for example, where the top quark Yukawa

ijk	λ_{ijk}	ijk	λ'_{ijk}	ijk	λ'_{ijk}	ijk	λ'_{ijk}	ijk	λ''_{ijk}
121	$0.05^{a\dagger}$	111	0.001^d	211	0.09^h	311	0.16^k	112	$10^{-6,\ell}$
122	$0.05^{a\dagger}$	112	$0.02^{a\dagger}$	212	0.09^h	312	0.16^k	113	$10^{-5,m}$
123	$0.05^{a\dagger}$	113	$0.02^{a\dagger}$	213	0.09^h	313	0.16^k	123	1.25^{**}
131	0.06^b	121	$0.035^{e\dagger}$	221	0.18^i	321	0.20^{f*}	212	1.25^{**}
132	0.06^b	122	0.06^c	222	0.18^i	322	0.20^{f*}	213	1.25^{**}
133	0.004^c	123	0.20^{f*}	223	0.18^i	323	0.20^{f*}	223	1.25^{**}
231	0.06^b	131	$0.035^{e\dagger}$	231	$0.22^{j\dagger}$	331	0.26^g	312	0.43^g
232	0.06^b	132	0.33^g	232	0.39^g	332	0.26^g	313	0.43^g
233	0.06^b	133	0.002^c	233	0.39^g	333	0.26^g	323	0.43^g

Table 1: Strictest bounds on $\not{R}_p$ Yukawa couplings for $\tilde{m} = 100\, GeV$. The physical processes from which they are obtained are summarized in the main text.

coupling is almost a factor 40 larger than the bottom Yukawa coupling. Since we do not know the origins of Yukawa couplings, we do not know whether this is a generic feature.

Before presenting the complete bounds, I shall discuss one example [22] to show how such bounds can be obtained. The operator $L_1 L_3 \bar{E}_k$ can contribute to the decay $\tau \to e\nu\bar{\nu}$ via the diagram in Figure 1b. For large slepton masses, $\tilde{m}(\tilde{e}^k_R)$, this interaction is described by an effective 4-fermion Lagrangian (after Fierz re-ordering) [22]

$$\mathcal{L}_{eff} = \frac{|\lambda_{13k}|^2}{2\tilde{m}^2}(\bar{e}_L \gamma^\mu \nu_{eL})(\bar{\nu}_{\tau L}\gamma_\mu \tau_L). \tag{12}$$

This has the same structure as the term in the effective SM Lagrangian and thus leads to an apparent shift in the Fermi constant for tau decays. Considering the ratio $R_\tau \equiv \Gamma(\tau \to e\nu\bar{\nu})/\Gamma(\tau \to \mu\nu\bar{\nu})$, the contribution from $\not{R}_p$ relative to the SM contribution is [22]

$$R_\tau = R_\tau(SM)\left[1 + 2\frac{M_W^2}{g^2}\left(\frac{|\lambda_{13k}|^2}{\tilde{m}^2(\tilde{e}^k_R)}\right)\right]. \tag{13}$$

Using the experimental value [23] $R_\tau/R_\tau(SM) = 1.0006 \pm 0.0103$ we obtain the bounds

$$|\lambda_{13k}| < 0.06\left(\frac{\tilde{m}(\tilde{e}^k_R)}{100\, GeV}\right), \quad k = 1, 2, 3, \tag{14}$$

which are given in Table 1. The strictest bounds on the remaining operators are also summarized in Table 1.

The bounds in Table 1 are obtained from the following physical processes: [a] charged current universality [22,10], [b] $\Gamma(\tau \to e\nu\bar{\nu})/\Gamma(\tau \to \mu\nu\bar{\nu})$ [22,10], [c] bound on the mass of ν_e [5,24,25], [d] neutrinoless double-beta decay [26,27], [e] atomic parity violation [28,29,30], [f] $D^0 - \bar{D}^0$ mixing [31,32,33], [g] $R_\ell = \Gamma_{had}(Z^0)/\Gamma_\ell(Z^0)$ [34,20], [h] $\Gamma(\pi \to e\bar{\nu})/\Gamma(\pi \to \mu\bar{\nu})$ [22], [i] $BR(D^+ \to \bar{K}^{0*}\mu^+\nu_\mu)/BR(D^+ \to \bar{K}^{0*}e^+\nu_\mu)$ [35,20], [j] ν_μ deep-inelastic scattering [22], [k] $BR(\tau \to \pi\nu_\tau)$ [35,20], [l] heavy nucleon decay [11], and [m] $n - \bar{n}$ oscillations [11].

The bounds are all given for $\tilde{m} = 100\,GeV$ and they become weaker with increasing $\tilde{m}$. They each depend on a specific scalar mass and have various functional dependences on this mass.

$$
\begin{array}{|c|c||c|c|}
\hline
\lambda_{ijk} & m_{\tilde{e}_{Rk}}/100\,GeV & \lambda'_{111} & (m_{\tilde{q}}/100\,GeV)^2(m_{\tilde{g}}/1\,TeV)^{1/2} \\
\lambda'_{11k},\lambda'_{21k} & m_{\tilde{d}_{Rk}}/100\,GeV & \lambda'_{1j1} & m_{\tilde{q}_{Lj}}/100\,GeV \\
\lambda_{133},\lambda'_{1jj} & \sqrt{m_{\tilde{\tau},\tilde{d}_j}/100\,GeV} & \lambda'_{123} & \sqrt{m_{\tilde{b}_R}/100\,GeV} \\
\lambda'_{231} & m_{\tilde{\nu}_{\tau L}}/100\,GeV & \lambda'_{32k} & \sqrt{m_{\tilde{d}_{Rk}}/100\,GeV} \\
\hline
\end{array}
\tag{15}
$$

For λ'_{111} the dependence can be on either $m_{\tilde{d}_{Rk}}$, or $m_{\tilde{u}_{Lk}}$. The bound in Table 1 is given for $\tilde{m}(\tilde{g}) = 1\,TeV$. For $\lambda'_{132}, \lambda'_{22k}, \lambda'_{23k}, \lambda'_{31k}$, and λ'_{32k} one must consult the appropriate references since the dependence is only given numerically. The bounds on $\lambda''_{112,113}$ from heavy nucleon decay and $n - \bar{n}$ oscillations have very strong mass dependences [11].

I have updated the previous bounds from charged current universality [22], from lepton-universality [22] and from [c] R_ℓ using more recent data [10,23]. For the bound from the electron neutrino mass I have used the upper bound [25] $m_{\nu_e} < 5\,,eV$. The PDG number is $10 - 15\,eV$ [10] and is very conservative [36]. The bound on λ from m_{ν_e} scales with the square root of the upper bound on m_{ν_e}. For the bound from atomic parity violation I have used the theory value [30]: $Q_W^{th} = -73.17 \pm 0.13$. The error includes the variations due to the unknown Higgs mass. I have also used the recent new experimental number [29]. For the bound from $D^0 - \bar{D}^0$ mixing, I have updated the bound from [33] to include a lattice calculation of [32] B_D and a more updated value of f_D [31]. I have also included a 10% error to account for the quenched approximation. The bounds denoted by $\dagger$ are 2σ bounds, the other bounds are at the 1 sigma level. The bounds denoted by [**] are not direct experimental bounds. They are obtained [37,11] from the requirement that the $\not{R}_p$-coupling remains within the unitarity bound up to the grand unified scale of $10^{16}\,GeV$. This need not be the case.

The bounds denoted by * are based on a further *assumption* about the

---

[c]I thank Gautam Bhattacharyya for providing me with updates on the bounds resulting from R_ℓ.

absolute mixing in the (SM) quark sector. As stated before, we do not know the physical origin of Yukawa couplings or superpotential terms. It is a reasonable (but not necessary) assumption that their structure is determined by some symmetry at an energy scale well above the electroweak scale, *e.g.* the GUT or the Planck scale. We would expect this symmetry to be in terms of the weak *current* eigenstates. Such a symmetry could then give us a single dominant operator, for example

$$L_1 Q_1 \bar{D}_1 = \lambda'_{111}(-\tilde{e}_L u_L \bar{d}_R + \tilde{\nu}_{eL} d_L \bar{d}_R \ldots). \tag{16}$$

Below the electroweak scale the quarks become massive and we must rotate them to their *mass* eigenstate basis. (The squarks must separately also be rotated by a different rotation but that is not relevant to these bounds.) In (16) there are then separate rotations: $d_L \to \mathcal{D}_{1j} d'_{jL}$ and $u_L \to \mathcal{U}_{1j} u'_{jL}$ which generate extra $\not{R}_p$ terms suppressed by mixing angles [38].

For the quarks we do not know the *absolute* mixing of the down-quark sector, $\mathcal{D}_{ij}$, or of the up-quark sector $\mathcal{U}_{ij}$ and thus do not know by how much to rotate the up- and down-quark current eigenstates. The *relative* mixing of these two sectors is given by the CKM-matrix [10] of the SM. If we assume the relative rotation is solely due to an absolute mixing in the up-quark sector ($\mathcal{D}_{ij} = 1$) the best bounds are those given in Table 1. Those denoted by * are specifically based on this mixing assumption. If however the relative mixing is solely due to absolute mixing in the down-quark sector ($\mathcal{U}_{ij} = 1$) the D^0-$\bar{D}^0$ mixing bounds no longer apply. There are then significantly stricter bounds on many couplings from measurements of $K^+ \to \pi^+ \nu\nu$ decays [33]

$$\lambda'_{ijk} < 0.012, \ (90\%CL), \quad j \neq 3. \tag{17}$$

For Table 1 we have adopted the conservative estimate that the mixing is solely due to the up-quark sector since we do not know the absolute mixing. We therefore did not include the bounds (17).[d]

6 Changes to R_p-MSSM

On the Lagrangian level the only change to the R_p-MSSM is the inclusion of the operators in $W_{\not{R}_p}$ which give new lepton- and baryon number violating Yukawa couplings. There are several changes in the phenomenology of supersymmetry due to these couplings [38].

---

[d]There is a possible loop-hole. The symmetry at the high energy scale could just produce such a combination of couplings that is rotated to one single dominant coupling at low energy. After all, it is possibly the same symmetry which produces the single dominant quark Yukawa coupling in the SM. However, I do not adopt this philosophy here.

470

1. Lepton- or baryon number is violated as discussed in Sect. 5.

2. The LSP is not stable and can decay in the detector. It is no longer a dark matter candidate.

3. The neutralino is not necessarily the LSP.

4. The single production of supersymmetric particles is possible.

2. If for example the neutralino is the LSP and the dominant $\not{R}_p$ operator is $L_1 Q_2 \bar{D}_1$ it can decay as shown in Figure 1c. For LSP= $\tilde{\gamma}$ the decay rate is [39,40]

$$\Gamma_{\tilde{\gamma}} = \frac{3\alpha\lambda_{121}'^2}{128\pi^2} \frac{M_{\chi_1^0}^5}{\tilde{m}^4}.$$ (18)

The decay occurs in the detector if $c\gamma_L \tau(\tilde{\gamma}) \lesssim 1\,m$, or

$$\lambda_{121}' > 1.4 \cdot 10^{-6} \sqrt{\gamma_L} \left(\frac{\tilde{m}}{200\,GeV}\right)^2 \left(\frac{100\,GeV}{M_{\tilde{\gamma}}}\right)^{5/2}.$$ (19)

where γ_L is the Lorentz boost factor. This is well below the bound of Table 1. Recall also for comparison, that in the SM Yukawa couplings can be very small: for the electron $h^e = 3 \cdot 10^{-6}$. We have presented these numerical results for a photino for simplicity and clarity. The full analysis with a neutralino LSP has been performed in [41,42]. It involves several subtleties due to the R_p-MSSM parameter space which can have significant effects on the lifetime. Due to the LSP decay, supersymmetry with broken R_p has no natural dark matter candidate.

3. In the R_p-MSSM the stable LSP must be charge and colour neutral for cosmological reasons (*cf.* Chapter 15). In the $\not{R}_p$-MSSM there is no preference for the nature of the unstable LSP. It can be any of the following [e]

$$LSP \quad \epsilon \quad \{\chi_1^0, \chi_1^\pm, \tilde{g}, \tilde{q}, \tilde{t}, \tilde{\ell}, \tilde{\nu}\}.$$ (20)

In each case the collider phenomenology can be quite distinct.

4. In the $\not{R}_p$-MSSM there are resonant and non-resonant single particle production mechanisms. The resonant production mechanisms are

$$e^+ + e^- \quad \rightarrow \quad \tilde{\nu}_{Lj}, \qquad L_1 L_j \bar{E}_1,$$ (21)

$$e^- + u_j \quad \rightarrow \quad \tilde{d}_{Rk}, \qquad L_1 Q_j \bar{D}_k,$$ (22)

$$e^- + \bar{d}_k \quad \rightarrow \quad \tilde{u}_{Lj}, \qquad L_1 Q_j \bar{D}_k,$$ (23)

---

[e]The stop is listed separately since it has a special theoretical motivation [43] and leads to quite distinct phenomenology given that the top quark is so heavy.

$$\begin{aligned}
\bar{u}_j + d_k &\rightarrow \tilde{e}_{Li}^{-}, & L_i Q_j \bar{D}_k, & \qquad (24)\\
d_j + \bar{d}_k &\rightarrow \tilde{\bar{\nu}}_{Li}, & L_1 Q_j \bar{D}_k, & \qquad (25)\\
\bar{u}_i + \bar{d}_j &\rightarrow \tilde{d}_{Rk}, & \bar{U}_i \bar{D}_j \bar{D}_k, & \qquad (26)\\
d_j + d_k &\rightarrow \tilde{\bar{u}}_{Ri}, & \bar{U}_i \bar{D}_j \bar{D}_k & \qquad (27)
\end{aligned}$$

These processes can be realized at e^+e^--colliders, at HERA, and at hadron colliders, respectively. There are many further t-channel single sparticle production processes. For example at an e^+e^--collider, we can have $e^+ + e^- \rightarrow \tilde{\chi}_1^0 + \nu_j$ via t-channel selectron exchange. The t-channel exchange of squarks (sleptons) can also contribute to $q\bar{q}$ ($\ell\tilde{\ell}$) pair production, leading to indirect bounds[44].

7 Collider Phenomenology

The supersymmetric signals for $\not{R}_p$ will be a combination of supersymmetric production and decay to R_p even final states. Supersymmetric particles can be produced in pairs via MSSM gauge couplings or singly as in (21)-(27). The former benefits from large couplings while being kinematically restricted to masses $< \sqrt{s}/2$. The latter case has double the kinematic reach but suffers from typically small Yukawa couplings. Combining the various production modes with the decays and the different dominant operators leads to a wide range of potential signals to search for. Instead of systematically listing them I shall focus on two examples. Throughout we shall assume a neutralino LSP.

7.1 Squark Pair Production at the Tevatron

Squark pair production at the Tevatron proceeds via the known gauge couplings of the R_p-MSSM

$$q\bar{q}, gg \rightarrow \tilde{q} + \tilde{\bar{q}}. \qquad (28)$$

In $\not{R}_p$, once produced the squarks decay to an R_p even final state. Let us consider a dominant $L_i L_j \bar{E}_k$ operator. The couplings λ_{ijk} are bounded to be smaller than gauge couplings. Thus we expect the squarks to cascade decay to LSPs as in the MSSM. The LSPs in turn will then decay via the operator $L_i L_j \bar{E}_k$ to two charged leptons and a neutrino each (*cf.* Figure 1c). If each squark decays directly to the LSP (assuming it is the second lightest)

$$q'\bar{q}', gg \rightarrow \tilde{q} + \tilde{\bar{q}} \rightarrow q\bar{q} + \tilde{\chi}_1^0 \tilde{\chi}_1^0 \rightarrow q\bar{q} + l^+ l^- l^+ l^- \nu\nu. \qquad (29)$$

We therefore have a multi-lepton signal which is detectable[45]. To date it has not been searched for with $\not{R}_p$ in mind. However, before the top quark discovery there was a bound from CDF on a di-lepton production cross section. Making

	$L_{1,2}Q_{1,2}\bar{D}_k,$	$L_{1,2}L_3\bar{E}_3$	$L_1L_2\bar{E}_3$	$L_{1,2}L_3\bar{E}_{1,2}$	$L_1L_2\bar{E}_{1,2}$
$m_{\tilde{q}}$	$100\,GeV$	$100\,GeV$	$140\,GeV$	$160\,GeV$	$175\,GeV$

Table 2: Squark mass bounds from the Tevatron for various dominant $\not{R}_p$-operators [45].

corresponding cuts and with some simple assumptions this can be translated into a bound on the rate of the process (29) and thus a lower bound on the squark mass [45]. The assumptions are: (i) $BR(\tilde{q} \to \tilde{\gamma}q) = 100\%$, $(m_{\tilde{q}} < m_{\tilde{g}})$, (ii) LSP$= \tilde{\gamma}$ with $M_{\tilde{\gamma}} = 30\,GeV$, (iii) λ, λ' satisfy the bound (19). For various dominant operators the bounds are given in Table 2. No attempt was made to consider final state τ's due to lack of data. These bounds are comparable to the R_p-MSSM squark mass bounds. Since, the theoretical analysis has been improved to allow for neutralino LSPs, more involved cascade decays and the operator $\bar{U}\bar{D}\bar{D}$ [38,46]. However, to date no experimental analysis has been performed.

7.2 Resonant Squark Production at HERA

HERA offers the possibility to test the operators $L_1Q_j\bar{D}_k$ via resonant squark production [47] ^{f}

$$e^+ + d_k \quad \to \quad \tilde{u}_j \to (e^+ + d_k, \; \tilde{\chi}_1^0 + u_j, \; \tilde{\chi}_1^+ + d_j), \qquad (30)$$

$$e^+ + \bar{u}_j \quad \to \quad \bar{\tilde{d}}_k \to (e^+ + \bar{u}_j, \; \bar{\nu}_e + \bar{d}_j, \; \tilde{\chi}_1^0 + \bar{d}_k). \qquad (31)$$

We have included what are most likely the dominant decay modes. The neutralino and chargino will decay as in Figure 1c

$$\tilde{\chi}_1^0 \to (e^\pm, \nu) + 2\,\text{jets}, \quad \tilde{\chi}_1^+ \to (e^+, \nu) + 2\,\text{jets}. \qquad (32)$$

The neutralino can decay to the electron or positron since it is a Majorana fermion. We are thus left with several distinct decay topologies. *(i)* If the squark is the LSP it will decay to $e^+ + q$ or $\bar{\nu}_e + q$ $(\bar{\tilde{d}}_k)$. The first looks just like neutral current DIS, except that for $x_{Bj} \approx \tilde{m}^2(\tilde{q})/s$ it results in a flat distribution in y_e whereas NC-DIS gives a $1/y_e^2$ distribution. The latter looks just like CC-DIS. *(ii)* If the gauginos are lighter than the squark the gaugino decay will dominate [42] ^{g}. The clearest signal is a high p_T electron which is essentially background free. The high p_T positron or the missing p_T of the neutrino can also be searched for.

---

^{f}HERA has accumulated most of its data as a positron proton collider.

^{g}The gaugino decays could be suppressed by phase space or by partial cancellations of the neutralino couplings [41,48].

Table 3: Exclusion upper limits at 95% CL on λ'_{1jk} for $\tilde{m}(\tilde{q}) = 150\,GeV$ and $\tilde{m}(\tilde{\chi}_1^0) = 40\,GeV$ for two different dominant admixtures of the neutralino.

	λ'_{111}	λ'_{112}	λ'_{113}	λ'_{121}	λ'_{122}	λ'_{123}	λ'_{131}	λ'_{132}	λ'_{133}
$\tilde{\gamma}$-like	0.056	0.14	0.18	0.058	0.19	0.30	0.06	0.22	0.55
$\tilde{Z}^0$-like	0.048	0.12	0.15	0.048	0.16	0.26	0.05	0.19	0.48

All five signals have been searched for by the H1 collaboration[49] in the 1994 e^+ data ($\mathcal{L} = 2.83\,pb^{-1}$). The observations were in excellent agreement with the SM. The resulting bounds on the couplings are summarized in Table 3. After rescaling the bounds of Table 1 we see that the direct search is an improvement for λ'_{121}, λ'_{131}, and λ'_{132}. In the more recent data, an excess has been observed in high Q^2 NC-DIS[50]. If this persists it can possibly be interpreted as the resonant production of a squark via an $L_1 Q_j \bar{D}_k$ operator[48,51].

8 Cosmology

8.1 Bounds from GUT-Scale Baryogenesis

There is a very strict bound on all $\not{R}_p$ Yukawa couplings assuming the presently observed matter-asymmetry was created above the electroweak scale, e.g. at the GUT-scale[52] which I briefly recount. Assume that at the GUT scale a baryon- and lepton-asymmetry was created with possibly both $B + L \neq 0$ and $B - L \neq 0$. The electroweak sector of the SM (and MSSM) has baryon-number and lepton-number violating "sphaleron" interactions which conserve $B - L$ but violate $B + L$. These are in equilibrium above the electroweak phase transition and they thus erase the $B + L \neq 0$ component of the matter asymmetry.

Consider now adding one additional $\not{R}_p$ operator, e.g. $\bar{U}\bar{D}\bar{D}$ which violates baryon number. If it is in thermal equilibrium during an epoch after the GUT epoch and together with the sphaleron-interactions then together they will erase the entire matter asymmetry. In order to avoid this scenario the $\not{R}_p$ interactions should not be in thermal equilibrium above the electroweak scale resulting in the bounds[52,53,54]

$$\lambda, \lambda', \lambda'' < 5 \cdot 10^{-7} \left(\frac{\tilde{m}}{1\,TeV} \right)^{1/2}. \tag{33}$$

It should be clear that the argument holds for $LL\bar{E}$ or $LQ\bar{D}$ operators as well if lepton flavour is universal[52]. This is an extremely strict bound on *all* the couplings. If it is valid then $\not{R}_p$ is irrelevant for collider physics and can only have cosmological effects.

There are two important loop-holes in this argument. The first and most obvious one is that the matter genesis occurred at the electroweak scale or below [55]. The second loop-hole has to do with the inclusion of all the symmetries and conserved quantum numbers [54,56,57]. The electroweak sphaleron interactions do not just conserve $B - L$. They conserve the three quantum numbers $B/3 - L_i$, one for each lepton flavour. These can also be written as $B - L$ and two independent combinations of $L_i - L_j$. First, again consider an additional $\bar{U}\bar{D}\bar{D}$ operator. If the matter genesis at the GUT scale is asymmetric in the lepton flavours, $(L_i - L_j)|_{M_{GUT}} \neq 0$, then this lepton-asymmetry is untouched by the sphalerons and by the $\Delta B \neq 0$ operators $\bar{U}\bar{D}\bar{D}$ operators. The baryon asymmetry is however erased. Below the electroweak scale, the lepton-asymmetry is partially converted into a baryon-asymmetry via (SM) leptonic *and* supersymmetric mass effects [54]. If now instead, we add a lepton-number violating operator, we will retain a matter asymmetry as long as one lepton flavour remains conserved. In order for L_τ for example to be conserved, all L_τ violating operators must remain out of thermal equilibrium above the elctroweak scale, *i.e.* satisfy the bound (33). From the low-energy point of view this is completely consistent with our Ansatz of considering only one large dominant coupling at a time. Thus in these simple scenarios the bounds (33) are evaded.

8.2 Long-Lived LSP

One can consider three distinct ranges for the lifetime of the LSP

$$(i)\ \tau_{LSP} \stackrel{<}{\sim} 10^{-8}s, \quad (ii)\ 10^{-8}s \stackrel{<}{\sim} \tau_{LSP} \stackrel{<}{\sim} 10^7 \tau_u, \quad (iii)\ \tau_{LSP} > 10^7 \tau_u, \quad (34)$$

where $\tau_u \sim 10^{10} yr$ is the present age of the universe. We have discussed the first case in detail in the previous chapters. The third case is indistinguishable from the R_p-MSSM with the LSP being a good dark matter candidate. In the second case, the LSP can provide a long-lived relic whose decays can potentially lead to observable effects in the universe. There are bounds excluding any such relic with lifetimes [58]

$$1s < \tau_{LSP} < 10^{17} yr. \qquad (35)$$

The lower end of the excluded region is due to the effects of hadron showers from LSP decays on the primordial abundances of light nuclei [59]. The upper bound is from searches for upward going muons in underground detectors which can result from ν_μ's in LSP decays [60]. Note that even if $\tau_{LSP} > \tau_u$ the relic abundance is so large [h] that the decay of only a small fraction can lead to observable effects.

---

[h]This is for most values of the MSSM parameters, *cf.* Chapter 15.

The above restrictions on decay lifetimes can be immediately applied to the case of $\not{R}_p$-MSSM. If we include LSP decays in collider experiments we are left with a gap of eight orders of magnitude in lifetimes $10^{-8}s < \tau_{LSP} < 1s$ where no observational tests are presently known. It is very important to find physical effects which could help to close this gap. Since the lifetime depends on the square of the $\not{R}_p$ Yukawa coupling this corresponds to a gap of four orders of magnitude in the coupling. For a photino LSP we can translate the above bounds into bounds on the $\not{R}_p$ Yukawa couplings[53]. Using Eq.(18) we obtain the excluded region for the couplings

$$10^{-22} < (\lambda, \lambda', \lambda'') \cdot \left(\frac{200\,GeV}{\tilde{m}}\right)^2 \left(\frac{M_{\tilde{\gamma}}}{100\,GeV}\right)^{5/2} < 10^{-10}. \tag{36}$$

Note that the lower range of these bounds extends well beyond the already strict bound from proton decay (8). For a generic neutralino LSP the lifetime depends strongly on the MSSM parameters[41,42] and the bounds can only be transferred with caution.

9 Outlook

Once we include the $\not{R}_p$ terms in the superpotential we are left with a bewildering set of possibilities. We have 45 new Yukawa couplings of which any could be dominant and we have a set of seven different potential LSPs, each possibly leading to quite different phenomenology. This situation requires a systematic approach.

I would here like to suggest a two-fold approach. The theoretically best motivated model is one based on universal soft breaking terms at the unification scale $\sim 10^{16}\,GeV$, completely analogous to the MSSM. To obtain the low-energy spectrum one then employs the renormalisation group equations *including* the $\not{R}_p$-Yukawa couplings and all the soft breaking terms. This program has yet to be completed[61,21]. However, since most of the $\not{R}_p$-Yukawa couplings are bounded to be relatively small we expect for large regions in parameter space the spectrum of the $\not{R}_p$-MSSM to look just like that of the R_p-MSSM . The only difference will be a decaying neutralino LSP. To this extent the program has been implemented in SUSYGEN[62], the supersymmetry Monte Carlo generator for e^+e^--colliders. $\not{R}_p$ has only been implemented partially in ISAJET, a generator for hadron colliders[63].

As a second step, I suggest a systematic listing of potential signal topologies which can arise for spectra not obtained in the simple unification approach. Any exotic topologies can easily be searched for on a qualitative level. These two approaches combined should ensure that we do not miss any signal for supersymmetry and also do not end up searching vigorously in the $\not{R}_p$-hat every time an experimental anomaly appears.

Acknowledgments

I would like to thank the many people I have collaborated with on R-parity violation: Ben Allanach, Jon Butterworth, A. Chamseddine, Manoranjan Guchait, Smaragda Lola, Peter Morawitz, Felicitas Pauss, Emmanuelle Perez, Roger Phillips, Heath Pois, D.P. Roy and Yves Sirois. In particular I would like to thank Graham Ross who got me started on the subject and continuously encouraged me. I would furthermore like to thank Sacha Davidson and Gautam Bhattacharyya for discussions on updating the indirect bounds. I would like to thank Subir Sarkar for very helpful discussions on the long-lived LSP. I thank Ben Allanach, Gautam Bhattacharyya, Sacha Davidson, Gian Giudice, and Subir Sarkar for reading the manuscript and their helpful comments.

References

1. G. Farrar, P. Fayet, Phys. Lett. B 76 (1978) 575.
2. D. Z. Freedman, Phys. Rev. D 15 (1977) 1173; S. Ferrara, L. Girardello, T. Kugo, A. van Proeyen, Nucl. Phys. B 223 (1983) 191.
3. A.H. Chamseddine, H. Dreiner, Nucl. Phys. B 458 (1996) 65, hep-ph/9504337.
4. S. Weinberg Phys. Rev. D 26 (1982) 287, N. Sakai, T. Yanagida, Nucl. Phys. B 197 (1982) 133.
5. L.J. Hall, M. Suzuki, Nucl. Phys. B 231 (1984) 419.
6. T. Banks, Y. Grossman, E. Nardi, Y. Nir, Phys. Rev. D 52 (1995) 5319, hep-ph/9505248.
7. B. de Carlos, P.L. White, Phys. Rev. D 54 (1996) 3427, hep-ph/9602381; E. Nardi, Phys. Rev. D 55 (1997) 5772, hep-ph/9610540.
8. R. Hempfling, Nucl. Phys. B 478 (1996) 3, hep-ph/9511288; H.-P. Nilles, N. Polonsky, Nucl. Phys. B 484 (1997) 33, hep-ph/9606388.
9. S. Davidson, J. Ellis, Phys. Lett. B 390 (1997) 210, hep-ph/9609451; and CERN-TH-97-14, hep-ph/9702247.
10. Particle Data Group, *Phys. Rev.* D **54**, 1 (1996).
11. J.L. Goity, Marc Sher, Phys. Lett. B 346 (1995) 69, erratum-ibid. B 385 (1996) 500, hep-ph/9412208.
12. A. Yu. Smirnov, F. Vissani, Phys. Lett. B 380 (1996) 317, hep-ph/9601387.

13. D. Brahm, L. Hall, Phys. Rev. D40 (1989) 2449; K. Tamvakis, Phys. Lett. B 382 (1996) 251, hep-ph/9604343.

14. G.F. Giudice, R. Rattazzi, CERN-TH-97-076, hep-ph/9704339.

15. R. Barbieri, A. Strumia, Z. Berezhiani, hep-ph/9704275; K. Tamvakis, Phys. Lett. B 383 (1996) 307, hep-ph/9602389; R. Hempfling, Nucl. Phys. B 478 (1996) 3, hep-ph/9511288, A. Yu. Smirnov, F. Vissani, Nucl. Phys. B460 (1996) 37, hep-ph/9506416.

16. M.C. Bento, L. Hall, G.G. Ross, Nucl. Phys. B 292 (1987) 400; N. Ganoulis, G. Lazarides, Q. Shafi, Nucl. Phys. B 323 (1989) 374.

17. L. M. Krauss, F. Wilczek, Phys. Rev. Lett. 62 (1989) 1221, T. Banks Nucl. Phys. B 323 (1989) 90.

18. L.E. Ibanez, G.G. Ross, Phys. Lett. B 260 (1991) 291; Nucl. Phys. B 368 (1992) 3.

19. T. Banks, M. Dine, Phys. Rev. D 45 (1992) 1424.

20. G. Bhattacharyya, Nucl. Phys. Proc. Suppl. 52A (1997) 83, hep-ph/9608415.

21. B. de Carlos, P.L. White, Phys. Rev. D 54 (1996) 3427, hep-ph/9602381.

22. V. Barger, G.F. Giudice, T. Han, Phys. Rev. D 40 (1989) 2987.

23. ALEPH Collaboration (D. Buskulic et al.). Z. Phys. C 70 (1996) 561.

24. R. M. Godbole, P. Roy, X. Tata, Nucl. Phys. B 401 (1993) 67, hep-ph/9209251.

25. A.I. Belesev et al, Phys. Lett. B 350 (1995) 263; C. Weinheimer, et al, Phys. Lett. B 300 (1993) 210 and update in http:// www.na.infn.it/ win97/second.html.

26. R. Mohapatra, Phys. Rev. D 34 (1986) 3457; J.D. Vergados Phys. Lett. B 184 (1987) 55; M. Hirsch, H.V. Klapdor-Kleingrothaus, S.G. Kovalenko, Phys. Lett. B 352 (1995) 1.

27. M. Hirsch, H.V. Klapdor-Kleingrothaus, S.G. Kovalenko, Phys. Rev. Lett. 75 (1995) 17; Phys. Rev. D 53 (1996) 1329, hep-ph/9502385.

28. S. Davidson, D. Bailey, B. A. Campbell, Z. Phys. C 61 (1994) 613, hep-ph/9309310.

29. C.S. Wood *et al.* , Science 275 (1997) 1759.

30. W. J. Marciano, J. L. Rosner, Phys. Rev. Lett. 65 (1990) 2963, ERRATUM-ibid. 68 (1992) 898; see also an article by W.J. Marciano in "Precision Tests of the Standard Electroweak Model", ed. P. Langacker, World Scientific, 1995. The number I present is an update of the last number by W. J. Marciano, talk given in the 1997 INT Summer Workshop.

31. I have updated the bound from [33] using better lattice data: H. Wittig, Oxford preprint, OUTP-97-20P, to be published in Int. J. Mod. Phys.

A; hep-lat/9705034.

32. R. Gupta, T. Bhattacharya, S. Sharpe, Phys. Rev. D 55 (1996) 4036, hep-lat/9611023.

33. K. Agashe, M. Graeser, Phys. Rev. D 54 (1996) 4445.

34. G. Bhattacharyya, J. Ellis, K. Sridhar, Mod. Phys. Lett. A10 (1995) 1583, hep-ph/9503264; G. Bhattacharyya, D. Choudhury, K. Sridhar, Phys. Lett. B 355 (1995) 193, hep-ph/9504314; J. Ellis, S. Lola, K. Sridhar, e-Print Archive: hep-ph/9705416.

35. G. Bhattacharyya, D. Choudhury, Mod. Phys. Lett. A10 (1995) 1699.

36. V.M. Lobashev, talk at the "XVI edition of the International Workshop on Weak Interactions and Neutrinos," Capri, Italy, June 1997. The transparencies can be found at http://www.na.infn.it/win97/second.html. See in particular page 20.

37. B. Brahmachari, Probir Roy, Phys. Rev. D 50 (1994) 39, erratum-ibid. D 51 (1995) 3974, hep-ph/9403350.

38. H. Dreiner, G.G. Ross, Nucl. Phys. B 365 (1991) 597.

39. S. Dawson, Nucl. Phys. B 261 (1985) 297.

40. The more general formula for a neutralino LSP can be found in [41].

41. H. Dreiner, P. Morawitz, Nucl. Phys. B 428 (1994) 31, hep-ph/9405253.

42. E. Perez, Y. Sirois, H. Dreiner, Published in "Workshop on Future Physics at HERA", DESY, May 1996, hep-ph/9703444.

43. J. Ellis, S. Rudaz, Phys. Lett. B 128 (1983) 248.

44. S. Komamiya, CERN seminar, Feb. 25, 1997, OPAL, internal report PN280.

45. D.P. Roy, Phys. Lett. B 283 (1992) 270.

46. S. Dimopoulos, R. Esmailzadeh, L. Hall, G. Starkman, Phys. Rev. D 41 (1990) 2099, V. Barger, M.S. Berger, P. Ohmann, Phys. Rev. D 50 (1994) 4299, H. Baer, C. Kao, X. Tata, Phys. Rev. D 51 (1995) 2180, hep-ph/9410283, M. Guchait, D.P. Roy Phys. Rev. D 54 (1996) 3276, hep-ph/9603219.

47. J. Hewett, Proceedings of the 1990 Study on High Energy Physics, Snowmass; J. Butterworth, H. Dreiner, proceedings of the 2nd HERA Workshop, DESY, March - Oct. 1991; J. Butterworth, H. Dreiner, Nucl. Phys. B 397 (1993) 3; T. Kon and T. Kobayashi, Phys. Lett. B 270 (1991) 81.

48. G. Altarelli, J. Ellis, G.F. Giudice, S. Lola, M. L. Mangano, CERN-TH-97-040, hep-ph/9703276.

49. H1 Collaboration (S. Aid et al.), Z. Phys. C 71 (1996) 211, hep-ex/9604006.

50. H1 Collaboration (C. Adloff et al.), DESY-97-024, hep-ex/9702012; ZEUS Collaboration (J. Breitweg et al.), DESY-97-025, hep-ex/9702015.

51. D. Choudhury, S. Raychaudhuri, CERN-TH-97-026, hep-ph/9702392; H. Dreiner, P. Morawitz, hep-ph/9703279; J. Kalinowski, R. Ruckl, H. Spiesberger, P.M. Zerwas, DESY-97-038, hep-ph/9703288.

52. B. A. Campbell, S. Davidson, J. Ellis, K. A. Olive, Phys. Lett. B 256 (1991) 457; W. Fischler, G.F. Giudice, R.G. Leigh, S. Paban, Phys. Lett. B 258 (1991) 45.

53. B. A. Campbell, S. Davidson, J. Ellis, K. A. Olive, Astropart. Phys. 1 (1992) 77.

54. H. Dreiner, published in Marseille EPS HEP (1993) 424, (QCD 161: I48: 1993), hep-ph/9311286; H. Dreiner, G.G. Ross, *Nucl. Phys.* B **410**, 183 (1993), hep-ph/9207221.

55. For a recent review of baryogenesis at the electroweak scale within supersymmetry see: M. Carena, C.E.M. Wagner, to appear in 'Perspectives on Higgs Physics II', ed. G.L. Kane, World Scientific, Singapore, hep-ph/9704347.

56. For a related argument on majorana neutrino masses see A. E. Nelson, S. M. Barr, Phys. Lett. B 246 (1990) 141.

57. B. A. Campbell, S. Davidson, J. Ellis, K. A. Olive Phys. Lett. B 297 (1992) 118, hep-ph/9302221.

58. John Ellis, G.B. Gelmini, Jorge L. Lopez, D.V. Nanopoulos, S. Sarkar, Nucl. Phys. B 373 (1992) 399.

59. M. H. Reno, D. Seckel, Phys. Rev. D 37 (1988) 3441.

60. P. Gondolo, G.B. Gelmini, S. Sarkar, Nucl. Phys. B 392 (1993) 111.

61. H. Dreiner, H. Pois, ETH-TH-95-30, hep-ph/9511444; V. Barger, M.S. Berger, R.J.N. Phillips, T. Wohrmann, Phys. Rev. D 53 (1996) 6407, hep-ph/9511473

62. S.Katsanevas, P.Morawitz, "SUSYGEN 2.0 - A Monte Carlo Event Generator for MSSM Sparticle Production at e+ e- Colliders", http:// ly-ohp5.in2p3.fr/ delphi/katsan/susygen.html.

63. H. Baer, F. Paige, S. Protopopescu, X. Tata in proceedings of the Workshop on Physics at Current Accelerators and Supercolliders, Eds. J. Hewett, A. White and D. Zeppenfeld, hep-ph/9305342. Single LSP decay modes can be implemented via the FORCE command. If there are several LSP decays the ISAJET decay table must be modified, *e.g.* H. Baer, M. Brhlik, Chih-hao Chen, X. Tata, Phys. Rev. D 55 (1997) 4463, hep-ph/9610358.

才 103、ﾚ0